THE WORLD'S CLASSICS

THE SMALL HOUSE AT ALLINGTON

NOTE ON THE TEXT

The Small House at Allington was first published serially in the *Cornhill Magazine*, between September 1862 and April 1864, and was first issued in book form in March 1864. The text reproduced here is based on that of the hardback World's Classic edition, first published by Oxford University Press in 1939. *The Small House* was not included by Trollope himself in the Barsetshire series of novels, but since it deals with many of the same characters it has generally been adopted as the fifth of the six *Chronicles of Barsetshire*.

THE WORLD'S CLASSICS

ANTHONY TROLLOPE

The Small House at Allington

Edited by
JAMES R. KINCAID

OXFORD UNIVERSITY PRESS
New York Oxford

Oxford University Press

Oxford New York Toronto
Delhi Bombay Calcutta Madras Karachi
Petaling Jaya Singapore Hong Kong Tokyo
Nairobi Dar es Salaam Cape Town
Melbourne Auckland

and associated companies in
Berlin Ibadan

First published in hardback World's Classics series by
Oxford University Press 1939
First issued as a World's Classics paperback 1980 and in a hardback edition 1981
World's Classics paperback reprinted 1983, 1984, 1985, 1986, 1987
This cloth edition issued in 1989 by Oxford University Press, Inc.
200 Madison Avenue, New York, New York 10016

British Library Cataloguing in Publication Data
Trollope, Anthony.
The Small House at Allington.–(World's classics)
I. Title
823'.8 PR5684.S655 80-40660
ISBN 0-19-281552-0

Library of Congress Cataloging-in-Publication Data
Trollope, Anthony, 1815–1882.
The small house at Allington / Anthony Trollope ;
edited by James R. Kincaid.
p. cm. — (The Barsetshire novels)
ISBN 0-19-520810-2 (cloth, U.S.)
I. Kincaid, James R. (James Russell) II. Title.
III. Series: Trollope, Anthony, 1815–1882. Barsetshire novels.
PR5684.S55 1989 823'.8—dc20 89-8717 CIP

Printed in the United States of America

The Barsetshire Novels

THE WARDEN (1855)
BARCHESTER TOWERS (1857)
DOCTOR THORNE (1858)
FRAMLEY PARSONAGE (1861)
THE SMALL HOUSE AT ALLINGTON (1864)
THE LAST CHRONICLE OF BARSET (1867)

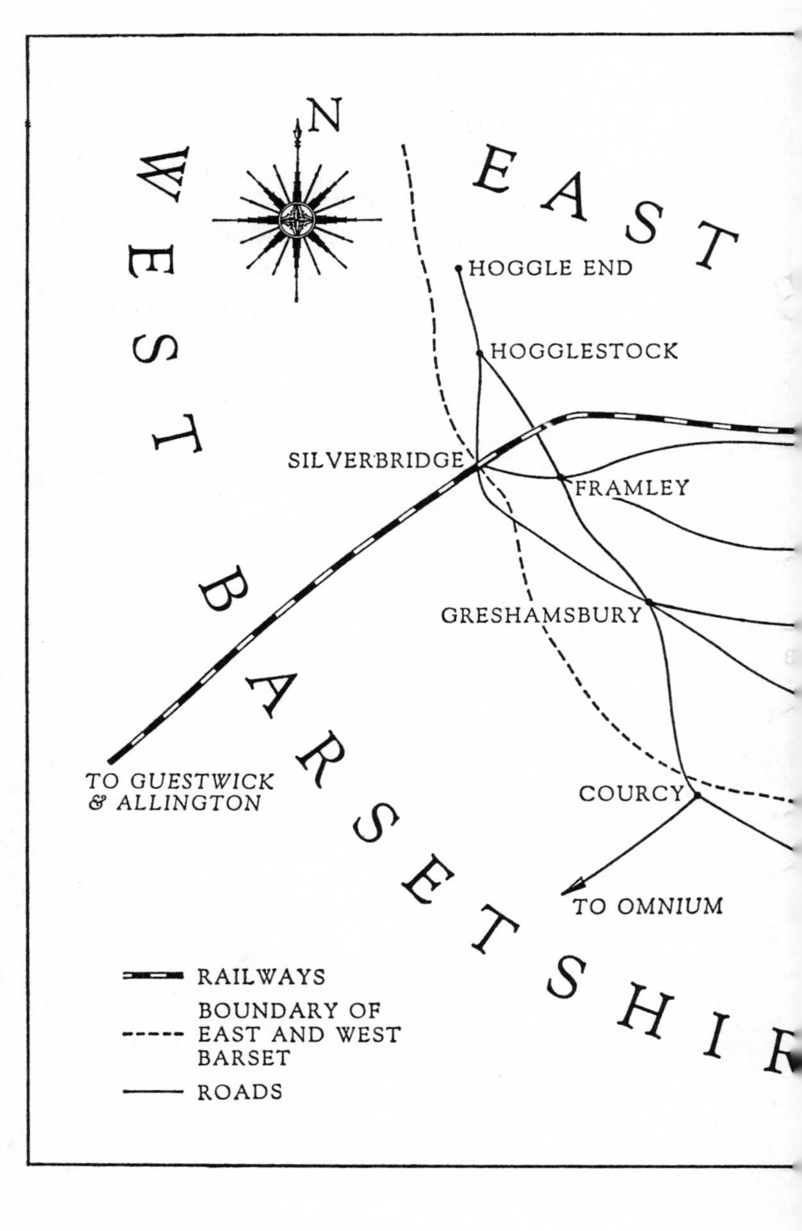
N
WEST BARSETSHIR
EAST
HOGGLE END
HOGGLESTOCK
SILVERBRIDGE
FRAMLEY
GRESHAMSBURY
COURCY
TO GUESTWICK
& ALLINGTON
TO OMNIUM
RAILWAYS
BOUNDARY OF
EAST AND WEST
BARSET
ROADS

BARSETSHIRE

after the reconstruction by

MGR. *RONALD KNOX*

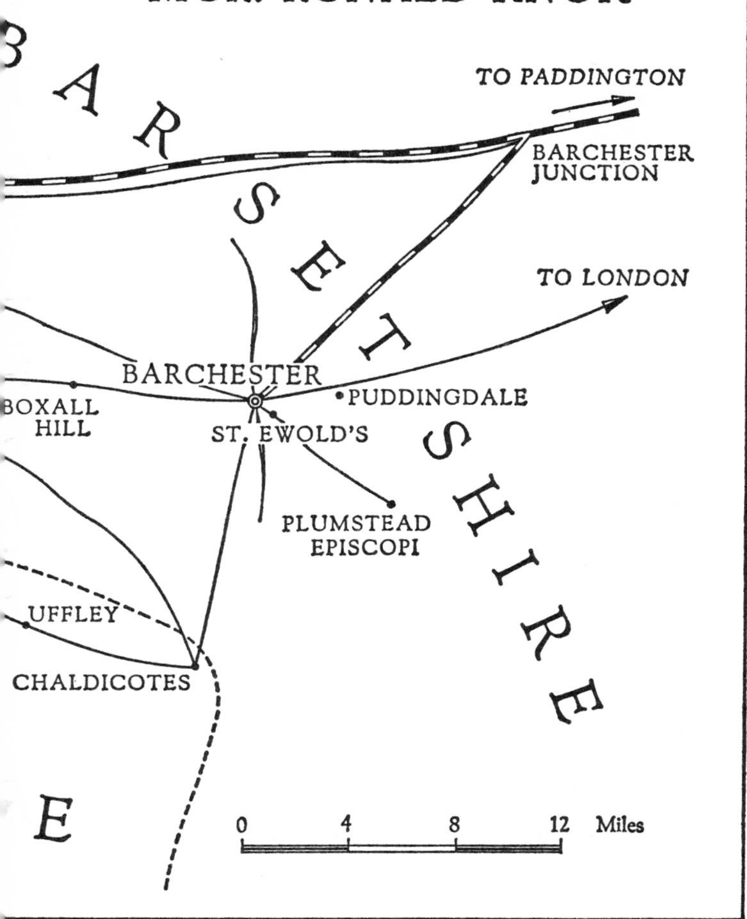

INTRODUCTION

The enormous popular success of *The Small House at Allington* appears to have struck Trollope as gratifying, bemusing, and somewhat annoying. He saw that it was Lily Dale who had 'made her way into the hearts of many readers', and he saw further that she had done so 'because she could not get over her troubles'. He steadfastly resisted the pleas of correspondents begging him to marry Lily to Johnny Eames. 'Had I done so,' he says, 'Lily would never have so endeared herself to these as to induce them to write letters to the author concerning her fate.' Perhaps understanding more than his readers or perhaps too close to her own peculiar mental state, he became irritated by her: 'In the love with which she has been greeted I have hardly joined with much enthusiasm, feeling that she is somewhat of a female prig.' In the succeeding novel in the Barsetshire series, *The Last Chronicle of Barset*, Trollope's annoyance with her becomes clear, but in the writing of *The Small House* he seems to have understood perfectly and presented with relentless unsentimentality Lily's peculiar blend of masochism and pathetic constancy to a world and values that nowhere exist.

'Outside of Lily Dale and the chief interest in the novel,' Trollope added, '*The Small House at Allington* is, I think, good.' A word now reserved for hopeless mediocrity, 'good' was, for Trollope, about the strongest expression of praise he allowed himself. And, despite his reservations as regards Lily, he was clearly not unsatisfied with the popularity of the novel. He received for it £3000, exactly triple the amount he had received for *Framley Parsonage*, which had established his reputation as a writer and as a commercial success. Beyond that, he tells us that *The Small House* 'redeemed my reputation' with the *Cornhill*, which, after the triumph of *Framley Parsonage*, had had to suffer through the tedious, pseudo-Dickensian *Struggles of Brown, Jones and Robinson*.

Confirmation of his ability and his earning power was

especially important to Trollope. He was, at the time the first serial instalment of *The Small House* appeared in the *Cornhill* (Sept. 1862), 47 years of age and had been writing at an astonishing pace for more than 15 years while pursuing actively and cantankerously a career with the General Post Office. It was as a Deputy Surveyor of rural letter deliveries in south-west England that he had visited Salisbury and laid claim to that imaginative landscape that was to become Barsetshire. He stayed on at the Post Office for another five years after *The Small House*, at which time the income from his novels no longer made such work necessary. By that time he had published 21 novels, most of them triple-deckers, two long travel books (on North America and on the West Indies), two volumes of short stories, three volumes of miscellaneous sketches, and many essays in periodicals.

By his own later standards, Trollope had been desultory and unproductive during this time. He was, no doubt, inspired by the image of his courageously industrious mother, who had taken up a literary career at the age of 50 in order to rescue her family from penury and had produced, he tells us, 114 volumes in the next 26 years, some of those written while she was nursing in the next room a dying husband and two dying children. 'Her career', Trollope quietly says, 'offers great encouragement to those who have not begun early in life.'

Great encouragement of a different sort came from Trollope's lifelong attempt to cover with success and popularity the terrible and shameful wounds he had received as a child. Pushed by a father who was both tragically unsuccessful in all he undertook and socially ambitious for his sons, Trollope was sent to a series of famous public schools – Harrow and Winchester most notably – each apparently offering more humiliation to him than the last. He was, he says, poor, ill-dressed, big, awkward, ugly, and sullen: a perfect target for the notorious and often vicious hazing. Even his older brother joined in, flogging him daily at one period with a big stick. 'I feel convinced in my mind that I have been flogged oftener than any human being alive,' he says. He was an outcast, taught to think of himself for many years after as 'an evil, an encumbrance, a useless thing, – as a creature of whom those connected with him had to be ashamed.' It is doubtful that he ever totally released himself from this sad self-portrait.

But he, like his mother, never gave up trying. He wrote in all spare hours when he wasn't chasing down postal problems or chasing after the hounds. Trollope's candid (or perhaps tongue-in-cheek) account of his writing practices is well-known but perhaps not so well-understood. He paid his groom £5 a year extra to wake him early so that he could begin work at 5.30 and complete the day's literary labours before dressing for breakfast. He wrote with his watch before him and asked from himself 250 words every 15 minutes – and got them. The literary world of 1883, committed to the austere doctrines of Henry James and to a self-congratulatory idea of 'genius' more extreme than anything the early Romantics would have dreamed of, seized on Trollope's bland accounting and used it as an indictment both of Trollope and of the tradition he represented. As a result of this and other factors, his reputation suffered greatly and has only very recently begun to recover.

The whole episode marks an interesting chapter in the history of taste, or perhaps only in the power of smugness and dogmatism. Trollope's position, and it is a venerable one, sees art primarily as a craft, the artist as a maker, not an inspired historian creating a naïve illusion of verisimilitude. Trollope had great disdain for the myth of the fastidious genius waiting for inspiration, arguing that a shoemaker might as well blame an off day on a reluctant muse. Trollope was perfectly willing to admit that he was no genius. He was not, however, merely the mechanic, producing so many words per minute on demand. He was, he says, able to write so quickly because he had mastered the craft. Further, he thought so intently during the day of what he was to write the next morning that there was no need to chew his pencil and stare at the wall. All this suggests an imagination that is not feeble but intense, disciplined, and focused. It is an imagination nursed, as he says, on a habit developed during his lonely schooldays, a habit of constructing in his head long narratives with himself as the hero. This habit he simply continued, with the single change of substituting other heroes and heroines for himself. Trollope had, then, a very long and rigorous apprenticeship. It should not surprise us that he was able to write so quickly and with such apparent effortlessness – once he had found a congenial mode.

But that took some doing. The early Irish novels are indeed interesting, but they are a series of false starts: tragedies, an excursion into historical romance, an extended diatribe after the manner of Carlyle, and a play. Even *The Warden* has a heavy vein of satire, a manner and point of view Trollope came more and more to dislike, and *Barchester Towers,* famous as it is, is not very Trollopean: its humour is often very broad indeed; its villains are pretty clearly marked; its moral dilemmas are muted and undramatic.

It is only with *Doctor Thorne* and *Framley Parsonage* that Trollope found his characteristic manner: a complex blend of the comedy of manners tradition he found in Jane Austen, the great nineteenth-century realist tradition, and the particular Trollopean complications of the disrupting, sly narrator and the contradictory subplots. It is in this Barchester series – *The Warden, Barchester Towers, Doctor Thorne, Framley Parsonage, The Small House at Allington,* and *The Last Chronicle of Barset* – that Trollope is able to find adequate space for exploring his truly radical distrust of plot and his reliance on an open, expansive exploration of situation and character.

While Trollope was for a time reluctant to include *The Small House* in the series, Allington not being in Barsetshire proper, he was clearly right to yield to the pressure of friends and publishers and put it there; for all six novels share the same basic pattern: an invasion, always from London, is launched against a small rural community and its values. The clash is between a world settled, conservative, easy in its morals and practices and a newer, more progressive, ruthless, and subtly inhuman one. The pattern is an ancient one reflected in the pastoral and its recall of a simpler, more stable world, one more kindly, if, in Trollope at least, possibly more corrupt. While Trollope does not idealize the fading world of old feudal England, neither does he regard with unalloyed delight the new, hasty, brash arena that he figures in London. He watches as the new reformers, usually well-meaning and armed with abstract and unassailable principles, spread out from London into the unsuspecting and virtually unarmed country. The early London Invaders, John Bold in *The Warden* and Mr. Slope in *Barchester Towers,* are energetic and, in an abstract sense, often right. They cause great furore, pangs of conscience among the natives; but in the end they are dismissed: John Bold conveniently dies and Mr. Slope is sent packing. By the time of *Framley Parsonage* there is hardly a threat and the pastoral world seems secure. The enemy is not only beaten back but forced to yield territory and rights. The novel ends with a perfect riot of marriages, and the assurances to us and to the pastoral world are very nearly absolute.

But in *The Small House* they are altogether gone. At a point that looks like a comic climax, the hero, John Eames, saves the Lord De Guest from an attack by a bull and wins the Lord's powerful support and friendship. The Lord is a great believer in pastoral values and in his own shepherding powers: 'Guided by faith in his own teaching the earl had taught himself to look upon his bull as a large, horned, innocent lamb of the flock.' The wonderful ability to deny empirical fact is not shaken by this experience at all: 'The gentlest creature alive; he's a lamb generally – just like a lamb. Perhaps he saw my red pocket-handkerchief.' Johnny thus seems to be protecting the latest chief shepherd, who will now confer pastoral blessings on him. Indeed, that is Lord De Guest's firm intention, and he goes to work with an open hand and a very warm heart to arrange the proper marriages and secure the proper alignments to support the old values. But none of it comes off. Suddenly the magic power is gone, and the major values are unrealized.

As a result, this novel is far and away the darkest of the series, so dark that it has sometimes been dismissed by lovers of Trollope who expected an uninterrupted idyllic series. Adolphus Crosbie is the first really powerful invader from London, and the pastoral world seems to collapse before him. The degeneration of the aristocracy, that is to say, the growth in power of liberal aristocrats, which had seemed to be checked in *Framley Parsonage,* is now out of control. The De Courcy people here operate like a nineteenth-century version of the Mafia, with equal power and equal terror. They tease Crosbie, whom they snatch hold of very quickly, about 'going about with a crook' at Allington, and soon teach him to distrust the comic and pastoral values. And he proceeds nearly to smash that world. It is scarcely redeeming that he is to some extent smashed himself.

In *The Small House,* as never before, the pastoral seems a small island of virtue surrounded by conditions which are, in their essence, incapable of resolution. The paradigmatic activity in this new ironic world is Mrs. Roper's. She runs a 'genteel' boarding house for miscellaneous sorts in London: 'Poor woman! Few positions in life could be harder to bear than hers! To be ever tugging at others for money that they could not pay; to desire respectability for its own sake but to be driven to confess that it was a luxury beyond her means; to put up with disreputable belongings for the sake of lucre, and then not to get the lucre.' Her daughter Amelia expresses even more succinctly the dominant and pointless immorality: she says she has been a knave and a fool, and 'both for nothing'.

In such a world nothing seems stable or connected. True virtue, therefore, is unsupported and can depend only upon itself. It is thus very likely to appear or to become perverse. It has none of the communal reliance which could make that virtue lie easily, unconsciously with the virtuous man. When, therefore, it is clutched firmly, as it must be, it becomes abstracted and unnaturally firm, removed from the rhythms of change and delicate modification that control comedy. Those, then, who would in a better world be chief actors in a natural comedy are now seen specifically as unnatural. Thus the problem of the novel is finally the problem of Lily Dale and the peculiar, twisted psychological position she finds herself in. At one point the narrator comments that 'it is the view which the mind takes of a thing which creates the sorrow that arises from it'. Lily's own view may seem so outrageously arbitrary that we often want to shake her. The temptation to attack her is almost irresistible. As we have seen, Trollope himself found it more than he could resist. But he also saw that her brilliantly portrayed suspension from the natural currents of comedy was at the heart of the book. There is no easy explanation for Lily's state in psychological terms. That she is attracted to pain is certain, but, as in Squire Dale's case, that attraction is partly based on a certain and generally accurate expectation of pain in any case. Neither Lily nor Trollope is purely masochistic or perverse, or, if we choose to think they are, a perception of the condition is less important than an understanding of its causes.

Lily is not the only character who is firm unto perversity. Firmness is a characteristic of all those whom we are asked to respect in this novel. 'When did you ever know Christopher Dale change his mind?' asks Mrs. Hearn. Or any other Dale, for that matter. The theme of constancy is kept alive by frequent reiteration and by parodies in such people as the Hon. John De Courcy, who declares, 'they'll find no change in me', and in his sister, Amelia Gazebee, who has 'done her duty in her new sphere of life with some constancy and a fixed purpose'. The fixed purpose is something close to legal gangsterism, but she is constant to it. In its serious reflections, such constancy represents the last grim stand of pastoral values. It is the necessary reaction to the fluidity of all bonds. The insistence on a constancy at all costs is the inevitable and very dangerous last assertion of permanence in an unstable world. The Dales and those about them have resisted the movements of the world at large, but their very resistance creates their vulnerability. The attempt to retain innocence in a fallen world leads finally to a mad fixity that displaces them from the natural world they sought to inhabit. It renders them unable to join the supple currents of a flexible nature: 'Was she not a Dale? And when did a Dale change his mind?'

Ironically, this is said of Lily's sister Bell, who is one Dale who does, in fact, change – and change radically. She is rewarded for her change by being allowed to participate in the novel's only fully comic action. Other plots move toward comedy, but none is allowed to reach its destination, and Lily's plot is derailed entirely. The basic rhetorical strategy here is to play off the lack of fulfilment and resolution in Lily's life and others' against very powerful currents of natural comedy. The novel makes us see clearly as we did in *Framley Parsonage* that a comic resolution is demanded, but here one is never presented. Everything in the novel moves toward comedy except the action. The major tension is thus established and the appropriate rhetoric of frustration produced. The formal conflict is arranged mostly through the intricate structural parallels in the four main actions: Bell's rejection of Bernard Dale and final marriage to Dr. Crofts, the movement of Mrs. Dale and the Squire toward greater understanding, the growth of John Eames into manhood, and the story of Lily and Crosbie.

The first plot is by far the least noticeable. Bell rejects the arranged love set up for her with the wooden Bernard, who 'had his feelings well under control' so well that his tenacity in clinging to her seems entirely impersonal, a light parody of the twisted constancy elsewhere. Bell has plenty of firmness of her own: 'If there was anything in this world as to which Isabella Dale was quite certain, it was this – that she was not in love with Dr. Crofts.' But of course that is what she is – in love with Dr. Crofts. Nature is allowed this one victory over unnatural, self-punishing rigidity. Such a triumph shows us what should, but does not, happen elsewhere.

One level more prominent but also one level more complex is the comic rejuvenation of Squire Christopher Dale. The Squire is one of those Trollope characters who is introduced with a long list of faults and a very short list of virtues, often even made up of spillovers from the vices: an idle man does not, at any rate, commit violent acts; a wrathful man is not idle. But there is always a quiet climax to such lists that renders the other traits superficial: 'And, moreover, our Mr. Christopher Dale was a gentleman.' We recognize immediately the signal intended here. Mr. Christopher Dale is the moral touchstone of the novel; he is a gentleman, and, moreover, his house is possessed of that Trollopean emblem of approval: Tudor windows. The symbolism is quite unmistakable. Dale is concerned with the future of his estate, the continuity of family, and he therefore becomes deeply involved in the affairs of young people. All this sounds like a score of Trollope's other secret heroes, representatives of the conscience of the county. Here, however, the Squire lives in constant expectation of being thwarted. No one even pays attention to what he says. He is misunderstood and alone, cut off both from his fellow squires and from his tenants, really from the entire world: 'It makes me feel that the world is changed, and that it is no longer worth a man's while to live in it.' He could stand that feeling – it might even grow into the sort of happy grievance Trollope's Tories love – but he is not able to live easily with the hostility of his own family, his sister-in-law and nieces at the Small House: 'You and the girls have been living here, close to me, for – how many years is it now? – and during all those years there has grown up for me no kindly feeling. Do you suppose that I am a fool and do not know?'

Mrs. Dale understands the full force of his complaint and begins to understand the full warmth of his heart. In doing so, she begins to come to life herself. She had vowed to 'bury herself in order that her daughters might live well above ground', another unnatural resolution virtue is forced to make, one the narrator flatly says is 'wrong'. Mrs. Dale secretly thinks that it is wrong too, finds no masochistic pleasure in self-denial, and frets about getting back into life. The pressure of this romantic and comic movement is so great that we are likely not to attend to the Squire's protestations: 'What, begin again at near seventy! No, Mary, there is no more beginning again for me.' But, though he does manage to come more to the surface, offering Lily money and working actively in the conspiracy to help her, his rebirth is never complete. He never entails the estate or arranges a marriage for Bernard and is troubled by his failure to provide for the property. Correspondingly, his psychological growth is also suspended and the narrator can only say at the end, 'he was a man for whom we may predicate some gentle sadness and continued despondency to the end of his life's chapter.'

But the Squire has a much more hopeful counterpart, the Earl De Guest, who refuses to give up on the pastoral world. When Dale tells him that the time for renewal 'has never come to you and me', his friend vigorously denies it: '"Yes it has", said the earl, with no slight touch of feeling and even of romance in what he said. "We have retricked our beams in our own ways, and our lives have not been desolate."' The similarity of the earl's life to that of Dale is stressed, but the earl lives in a different world altogether. He is purely of the country, living with a cosy disdain for London. He has never abandoned the comic premises his life has, on the whole, affirmed. He was poor, but now he is comfortable. He has, unlike Dale, solidified the estate and become a part of it: 'He knew every acre of his own estate, and every tree upon it.' Because he believes so firmly in innocent and beneficent change, he can himself practise a healthy constancy, not the one that is steadfast to pain but one that is loyal to happy alterations, satisfactory endings. He is, potentially, another Prospero, and when Johnny Eames saves him from the bull – more precisely, saves him from having to readjust his principles – the earl vows to support his young friend with all his comic constancy: 'Now, good-night, my dear fellow, and remember this – when I say a thing I mean it. I think I may boast that I never yet went back from my word.'

Johnny is the perfect natural hero: generous, open, and imaginative. He is Lord De Guest in an earlier stage of development, as the earl clearly recognizes. Johnny's faults are purely those which easy, natural education will remedy. He comes straight out of an irresistible tradition that rewards the gentle, the meek, the good-hearted. But here the tradition is resisted, just at the last. Johnny's education is, of course, conducted along the standard lines. He learns the first lesson of a Trollope gentleman, his comparative insignificance: 'I made a fool of myself, and have been a fool all along. I am foolish now to tell you this, but I cannot help it.' His insight and his impulsiveness are, according to the tradition, sure signs that in the very process of acknowledging himself to be a fool, saying so because he 'cannot help it', he is actively demonstrating his wisdom and his good heart. He can, therefore, survive a rough training period in London, the territory of the enemy. He plunges directly into a hellish world. There everyone struggles to hold on to connections, to bind people by force to vows that are always being broken. Every motive seems perverse, truly as masochistic as poor Cradell's 'moth-like weakness' for Mrs. Lupex's candle. Trollope specifically refers to this period as John's initiation and makes it seem all the more real by making it so very unsentimental. John escapes without cost to himself, but others are made to pay, especially the pathetic Amelia Roper: 'But the world had been hard to her; knocking her about hither and thither unmercifully; threatening, as it now threatened, to take from her what few good things she enjoyed.' John tries to slither out of this fluid world of the boarding-house with a few platitudes to Amelia about how it is all for the best, how 'we should never be happy'. But Amelia's startling response brings into focus for a moment the dark world from which for so many there is no escape: 'I should be happy – very happy indeed.' But 'John Eames becomes a man' and manages to 'come out of the fire comparatively unharmed'. He has so much on his side: 'You have everybody in your favour – the squire, her mother, and all.' The 'all' includes here not only the earl, who is speaking, but the whole tradition of romantic comedy. But though he can thrash the villain and win the heart of Lady Julia De Guest, Johnny cannot win the heart of Lily. The energies of the tradition are thus allowed full rein and are suddenly blocked, to our great discomfort.

Trollope's narrator calls attention to this countering of tradition at the very end of the novel with a mock apology: 'I feel that I have been in fault in giving such prominence to a hobbledehoy, and that I should have told my story better had I brought Mr. Crosbie more conspicuously forward on my canvas. He at any rate has gotten to himself a wife – as a hero always should do.' Crosbie, who is, the narrator insists, 'not altogether a villain', gets what he perhaps deserves – nothing at all. He is punished somewhat by his marriage, but his wife soon flees, and he is liberated from definite punishment into a more appropriate emptiness. The form properly resists either punishing or rewarding him. Here, as elsewhere, resolution is denied. The novel makes it difficult to respond to Crosbie in any simple way. Though a genuine scoundrel, he really never meant harm. And he is a victim of Courcy Castle, which 'had tended to destroy all that was good and true within him.' Ironically, he finds it much easier to break the oath he has made to the constant Lily than that he has made to the slippery Courcy clan. There is a much subtler sense too in which we recognize that he is running to the Courcy people to escape another kind of victimization from Lily. After the engagement, Lily puts a sort of pressure on him that makes him feel caged and on display: 'And then she exacted from him the repetition of the promise which he had so often given her.' Surely this is Amelia Roper on a more advanced or just less self-conscious level. Lily throws herself, as it were, into Crosbie's arms and then looks up beaming, 'Yes, your own, to take when you please, and leave untaken while you please; and as much your own in one way as in the other.' He is understandably a bit uncomfortable with the burden and the sly trap it creates for him; the possession threatens to possess the owner. Lily says she desires to 'do everything for you. I sometimes think that a very poor man's wife is the happiest, because she does do everything.' There is a desire for power here that exposes how much of her excessive self-effacement, her exaggerated submission to Crosbie, is really a cry of triumph. Crosbie hears the bray and retreats. There is, then, a cutting sarcasm at work when Lily's sentimental and deliberately cute resolutions to punish herself for forgetting how much Crosbie is giving up by marrying her are echoed seriously a page or two later by Crosbie, who comes to believe her. Perhaps he *is* giving up too much.

But Lily's sentimental, mock desire for punishment becomes, in her painful humiliation, genuine perversity. She recognizes that she can discover no reason for her tenacity. At first she declares, 'I believe, in my heart, that he still loves me', but her firmness is not shaken by clear evidence that he does not. Like a parody of Lord De Guest, she turns away from all evidence and embraces a world of absolutisms: 'I have made up my mind about it all clearly and with an absolute certainty.' Lily is not, then, just a masochist but a sentimental idealist, one who, unlike her sister, prefers novels whose capacity to minister to wish-fulfilment is greatest. Her pride contributes to her firmness, too, but Lily represents the attempt of the pressured pastoral world to reach out desperately for some stability. The great agent of comic fulfilment, Lord De Guest, says at the end that time will cure all, that Lily, like other girls, will change in accord with the gentle pressures of love, sex, growth. But the earl is wrong, and his hope for an innocent comic world where all bulls are really lambs is never realized.

JAMES R. KINCAID

THE WORLD'S CLASSICS

THE SMALL HOUSE AT ALLINGTON

THE SMALL HOUSE AT ALLINGTON

*

CHAPTER I

THE SQUIRE OF ALLINGTON

OF course there was a Great House at Allington. How otherwise should there have been a Small House? Our story will, as its name imports, have its closest relations with those who lived in the less dignified domicile of the two; but it will have close relations also with the more dignified, and it may be well that I should, in the first instance, say a few words as to the Great House and its owner.

The squires of Allington had been squires of Allington since squires, such as squires are now, were first known in England. From father to son, from uncle to nephew, and, in one instance, from second cousin to second cousin, the sceptre had descended in the family of the Dales; and the acres had remained intact, growing in value and not decreasing in number, though guarded by no entail* and protected by no wonderful amount of prudence or wisdom. The estate of Dale of Allington had been coterminous with the parish of Allington for some hundreds of years; and though, as I have said, the race of squires had possessed nothing of superhuman discretion, and had perhaps been guided in their walks through life by no very distinct principles, still there had been with them so much of adherence to a sacred law, that no acre of the property had ever been parted from the hands of the existing squire. Some futile attempts had been made to increase the territory, as indeed had been done by Kit Dale, the father of Christopher Dale, who will appear as our squire of Allington when the persons of our drama are introduced. Old Kit Dale, who had married money, had bought outlying farms, – a bit of ground here and a bit there, – talking, as he did so, much of political influence and of the good old Tory cause. But these farms and bits of ground had gone again before our time. To them had been attached no religion. When old Kit had found himself pressed in that matter of the majority of the Nineteenth Dragoons, in which crack regiment his second son made for himself quite a career, he found it easier to sell than to save – seeing that that which he sold was his own and not the patrimony of the Dales. At his death the remainder of these purchases had gone. Family arrangements required completion, and Christopher Dale required ready money. The outlying farms flew away, as such new purchases had flown before; but the old patrimony of the Dales remained untouched, as it had ever remained.

It had been a religion among them; and seeing that the worship had been carried on without fail, that the vestal fire had never gone down upon the hearth, I should not have said that the Dales had walked their ways without high principle. To this religion they had all adhered, and the new heir had ever entered in upon his domain without other encumbrances than those with which he himself was already burdened. And yet there had been no entail. The idea of an entail was not in accordance with the peculiarities of the Dale mind. It was necessary to the Dale religion that each squire should have the power of wasting the acres of Allington, – and that he should abstain from wasting them. I remember to have dined at a house, the whole glory and fortune of which depended on the safety of a glass goblet. We all know the story. If the luck of Edenhall* should be shattered, the doom of the family would be sealed. Nevertheless I was bidden to drink out of the fatal glass, as were all guests in that house. It would not have contented the chivalrous mind of the master to protect his doom by lock and key and padded chest. And so it was with the Dales of Allington. To them an entail would have been a lock and key and a padded chest; but the old chivalry of their house denied to them the use of such protection.

I have spoken something slightingly of the acquirements and doings of the family; and indeed, their acquirements had been few and their doings little. At Allington, Dale of

Allington had always been known as a king. At Guestwick, the neighbouring market town, he was a great man – to be seen frequently on Saturdays, standing in the market-place, and laying down the law as to barley and oxen among men who knew usually more about barley and oxen than did he. At Hamersham, the assize town, he was generally in some repute, being a constant grand juror for the county, and a man who paid his way. But even at Hamersham the glory of the Dales had, at most periods, begun to pale, for they had seldom been widely conspicuous, in the county, and had earned no great reputation by their knowledge of jurisprudence in the grand jury room. Beyond Hamersham their fame had not spread itself.

They had been men generally built in the same mould, inheriting each from his father the same virtues and the same vices, – men who would have lived, each, as his father had lived before him, had not the new ways of the world gradually drawn away with them, by an invisible magnetism, the up-coming Dale of the day, – not indeed in any case so moving him as to bring him up to the spirit of the age in which he lived, but dragging him forward to a line in advance of that on which his father had trodden. They had been obstinate men; believing much in themselves; just according to their ideas of justice; hard to their tenants – but not known to be hard even by the tenants themselves, for the rules followed had ever been the rules on the Allington estate; imperious to their wives and children, but imperious within bounds, so that no Mrs. Dale had fled from her lord's roof, and no loud scandals had existed between father and sons; exacting in their ideas as to money, expecting that they were to receive much and to give little, and yet not thought to be mean, for they paid their way, and gave money in parish charity and in county charity. They had ever been steady supporters of the Church, graciously receiving into their parish such new vicars as, from time to time, were sent to them from King's College, Cambridge, to which establishment the gift of the living belonged; – but, nevertheless, the Dales had ever carried on some unpronounced warfare against the clergyman, so that the intercourse between the lay family and the clerical had seldom been in all respects pleasant.

Such had been the Dales of Allington, time out of mind, and such in all respects would have been the Christopher Dale of our time, had he not suffered two accidents in his youth. He had fallen in love with a lady who obstinately refused his hand, and on her account he had remained single; that was his first accident. The second had fallen upon him with reference to his father's assumed wealth. He had supposed himself to be richer than other Dales of Allington when coming in upon his property, and had consequently entertained an idea of sitting in Parliament for his county. In order that he might attain this honour, he had allowed himself to be talked by the men of Hamersham and Guestwick out of his old family politics, and had declared himself a Liberal. He had never gone to the poll, and, indeed, had never actually stood for the seat. But he had come forward as a Liberal politician, and had failed; and, although it was well known to all around that Christopher Dale was in heart as thoroughly Conservative as any of his forefathers, this accident had made him sour and silent on the subject of politics, and had somewhat estranged him from his brother squires.

In other respects our Christopher Dale was, if anything, superior to the average of the family. Those whom he did love he loved dearly. Those whom he hated he did not ill-use beyond the limits of justice. He was close in small matters of money, and yet in certain family arrangements he was, as we shall see, capable of much liberality. He endeavoured to do his duty in accordance with his lights, and had succeeded in weaning himself from personal indulgences, to which during the early days of his high hopes he had become accustomed. And in that matter of his unrequited love he had been true throughout. In his hard, dry, unpleasant way he had loved the woman; and when at last he learned to know that she would not have his love, he had been unable to transfer his heart to another. This had happened just at the period of his father's death, and he had endeavoured to console himself with politics, with what fate we have already seen. A constant, upright, and by no means insincere man was our Christopher Dale, – thin and meagre in his mental attributes, by no means even understanding the fulness of a full man, with power of eye-sight very limited in seeing aught which was above him, but yet worthy of regard in that he had realized a path of duty and did endeavour to walk therein. And, moreover, our Mr. Christopher Dale was a gentleman.

Such in character was the squire of Allington, the only regular inhabitant of the Great House. In person, he was a plain, dry man, with short grizzled hair and thick grizzled eyebrows. Of beard, he had very little, carrying the smallest possible grey whiskers, which hardly fell below the points of his ears. His eyes were sharp and expressive, and his nose was straight and well formed, – as was also his chin. But the nobility of his face was destroyed by a mean mouth with thin lips; and his forehead, which was high and narrow, though it forbad you to take Mr. Dale for a fool, forbad you also to take him for a man of great parts, or of a wide capacity. In height, he was about five feet ten; and at the time of our story was as near to seventy as he was to sixty. But years had treated him very lightly, and he bore few signs of age. Such in person was Christopher Dale, Esq., the squire of Allington, and owner of some three thousand a year, all of which proceeded from the lands of that parish.

And now I will speak of the Great House of Allington. After all, it was not very great; nor was it surrounded by much of that exquisite nobility of park appurtenance which graces the habitations of most of our landed proprietors. But the house itself was very graceful. It had been built in the days of the early Stuarts, in that style of architecture to which we give the name of the Tudors. On its front it showed three pointed roofs, or gables, as I believe they should be called; and between each gable a thin tall chimney stood, the two chimneys thus raising themselves just above the three peaks I have mentioned. I think that the beauty of the house depended much on those two chimneys; on them, and on the mullioned windows with which the front of the house was closely filled. The door, with its jutting porch, was by no means in the centre of the house. As you entered, there was but one window on your right hand, while on your left there were three. And over these there was a line of five windows, one taking its place above the porch. We all know the beautiful old Tudor window, with its stout stone mullions and its stone transoms, crossing from side to side at a point much nearer to the top than to the bottom. Of all windows ever invented it is the sweetest. And here, at Allington, I think their beauty was enhanced by the fact that they were not regular in their shape. Some of these windows were long windows, while some of them were high. That to the right of the door, and that at the other extremity of the house, were among the former. But the others had been put in without regard to uniformity, a long window here, and a high window there, with a general effect which could hardly have been improved. Then above, in the three gables, were three other smaller apertures. But these also were mullioned, and the entire frontage of the house was uniform in its style.

Round the house there were trim gardens, not very large, but worthy of much note in that they were so trim, – gardens with broad gravel paths, with one walk running in front of the house so broad as to be fitly called a terrace. But this, though in front of the house, was sufficiently removed from it to allow of a coach-road running inside it to the front door. The Dales of Allington had always been gardeners, and their garden was perhaps more noted in the county than any other of their properties. But outside the gardens no pretensions had been made to the grandeur of a domain. The pastures round the house were but pretty fields, in which timber was abundant. There was no deer-park at Allington; and though the Allington woods were well known, they formed no portion of the whole of which the house was a part. They lay away, out of sight, a full mile from the back of the house; but not on that account of less avail for the fitting preservation of foxes.

And the house stood much too near the road for purposes of grandeur, had such purposes ever swelled the breast of any of the squires of Allington. But I fancy that our ideas of rural grandeur have altered since many of our older country seats were built. To be near the village, so as in some way to afford comfort, protection, and patronage, and perhaps also with some view to the pleasantness of neighbourhood for its own inmates, seemed to be the object of a gentleman when building his house in the old days. A solitude in the centre of a wide park is now the only site that can be recognized as eligible. No cottage must be seen, unless the cottage orné of the gardener. The village, if it cannot be abolished, must be got out of sight. The sound of the church bells is not desirable, and the road on which the profane vulgar travel by their own right must be at a distance. When some old Dale of Allington built his house, he thought differently. There stood the church and there the village, and, pleased with such vicinity, he sat himself down close to his God and to his tenants.

As you pass along the road from Guestwick into the village you see the church near to you on your left hand; but the house is hidden from the road. As you approach the church, reaching the gate of it which is not above two hundred yards from the high road, you see the full front of the Great House. Perhaps the best view of it is from the churchyard. The lane leading up to the church ends in a gate, which is the entrance into Mr. Dale's place. There is no lodge there, and the gate generally stands open, – indeed, always does so, unless some need of cattle grazing within requires that it should be closed. But there is an inner gate, leading from the home paddock, through the gardens to the house, and another inner gate, some thirty yards further on, which will take you into the farm-yard. Perhaps it is a defect at Allington that the farm-yard is very close to the house. But the stables, and the straw-yards, and the unwashed carts, and the lazy lingering cattle of the homestead, are screened off by a row of chestnuts, which, when in its glory of flower, in the early days of May, no other row in England can surpass in beauty. Had any one told Dale of Allington – this Dale or any former Dale – that his place wanted wood, he would have pointed with mingled pride and disdain to his belt of chestnuts.

Of the church itself I will say the fewest possible number of words. It was a church such as there are, I think, thousands in England – low, incommodious, kept with difficulty in repair, too often pervious to the wet, and yet strangely picturesque, and correct too, according to great rules of architecture. It was built with a nave and aisles, visibly in the form of a cross, though with its arms clipped down to the trunk, with a separate chancel, with a large square short tower, and with a bell-shaped spire, covered with lead and irregular in its proportions. Who does not know the low porch, the perpendicular Gothic window, the flat-roofed aisles, and the noble old gray tower of such a church as this? As regards its interior, it was dusty; it was blocked up with high-backed ugly pews; the gallery in which the children sat at the end of the church, and in which two ancient musicians blew their bassoons, was all awry, and looked as though it would fall; the pulpit was an ugly useless edifice, as high nearly as the roof would allow, and the reading-desk under it hardly permitted the parson to keep his head free from the dangling tassels of the cushion above him. A clerk also was there beneath him, holding a third position somewhat elevated; and upon the whole things there were not quite as I would have had them. But, nevertheless, the place looked like a church, and I can hardly say so much for all the modern edifices which have been built in my days towards the glory of God. It looked like a church, and not the less so because in walking up the passage between the pews, the visitor trod upon the brass plates which dignified the resting-places of the departed Dales of old.

Below the church, and between that and the village, stood the vicarage, in such position that the small garden of the vicarage stretched from the churchyard down to the backs of the village cottages. This was a pleasant residence, newly built within the last thirty years, and creditable to the ideas of comfort entertained by the rich collegiate body from which the vicars of Allington always came. Doubtless we shall in the course of our sojourn at Allington visit the vicarage now and then, but I do not know that any further detailed account of its comforts will be necessary to us.

Passing by the lane leading to the vicarage, the church and to the house, the high road descends rapidly to a little brook which runs through the village. On the right as you descend, you will have seen, the 'Red Lion', and will have seen no other house conspicuous in any way. At the bottom, close to the brook, is the post-office, kept surely by the crossest old woman in all those parts. Here the road passes through the water, the accommodation of a narrow wooden bridge having been afforded for those on foot. But before passing the stream, you will see a cross street, running to the left, as had run that other lane leading to the house. Here, as this cross street rises the hill, are the best houses in the village. The baker lives here, and that respectable woman, Mrs. Frummage, who sells ribbons, and toys, and soap, and straw bonnets, with many other things too long to mention. Here, too, lives an apothecary, whom the veneration of this and neighbouring parishes has raised to the dignity of a doctor. And here also, in the smallest but prettiest cottage that can be imagined, lives Mrs. Hearn, the widow of a former vicar, on terms, however, with her neighbour the squire which I regret to say are not as friendly as they should be. Beyond this lady's modest residence, Allington Street, for so the road is called, turns suddenly round towards the church, and at the point of the turn is a pretty low iron railing with a gate, and with a covered way, which leads up to the front door of the house which stands there. I will only say here, at this fag end of a chapter, that it is the Small House at Allington. Allington Street, as I have said, turns short round towards the church at this point, and there ends at a white gate, leading into the churchyard by a second entrance.

So much it was needful that I should say of Allington Great House, of the Squire, and of the village. Of the Small House, I will speak separately in a further chapter.

CHAPTER II

THE TWO PEARLS OF ALLINGTON

'BUT Mr. Crosbie is only a mere clerk.'

This sarcastic condemnation was spoken by Miss Lilian Dale to her sister Isabella, and referred to a gentleman with whom we shall have much concern in these pages. I do not say that Mr. Crosbie will be our hero, seeing that that part in the drama will be cut up, as it were, into fragments. Whatever of the magnificent may be produced will be diluted and apportioned out in very moderate quantities among two or more, probably among three or four, young gentlemen – to none of whom will be vouchsafed the privilege of much heroic action.

'I don't know what you call a mere clerk, Lily. Mr. Fanfaron is a mere barrister, and Mr. Boyce is a mere clergyman.' Mr. Boyce was the vicar of Allington, and Mr. Fanfaron was a lawyer who had made his way over to Allington during the last assizes. 'You might as well say that Lord De Guest is a mere earl.'

'So he is – only a mere earl. Had he ever done anything except have fat oxen, one wouldn't say so. You know what I mean by a mere clerk? It isn't much in a man to be in a public office, and yet Mr. Crosbie gives himself airs.'

'You don't suppose that Mr. Crosbie is the same as John Eames,' said Bell, who, by her tone of voice, did not seem inclined to undervalue the qualifications of Mr. Crosbie. Now John Eames was a young man from Guestwick, who had been appointed to a clerkship in the Income-tax Office, with eighty pounds a year, two years ago.

'Then Johnny Eames is a mere clerk,' said Lily; 'and Mr. Crosbie is— After all, Bell, what is Mr. Crosbie, if he is not a mere clerk? Of course, he is older than John Eames; and, as he has been longer at it, I suppose he has more than eighty pounds a year.'

'I am not in Mr. Crosbie's confidence. He is in the General Committee Office, I know; and, I believe, has pretty nearly the management of the whole of it. I have heard Bernard say that he has six or seven young men under him, and that—; but, of course, I don't know what he does at his office.'

'I'll tell you what he is, Bell; Mr. Crosbie is a swell.' And Lilian Dale was right; Mr. Crosbie was a swell.

And here I may perhaps best explain who Bernard was, and who was Mr. Crosbie. Captain Bernard Dale was an officer in the corps of Engineers, was the first cousin of the two girls who have been speaking, and was nephew and heir presumptive to the squire. His father, Colonel Dale, and his mother, Lady Fanny Dale, were still living at Torquay – an effete, invalid, listless couple, pretty well dead to all the world beyond the region of the Torquay card-tables. He it was who had made for himself quite a career in the Nineteenth Dragoons. This he did by eloping with the penniless daughter of that impoverished earl, the Lord De Guest. After the conclusion of that event, circumstances had not afforded him the opportunity of making himself conspicuous; and he had gone on declining gradually in the world's esteem – for the world had esteemed him when he first made good his running with the Lady Fanny – till now, in his slippered years, he and his Lady Fanny were unknown except among those Torquay Bath chairs and card-tables. His elder brother was still a hearty man, walking in thick shoes, and constant in his saddle; but the colonel, with nothing beyond his wife's title to keep his body awake, had fallen asleep somewhat prematurely among his slippers. Of him and of Lady Fanny, Bernard Dale was the only son. Daughters they had had; some were dead, some married, and one living with them among the card-tables. Of his parents Bernard had latterly not seen much; not more, that is, than duty and a due attention to the fifth commandment required of him. He also was making a career for himself, having obtained a commission in the Engineers, and being known to all his compeers as the nephew of an earl, and as the heir to a property of three thousand a year. And when I say that Bernard Dale was not inclined to throw away any of these advantages, I by no means intend to speak in his dispraise. The advantage of being heir to a good property is so manifest – the advantages over and beyond those which are merely fiscal – that no man thinks of throwing them away, or expects another man to do so. Moneys in possession or in expectation do give a set to the head, and a confidence to the voice, and an assurance to the man, which will help him much in his walk in life – if the owner of them will simply use them, and not abuse them. And for Bernard Dale I will say that he did not often talk of his uncle the earl. He was conscious that his uncle was an earl, and that other men knew the fact. He knew that he would not otherwise have been elected at the Beaufort, or at that most aristocratic of little clubs, called Sebright's. When noble blood was called in question, he never alluded specially to his own, but he knew how to speak as one of whom all the world was aware on which side he had been placed by the circumstances of his birth. Thus he used his advantage, and did not abuse it. And in his profession he had been equally fortunate. By industry, by a small but wakeful intelligence, and by some aid from patronage, he had got on till he had almost achieved the reputation of talent. His name had become known among scientific experimentalists, not as that of one who had himself invented a cannon or an antidote to a cannon, but as of a man understanding in cannons, and well fitted to look at those invented by others; who would honestly test this or that antidote; or, if not honestly, seeing that such thin-minded men can hardly go to the proof of any matter without some pre-judgment in their minds, at any rate with such appearance of honesty that the world might be satisfied. And in this way Captain Dale was employed much at home, about London; and was not called on to build barracks in Nova Scotia, or to make roads in the Punjaub.

He was a small slight man, smaller than his uncle, but in face very like him. He had the same eyes, and nose, and chin, and the same mouth; but his forehead was better, – less high and pointed, and better formed about the brows. And then he wore moustaches, which somewhat hid the thinness of his mouth. On the whole, he was not ill-looking; and, as I have said before, he carried with him an air of self-assurance and a confident balance, which in itself gives a grace to a young man.

He was staying at the present time in his uncle's house, during the delicious warmth of the summer, – for, as yet, the month of July was not all past; and his intimate friend, Adolphus Crosbie, who was or was not a mere clerk as my readers may choose to form their own opinions on that matter, was a guest in the house with him. I am inclined to say that Adolphus Crosbie was not a mere clerk; and I do not think that he would have been so called, even by Lily Dale, had he not given signs to her that he was a 'swell'. Now a man in becoming a swell, – a swell of such an order as could possibly be known to Lily Dale, – must have ceased to be a mere clerk in that very process. And, moreover, Captain Dale would not have been Damon to any Pythias* of whom it might fairly be said that he was a mere clerk. Nor could any mere clerk have got himself in either at the Beaufort or at Sebright's. The evidence against that former assertion made by Lily Dale is very strong; but then the evidence as to her latter assertion is as strong. Mr. Crosbie certainly was a swell. It is true that he was a clerk in the General Committee Office. But then, in the first place, the General Committee Office is situated in Whitehall; whereas poor John Eames was forced to travel daily from his lodgings in Burton Crescent, ever so far beyond Russell Square, to his dingy room in Somerset House. And Adolphus Crosbie, when very young, had been a private secretary, and had afterwards mounted up in his office to some quasi authority and senior-clerkship, bringing him in seven hundred a year, and giving him a status among assistant secretaries and the like, which even in an official point of view was something. But the triumphs of Adolphus Crosbie had been other than these. Not because he had been intimate with assistant secretaries, and was allowed in Whitehall a room to himself with an arm-chair, would he have been entitled to stand upon the rug at Sebright's, and speak while rich men listened, – rich men, and men also who had handles to their names! Adolphus Crosbie had done more than make minutes with discretion on the papers of the General Committee Office. He had set himself down before the gates of the city of fashion, and had taken them by storm; or, perhaps, to speak with more propriety, he had picked the locks and let himself in. In his walks of life he was somebody in London. A man at the West End who did not know who was Adolphus Crosbie knew nothing. I do not say that he was the intimate friend of many great men; but even great men acknowledged the acquaintance of Adolphus Crosbie, and he was to be seen in the drawing-rooms, or at any rate on the staircases, of Cabinet Ministers.

Lilian Dale, dear Lily Dale – for my reader must know that she is to be very dear, and that my story will be nothing to him if he do not love Lily Dale – Lilian Dale had discovered that Mr. Crosbie was a swell. But I am bound to say that Mr. Crosbie did not habitually proclaim the fact in any offensive manner; nor in becoming a swell had he become altogether a bad fellow. It was not to be expected that a man who was petted at Sebright's should carry himself in the Allington drawing-room as would Johnny Eames, who had never been petted by any one but his mother. And this fraction of a hero of ours had other advantages to back him, over and beyond those which fashion had given him. He was a tall, well-looking man, with pleasant eyes and an expressive mouth, – a man whom you would probably observe in whatever room you might meet him. And he knew how to talk, and had in him something which justified talking. He was no butterfly or dandy, who flew about in the world's sun, warmed into prettiness by a sunbeam. Crosbie had his opinion on things, – on politics, on religion, on the philanthropic tendencies of the age, and had read something here and there as he formed his opinion. Perhaps he might have done better in the world had he not been placed so early in life in that Whitehall public office. There was that in him which might have earned better bread for him in an open profession.

But in that matter of his bread the fate of Adolphus Crosbie had by this time been decided for him, and he had reconciled himself to fate that was now inexorable. Some very slight patrimony, a hundred a year or so, had fallen to his share. Beyond that he had his salary from his office, and nothing else; and on his income, thus made up, he had lived as a bachelor in London, enjoying all that London could give him as a man in moderately easy circumstances, and looking forward to no costly luxuries, – such as a wife, a house of his own, or a stable full of horses. Those which he did enjoy of the good things of the world would, if known to John Eames, have made him appear fabulously rich in the eyes of that brother clerk. His lodgings in Mount Street were elegant in their belongings. During three months of the season in London he called himself the master of a very neat hack. He was always well dressed, though never over-dressed. At his clubs he could live on equal terms with men having ten times his income. He was not married. He had acknowledged to himself that he could not marry without money; and he would not marry for money. He had put aside from him, as not within his reach, the comforts of marriage. But— We will not, however, at the present moment inquire more curiously into the private life and circumstances of our new friend Adolphus Crosbie.

After the sentence pronounced against him by Lilian, the two girls remained silent for awhile. Bell was, perhaps, a little angry with her sister. It was not often that she allowed herself to say much in praise of any gentleman; and, now that she had spoken a word or two in favour of Mr. Crosbie, she felt herself to be rebuked by her sister for this unwonted enthusiasm. Lily was at work on a drawing, and in a minute or two had forgotten all about Mr. Crosbie; but the injury remained on Bell's mind, and prompted her to go back to the subject. 'I don't like those slang words, Lily.'

'What slang words?'

'You know what you called Bernard's friend.'

'Oh; a swell. I fancy I do like slang. I think it's awfully jolly to talk about things being jolly. Only that I was afraid of your nerves I should have called him stunning. It's so slow, you know, to use nothing but words out of a dictionary.'

'I don't think it's nice in talking of gentlemen.'

'Isn't it? Well, I'd like to be nice – if I knew how.'

If she knew how! There is no knowing how, for a girl, in that matter. If nature and her mother have not done it for her, there is no hope for her on that head. I think I may say that nature and her mother had been sufficiently efficacious for Lilian Dale in this respect.

'Mr. Crosbie is, at any rate, a gentleman, and knows how to make himself pleasant. That was all that I meant. Mamma said a great deal more about him than I did.'

'Mr. Crosbie is an Apollo; and I always look upon Apollo as the greatest – you know what – that ever lived. I mustn't say the word, because Apollo was a gentleman.'

At this moment, while the name of the god was still on her lips, the high open window of the drawing-room was darkened, and Bernard entered, followed by Mr. Crosbie.

'Who is talking about Apollo?' said Captain Dale.

The girls were both stricken dumb. How would it be with them if Mr. Crosbie had heard himself spoken of in those last words of poor Lily's? This was the rashness of which Bell was ever accusing her sister, and here was the result! But, in truth, Bernard had heard nothing more than the name, and Mr. Crosbie, who had been behind him, had heard nothing.

'"As sweet and musical as bright Apollo's lute, strung with his hair,"' said Mr. Crosbie, not meaning much by the quotation, but perceiving that the two girls had been in some way put out and silenced.

'What very bad music it must have made,' said Lily; 'unless, indeed, his hair was very different from ours.'

'It was all sunbeams,' suggested Bernard. But by that time Apollo had served his turn, and the ladies welcomed their guests in the proper form.

'Mamma is in the garden,' said Bell, with that hypocritical pretence so common with young ladies when young gentlemen call; as though they were aware that mamma was the object specially sought.

'Picking peas, with a sun-bonnet on,' said Lily.

'Let us by all means go and help her,' said Mr. Crosbie; and then they issued out into the garden.

The gardens of the Great House of Allington and those

of the Small House open on to each other. A proper boundary of thick laurel hedge, and wide ditch, and of iron spikes guarding the ditch, there is between them; but over the wide ditch there is a foot-bridge, and at the bridge there is a gate which has no key; and for all purposes of enjoyment the gardens of each house are open to the other. And the gardens of the Small House are very pretty. The Small House itself is so near the road that there is nothing between the dining-room windows and the iron rail but a narrow edge rather than border, and a little path made with round fixed cobble stones, not above two feet broad, into which no one but the gardener ever makes his way. The distance from the road to the house is not above five or six feet, and the entrance from the gate is shut in by a covered way. But the garden behind the house, on to which the windows from the drawing-room open, is to all the senses as private as though there were no village of Allington, and no road up to the church within a hundred yards of the lawn. The steeple of the church, indeed, can be seen from the lawn, peering, as it were, between the yew-trees which stand in the corner of the churchyard adjoining to Mrs. Dale's wall. But none of the Dale family have any objection to the sight of that steeple. The glory of the Small House at Allington certainly consists in its lawn, which is as smooth, as level, and as much like velvet as grass has ever yet been made to look. Lily Dale, taking pride in her own lawn, has declared often that it is no good attempting to play croquet up at the Great House. The grass, she says, grows in tufts, and nothing that Hopkins, the gardener, can or will do has any effect upon the tufts. But there are no tufts at the Small House. As the squire himself has never been very enthusiastic about croquet, the croquet implements have been moved permanently down to the Small House, and croquet there has become quite an institution.

And while I am on the subject of the garden I may also mention Mrs. Dale's conservatory, as to which Bell was strenuously of opinion that the Great House had nothing to offer equal to it – 'For flowers, of course, I mean,' she would say, correcting herself; for at the Great House there was a grapery very celebrated. On this matter the squire would be less tolerant than as regarded the croquet, and would tell his niece that she knew nothing about flowers. 'Perhaps not, uncle Christopher,' she would say. 'All the same, I like our geraniums best;' for there was a spice of obstinacy about Miss Dale, – as, indeed, there was in all the Dales, male and female, young and old.

It may be as well to explain that the care of this lawn and of this conservatory, and, indeed, of the entire garden belonging to the Small House, was in the hands of Hopkins, the head gardener to the Great House; and it was so simply for this reason, that Mrs. Dale could not afford to keep a gardener herself. A working lad, at ten shillings a week, who cleaned the knives and shoes, and dug the ground, was the only male attendant on the three ladies. But Hopkins, the head gardener of Allington, who had men under him, was as widely awake to the lawn and the conservatory of the humbler establishment as he was to the grapery, peach-walls, and terraces of the grander one. In his eyes it was all one place. The Small House belonged to his master, as indeed did the very furniture within it; and it was lent, not let, to Mrs. Dale. Hopkins, perhaps, did not love Mrs. Dale, seeing that he owed her no duty as one born a Dale. The two young ladies he did love, and also snubbed in a very peremptory way sometimes. To Mrs. Dale he was coldly civil, always referring to the squire if any direction worthy of special notice as concerning the garden was given to him.

All this will serve to explain the terms on which Mrs. Dale was living at the Small House, – a matter needful of explanation sooner or later. Her husband had been the youngest of three brothers, and in many respects the brightest. Early in life he had gone up to London, and there had done well as a land surveyor. He had done so well that Government had employed him, and for some three or four years he had enjoyed a large income, but death had come suddenly on him, while he was only yet ascending the ladder; and, when he died, he had hardly begun to realize the golden prospects which he had seen before him. This had happened some fifteen years before our story commenced, so that the two girls hardly retained any memory of their father. For the first five years of her widowhood, Mrs. Dale, who had never been a favourite of the squire's, lived with her two little girls in such modest way as her very limited means allowed. Old Mrs. Dale, the squire's mother, then occupied the Small House. But when old Mrs. Dale died, the squire offered the place rent-free to his sister-in-law, intimating to her that her daughters would obtain considerable social advantages by living at Allington. She had accepted the offer, and the social advantages had certainly followed. Mrs. Dale was poor, her whole income not exceeding three hundred a year, and therefore her own style of living was of necessity very unassuming; but she saw her girls becoming popular in the county, much liked by the families around them, and enjoying nearly all the advantages which would have accrued to them had they been the daughters of Squire Dale of Allington. Under such circumstances it was little to her whether or no she were loved by her brother-in-law, or respected by Hopkins. Her own girls loved her, and respected her, and that was pretty much all that she demanded of the world on her own behalf.

And uncle Christopher had been very good to the girls in his own obstinate and somewhat ungracious manner. There were two ponies in the stables of the Great House, which they were allowed to ride, and which, unless on occasions, nobody else did ride. I think he might have given the ponies to the girls, but he thought differently. And he contributed to their dresses, sending them home now and again things which he thought necessary, not in the pleasantest way in the world. Money he never gave them, nor did he make them any promises. But they were Dales, and he loved them; and with Christopher Dale to love once was to love always. Bell was his chief favourite, sharing with his nephew Bernard the best warmth of his heart. About these two he had his projects, intending that Bell should be the future mistress of the Great House of Allington; as to which project however, Miss Dale was as yet in very absolute ignorance.

We may now, I think, go back to our four friends, as they walked out upon the lawn. They were understood to be on a mission to assist Mrs. Dale in the picking of the peas; but pleasure intervened in the way of business, and the young people, forgetting the labours of their elder, allowed themselves to be carried away by the fascinations of croquet. The iron hoops and the sticks were fixed. The mallets and the balls were lying about; and then the party was so nicely made up! 'I haven't had a game of croquet yet,' said Mr. Crosbie. It cannot be said that he had lost much time, seeing that he had only arrived before dinner on the preceding day. And then the mallets were in their hands in a moment.

'We'll play sides, of course,' said Lily. 'Bernard and I'll play together.' But this was not allowed. Lily was well known to be the queen of the croquet ground; and as Bernard was supposed to be more efficient than his friend, Lily had to take Mr. Crosbie as her partner. 'Apollo can't get through the hoops,' Lily said afterwards to her sister; 'but then how gracefully he fails to do it!' Lily, however, had been beaten, and may therefore be excused for a little spite against her partner. But it so turned out that before Mr. Crosbie took his final departure from Allington he could get through the hoops; and Lily, though she was still queen of the croquet ground, had to acknowledge a male sovereign in that dominion.

'That's not the way we played at—,' said Crosbie, at one point of the game, and then stopped himself.

'Where was that?' said Bernard.

'A place I was at last summer, – in Shropshire.'

'Then they don't play the game, Mr. Crosbie, at the place you were at last summer, – in Shropshire,' said Lily.

'You mean Lady Hartletop's,' said Bernard. Now, the Marchioness of Hartletop was a very great person indeed, and a leader in the fashionable world.

'Oh! Lady Hartletop's!' said Lily. 'Then I suppose we must give in;' which little bit of sarcasm was not lost upon Mr. Crosbie, and was put down by him in the tablets of his mind as quite undeserved. He had endeavoured to avoid any mention of Lady Hartletop and her croquet ground, and her ladyship's name had been forced upon him. Nevertheless, he liked Lily Dale through it all. But he thought he liked Bell the best, though she said little; for Bell was the beauty of the family.

During the game Bernard remembered that they had especially come over to bid the three ladies to dinner at the house on that day. They had all dined there on the day before, and the girls' uncle had now sent directions to them to come again. 'I'll go and ask mamma about it,' said Bell, who was out first. And then she returned, saying, that she and her sister would obey their uncle's behest; but that her mother would prefer to remain at home. 'There are the peas to be eaten, you know,' said Lily.

'Send them up to the Great House,' said Bernard.

'Hopkins would not allow it,' said Lily. 'He calls that a mixing of things. Hopkins doesn't like mixings.' And then when the game was over, they sauntered about, out of the small garden into the larger one, and through the shrubberies, and out upon the fields, where they found the still lingering remnants of the haymaking. And Lily took a rake, and raked for two minutes; and Mr. Crosbie, making an attempt to pitch the hay into the cart, had to pay half-a-crown for his footing to the haymakers; and Bell sat quiet under a tree, mindful of her complexion; whereupon Mr. Crosbie, finding the hay-pitching not much to his taste, threw himself under the same tree also, quite after the manner of Apollo, as Lily said to her mother late in the evening. Then Bernard covered Lily with hay, which was a great feat in the jocose way for him; and Lily, in returning the compliment almost smothered Mr. Crosbie, – by accident.

'Oh, Lily,' said Bell.

'I'm sure I beg your pardon, Mr. Crosbie. It was Bernard's fault. Bernard, I never will come into a hayfield with you again.' And so they all became very intimate; while Bell sat quietly under the tree, listening to a word or two now and then as Mr. Crosbie chose to speak them. There is a kind of enjoyment to be had in society, in which very few words are necessary. Bell was less vivacious than her sister Lily; and when, an hour after this, she was dressing herself for dinner, she acknowledged that she had passed a pleasant afternoon, though Mr. Crosbie had not said very much.

CHAPTER III

THE WIDOW DALE OF ALLINGTON

As Mrs. Dale, of the Small House, was not a Dale by birth, there can be no necessity for insisting on the fact that none of the Dale peculiarities should be sought for in her character. These peculiarities were not, perhaps, very conspicuous in her daughters, who had taken more in that respect from their mother than from their father; but a close observer might recognize the girls as Dales. They were constant, perhaps obstinate, occasionally a little uncharitable in their judgment, and prone to think that there was a great deal in being a Dale, though not prone to say much about it. But they had also a better pride than this, which had come to them as their mother's heritage.

Mrs. Dale was certainly a proud woman, – not that there was anything appertaining to herself in which she took a pride. In birth she had been much lower than her husband, seeing that her grandfather had been almost nobody. Her fortune had been considerable for her rank in life, and on its proceeds she now mainly depended; but it had not been sufficient to give any of the pride of wealth. And she had been a beauty; according to my taste, was still very lovely; but certainly at this time of life, she, a widow of fifteen years' standing, with two grown-up daughters, took no pride in her beauty. Nor had she any conscious pride in the fact that she was a lady. That she was a lady, inwards and outwards, from the crown of her head to the sole of her feet, in head, in heart, and in mind, a lady by education and a lady by nature, a lady also by birth, in spite of that deficiency respecting her grandfather, I hereby state as a fact – meo periculo.* And the squire, though he had no special love for her, had recognized this, and in all respects treated her as his equal.

But her position was one which required that she should either be very proud or else very humble. She was poor, and yet her daughters moved in a position which belongs, as a rule, to the daughters of rich men only. This they did as nieces of the childless squire of Allington, and as his nieces she felt that they were entitled to accept his countenance and kindness, without loss of self-respect either to her or to them. She would have ill done her duty as a mother to them had she allowed any pride of her own to come between them and such advantage in the world as their uncle might be able to give them. On their behalf she had accepted the loan of the house in which she lived, and the use of many of the appurtenances belonging to her brother-in-law; but on her own account she had accepted nothing. Her marriage with Philip Dale had been disliked by his brother the squire, and the squire, while Philip was still living, had continued to show that his feelings in this respect were not to be overcome. They never had been overcome; and now, though the brother-in-law and sister-in-law had been close neighbours for years, living as one may say almost in the same family, they had never become friends. There had not been a word of quarrel between them. They met constantly. The squire had unconsciously come to entertain a profound respect for his brother's widow. The widow had acknowledged to herself the truth of the affection shown by the uncle to her daughters. But yet they had never come together as friends. Of her own money matters Mrs. Dale had never spoken a word to the squire. Of his intention respecting the girls the squire had never spoken a word to the mother. And in this way they had lived and were living at Allington.

The life which Mrs. Dale led was not altogether an easy life, – was not devoid of much painful effort on her part. The theory of her life one may say was this – that she should bury herself in order that her daughters might live well above ground. And in order to carry out this theory, it was necessary that she should abstain from all complaint or show of uneasiness before her girls. Their life above ground would not be well if they understood that their mother, in this underground life of hers, was enduring any sacrifice on their behalf. It was needful that they should think that the picking of peas in a sun-bonnet, or long readings by her own fire-side, and solitary hours spent in thinking, were specially to her mind. 'Mamma doesn't like going out.' 'I don't think mamma is happy anywhere out of her own drawing-room.' I do not say that the girls were taught to say such words, but they were taught to have thoughts which led to such words, and in the early days of their going out into the world used so to speak of their mother. But a time came to them before long, – to one first and then to the other, in which they knew that it was not so, and knew also that their mother had suffered for their sakes.

And in truth Mrs. Dale could have been as young in heart as they were. She, too, could have played croquet, and have coquetted with a haymaker's rake, and have delighted in her pony, ay, and have listened to little nothings from this and that Apollo, had she thought that things had been conformable thereto. Women at forty do not become ancient misanthropes, or stern Rhadamanthine moralists,* indifferent to the world's pleasures – no, not even though they be widows. There are those who think that such should be the phase of their minds. I profess that I do not so think. I would have women, and men also, young as long as they can be young. It is not that a woman should call herself in years younger than her father's family Bible will have her to be. Let her who is forty call herself forty; but if she can be young in spirit at forty, let her show that she is so.

I think that Mrs. Dale was wrong. She would have joined that party on the croquet ground, instead of remaining among the pea-sticks in her sun-bonnet, had she done as I would have counselled her. Not a word was spoken among the four that she did not hear. Those pea-sticks were only removed from the lawn by a low wall and a few shrubs. She listened, not as one suspecting, but simply as one loving. The voices of her girls were very dear to her, and the silver ringing tones of Lily's tongue were as sweet to her ears as the music of the gods. She heard all that about Lady Hartletop, and shuddered at Lily's bold sarcasm. And she heard Lily say that mamma would stay at home and eat the peas, and said to herself sadly that that was now her lot in life.

'Dear, darling girl – and so it should be!'

It was thus her thoughts ran. And then, when her ear had traced them, as they passed across the little bridge into the other grounds, she returned across the lawn to the house with her burden on her arm, and sat herself down on the step of the drawing-room window, looking out on the sweet summer flowers and the smooth surface of the grass before her.

Had not God done well for her to place her where she was? Had not her lines been set for her in pleasant places? Was she not happy in her girls, – her sweet, loving, trusting, trusty children? As it was to be that her lord, the best half of herself, was to be taken from her in early life, and that the springs of all the lighter pleasures were to be thus stopped for her, had it not been well that in her bereavement so much had been done to soften her lot in life and give it grace and beauty? 'Twas so, she argued with herself, and yet she acknowledged to herself that she was not happy. She had resolved, as she herself had said often, to put away childish things, and now she pined for those things which she so put from her. As she sat she could still hear Lily's voice as they went through the shrubbery, – hear it when none but a mother's ears would have distinguished the sound. Now that those young men were at the Great House, it was natural that her girls should be there too. The squire would not have had young men to stay with him had there been no ladies to grace his table. But for her, – she knew that no one would want her there. Now and again she must go, as otherwise her very existence, without going, would be a thing disagreeably noticeable. But there was no other reason why she should join the party; nor in joining it would she either give or receive pleasure. Let her daughters eat from her brother's table and drink of his cup. They were made welcome to do so from the heart. For her there was no such welcome as that at the Great House, – nor at any other house, or any other table!

'Mamma will stay at home to eat the peas.'

And then she repeated to herself the words which Lily had spoken, sitting there, leaning with her elbow on her knee, and her head upon her hand.

'Please, ma'am, cook says can we have the peas to shell?' and then her reverie was broken.

Whereupon Mrs. Dale got up and gave over her basket. 'Cook knows that the young ladies are going to dine at the Great House?'

'Yes, ma'am.'

'She needn't mind getting dinner for me. I will have tea early.' And so, after all, Mrs. Dale did not perform that special duty appointed for her.

But she soon set herself to work upon another duty. When a family of three persons has to live upon an income of three hundred a year, and, nevertheless, makes some pretence of going into society, it has to be very mindful of small details, even though that family may consist only of ladies. Of this Mrs. Dale was well aware, and as it pleased her that her daughters should be nice and fresh, and pretty in their attire, many a long hour was given up to that care. The squire would send them shawls in winter, and had given them riding habits, and had sent them down brown silk dresses from London, – so limited in quantity that the due manufacture of two dresses out of the material had been found to be beyond the art of woman, and the brown silk garments had been a difficulty from that day to this, – the squire having a good memory in such matters, and being anxious to see the fruits of his liberality. All this was doubtless of assistance, but had the squire given the amount which he so expended in money to his nieces, the benefit would have been greater. As it was, the girls were always nice and fresh and pretty, they themselves not being idle in that matter; but their tire-woman* in chief was their mother. And now she went up to their room and got out their muslin frocks, and – but, perhaps, I should not tell such tales! – She, however, felt no shame in her work, as she sent for a hot iron, and with her own hands smoothed out the creases, and gave the proper set to the crimp flounces, and fixed a new ribbon where it was wanted, and saw that all was as it should be. Men think but little how much of this kind is endured that their eyes may be pleased, even though it be but for an hour.

'Oh! mamma, how good you are,' said Bell, as the two girls came in, only just in time to make themselves ready for returning to dinner.

'Mamma is always good,' said Lily. 'I wish, mamma, I could do the same for you oftener,' and then she kissed her mother. But the squire was exact about dinner, so they dressed themselves in haste, and went off again through the garden, their mother accompanying them to the little bridge.

'Your uncle did not seem vexed at my not coming?' said Mrs. Dale.

'We have not seen him, mamma,' said Lily. 'We have been ever so far down the fields, and forgot altogether what o'clock it was.'

'I don't think uncle Christopher was about the place, or we should have met him,' said Bell.

'But I am vexed with you, mamma. Are not you, Bell? It is very bad of you to stay here all alone, and not come.'

'I suppose mamma likes being at home better than up at the Great House,' said Bell, very gently; and as she spoke she was holding her mother's hand.

'Well; good-bye, dears. I shall expect you between ten and eleven. But don't hurry yourselves if anything is going on.' And so they went, and the widow was again alone. The path from the bridge ran straight up towards the back of the Great House, so that for a moment or two she could see them as they tripped on almost in a run. And then she saw their dresses flutter as they turned sharp round, up the terrace steps. She would not go beyond the nook among the laurels by which she was surrounded, lest any one should see her as she looked after her girls. But when the last flutter of the pink muslin had been whisked away from her sight, she felt it hard that she might not follow them. She stood there, however, without advancing a step. She would not have Hopkins telling how she watched her daughters as they went from her own home to that of her brother-in-law. It was not within the capacity of Hopkins to understand why she watched them.

'Well, girls, you're not much too soon. I think your mother might have come with you,' said uncle Christopher. And this was the manner of the man. Had he known his own wishes he must have acknowledged to himself that he was better pleased that Mrs. Dale should stay away. He felt himself more absolutely master and more comfortably at home at his own table without her company than with it. And yet he frequently made a grievance of her not coming, and himself believed in that grievance.

'I think mamma was tired,' said Bell.

'Hem. It's not very far across from one house to the other. If I were to shut myself up whenever I'm tired— But never mind. Let's go to dinner. Mr. Crosbie, will you take my niece Lilian.' And then, offering his own arm to Bell, he walked off to the dining-room.

'If he scolds mamma any more, I'll go away,' said Lily to her companion; by which it may be seen that they had all become very intimate during the long day that they had passed together.

Mrs. Dale, after remaining for a moment on the bridge, went in to her tea. What succedaneum* of mutton chop or broiled ham she had for the roast duck and green peas which were to have been provided for the family dinner we will not particularly inquire. We may, however, imagine that she did not devote herself to her evening repast with any peculiar energy of appetite. She took a book with her as she sat herself down, – some novel, probably, for Mrs. Dale was not above novels, – and read a page or two as she sipped her tea. But the book was soon laid on one side, and the tray on which the warm plate had become cold was neglected, and she threw herself back in her own familiar chair, thinking of herself, and of her girls, and thinking also what might have been her lot in life had he lived who had loved her truly during the few years that they had been together.

It is especially the nature of a Dale to be constant in his likings and his dislikings. Her husband's affection for her had been unswerving, – so much so that he had quarrelled with his brother because his brother would not express himself in brotherly terms about his wife; but, nevertheless, the two brothers had loved each other always. Many years had now gone by since these things had occurred, but still the same feelings remained. When she had first come down to Allington she had resolved to win the squire's regard, but she had now long known that any such winning was out of the question; indeed, there was no longer a wish for it. Mrs.

Dale was not one of those soft-hearted women who sometimes thank God that they can love any one. She could once have felt affection for her brother-in-law, – affection, and close, careful, sisterly friendship; but she could not do so now. He had been cold to her, and had with perseverance rejected her advances. That was now seven years since; and during those years Mrs. Dale had been, at any rate, as cold to him as he had been to her.

But all this was very hard to bear. That her daughters should love their uncle was not only reasonable, but in every way desirable. He was not cold to them. To them he was generous and affectionate. If she were only out of the way, he would have taken them to his house as his own, and they would in all respects have stood before the world as his adopted children. Would it not be better if she were out of the way?

It was only in her most dismal moods that this question would get itself asked within her mind, and then she would recover herself, and answer it stoutly with an indignant protest against her own morbid weakness. It would not be well that she should be away from her girls, – not though their uncle should have been twice a better uncle; not though, by her absence, they might become heiresses of all Allington. Was it not above everything to them that they should have a mother near them? And as she asked of herself that morbid question, – wickedly asked it, as she declared to herself, – did she not know that they loved her better than all the world beside, and would prefer her caresses and her care to the guardianship of any uncle, let his house be ever so great? As yet they loved her better than all the world beside. Of other love, should it come, she would not be jealous. And if it should come, and should be happy, might there not yet be a bright evening of life for herself? If they should marry, and if their lords would accept her love, her friendship, and her homage, she might yet escape from the deathlike coldness of that Great House, and be happy in some tiny cottage, from which she might go forth at times among those who would really welcome her. A certain doctor there was, living not very far from Allington, at Guestwick, as to whom she had once thought that he might fill that place of son-in-law, – to be well-beloved. Her quiet, beautiful Bell had seemed to like the man; and he had certainly done more than seem to like her. But now, for some weeks past, this hope, or rather this idea, had faded away. Mrs. Dale had never questioned her daughter on the matter; she was not a woman prone to put such questions. But during the month or two last past, she had seen with regret that Bell looked almost coldly on the man whom her mother favoured.

In thinking of all this the long evening passed away, and at eleven o'clock she heard the coming steps across the garden. The young men had, of course, accompanied the girls home; and as she stepped out from the still open window of her own drawing-room, she saw them all on the centre of the lawn before her.

'There's mamma,' said Lily. 'Mamma, Mr. Crosbie wants to play croquet by moonlight.'

'I don't think there is light enough for that,' said Mrs. Dale.

'There is light enough for him,' said Lily, 'for he plays quite independently of the hoops; don't you, Mr. Crosbie?'

'There's a very pretty croquet light, I should say,' said Mr. Crosbie, looking up at the bright moon; 'and then it is so stupid going to bed.'

'Yes, it is stupid going to bed,' said Lily; but people in the country are stupid, you know. Billiards, that you can play all night by gas, is much better, isn't it?'

'Your arrows fall terribly astray there, Miss Dale, for I never touch a cue; you should talk to your cousin about billiards.'

'Is Bernard a great billiard player?' asked Bell.

'Well, I do play now and again; about as well as Crosbie does croquet. Come, Crosbie, we'll go home and smoke a cigar.'

'Yes,' said Lily; 'and then, you know, we stupid people can go to bed. Mamma, I wish you had a little smoking-room here for us. I don't like being considered stupid.' And then they parted, – the ladies going into the house, and the two men returning across the lawn.

'Lily, my love,' said Mrs. Dale, when they were all together in her bedroom, 'it seems to me that you are very hard upon Mr. Crosbie.'

'She has been going on like that all the evening,' said Bell.

'I'm sure we are very good friends,' said Lily.

'Oh, very!' said Bell.

'Now, Bell, you're jealous; you know you are.' And then, seeing that her sister was in some slight degree vexed, she went up to her and kissed her. 'She shan't be called jealous; shall she, mamma?'

'I don't think she deserves it,' said Mrs. Dale.

'Now, you don't mean to say that you think I meant anything?' said Lily. 'As if I cared a buttercup about Mr. Crosbie.'

'Or I either, Lily.'

'Of course you don't. But I do care for him very much, mamma. He is such a duck of an Apollo. I shall aways call him Apollo; Phoebus Apollo! And when I draw his picture he shall have a mallet in his hand instead of a bow. Upon my word I am very much obliged to Bernard for bringing him down here; and I do wish he was not going away the day after to-morrow.'

'The day after to-morrow!' said Mrs. Dale. 'It was hardly worth coming for two days.'

'No, it wasn't, – disturbing us all in our quiet little ways just for such a spell as that, – not giving one time even to count his rays.'

'But he says he shall perhaps come again,' said Bell.

'There is that hope for us,' said Lily. 'Uncle Christopher asked him to come down when he gets his long leave of absence. This is only a short sort of leave. He is better off than poor Johnny Eames. Johnny Eames only has a month, but Mr. Crosbie has two months just whenever he likes it; and seems to be pretty much his own master all the year round besides.'

'And uncle Christopher asked him to come down for the shooting in September,' said Bell.

'And though he didn't say he'd come I think he meant it,' said Lily. 'There is that hope for us, mamma.'

'Then you'll have to draw Apollo with a gun instead of a mallet.'

'That is the worst of it, mamma. We shan't see much of him or of Bernard either. They wouldn't let us go out into the woods as beaters, would they?'

'You'd make too much noise to be of any use.'

'Should I? I thought the beaters had to shout at the birds. I should get very tired of shouting at birds, so I think I'll stay at home and look after my clothes.'

'I hope he will come, because uncle Christopher seems to like him so much,' said Bell.

'I wonder whether a certain gentleman at Guestwick will like his coming,' said Lily. And then, as soon as she had spoken the words, she looked at her sister, and saw that she had grieved her.

'Lily, you let your tongue run too fast,' said Mrs. Dale.

'I didn't mean anything, Bell,' said Lily. 'I beg your pardon.'

'It doesn't signify,' said Bell. 'Only Lily says things without thinking.' And then that conversation came to an end, and nothing more was said among them beyond what appertained to their toilet, and a few last words at parting. But the two girls occupied the same room, and when their own door was closed upon them, Bell did allude to what had passed with some spirit.

'Lily, you promised me,' she said, 'that you would not say anything more to me about Dr. Crofts.'

'I know I did, and I was very wrong. I beg your pardon, Bell; and I won't do it again, – not if I can help it.'

'Not help it, Lily!'

'But I'm sure I don't know why I shouldn't speak of him, – only not in the way of laughing at you. Of all the men I ever saw in my life I like him best. And only that I love you better than I love myself I could find it in my heart to grudge you his—'

'Lily, what did you promise just now?'

'Well; after to-night. And I don't know why you should turn against him.'

'I have never turned against him or for him.'

'There's no turning about him. He'd give his left hand if you'd only smile on him. Or his right either, – and that's what I should like to see; so now you've heard it.'

'You know you are talking nonsense.'

'So I should like to see it. And so would mamma, too, I'm sure; though I never heard her say a word about him. In my mind he's the finest fellow I ever saw. What's Mr. Apollo Crosbie to him? And now, as it makes you unhappy, I'll never say another word about him.'

As Bell wished her sister good-night with perhaps more than her usual affection, it was evident that Lily's words and eager tone had in some way pleased her, in spite of their opposition to the request which she had made. And Lily was aware that it was so.

CHAPTER IV

MRS. ROPER'S BOARDING-HOUSE

I HAVE said that John Eames had been petted by none but his mother, but I would not have it supposed, on this account, that John Eames had no friends. There is a class of young men who never get petted, though they may not be the less esteemed, or perhaps loved. They do not come forth to the world as Apollos, nor shine at all, keeping what light they may have for inward purposes. Such young men are often awkward, ungainly, and not yet formed in their gait; they straggle with their limbs, and are shy; words do not come to them with ease, when words are required, among any but their accustomed associates. Social meetings are periods of penance to them, and any appearance in public will unnerve them. They go much about alone, and blush when women speak to them. In truth, they are not as yet men, whatever the number may be of their years; and, as they are no longer boys, the world has found for them the ungraceful name of hobbledehoy.

Such observations, however, as I have been enabled to make on this matter have led me to believe that the hobbledehoy is by no means the least valuable species of the human race. When I compare the hobbledehoy of one or two and twenty to some finished Apollo of the same age, I regard the former as unripe fruit, and the latter as fruit that is ripe. Then comes the question as to the two fruits. Which is the better fruit, that which ripens early – which is, perhaps, favoured with some little forcing apparatus, or which, at least, is backed by the warmth of a southern wall; or that fruit of slower growth, as to which nature works without assistance, on which the sun operates in its own time, – or perhaps never operates if some ungenial shade has been allowed to interpose itself? The world, no doubt, is in favour of the forcing apparatus or of the southern wall. The fruit comes certainly, and at an assured period. It is spotless, speckless, and of a certain quality by no means despicable. The owner has it when he wants it, and it serves its turn. But, nevertheless, according to my thinking, the fullest flavour of the sun is given to that other fruit, – is given in the sun's own good time, if so be that no ungenial shade has interposed itself. I like the smack of the natural growth, and like it, perhaps, the better because that which has been obtained has been obtained without favour.

But the hobbledehoy, though he blushes when women address him, and is uneasy even when he is near them, though he is not master of his limbs in a ball-room, and is hardly master of his tongue at any time, is the most eloquent of beings, and especially eloquent among beautiful women. He enjoys all the triumphs of a Don Juan, without any of Don Juan's heartlessness, and is able to conquer in all encounters, through the force of his wit and the sweetness of his voice. But his eloquence is heard only by his own inner ears, and these triumphs are the triumphs of his imagination.

The true hobbledehoy is much alone, not being greatly given to social intercourse even with other hobbledehoys – a trait in his character which I think has hardly been sufficiently observed by the world at large. He has probably become a hobbledehoy instead of an Apollo, because circumstances have not afforded him much social intercourse; and, therefore, he wanders about in solitude, taking long walks, in which he dreams of those successes which are so far removed from his powers of achievement. Out in the fields, with his stick in his hand, he is very eloquent, cutting off the heads of the springing summer weeds, as he practises his oratory with energy. And thus he feeds an imagination for which those who know him give him but scanty credit, and unconsciously prepares himself for that later ripening, if only the ungenial shade will some day cease to interpose itself.

Such hobbledehoys receive but little petting, unless it be from a mother; and such a hobbledehoy was John Eames when he was sent away from Guestwick to begin his life in the big room of a public office in London. We may say that there was nothing of the young Apollo about him. But yet he was not without friends – friends who wished him well, and thought much of his welfare. And he had a younger sister who loved him dearly, who had no idea that he was a hobbledehoy, being somewhat of a hobbledehoya herself. Mrs. Eames, their mother, was a widow, living in a small house in Guestwick, whose husband had been throughout his whole life an intimate friend of our squire. He had been a man of many misfortunes, having begun the world almost with affluence, and having ended it in poverty. He had lived all his days in Guestwick, having at one time occupied a large tract of land, and lost much money in experimental farming; and late in life he had taken a small house on the outskirts of the town, and there had died, some two years previously to the commencement of this story. With no other man had Mr. Dale lived on terms so intimate; and when Mr. Eames died Mr. Dale acted as executor under his will, and as guardian to his children. He had, moreover, obtained for John Eames that situation under the Crown which he now held.

And Mrs. Eames had been and still was on very friendly terms with Mrs. Dale. The squire had never taken quite kindly to Mrs. Eames, whom her husband had not met till he was already past forty years of age. But Mrs. Dale had made up by her kindness to the poor forlorn woman for any lack of that cordiality which might have been shown to her from the Great House. Mrs. Eames was a poor forlorn woman – forlorn even during the time of her husband's life, but very wobegone now in her widowhood. In matters of importance the squire had been kind to her; arranging for her her little money affairs, advising her about her house and income, also getting for her that appointment for her son. But he snubbed her when he met her, and poor Mrs. Eames held him in great awe. Mrs. Dale held her brother-in-law in no awe, and sometimes gave to the widow from Guestwick advice quite at variance to that given by the squire. In this way there had grown up an intimacy between Bell and Lily and the young Eames, and either of the girls was prepared to declare that Johnny Eames was her own and well-loved friend. Nevertheless, they spoke of him occasionally with some little dash of merriment – as is not unusual with prety girls who have hobbledehoys among their intimate friends, and who are not themselves unaccustomed to the grace of an Apollo.

I may as well announce at once that John Eames, when he went up to London, was absolutely and irretrievably in love with Lily Dale. He had declared his passion in the most moving language a hundred times; but he had declared it only to himself. He had written much poetry about Lily, but he kept his lines safe under double lock and key. When he gave the reins to his imagination, he flattered himself that he might win not only her but the world at large also by his verses; but he would have perished rather than exhibit them to human eye. During the last ten weeks of his life at Guestwick, while he was preparing for his career in London, he hung about Allington, walking over frequently and then walking back again; but all in vain. During these visits he would sit in Mrs. Dale's drawing-room, speaking but little, and addressing himself usually to the mother; but on each occasion, as he started on his long, hot walk, he resolved that he would say something by which Lily might know of his love. When he left for London that something had not been said.

He had not dreamed of asking her to be his wife. John Eames was about to begin the world with eighty pounds a year, and an allowance of twenty more from his mother's purse. He was well aware that with such an income he could not establish himself as a married man in London, and he also felt that the man who might be fortunate enough to win Lily for his wife should be prepared to give her every soft luxury that the world could afford. He knew well that he ought not to expect any assurance of Lily's love; but, nevertheless, he thought it possible that he might give her an assurance of his love. It would probably be in vain. He had no real hope, unless when he was in one of those poetic moods. He had acknowledged to himself, in some indistinct way, that he was no more than a hobbledehoy, awkward, silent, ungainly, with a face unfinished, as it were, or unripe. All this he knew, and knew also that there were Apollos in the world who would be only too ready to carry off Lily in their splendid cars. But not the less did he make up his mind that having loved her once, it behoved him, as a true man, to love her on to the end.

One little word he had said to her when they parted, but it had been a word of friendship rather than of love. He had strayed out after her on to the lawn, leaving Bell alone in the drawing-room. Perhaps Lily, had understood something of the boy's feeling, and had wished to speak kindly to him at parting, or almost more than kindly. There is a silent love which women recognize, and which in some silent way they acknowledge, – giving gracious but silent thanks for the respect which accompanies it.

'I have come to say good-bye, Lily,' said Johnny Eames, following the girl down one of the paths.

'Good-bye, John,' said she, turning round. 'You know how sorry we are to lose you. But it's a great thing for you to be going up to London.'

'Well; yes. I suppose it is. I'd sooner remain here, though.'

'What! stay here, doing nothing! I am sure you would not.'

'Of course, I should like to do something. I mean—'

'You mean that it is painful to part with old friends; and I'm sure that we all feel that at parting with you. But you'll have a holiday sometimes, and then we shall see you.'

'Yes; of course, I shall see you then. I think, Lily, I shall care more about seeing you than anybody.'

'Oh, no, John. There'll be your own mother and sister.'

'Yes, there'll be mother and Mary, of course. But I will come over here the very first day, – that is, if you'll care to see me?'

'We shall care to see you very much. You know that. And – dear John, I do hope you'll be happy.'

There was a tone in her voice as she spoke which almost upset him; or, I should rather say, which almost put him up upon his legs and made him speak; but its ultimate effect was less powerful. 'Do you?' said he, as he held her hand for a few happy seconds. 'And I'm sure I hope you'll always be happy. Good-bye, Lily.' Then he left her, returning to the house, and she continued her walk, wandering down among the trees in the shrubbery, and not showing herself for the next half-hour. How many girls have some such lover as that, – a lover who says no more to them than Johnny Eames then said to Lily Dale, who never says more than that? And yet when, in after years, they count over the names of all who have loved them, the name of that awkward youth is never forgotten.

That farewell had been spoken nearly two years since, and Lily Dale was then seventeen. Since that time, John Eames had been home once, and during his month's holiday had often visited Allington. But he had never improved upon that occasion of which I have told. It had seemed to him that Lily was colder to him than in old days, and he had become, if anything, more shy in his ways with her. He was to return to Guestwick again during this autumn; but, to tell honestly the truth in the matter, Lily Dale did not think or care very much for his coming. Girls of nineteen do not care for lovers of one-and-twenty, unless it be when the fruit has had the advantage of some forcing apparatus or southern wall.

John Eames's love was still as hot as ever, having been sustained on poetry, and kept alive, perhaps, by some close confidence in the ears of a brother clerk; but it is not to be supposed that during these two years he had been a melancholy lover. It might, perhaps, have been better for him had his disposition led him to that line of life. Such, however, had not been the case. He had already abandoned the flute on which he had learned to sound three sad notes before he left Guestwick, and, after the fifth or sixth Sunday, he had relinquished his solitary walks along the towing-path of the Regent's Park Canal. To think of one's absent love is very sweet; but it becomes monotonous after a mile or two of a towing-path, and the mind will turn away to Aunt Sally, the Cremorne Gardens, and financial questions. I doubt whether any girl would be satisfied with her lover's mind if she knew the whole of it.

'I say, Caudle, I wonder whether a fellow could get into a club?'

This proposition was made, on one of those Sunday walks, by John Eames to the friend of his bosom, a brother clerk, whose legitimate name was Cradell, and who was therefore called Caudle by his friends.

'Get into a club? Fisher in our room belongs to a club.'

'That's only a chess-club. I mean a regular club.'

'One of the swell ones at the West End?' said Cradell, almost lost in admiration at the ambition of his friend.

'I shouldn't want it to be particularly swell. If a man isn't a swell, I don't see what he gets by going among those who are. But it is so uncommon slow at Mother Roper's.' Now Mrs. Roper was a respectable lady, who kept a boarding-house in Burton Crescent, and to whom Mrs. Eames had been strongly recommended when she was desirous of finding a specially safe domicile for her son. For the first year of his life in London John Eames had lived alone in lodgings; but that had resulted in discomfort, solitude, and, alas! in some amount of debt, which had come heavily on the poor widow. Now, for the second year, some safer mode of life was necessary. She had learned that Mrs. Cradell, the widow of a barrister, who had also succeeded in getting her son into the Income-tax Office, had placed him in charge of Mrs. Roper; and she, with many injunctions to that motherly woman, submitted her own boy to the same custody.

'And about going to church?' Mrs. Eames had said to Mrs. Roper.

'I don't suppose I can look after that, ma'am,' Mrs. Roper had answered, conscientiously. 'Young gentlemen choose mostly their own churches.'

'But they do go?' asked the mother, very anxious in her heart as to this new life in which her boy was to be left to follow in so many things the guidance of his own lights.

'They who have been brought up steady do so, mostly.'

'He has been brought up steady, Mrs. Roper. He has, indeed. And you won't give him a latch-key?'

'Well, they always do ask for it.'

'But he won't insist, if you tell him that I had rather that he shouldn't have one.'

Mrs. Roper promised accordingly, and Johnny Eames was left under her charge. He did ask for the latch-key, and Mrs. Roper answered as she was bidden. But he asked again, having been sophisticated by the philosophy of Cradell, and then Mrs. Roper handed him the key. She was a woman who plumed herself on being as good as her word, not understanding that any one could justly demand from her more than that. She gave Johnny Eames the key, as doubtless she had intended to do; for Mrs. Roper knew the world, and understood that young men without latch-keys would not remain with her.

'I thought you didn't seem to find it so dull since Amelia came home,' said Cradell.

'Amelia! What's Amelia to me? I have told you everything, Cradell, and yet you can talk to me about Amelia Roper!'

'Come now, Johnny—.' He had always been called Johnny, and the name had gone with him to his office. Even Amelia Roper had called him Johnny on more than one occasion before this. 'You were as sweet to her the other night as though there were no such person as L. D. in existence.' John Eames turned away and shook his head. Nevertheless, the words of his friend were grateful to him. The character of a Don Juan was not unpleasant to his imagination, and he liked to think that he might amuse Amelia Roper with a passing word, though his heart was true to Lilian Dale. In truth, however, many more of the passing words had been spoken by the fair Amelia than by him.

Mrs. Roper had been quite as good as her word when she told Mrs. Eames that her household was composed of herself, of a son who was in an attorney's office, of an ancient maiden cousin, named Miss Spruce, who lodged with her, and of Mr. Cradell. The divine Amelia had not then been living with her, and the nature of the statement which she was making by no means compelled her to inform Mrs. Eames that the young lady would probably return home in the following winter. A Mr. and Mrs. Lupex had also joined the family lately, and Mrs. Roper's house was now supposed to be full.

And it must be acknowledged that Johnny Eames had, in certain unguarded moments, confided to Cradell the secret of a second weaker passion for Amelia. 'She is a fine girl, – a deuced fine girl!' Johnny Eames had said, using a style of language which he had learned since he left Guestwick and Allington. Mr. Cradell, also, was an admirer of the fair sex; and, alas! that I should say so, Mrs. Lupex, at the present moment, was the object of his admiration. Not that he entertained the slightest idea of wronging Mr. Lupex, – a man who was a scene-painter, and knew the world. Mr. Cradell admired Mr. Lupex as a connoisseur, not simply as a man. 'By heavens! Johnny, what a figure that woman has!' he said, one morning, as they were walking to their office.

'Yes; she stands well on her pins.'

'I should think she did. If I understand anything of form,' said Cradell, 'that woman is nearly perfect. What a torso she has!'

From which expression, and from the fact that Mrs. Lupex depended greatly upon her stays and crinoline for such figure as she succeeded in displaying, it may, perhaps, be understood that Mr. Cradell did not understand much about form.

'It seems to me that her nose isn't quite straight,' said Johnny Eames. Now, it undoubtedly was the fact that the nose on Mrs. Lupex's face was a little awry. It was a long, thin nose, which, as it progressed forward into the air, certainly had a preponderating bias towards the left side.

'I care more for figure than face,' said Cradell. 'But Mrs. Lupex has fine eyes – very fine eyes.'

'And knows how to use them, too,' said Johnny.

'Why shouldn't she? And then she has lovely hair.'

'Only she never brushes it in the morning.'

'Do you know, I like that kind of deshabille,'* said Cradell. 'Too much care always betrays itself.'

'But a woman should be tidy.'

'What a word to apply to such a creature as Mrs. Lupex! I call her a splendid woman. And how well she was got up last night. Do you know, I've an idea that Lupex treats her very badly. She said a word or two to me yesterday that—,' and then he paused. There are some confidences which a man does not share even with his dearest friend.

'I rather fancy it's quite the other way,' said Eames.

'How the other way?'

'That Lupex has quite as much as he likes of Mrs. L. The sound of her voice sometimes makes me shake in my shoes, I know.'

'I like a woman with spirit,' said Cradell.

'Oh, so do I. But one may have too much of a good thing. Amelia did tell me; – only you won't mention it.'

'Of course I won't.'

'She told me that Lupex sometimes was obliged to run away from her. He goes down to the theatre, and remains there two or three days at a time. Then she goes to fetch him, and there is no end of a row in the house.'

'The fact is, he drinks,' said Cradell. 'By George, I pity a woman whose husband drinks – and such a woman as that, too!'

'Take care, old fellow, or you'll find yourself in a scrape.'

'I know what I'm at. Lord bless you, I'm not going to lose my head because I see a fine woman.'

'Or your heart either?'

'Oh, heart! There's nothing of that kind of thing about me. I regard a woman as a picture or a statue. I dare say I shall marry some day, because men do; but I've no idea of losing myself about a woman.'

'I'd lose myself ten times over for—'

'L.D.,' said Cradell.

'That I would. And yet I know I shall never have her. I'm a jolly, laughing sort of fellow; and yet, do you know, Caudle, when that girl marries, it will be all up with me. It will, indeed.'

'Do you mean that you'll cut your throat?'

'No; I shan't do that. I shan't do anything of that sort; and yet it will be all up with me.'

'You are going down there in October; – why don't you ask her to have you?'

'With ninety pounds a year!' His grateful country had twice increased his salary at the rate of five pounds each year. 'With ninety pounds a year, and twenty allowed me by my mother!'

'She could wait, I suppose. I should ask her, and no mistake. If one is to love a girl, it's no good one going on in that way!'

'It isn't much good, certainly,' said Johnny Eames. And then they reached the door of the Income-tax Office, and each went away to his own desk.

From this little dialogue, it may be imagined that though Mrs. Roper was as good as her word, she was not exactly the woman Mrs. Eames would have wished to select as a protecting angel for her son. But the truth I take to be this, that protecting angels for widows' sons, at forty-eight pounds a year, paid quarterly, are not to be found very readily in London. Mrs. Roper was not worse than others of her class. She would much have preferred lodgers who were respectable to those who were not so, – if she could only have found respectable lodgers as she wanted them. Mr. and Mrs. Lupex hardly came under that denomination; and when she gave them up her big front bedroom at a hundred a year, she knew she was doing wrong. And she was troubled, too, about her own daughter Amelia, who was already over thirty years of age. Amelia was a very clever young woman, who had been, if the truth must be told, first young lady at a millinery establishment in Manchester. Mrs. Roper knew that Mrs. Eames and Mrs. Cradell would not wish their sons to associate with her daughter. But what could she do? She could not refuse the shelter of her own house to her own child, and yet her heart misgave her when she saw Amelia flirting with young Eames.

'I wish, Amelia, you wouldn't have so much to say to that young man.'

'Laws, mother.'

'So I do. If you go on like that, you'll put me out of both of my lodgers.'

'Go on like what, mother? If a gentleman speaks to me, I suppose I'm to answer him? I know how to behave myself, I believe.' And then she gave her head a toss. Whereupon her mother was silent; for her mother was afraid of her.

CHAPTER V

ABOUT L. D.

APOLLO CROSBIE left London for Allington on the 31st of August, intending to stay there four weeks, with the declared intention of recruiting his strength by an absence of two months from official cares, and with no fixed purpose as to his destiny for the last of those two months. Offers of hospitality had been made to him by the dozen. Lady Hartletop's doors, in Shropshire, were open to him, if he chose to enter them. He had been invited by the Countess de Courcy to join her suite at Courcy Castle. His special friend, Montgomerie Dobbs, had a place in Scotland, and then there was a yachting party by which he was much wanted. But Mr. Crosbie had as yet knocked himself down to none of these biddings, having before him when he left London no other fixed engagement than that which took him to Allington. On the 1st of October we shall also find ourselves at Allington in company with Johnny Eames; and Apollo Crosbie will still be there, – by no means to the comfort of our friend from the Income-tax Office.

Johnny Eames cannot be called unlucky in that matter of his annual holiday, seeing that he was allowed to leave London in October, a month during which few choose to own that they remain in town. For myself, I always regard May as the best month for holiday-making; but then no Londoner cares to be absent in May. Young Eames, though he lived in Burton Crescent, and had as yet no connection with the West End, had already learned his lesson in this respect. 'Those fellows in the big room want me to take May,' he had said to his friend Cradell. 'They must think I'm uncommon green.'

'It's too bad,' said Cradell. 'A man shouldn't be asked to take his leave in May. I never did, and what's more, I never will. I'd go to the Board first.'

Eames had escaped this evil without going to the Board, and had succeeded in obtaining for himself for his own holiday that month of October, which, of all months, is perhaps the most highly esteemed for holiday purposes. 'I shall go down by the mail-train to-morrow night,' he said to Amelia Roper, on the evening before his departure. At that moment he was sitting alone with Amelia in Mrs. Roper's back drawing-room. In the front room Cradell was talking to Mrs. Lupex; but as Miss Spruce was with them, it may be presumed that Mr. Lupex need have had no cause for jealousy.

'Yes,' said Amelia; 'I know how great is your haste to get down to that fascinating spot. I could not expect that you would lose one single hour in hurrying away from Burton Crescent.'

Amelia Roper was a tall, well-grown young woman, with dark hair and dark eyes; – not handsome, for her nose was thick, and the lower part of her face was heavy, but yet not without some feminine attractions. Her eyes were bright; but then, also, they were mischievous. She could talk fluently enough; but then, also, she could scold. She could assume sometimes the plumage of a dove; but then again she could occasionally ruffle her feathers like an angry kite. I am quite prepared to acknowledge that John Eames should have kept himself clear of Amelia Roper; but then young men so frequently do those things which they should not do!

'After twelve months up here in London one is glad to get away to one's own friends,' said Johnny.

'Your own friends, Mr. Eames! What sort of friends? Do you suppose I don't know?'

'Well, no. I don't think you do know.'

'L. D.!' said Amelia, showing that Lily had been spoken of among people who should never have been allowed to hear her name. But perhaps, after all, no more than those two initials were known in Burton Crescent. From the tone which was now used in naming them, it was sufficiently manifest that Amelia considered herself to be wronged by their very existence.

'L. S. D.,'* said Johnny, attempting the line of a witty, gay young spendthrift. 'That's my love – pounds, shillings, and pence; and a very coy mistress she is.'

'Nonsense, sir. Don't talk to me in that way. As if I didn't know where your heart was. What right had you to speak to me if you had an L. D. down in the country?'

It should be here declared on behalf of poor John Eames that he had not ever spoken to Amelia – he had not spoken to her in any such phrase as her words seemed to imply. But then he had written to her a fatal note of which we will speak further before long, and that perhaps was quite as bad, – or worse.

'Ha, ha, ha!' laughed Johnny. But the laugh was assumed, and not assumed with ease.

'Yes, sir; it's a laughing matter to you, I daresay. It is very easy for a man to laugh under such circumstances; – that is to say, if he is perfectly heartless, – if he's got a stone inside of his bosom instead of flesh and blood. Some men are made of stone, I know, and are troubled with no feelings.'

'What is it you want me to say? You pretend to know all about it, and it wouldn't be civil in me to contradict you.'

'What is it I want? You know very well what I want; or rather, I do not want anything. What is it to me? It is nothing to me about L.D. You can go down to Allington and do what you like for me. Only I hate such ways.'

'What ways, Amelia?'

'What ways! Now, look here, Johnny: I'm not going to make a fool of myself for any man. When I came home here three months ago – and I wish I never had;' – she paused here a moment, waiting for a word of tenderness; but as the word of tenderness did not come, she went on – 'but when I did come home, I didn't think there was a man in all London could make me care for him, – that I didn't. And now you're going away, without so much as hardly saying a word to me.' And then she brought out her handkerchief.

'What am I to say, when you keep on scolding me all the time?'

'Scolding you! – And me too! No, Johnny, I ain't scolding you, and don't mean to. If it's to be all over between us, say the word, and I'll take myself away out of the house before you come back again. I've had no secrets from you. I can go back to my business in Manchester, though it is beneath my birth, and not what I've been used to. If L.D. is more to you than I am, I won't stand in your way. Only say the word.'

L. D. was more to him than Amelia Roper, – ten times more to him. L. D. would have been everything to him, and Amelia Roper was worse than nothing. He felt all this at the moment, and struggled hard to collect an amount of courage that would make him free.

'Say the word,' said she, rising on her feet before him, 'and all between you and me shall be over. I have got your promise, but I'd scorn to take advantage. If Amelia hasn't got your heart, she'd despise to take your hand. Only I must have an answer.'

It would seem that an easy way of escape was offered to him; but the lady probably knew that the way as offered by her was not easy to such a one as John Eames.

'Amelia,' he said, still keeping his seat.

'Well, sir?'

'You know I love you.'

'And about L. D.?'

'If you choose to believe all the nonsense that Cradell puts into your head, I can't help it, if you like to make yourself jealous about two letters, it isn't my fault.'

'And you love me?' said she.

'Of course I love you.' And then, upon hearing these words, Amelia threw herself into his arms.

As the folding doors between the two rooms were not closed, and as Miss Spruce was sitting in her easy chair immediately opposite to them, it was probable that she saw what passed. But Miss Spruce was a taciturn old lady, not easily excited to any show of surprise or admiration; and as she had lived with Mrs. Roper for the last twelve years, she was probably well acquainted with her daughter's ways.

'You'll be true to me?' said Amelia, during the moment of that embrace – 'true to me for ever?'

'Oh, yes; that's a matter of course,' said John Eames. And then she liberated him; and the two strolled into the front sitting-room.

'I declare, Mr. Eames,' said Mrs. Lupex, 'I'm glad you've come. Here's Mr. Cradell does say such queer things.'

'Queer things!' said Cradell. 'Now, Miss Spruce, I appeal to you – Have I said any queer things?'

'If you did, sir, I didn't notice them,' said Miss Spruce.

'I noticed them, then,' said Mrs. Lupex. 'An unmarried man like Mr. Cradell has no business to know whether a married lady wears a cap or her own hair – has he, Mr. Eames?'

'I don't think I ever know,' said Johnny, not intending any sarcasm on Mrs. Lupex.

'I dare say not, sir,' said the lady. 'We all know where your attention is riveted. If you were to wear a cap, my dear, somebody would see the difference very soon – wouldn't they, Miss Spruce?'

'I dare say they would,' said Miss Spruce.

'If I could look as nice in a cap as you do, Mrs. Lupex, I'd wear one to-morrow,' said Amelia, who did not wish to quarrel with the married lady at the present moment. There were occasions, however, on which Mrs. Lupex and Miss Roper were by no means so gracious to each other.

'Does Lupex like caps?' asked Cradell.

'If I wore a plumed helmet on my head, it's my belief he wouldn't know the difference; nor yet if I had got no head at all. That's what comes of getting married. If you'll take my advice, Miss Roper, you'll stay as you are; even though somebody should break his heart about it. Wouldn't you, Miss Spruce?'

'Oh, as for me, I'm an old woman, you know,' said Miss Spruce, which was certainly true.

'I don't see what any woman gets by marrying,' continued Mrs. Lupex. 'But a man gains everything. He don't know how to live, unless he's got a woman to help him.'

'But is love to go for nothing?' said Cradell.

'Oh, love! I don't believe in love. I suppose I thought I loved once, but what did it come to after all? Now, there's Mr. Eames – we all know he's in love.'

'It comes natural to me, Mrs. Lupex. I was born so,' said Johnny.

'And there's Miss Roper – one never ought to speak free about a lady, but perhaps she's in love too.'

'Speak for yourself, Mrs. Lupex,' said Amelia.

'There's no harm in saying that, is there? I'm sure, if you ain't, you're very hard-hearted; for, if ever there was a true lover, I believe you've got one of your own. My! – if there's not Lupex's step on the stair? What can bring him home at this hour? If he's been drinking, he'll come home as cross as anything.' Then Mr. Lupex entered the room, and the pleasantness of the party was destroyed.

It may be said that neither Mrs. Cradell nor Mrs. Eames would have placed their sons in Burton Crescent if they had known the dangers into which the young men would fall. Each, it must be acknowledged, was imprudent; but each clearly saw the imprudence of the other. Not a week before this, Cradell had seriously warned his friend against the arts of Miss Roper. 'By George, Johnny, you'll get yourself entangled with that girl.'

'One always has to go through that sort of thing,' said Johnny.

'Yes; but those who go through too much of it never get

out again. Where would you be if she got a written promise of marriage from you?'

Poor Johnny did not answer this immediately, for in very truth Amelia Roper had such a document in her possession.

'Where should I be?' said he. 'Among the breaches of promise, I suppose.'

'Either that, or else among the victims of matrimony. My belief of you is, that if you gave such a promise, you'd carry it out.'

'Perhaps I should,' said Johnny; 'but I don't know. It's a matter of doubt what a man ought to do in such a case.'

'But there's been nothing of that kind yet?'

'Oh dear, no!'

'If I was you, Johnny, I'd keep away from her. It's very good fun, of course, that sort of thing; but it is so uncommon dangerous! Where would you be now with such a girl as that for your wife?'

Such had been the caution given by Cradell to his friend. And now, just as he was starting for Allington, Eames returned the compliment. They had gone together to the Great Western station at Paddington, and Johnny tendered his advice at they were walking together up and down the platform.

'I say, Caudle, old boy, you'll find yourself in trouble with that Mrs. Lupex, if you don't take care of yourself.'

'But I shall take care of myself. There's nothing so safe as a little nonsense with a married woman. Of course, it means nothing, you know, between her and me.'

'I don't suppose it does mean anything. But she's always talking about Lupex being jealous; and if he was to cut up rough, you wouldn't find it pleasant.'

Cradell, however, seemed to think that there was no danger. His little affair with Mrs. Lupex was quite platonic and safe. As for doing any real harm, his principles, as he assured his friend, were too high. Mrs. Lupex was a woman of talent, whom no one seemed to understand, and, therefore, he had taken some pleasure in studying her character. It was merely a study of character, and nothing more. Then the friends parted, and Eames was carried away by the night mail-train down to Guestwick.

How his mother was up to receive him at four o'clock in the morning, how her maternal heart was rejoicing at seeing the improvement in his gait, and the manliness of appearance imparted to him by his whiskers, I need not describe at length. Many of the attributes of a hobbledehoy had fallen from him, and even Lily Dale might now probably acknowledge that he was no longer a boy. All which might be regarded as good, if only in putting off childish things he had taken up things which were better than childish.

On the very first day of his arrival he made his way over to Allington. He did not walk on this occasion as he had used to do in the old happy days. He had an idea that it might not be well for him to go into Mrs. Dale's drawing-room with the dust of the road on his boots, and the heat of the day on his brow. So he borrowed a horse and rode over, taking some pride in a pair of spurs which he had bought in Piccadilly, and in his kid gloves, which were brought out new for the occasion. Alas, alas! I fear that those two years in London have not improved John Eames; and yet I have to acknowledge that John Eames is one of the heroes of my story.

On entering Mrs. Dale's drawing-room he found Mrs. Dale and her eldest daughter. Lily at the moment was not there, and as he shook hands with the other two, of course, he asked for her.

'She is only in the garden,' said Bell. 'She will be here directly.'

'She has walked across to the Great House with Mr. Crosbie,' said Mrs. Dale; 'but she is not going to remain. She will be so glad to see you, John! We all expected you to-day.'

'Did you?' said Johnny, whose heart had been plunged into cold water at the mention of Mr. Crosbie's name. He had been thinking of Lilian Dale ever since his friend had left him on the railway platform; and, as I beg to assure all ladies who may read my tale, the truth of his love for Lily had moulted no feather through that unholy liaison between him and Miss Roper. I fear that I shall be disbelieved in this; but it was so. His heart was and ever had been true to Lilian, although he had allowed himself to be talked into declarations of affection by such a creature as Amelia Roper. He had been thinking of his meeting with Lily all the night and throughout the morning, and now he heard that she was walking alone about the gardens with a strange gentleman. That Mr. Crosbie was very grand and very fashionable he had heard, but he knew no more of him. Why should Mr. Crosbie be allowed to walk with Lily Dale? And why should Mrs. Dale mention the circumstance as though it were quite a thing of course? Such mystery as there was in this was solved very quickly.

'I'm sure Lily won't object to my telling such a dear friend as you what has happened,' said Mrs. Dale. 'She is engaged to be married to Mr. Crosbie.'

The water into which Johnny's heart had been plunged now closed over his head and left him speechless. Lily Dale was engaged to be married to Mr. Crosbie! He knew that he should have spoken when he heard the tidings. He knew that the moments of silence as they passed by told his secret to the two women before him, – that secret which it would now behove him to conceal from all the world. But yet he could not speak.

'We are all very well pleased at the match,' said Mrs. Dale, wishing to spare him.

'Nothing can be nicer than Mr. Crosbie,' said Bell. 'We have often talked about you, and he will be so happy to know you.'

'He won't know much about me,' said Johnny; and even in speaking these few senseless words – words which he uttered because it was necessary that he should say something – the tone of his voice was altered. He would have given the world to have been master of himself at this moment, but he felt that he was utterly vanquished.

'There is Lily coming across the lawn,' said Mrs. Dale.

'Then I'd better go,' said Eames. 'Don't say anything about it; pray don't.' And then, without waiting for another word, he escaped out of the drawing-room.

CHAPTER VI

BEAUTIFUL DAYS

I AM well aware that I have not as yet given any description of Bell and Lilian Dale, and equally well aware that the longer the doing so is postponed the greater the difficulty becomes. I wish it could be understood without any description that they were two pretty, fair-haired girls, of whom Bell was the tallest and the prettiest, whereas Lily was almost as pretty as her sister, and perhaps was more attractive.

They were fair-haired girls, very like each other, of whom I have before my mind's eye a distinct portrait, which I fear I shall not be able to draw in any such manner as will make it distinct to others. They were something below the usual height, being slight and slender in all their proportions. Lily was the shorter of the two, but the difference was so trifling that it was hardly remembered unless the two were together. And when I said that Bell was the prettier, I should, perhaps, have spoken more justly had I simply declared that her features were more regular than her sister's. The two girls were very fair, so that the soft tint of colour which relieved the whiteness of their complexion was rather acknowledged than distinctly seen. It was there, telling its own tale of health, as its absence would have told a tale of present or coming sickness; and yet nobody could ever talk about the colour in their cheeks. The hair of the two girls was so alike in hue and texture, that no one, not even their mother, could say that there was a difference. It was not flaxen hair, and yet it was very light. Nor did it approach to auburn; and yet there ran through it a golden tint that gave it a distinct brightness of its own. But with Bell it was more plentiful than with Lily, and therefore Lily would always talk of her own scanty locks, and tell how beautiful were those belonging to her sister. Nevertheless Lily's head was quite as lovely as her sister's; for its form was perfect, and the simple braids in which they both wore their hair did not require any great exuberance in quantity. Their eyes were brightly blue; but

Bell's were long, and soft, and tender, often hardly daring to raise themselves to your face; while those of Lily were rounder, but brighter, and seldom kept by want of courage from fixing themselves where they pleased. And Lily's face was perhaps less oval in its form – less perfectly oval – than her sister's. The shape of the forehead was, I think, the same, but with Bell the chin was something more slender and delicate. But Bell's chin was unmarked, whereas on her sister's there was a dimple which amply compensated for any other deficiency in its beauty. Bell's teeth were more even than her sister's; but then she showed her teeth more frequently. Her lips were thinner, and, as I cannot but think, less expressive. Her nose was decidedly more regular in its beauty, for Lily's nose was somewhat broader than it should have been. It may, therefore, be understood that Bell would be considered the beauty by the family.

But there was, perhaps, more in the general impression made by these girls, and in the whole tone of their appearance, than in the absolute loveliness of their features or the grace of their figures. There was about them a dignity of demeanour devoid of all stiffness or pride, and a maidenly modesty which gave itself no airs. In them was always apparent that sense of security which women should receive from an unconscious dependence on their own mingled purity and weakness. These two girls were never afraid of men, – never looked as though they were so afraid. And I may say that they had little cause for that kind of fear to which I allude. It might be the lot of either of them to be ill-used by a man, but it was hardly possible that either of them should ever be insulted by one. Lily, as may, perhaps, have been already seen, could be full of play, but in her play she never so carried herself that any one could forget what was due to her.

And now Lily Dale was engaged to be married, and the days of her playfulness were over. It sounds sad, this sentence against her, but I fear that it must be regarded as true. And when I think that it is true, – when I see that the sportiveness and kitten-like gambols of girlhood should be over, and generally are over, when a girl has given her troth, it becomes a matter of regret to me that the feminine world should be in such a hurry after matrimony. I have, however, no remedy to offer for the evil; and indeed, am aware that the evil, if there be an evil, is not well expressed in the words I have used. The hurry is not for matrimony, but for love. Then, the love once attained, matrimony seizes it for its own, and the evil is accomplished.

And Lily Dale was engaged to be married to Adolphus Crosbie, – to Apollo Crosbie, as she still called him, confiding her little joke to his own ears. And to her he was an Apollo, as a man who is loved should be to the girl who loves him. He was handsome, graceful, clever, self-confident, and always cheerful when she asked him to be cheerful. But he had also his more serious moments, and could talk to her of serious matters. He would read to her, and explain to her things which had hitherto been too hard for her young intelligence. His voice, too, was pleasant, and well under command. It could be pathetic if pathos were required, or ring with laughter as merry as her own. Was not such a man fit to be an Apollo to such a girl, when once the girl had acknowledged to herself that she loved him?

She had acknowledged it to herself, and had acknowledged it to him, – as the reader will perhaps say without much delay. But the courtship had so been carried on that no delay had been needed. All the world had smiled upon it. When Mr. Crosbie had first come among them at Allington, as Bernard's guest, during those few days of his early visit, it had seemed as though Bell had been chiefly noticed by him. And Bell in her own quiet way had accepted his admiration, saying nothing of it and thinking but very little. Lily was heart-free at the time, and had ever been so. No first shadow from Love's wing had as yet been thrown across the pure tablets of her bosom. With Bell it was not so, – not so in absolute strictness. Bell's story, too, must be told, but not on this page. But before Crosbie had come among them, it was a thing fixed in her mind that such love as she had felt must be overcome and annihilated. We may say that it had been overcome and annihilated, and that she would have sinned in no way had she listened to vows from this new Apollo. It is almost sad to think that such a man might have had the love of either of such girls, but I fear that I must acknowledge that it was so. Apollo, in the plenitude of his power, soon changed his mind; and before the end of his first visit, had transferred the distant homage which he was then paying from the elder to the younger sister. He afterwards returned, as the squire's guest, for a longer sojourn among them, and at the end of the first month had already been accepted as Lily's future husband.

It was beautiful to see how Bell changed in her mood towards Crosbie and towards her sister as soon as she perceived how the affair was going. She was not long in perceiving it, having caught the first glimpses of the idea on that evening when they both dined at the Great House, leaving their mother alone to eat or to neglect the peas. For some six or seven weeks Crosbie had been gone, and during that time Bell had been much more open in speaking of him than her sister. She had been present when Crosbie had bid them good-bye, and had listened to his eagerness as he declared to Lily that he should soon be back again at Allington. Lily had taken this very quietly, as though it had not belonged at all to herself; but Bell had seen something of the truth, and, believing in Crosbie as an earnest, honest man, had spoken kind words of him, fostering any little aptitude for love which might already have formed itself in Lily's bosom.

'But he is such an Apollo, you know,' Lily had said.

'He is a gentleman; I can see that.'

'Oh, yes; a man can't be an Apollo unless he's a gentleman.'

'And he's very clever.'

'I suppose he is clever.' There was nothing more said about his being a mere clerk. Indeed, Lily had changed her mind on that subject. Johnny Eames was a mere clerk; whereas Crosbie, if he was to be called a clerk at all, was a clerk of some very special denomination. There may be a great difference between one clerk and another! A Clerk of the Council and a parish clerk are very different persons. Lily had got some such idea as this into her head as she attempted in her own mind to rescue Mr. Crosbie from the lower orders of the Government service.

'I wish he were not coming,' Mrs. Dale had said to her eldest daughter.

'I think you are wrong, mamma.'

'But if she should become fond of him, and then—'

'Lily will never become really fond of any man till he shall have given her proper reason. And if he admires her, why should they not come together?'

'But she is so young, Bell.'

'She is nineteen; and if they were engaged, perhaps, they might wait for a year or so. But it's no good talking in that way, mamma. If you were to tell Lily not to give him encouragement, she would not speak to him.'

'I should not think of interfering.'

'No, mamma; and therefore it must take its course. For myself, I like Mr. Crosbie very much.'

'So do I, my dear.'

'And so does my uncle. I wouldn't have Lily take a lover of my uncle's choosing.'

'I should hope not.'

'But it must be considered a good thing if she happens to choose one of his liking.'

In this way the matter had been talked over between the mother and her elder daughter. Then Mr. Crosbie had come; and before the end of the first month his declared admiration for Lily had proved the correctness of her sister's foresight. And during that short courtship all had gone well with the lovers. The squire from the first had declared himself satisfied with the match, informing Mrs. Dale, in his cold manner, that Mr. Crosbie was a gentleman with an income sufficient for matrimony.

'It would be close enough in London,' Mrs. Dale had said.

'He has more than my brother had when he married,' said the squire.

'If he will only make her as happy as your brother made me, – while it lasted!' said Mrs. Dale, as she turned away her face to conceal a tear that was coming. And then there was nothing more said about it between the squire and his sister-in-law. The squire spoke no word as to assistance in money matters, – did not even suggest that he would lend a hand to the young people at starting, as an uncle in such a position might surely have done. It may well be conceived that Mrs. Dale herself said nothing on the subject. And, indeed, it may be conceived, also, that the squire, let his intentions be what they might, would not divulge them to Mrs. Dale. This was uncomfortable, but the position was one that was well understood between them.

Bernard Dale was still at Allington, and had remained there through the period of Crosbie's absence. Whatever words Mrs. Dale might choose to speak on the matter would probably be spoken to him; but, then, Bernard could be quite as close as his uncle. When Crosbie returned, he and Bernard had, of course, lived much together; and, as was natural, there came to be close discussion between them as to the two girls, when Crosbie allowed it to be understood that his liking for Lily was becoming strong.

'You know, I suppose, that my uncle wishes me to marry the elder one,' Bernard had said.

'I have guessed as much.'

'And I suppose the match will come off. She's a pretty girl, and as good as gold.'

'Yes, she is.'

'I don't pretend to be very much in love with her. It's not my way, you know. But, some of these days, I shall ask her to have me, and I suppose it'll all go right. The governor has distinctly promised to allow me eight hundred a year off the estate, and to take us in for three months every year if we wish it. I told him simply that I couldn't do it for less, and he agreed with me.'

'You and he get on very well together.'

'Oh, yes! There's never been any fal-lal between us about love, and duty, and all that. I think we understand each other, and that's everything. He knows the comfort of standing well with the heir, and I know the comfort of standing well with the owner.' It must be admitted, I think, that there was a great deal of sound, common sense about Bernard Dale.

'What will he do for the younger sister?' asked Crosbie; and, as he asked the important question, a close observer might have perceived that there was some slight tremor in his voice.

'Ah! that's more than I can tell you. If I were you, I should ask him. The governor is a plain man, and likes plain business.'

'I suppose you couldn't ask him?'

'No; I don't think I could. It is my belief that he will not let her go by any means empty-handed.'

'Well, I should suppose not.'

'But remember this, Crosbie, – I can say nothing to you on which you are to depend. Lily, also, is as good as gold; and, as you seem to be fond of her, I should ask the governor, if I were you, in so many words, what he intends to do. Of course, it's against my interest, for every shilling he gives Lily will ultimately come out of my pocket. But I'm not the man to care about that, as you know.'

What might be Crosbie's knowledge on this subject we will not here inquire; but we may say that it would have mattered very little to him out of whose pocket the money came, so long as it went into his own. When he felt quite sure of Lily, – having, in fact, received Lily's permission to speak to her uncle, and Lily's promise that she would herself speak to her mother, – he did tell the squire what was his intention. This he did in an open, manly way, as though he felt that in asking for much he also offered to give much.

'I have nothing to say against it,' said the squire.

'And I have your permission to consider myself as engaged to her?'

'If you have hers and her mother's. Of course you are aware that I have no authority over her.'

'She would not marry without your sanction.'

'She is very good to think so much of her uncle,' said the squire; and his words as he spoke them sounded very cold in Crosbie's ears. After that Crosbie said nothing about money, having to confess to himself that he was afraid to do so. 'And what would be the use?' said he to himself, wishing to make excuses for what he felt to be weak in his own conduct. 'If he should refuse to give her a shilling I could not go back from it now.' And then some ideas ran across his mind as to the injustice to which men are subjected in this matter of matrimony. A man has to declare himself before it is fitting that he should make any inquiry about a lady's money; and then, when he has declared himself, any such inquiry is unavailing. Which consideration somewhat cooled the ardour of his happiness. Lily Dale was very pretty, very nice, very refreshing in her innocence, her purity, and her quick intelligence. No amusement could be more deliciously amusing than that of making love to Lily Dale. Her way of flattering her lover without any intention of flattery on her part, had put Crosbie into a seventh heaven. In all his experience he had known nothing like it. 'You may be sure of this,' she had said, – 'I shall love you with all my heart and all my strength.' It was very nice; – but then what were they to live upon? Could it be that he, Adolphus Crosbie, should settle down on the north side of the New Road, as a married man, with eight hundred a year? If indeed the squire would be as good to Lily as he had promised to be to Bell, then indeed things might be made to arrange themselves.

But there was no such drawback on Lily's happiness. Her ideas about money were rather vague, but they were very honest. She knew she had none of her own, but supposed it was a husband's duty to find what would be needful. She knew she had none of her own, and was therefore aware that she ought not to expect luxuries in the little household that was to be prepared for her. She hoped, for his sake, that her uncle might give some assistance, but was quite prepared to prove that she could be a good poor man's wife. In the old colloquies on such matters between her and her sister, she had always declared that some decent income should be considered as indispensable before love could be entertained. But eight hundred a year had been considered as doing much more than fulfilling this stipulation. Bell had had high-flown notions as to the absolute glory of poverty. She had declared that income should not be considered at all. If she had loved a man, she could allow herself to be engaged to him, even though he had no income. Such had been their theories; and as regarded money, Lily was quite contented with the way in which she had carried out her own.

In these beautiful days there was nothing to check her happiness. Her mother and sister united in telling her that she had done well, – that she was happy in her choice, and justified in her love. On that first day, when she told her mother all, she had been made exquisitely blissful by the way in which her tidings had been received.

'Oh! mamma, I must tell you something,' she said, coming up to her mother's bedroom, after a long ramble with Mr. Crosbie through those Allington fields.

'Is it about Mr. Crosbie?'

'Yes, mamma.' And then the rest had been said through the medium of warm embraces and happy tears rather than by words.

As she sat in her mother's room, hiding her face on her mother's shoulders, Bell had come, and had knelt at her feet.

'Dear Lily,' she had said, 'I am so glad.' And then Lily remembered how she had, as it were, stolen her lover from her sister, and she put her arms round Bell's neck and kissed her.

'I knew how it was going to be from the very first,' said Bell. 'Did I not, mamma?'

'I'm sure I didn't,' said Lily. 'I never thought such a thing was possible.'

'But we did, – mamma and I.'

'Did you?' said Lily.

'Bell told me that it was to be so,' said Mrs. Dale. 'But I could hardly bring myself at first to think that he was good enough for my darling.'

'Oh, mamma! you must not say that. You must think that he is good enough for anything.'

'I will think that he is very good.'

'Who could be better? And then, when you remember all that he is to give up for my sake! – And what can I do for him in return? What have I got to give him?'

Neither Mrs. Dale nor Bell could look at the matter in this light, thinking that Lily gave quite as much as she received. But they both declared that Crosbie was perfect, knowing that by such assurances only could they now administer to Lily's happiness; and Lily, between them, was made perfect in her happiness, receiving all manner of encouragement in her love, and being nourished in her passion by the sympathy and approval of her mother and sister.

And then had come that visit from Johnny Eames. As the poor fellow marched out of the room, giving them no time to say farewell, Mrs. Dale and Bell looked at each other sadly; but they were unable to concoct any arrangement, for Lily had run across the lawn, and was already on the ground before the window.

'As soon as we got to the end of the shrubbery there were uncle Christopher and Bernard close to us; so I told Adolphus he might go on by himself.'

'And who do you think has been here?' said Bell. But Mrs. Dale said nothing. Had time been given to her to use her own judgment, nothing should have been said at that moment as to Johnny's visit.

'Has anybody been here since I went? Whoever it was didn't stay very long.'

'Poor Johnny Eames,' said Bell. Then the colour came up into Lily's face, and she bethought herself in a moment that the old friend of her young days had loved her, that he, too, had had hopes as to his love, and that now he had heard tidings which would put an end to such hopes. She understood it all in a moment, but understood also that it was necessary that she should conceal such understanding.

'Dear Johnny!' she said. 'Why did he not wait for me?'

'We told him you were out,' said Mrs. Dale. 'He will be here again before long, no doubt.'

'And he knows—?'

'Yes; I thought you would not object to my telling him.'

'No, mamma; of course not. And he has gone back to Guestwick?'

There was no answer given to this question, nor were there any further words then spoken about Johnny Eames. Each of these women understood exactly how the matter stood, and each knew that the others understood it. The young man was loved by them all, but not loved with that sort of admiring affection which had been accorded to Mr. Crosbie. Johnny Eames could not have been accepted as a suitor by their pet. Mrs. Dale and Bell both felt that. And yet they loved him for his love, and for that distant, modest respect which had restrained him from any speech regarding it. Poor Johnny! But he was young, – hardly as yet out of his hobbledehoyhood, – and he would easily recover this blow, remembering, and perhaps feeling to his advantage, some slight touch of its passing romance. It is thus women think of men who love young and love in vain.

But Johnny Eames himself, as he rode back to Guestwick, forgetful of his spurs, and with his gloves stuffed into his pocket, thought of the matter very differently. He had never promised to himself any success as to his passion for Lily, and had, indeed, always acknowledged that he could have no hope; but, now that she was actually promised to another man, and as good as married, he was not the less broken-hearted because his former hopes had not been high. He had never dared to speak to Lily of his love, but he was conscious that she knew it, and he did not now dare to stand before her as one convicted of having loved in vain. And then, as he rode back, he thought also of his other love, not with many of those pleasant thoughts which Lotharios * and Don Juans may be presumed to enjoy when they contemplate their successes. 'I suppose I shall marry her, and there'll be an end of me,' he said to himself, as he remembered a short note which he had once written to her in his madness. There had been a little supper at Mrs. Roper's, and Mrs. Lupex and Amelia had made the punch. After supper, he had been by some accident alone with Amelia in the dining-parlour; and when, warmed by the generous god, he had declared his passion, she had shaken her head mournfully, and had fled from him to some upper region, absolutely refusing his proffered embrace. But on the same night, before his head had found its pillow, a note had come to him, half repentant, half affectionate, half repellent, – 'If, indeed, he would swear to her that his love was honest and manly, then, indeed, she might even yet, – see him through the chink of the door-way with the purport of telling him that he was forgiven.' Whereupon, a perfidious pencil being near to his hand, he had written the requisite words. 'My only object in life is to call you my own for ever.' Amelia had her misgivings whether such a promise, in order that it might be used as legal evidence, should not have been written in ink. It was a painful doubt; but nevertheless she was as good as her word, and saw him through the chink, forgiving him for his impetuosity in the parlour with, perhaps, more clemency than a mere pardon required. 'By George! how well she looked with her hair all loose,' he said to himself, as he at last regained his pillow, still warm with the generous god. But now, as he thought of that night, returning on his road from Allington to Guestwick, those loose, floating locks were remembered by him with no strong feeling as to their charms. And he thought also of Lily Dale, as she was when he had said farewell to her on that day before he first went up to London. 'I shall care more about seeing you than anybody,' he had said; and he had often thought of the words since, wondering whether she had understood them as meaning more than an assurance of ordinary friendship. And he remembered well the dress she had then worn. It was an old brown merino,* which he had known before, and which, in truth, had nothing in it to recommend it specially to a lover's notice. 'Horrid old thing!' had been Lily's own verdict respecting the frock, even before that day. But she had hallowed it in his eyes, and he would have been only too happy to have worn a shred of it near his heart, as a talisman. How wonderful in its nature is that passion of which men speak when they acknowledge to themselves that they are in love. Of all things, it is, under one condition, the most foul, and under another, the most fair. As that condition is, a man shows himself either as a beast or as a god! And so we will let poor Johnny Eames ride back to Guestwick, suffering much in that he had loved basely – and suffering much, also, in that he had loved nobly.

Lily, as she had tripped along through the shrubbery, under her lover's arm, looking up, every other moment, into his face, had espied her uncle and Bernard. 'Stop,' she had said, giving him a little pull at the arm; 'I won't go on. Uncle is always teasing me with some old-fashioned wit. And I've had quite enough of you to-day, sir. Mind you come over to-morrow before you go to your shooting.' And so she had left him.

We may as well learn here what was the question in dispute between the uncle and cousin, as they were walking there on the broad gravel path behind the Great House. 'Bernard,' the old man had said, 'I wish this matter could be settled between you and Bell.'

'Is there any hurry about it, sir?'

'Yes, there is hurry; or, rather, as I hate hurry in all things, I would say that there is ground for despatch. Mind, I do not wish to drive you. If you do not like your cousin, say so.'

'But I do like her; only I have a sort of feeling that these things grow best by degrees. I quite share your dislike to being in a hurry.'

'But time enough has been taken now. You see, Bernard, I am going to make a great sacrifice of income on your behalf.'

'I am sure I am very grateful.'

'I have no children, and have therefore always regarded you as my own. But there is no reason why my brother Philip's daughter should not be as dear to me as my brother Orlando's son.'

'Of course not, sir; or, rather, his two daughters.'

'You may leave that matter to me, Bernard. The younger girl is going to marry this friend of yours, and as he has a sufficient income to support a wife, I think that my sister-in-law has good reason to be satisfied by the match. She will not be expected to give up any part of her small income, as she must have done had Lily married a poor man.'

'I suppose she could hardly give up much.'

'People must be guided by circumstances. I am not disposed to put myself in the place of a parent to them both. There is no reason why I should, and I will not encourage false hopes. If I knew that this matter between you and Bell was arranged, I should have reason to feel satisfied with what I was doing.' From all which Bernard began to perceive that poor Crosbie's expectations in the matter of money would not probably receive much gratification. But he also perceived – or thought that he perceived – a kind of threat in this warning from his uncle. 'I have promised you eight hundred a year with your wife,' the warning seemed to say. 'But if you do not at once accept it, or let me feel that it will be accepted, it may be well for me to change my mind – especially as this other niece is about to be married. If I am to give you so large a fortune with Bell, I need do nothing for Lily. But if you do not choose to take Bell and the fortune, why then—' And so on. It was thus that Bernard read his uncle's caution, as they walked together on the broad gravel path.

'I have no desire to postpone the matter any longer,' said Bernard. 'I will propose to Bell at once, if you wish it.'

'If your mind be quite made up, I cannot see why you should delay it.'

And then, having thus arranged that matter, they received their future relative with kind smiles and soft words.

CHAPTER VII

THE BEGINNING OF TROUBLES

LILY, as she parted with her lover in the garden, had required of him to attend upon her the next morning as he went to his shooting, and in obedience to this command he appeared on Mrs. Dale's lawn after breakfast, accompanied by Bernard and two dogs. The men had guns in their hands, and were got up with all proper sporting appurtenances, but it so turned out that they did not reach the stubble-fields on the farther side of the road until after luncheon. And may it not be fairly doubted whether croquet is not as good as shooting when a man is in love?

It will be said that Bernard Dale was not in love; but they who bring such accusation against him, will bring it falsely. He was in love with his cousin Bell according to his manner and fashion. It was not his nature to love Bell as John Eames loved Lily; but then neither would his nature bring him into such a trouble as that which the charms of Amelia Roper had brought upon the poor clerk from the Income-tax Office. Johnny was susceptible, as the word goes; whereas Captain Dale was a man who had his feelings well under control. He was not one to make a fool of himself about a girl, or to die of a broken heart; but, nevertheless, he would probably love his wife when he got a wife, and would be a careful father to his children.

They were very intimate with each other now, – these four. It was Bernard and Adolphus, or sometimes Apollo, and Bell and Lily among them; and Crosbie found it to be pleasant enough. A new position of life had come upon him, and one exceeding pleasant; but, nevertheless, there were moments in which cold fits of a melancholy nature came upon him. He was doing the very thing which throughout all the years of his manhood he had declared to himself that he would not do. According to his plan of life he was to have eschewed marriage, and to have allowed himself to regard it as a possible event only under the circumstances of wealth, rank, and beauty all coming in his way together. As he had expected no such glorious prize, he had regarded himself as a man who would reign at the Beaufort and be potent at Sebright's to the end of his chapter. But now—.

It was the fact that he had fallen from his settled position, vanquished by a silver voice, a pretty wit, and a pair of moderately bright eyes. He was very fond of Lily, having in truth a stronger capability for falling in love than his friend Captain Dale; but was the sacrifice worth his while? This was the question which he asked himself in those melancholy moments; while he was lying in bed, for instance, awake in the morning, when he was shaving himself, and sometimes also when the squire was prosy after dinner. At such times as these, while he would be listening to Mr. Dale, his self-reproaches would sometimes be very bitter. Why should he undergo this, he, Crosbie of Sebright's, Crosbie of the General Committee Office, Crosbie who would allow no one to bore him between Charing Cross and the far end of Bayswater, – why should he listen to the long-winded stories of such a one as Squire Dale? If, indeed, the squire intended to be liberal to his niece, then it might be very well. But as yet the squire had given no sign of such intention, and Crosbie was angry with himself in that he had not had the courage to ask a question on that subject.

And thus the course of love was not all smooth to our Apollo. It was still pleasant for him when he was there on the croquet ground, or sitting in Mrs. Dale's drawing-room with all the privileges of an accepted lover. It was pleasant to him also as he sipped the squire's claret, knowing that his coffee would soon be handed to him by a sweet girl who would have tripped across the two gardens on purpose to perform for him this service. There is nothing pleasanter than all this, although a man when so treated does feel himself to look like a calf at the altar, ready for the knife, with blue ribbons round his horns and neck. Crosbie felt that he was such a calf, – and the more calf-like, in that he had not as yet dared to ask a question about his wife's fortune. 'I will have it out of the old fellow this evening,' he said to himself, as he buttoned on his dandy shooting gaiters * that morning.

'How nice he looks in them,' Lily said to her sister afterwards, knowing nothing of the thoughts which had troubled her lover's mind while he was adorning his legs.

'I suppose we shall come back this way,' Crosbie said, as they prepared to move away on their proper business when lunch was over.

'Well, not exactly!' said Bernard. 'We shall make our way round by Darvell's farm, and so back by Gruddock's. Are the girls going to dine up at the Great House to-day?'

The girls declared that they were not going to dine up at the Great House, – that they did not intend going to the Great House at all that evening.

'Then, as you won't have to dress, you might as well meet us at Gruddock's gate, at the back of the farmyard. We'll be there exactly at half-past five.'

'That is to say, we're to be there at half-past five, and you'll keep us waiting for three-quarters of an hour,' said Lily. Nevertheless, the arrangement as proposed was made, and the two ladies were not at all unwilling to make it. It is thus that the game is carried on among unsophisticated people who really live in the country. The farmyard gate at Farmer Gruddock's has not a fitting sound as a trysting-place in romance, but for people who are in earnest it does as well as any oak in the middle glade of a forest. Lily Dale was quite in earnest – and so indeed was Adolphus Crosbie, – only with him the earnest was beginning to take that shade of brown which most earnest things have to wear in this vale of tears. With Lily it was as yet all rose-coloured. And Bernard Dale was also in earnest. Throughout this morning he had stood very near to Bell on the lawn, and had thought that his cousin did not receive his little whisperings with any aversion. Why should she? Lucky girl that she was, thus to have eight hundred a year pinned to her skirt!

'I say, Dale,' Crosbie said, as in the course of their day's work they had come round upon Gruddock's ground, and were preparing to finish off his turnips before they reached the farmyard gate. And now, as Crosbie spoke, they stood leaning on the gate, looking at the turnips while the two dogs squatted on their haunches. Crosbie had been very silent for the last mile or two, and had been making up his mind for this conversation. 'I say, Dale, – your uncle has never said a word to me yet as to Lily's fortune.'

'As to Lily's fortune! The question is whether Lily has got a fortune.'

'He can hardly expect that I am to take her without something. Your uncle is a man of the world and he knows—'

'Whether or no my uncle is a man of the world, I will not say; but you are, Crosbie, whether he is or not. Lily, as you have always known, has nothing of her own.'

'I am not talking of Lily's own. I'm speaking of her uncle. I have been straightforward with him; and when I became attached to your cousin I declared what I meant at once.'

'You should have asked him the question, if you thought there was any room for such a question.'

'Thought there was any room! Upon my word, you are a cool fellow.'

'Now look here, Crosbie; you may say what you like about my uncle, but you must not say a word against Lily.'

'Who is going to say a word against her? You can little understand me if you don't know that the protection of her name against evil words is already more my care than it is yours. I regard Lily as my own.'

'I only meant to say, that any discontent you may feel as to her money, or want of money, you must refer to my uncle, and not to the family at the Small House.'

'I am quite well aware of that.'

'And though you are quite at liberty to say what you like to me about my uncle, I cannot say that I can see that he has been to blame.'

'He should have told me what her prospects are.'

'But if she have got no prospects! It cannot be an uncle's duty to tell everybody that he does not mean to give his niece a fortune. In point of fact, why should you suppose that he has such an intention?'

'Do you know that he has not? because you once led me to believe that he would give his niece money.'

'Now, Crosbie, it is necessary that you and I should understand each other in this matter—'

'But did you not?'

'Listen to me for a moment. I never said a word to you about my uncle's intentions in any way, until after you had become fully engaged to Lily with the knowledge of us all. Then, when my belief on the subject could make no possible difference in your conduct, I told you that I thought my uncle would do something for her. I told you so because I did think so; – and as your friend, I should have told you what I thought in any matter that concerned your interest.'

'And now you have changed your opinion?'

'I have changed my opinion; but very probably without sufficient ground.'

'That's hard upon me.'

'It may be hard to bear disappointment; but you cannot say that anybody has ill-used you.'

'And you don't think he will give her anything?'

'Nothing that will be of much moment to you.'

'And I'm not to say that that's hard? I think it confounded hard. Of course I must put off my marriage.'

'Why do you not speak to my uncle?'

'I shall do so. To tell the truth, I think it would have come better from him; but that is a matter of opinion. I shall tell him very plainly what I think about it; and if he is angry, why, I suppose I must leave his house; that will be all.'

'Look here, Crosbie; do not begin your conversation with the purpose of angering him. He is not a bad-hearted man, but is very obstinate.'

'I can be quite as obstinate as he is.' And, then, without further parley, they went in among the turnips, and each swore against his luck as he missed his birds. There are certain phases of mind in which a man can neither ride nor shoot, nor play a stroke at billiards, nor remember a card at whist, – and to such a phase of mind had come both Crosbie and Dale after their conversation over the gate.

They were not above fifteen minutes late at the trysting-place, but, nevertheless, punctual though they had been, the girls were there before them. Of course the first inquiries were made about the game, and of course the gentlemen declared that the birds were scarcer than they had ever been before, that the dogs were wilder, and their luck more excruciatingly bad, – to all which apologies very little attention was paid. Lily and Bell had not come there to inquire after partridges, and would have forgiven the sportsmen even though no single bird had been killed. But they could not forgive the want of good spirits which was apparent.

'I declare I don't know what's the matter with you,' Lily said to her lover.

'We have been over fifteen miles of ground, and—'

'I never knew anything so lackadaisical as you gentlemen from London. Been over fifteen miles of ground! Why, uncle Christopher would think nothing of that.'

'Uncle Christopher is made of sterner stuff than we are,' said Crosbie. 'They used to be born so sixty or seventy years ago.' And then they walked on through Gruddock's fields, and the home paddocks, back to the Great House, where they found the squire standing in the front of the porch.

The walk had not been so pleasant as they had all intended that it should be when they made their arrangements for it. Crosbie had endeavoured to recover his happy state of mind, but had been unsuccessful; and Lily, fancying that her lover was not all that he should be, had become reserved and silent. Bernard and Bell had not shared this discomfiture, but then Bernard and Bell were, as a rule, much more given to silence than the other two.

'Uncle,' said Lily, 'these men have shot nothing, and you cannot conceive how unhappy they are in consequence. It's all the fault of the naughty partridges.'

'There are plenty of partridges if they knew how to get them,' said the squire.

'The dogs are uncommonly wild,' said Crosbie.

'They are not wild with me,' said the squire; 'nor yet with Dingles.' Dingles was the squire's gamekeeper. 'The fact is, you young man, nowadays, expect to have dogs trained to do all the work for you. It's too much labour for you to walk up to your game. You'll be late for dinner, girls, if you don't look sharp.'

'We're not coming up this evening, sir,' said Bell.

'And why not?'

'We're going to stay with mamma.'

'And why will not your mother come with you? I'll be whipped if I can understand it. One would have thought that under the present circumstances she would have been glad to see you all as much together as possible.'

'We're together quite enough,' said Lily. 'And as for mamma, I suppose she thinks—' And then she stopped herself, catching the glance of Bell's imploring eye. She was going to make some indignant excuse for her mother, – some excuse which would be calculated to make her uncle angry. It was her practice to say such sharp words to him, and consequently he did not regard her as warmly as her more silent and more prudent sister. At the present moment he turned quickly round and went into the house; and then, with a very few words of farewell the two young men followed him. The girls went back over the little bridge by themselves, feeling that the afternoon had not gone off altogether well.

'You shouldn't provoke him, Lily,' said Bell.

'And he shouldn't say those things about mamma. It seems to me that you don't mind what he says.'

'Oh, Lily.'

'No more you do. He makes me so angry that I cannot hold my tongue. He thinks that because all the place is his, he is to say just what he likes. Why should mamma go up there to please his humours?'

'You may be sure that mamma will do what she thinks best. She is stronger-minded than uncle Christopher, and does not want any one to help her. But, Lily, you shouldn't speak as though I were careless about mamma. You didn't mean that, I know.'

'Of course I didn't.' Then the two girls joined their mother in their own little domain; but we will return to the men at the Great House.

Crosbie, when he went up to dress for dinner, fell into one of those melancholy fits of which I have spoken. Was he absolutely about to destroy all the good that he had done for himself throughout the past years of his hitherto successful life? or rather, as he at last put the question to himself more strongly, – was it not the case that he had already destroyed all that success? His marriage with Lily, whether it was to be for good or bad, was now a settled thing, and was not regarded as a matter admitting of any doubt. To do the man justice, I must declare that in all these moments of misery he still did the best he could to think of Lily herself as of a great treasure which he had won, – as of a treasure which should, and perhaps would, compensate him for his misery. But there was the misery very plain. He must give up his clubs, and his fashion, and all that he had hitherto gained, and be content to live a plain, humdrum, domestic life, with eight hundred a year, and a small house, full of babies. It was not the kind of Elysium for which he had tutored himself. Lily was very nice, very nice indeed. She was, as he said to himself, 'by odds, the nicest girl that he had ever seen.' Whatever might now turn up, her happiness should be his first care. But as for his own, – he began to fear that the compensation would hardly be perfect. 'It is my own doing,' he said to himself, intending to be rather noble in the purport of his soliloquy, 'I have trained myself for other things, – very foolishly. Of course I must suffer, – suffer damnably. But she shall never know it. Dear, sweet, innocent, pretty little thing!' And then he went on about the squire, as to whom he felt himself entitled to be indignant by his own disinterested and manly line of conduct towards the niece. 'But I will let him know what I think about it,' he said. 'It's all very well for Dale to say that I have been treated fairly. It isn't fair for a man to put forward his niece under false pretences. Of course I thought that he intended to provide for her.' And then, having made up his mind in a very manly way that he would not desert Lily altogether after having promised to marry her, he endeavoured to find consolation in the reflection that he might, at any rate, allow himself two years' more run as a bachelor in London. Girls who have to get themselves married without fortunes always know that they will have to wait. Indeed, Lily had already told him, that as far as she was concerned, she was in no hurry. He need not, therefore, at once withdraw his name from Sebright's. Thus he endeavoured to console himself, still, however, resolving that he would have a little serious conversation with the squire that very evening as to Lily's fortune.

And what was the state of Lily's mind at the same moment, while she, also, was performing some slight toilet changes preparatory to their simple dinner at the Small House?

'I didn't behave well to him,' she said to herself; 'I never do. I forget how much he is giving up for me; and then, when anything annoys him, I make it worse instead of comforting him.' And upon that she made accusation against herself that she did not love him half enough, – that she did not let him see how thoroughly and perfectly she loved him. She had an idea of her own, that as a girl should never show any preference for a man till circumstances should have fully entitled him to such manifestation, so also should she make no drawback on her love, but pour it forth for his benefit with all her strength, when such circumstances had come to exist. But she was ever feeling that she was not acting up to her theory, now that the time for such practice had come. She would unwittingly assume little reserves, and make small pretences of indifference in spite of her own judgment. She had done so on this afternoon, and had left him without giving him her hand to press, without looking up into his face with an assurance of love, and therefore she was angry with herself. 'I know I shall teach him to hate me,' she said out loud to Bell.

'That would be very sad,' said Bell; 'but I don't see it.'

'If you were engaged to a man you would be much better to him. You would not say so much, but what you did say would be all affection. I am always making horrid little speeches, for which I should like to cut out my tongue afterwards.'

'Whatever sort of speeches they are, I think that he likes them.'

'Does he? I'm not all so sure of that, Bell. Of course I don't expect that he is to scold me, – not yet, that is. But I know by his eye when he is pleased and when he is displeased.'

And then they went down to dinner.

Up at the Great House the three gentlemen met together in apparent good humour. Bernard Dale was a man of an equal temperament, who rarely allowed any feeling, or even any annoyance, to interfere with his usual manner, – a man who could always come to table with a smile, and meet either his friend or his enemy with a properly civil greeting. Not that he was especially a false man. There was nothing of deceit in his placidity of demeanour. It arose from true equanimity; but it was the equanimity of a cold disposition rather than of one well ordered by discipline. The squire was aware that he had been unreasonably petulant before dinner, and having taken himself to task in his own way, now entered the dining-room with the courteous greeting of a host. 'I find that your bag was not so bad after all,' he said, 'and I hope that your appetite is at least as good as your bag.'

Crosbie smiled, and made himself pleasant, and said a few flattering words. A man who intends to take some very decided step in an hour or two generally contrives to bear himself in the meantime as though the trifles of the world were quite sufficient for him. So he praised the squire's game; said a good-natured word as to Dingles, and bantered himself as to his own want of skill. Then all went merry, – not quite as a marriage bell; but still merry enough for a party of three gentlemen.

But Crosbie's resolution was fixed; and as soon therefore, as the old butler was permanently gone, and the wine steadily in transit upon the table, he began his task, not without some apparent abruptness. Having fully considered the matter, he had determined that he would not wait for Bernard Dale's absence. He thought it possible that he might be able to fight his battle better in Bernard's presence than he could do behind his back.

'Squire,' he began. They all called him squire when they were on good terms together, and Crosbie thought it well to begin as though there was nothing amiss between them. 'Squire, of course I am thinking a good deal at the present moment as to my intended marriage.'

'That's natural enough,' said the squire.

'Yes, by George! sir, a man doesn't make a change like that without finding that he has got something to think of.'

'I suppose not,' said the squire. 'I never was in the way of getting married myself, but I can easily understand that.'

'I've been the luckiest fellow in the world in finding such a girl as your niece—' Whereupon the squire bowed, intending to make a little courteous declaration that the luck in the matter was on the side of the Dales. 'I know that,' continued Crosbie. 'She is exactly everything that a girl ought to be.'

'She is a good girl,' said Bernard.

'Yes; I think she is,' said the squire.

'But it seems to me,' said Crosbie, finding that it was necessary to dash at once headlong into the water, 'that something ought to be said as to my means of supporting her properly.'

Then he paused for a moment, expecting that the squire would speak. But the squire sat perfectly still, looking intently at the empty fireplace and saying nothing. 'Of supporting her,' continued Crosbie, 'with all those comforts to which she has been accustomed.'

'She has never been used to expense,' said the squire. 'Her mother, as you doubtless know, is not a rich woman.'

'But living here, Lily has had great advantages, – a horse to ride, and all that sort of thing.'

'I don't suppose she expects a horse in the park,' said the squire, with a very perceptible touch of sarcasm in his voice.

'I hope not,' said Crosbie.

'I believe she has had the use of one of the ponies here sometimes, but I hope that has not made her extravagant in her ideas. I did not think that there was anything of that nonsense about either of them.'

'Nor is there, – as far as I know.'

'Nothing of the sort,' said Bernard.

'But the long and the short of it is this, sir!' and Crosbie, as he spoke, endeavoured to maintain his ordinary voice and usual coolness, but his heightened colour betrayed that he was nervous. 'Am I to expect any accession of income with my wife?'

'I have not spoken to my sister-in-law on the subject,' said the squire; 'but I should fear that she cannot do much.'

'As a matter of course, I would not take a shilling from her,' said Crosbie.

'Then that settles it,' said the squire.

Crosbie paused a moment, during which his colour became very red. He unconsciously took up an apricot and ate it, and then he spoke out. 'Of course I was not alluding to Mrs. Dale's income; I would not, on any account, disturb her arrangements. But I wished to learn, sir, whether you intend to do anything for your niece.'

'In the way of giving her a fortune? Nothing at all. I intend to do nothing at all.'

'Then I suppose we understand each other, – at last,' said Crosbie.

'I should have thought that we might have understood each other at first,' said the squire. 'Did I ever make you any promise, or give you any hint that I intended to provide for my niece? Have I ever held out to you any such hope? I don't know what you mean by that word "at last" – unless it be to give offence.'

'I meant the truth, sir; – I meant this – that seeing the manner in which your nieces lived with you, I thought it probable that you would treat them both as though they were your daughters. Now I find out my mistake; – that is all!'

'You have been mistaken, – and without a shadow of excuse for your mistake.'

'Others have been mistaken with me,' said Crosbie, forgetting, on the spur of the moment, that he had no right to drag the opinion of any other person into the question.

'What others?' said the squire, with anger; and his mind immediately betook itself to his sister-in-law.

'I do not want to make any mischief,' said Crosbie.

'If anybody connected with my family has presumed to tell you that I intended to do more for my niece Lilian than I have already done, such person has not only been false, but ungrateful. I have given to no one any authority to make any promise on behalf of my niece.'

'No such promise has been made. It was only a suggestion,' said Crosbie.

He was not in the least aware to whom the squire was alluding in his anger; but he perceived that his host was angry, and having already reflected that he should not have alluded to the words which Bernard Dale had spoken in his friendship, he resolved to name no one. Bernard, as he sat by listening, knew exactly how the matter stood; but, as he thought, there could be no reason why he should subject himself to his uncle's ill-will, seeing that he had committed no sin.

'No such suggestion should have been made,' said the squire. 'No one has had a right to make such a suggestion. No one has been placed by me in a position to make such a suggestion to you without manifest impropriety. I will ask no further questions about it; but it is quite as well that you should understand at once that I do not consider it to be my duty to give my niece Lilian a fortune on her marriage. I trust that your offer to her was not made under any such delusion.'

'No, sir; it was not,' said Crosbie.

'Then I suppose that no great harm has been done. I am sorry if false hopes have been given to you; but I am sure you will acknowledge that they were not given to you by me.'

'I think you have misunderstood me, sir. My hopes were never very high; but I thought it right to ascertain your intentions.'

'Now you know them. I trust, for the girl's sake, that it will make no difference to her. I can hardly believe that she has been to blame in the matter.'

Crosbie hastened at once to exculpate Lily; and then, with more awkward blunders than a man should have made who was so well acquainted with fashionable life as the Apollo of the Beaufort, he proceeded to explain that, as Lily was to have nothing, his own pecuniary arrangements would necessitate some little delay in their marriage.

'As far as I myself am concerned,' said the squire, 'I do not like long engagements. But I am quite aware that in this matter I have no right to interfere, unless, indeed—' and then he stopped himself.

'I suppose it will be well to fix some day; eh, Crosbie?' said Bernard.

'I will discuss that matter with Mrs. Dale,' said Crosbie.

'If you and she understand each other,' said the squire, 'that will be sufficient. Shall we go into the drawing-room now, or out upon the lawn?'

That evening, as Crosbie went to bed, he felt that he had not gained the victory in his encounter with the squire.

CHAPTER VIII

IT CANNOT BE

On the following morning at breakfast each of the three gentlemen at the Great House received a little note on pink paper, nominally from Mrs. Dale, asking them to drink tea at the Small House on that day week. At the bottom of the note which Lily had written for Mr. Crosbie was added: 'Dancing on the lawn, if we can get anybody to stand up. Of course you must come, whether you like it or not. And Bernard also. Do your possible to talk my uncle into coming.' And this note did something towards re-creating good-humour among them at the breakfast-table. It was shown to the squire, and at last he was brought to say that he would perhaps go to Mrs. Dale's little evening-party.

It may be well to explain that this promised entertainment had been originated with no special view to the pleasure of Mr. Crosbie, but altogether on behalf of poor Johnny Eames. What was to be done in that matter? This question had been fully discussed between Mrs. Dale and Bell, and they had come to the conclusion that it would be best to ask Johnny over to a little friendly gathering, in which he might be able to meet Lily with some strangers around them. In this way his embarrassment might be overcome. It would never do, as Mrs. Dale said, that he should be suffered to stay away, unnoticed by them. 'When the ice is once broken he won't mind it,' said Bell. And, therefore, early in the day, a messenger was sent over to Guestwick, who returned with a note from Mrs. Eames, saying that she would come on the evening in question, with her son and daughter. They would keep the fly and get back to Guestwick the same evening. This was added, as an offer had been made of beds for Mrs. Eames and Mary.

Before the evening of the party another memorable occurrence had taken place at Allington, which must be described, in order that the feelings of the different people on that evening may be understood. The squire had given his nephew to understand that he wished to have that matter settled as to his niece Bell; and as Bernard's views were altogether in accordance with the squire's, he resolved to comply with his uncle's wishes. The project with him was not a new thing. He did love his cousin quite sufficiently for purposes of matrimony, and was minded that it would be a good thing for him to marry. He could not marry without money, but this marriage would give him an income without the trouble of intricate settlements, or the interference of lawyers hostile to his own interests. It was possible that he might do better; but then it was possible also that he might do much worse; and, in addition to this, he was fond of his cousin. He discussed the matter within himself, very calmly; made some excellent resolutions as to the kind of life which it would behove him to live as a married man; settled on the street in London in which he would have his house, and behaved very prettily to Bell for four or five days running. That he did not make love to her, in the ordinary sense of the word, must, I suppose, be taken for granted, seeing that Bell herself did not recognize the fact. She had always liked her cousin, and thought that in these days he was making himself particularly agreeable.

On the evening before the party the girls were at the Great House, having come up nominally with the intention of discussing the expediency of dancing on the lawn. Lily had made up her mind that it was to be so, but Bell had objected that it would be cold and damp, and that the drawing-room would be nicer for dancing.

'You see we've only got four young gentlemen and one ungrown,' said Lily; 'and they will look so stupid standing up all properly in a room, as though we had a regular party.'

'Thank you for the compliment,' said Crosbie, taking off his straw hat.

'So you will; and we girls will look more stupid still. But out on the lawn it won't look stupid at all. Two or three might stand up on the lawn, and it would be jolly enough.'

'I don't quite see it,' said Bernard.

'Yes, I think I see it,' said Crosbie. 'The unadaptability of the lawn for the purpose of a ball—'

'Nobody is thinking of a ball,' said Lily, with mock petulance.

'I'm defending you, and yet you won't let me speak. The unadaptability of the lawn for the purposes of a ball will conceal the insufficiency of four men and a boy as a supply of male dancers. But, Lily, who is the ungrown gentleman? Is it your old friend Johnny Eames?'

Lily's voice became sobered as she answered him,

'Oh, no; I did not mean Mr. Eames. He is coming, but I

did not mean him. Dick Boyce, Mr. Boyce's son, is only sixteen. He is the ungrown gentleman.'

'And who is the fourth adult?'

'Dr. Crofts, from Guestwick. I do hope you will like him, Adolphus. We think he is the very perfection of a man.'

'Then of course I shall hate him; and be very jealous, too!'

And then that pair went off together, fighting their own little battle on that head, as turtle-doves will sometimes do. They went off, and Bernard was left with Bell standing together over the ha-ha* fence which divides the garden at the back of the house from the field.

'Bell,' he said, 'they seem very happy, don't they?'

'And they ought to be happy now, oughtn't they? Dear Lily! I hope he will be good to her. Do you know, Bernard, though he is your friend, I am very, very anxious about it. It is such a vast trust to put in a man when we do not quite know him.'

'Yes, it is; but they'll do very well together. Lily will be happy enough.'

'And he?'

'I suppose he'll be happy, too. He'll feel himself a little straitened as to income at first, but that will all come round.'

'If he is not, she will be wretched.'

'They will do very well. Lily must be prepared to make the money go as far as she can, that's all.'

'Lily won't feel the want of money. It is not that. But if he lets her know that she has made him a poor man, then she will be unhappy. Is he extravagant, Bernard?'

But Bernard was anxious to discuss another subject, and therefore would not speak such words of wisdom as to Lily's engagement as might have been expected from him had he been in a different frame of mind.

'No, I should say not,' said he. 'But, Bell—'

'I do not know that we could have acted otherwise than we have done, and yet I fear that we have been rash. If he makes her unhappy, Bernard, I shall never forgive you.'

But as she said this she put her hand lovingly upon his arm, as a cousin might do, and spoke in a tone which divested her threat of its acerbity.

'You must not quarrel with me, Bell, whatever may happen. I cannot afford to quarrel with you.'

'Of course I was not in earnest as to that.'

'You and I must never quarrel, Bell; at least, I hope not. I could bear to quarrel with any one rather than with you.' And then, as he spoke, there was something in his voice which gave the girl some slight, indistinct warning of what might be his intention. Not that she said to herself at once, that he was going to make her an offer of his hand, – now, on the spot; but she felt that he intended something beyond the tenderness of ordinary cousinly affection.

'I hope we shall never quarrel,' she said. But as she spoke, her mind was settling itself, – forming its resolution, and coming to a conclusion as to the sort of love which Bernard might, perhaps, expect. And it formed another conclusion; as to the sort of love which might be given in return.

'Bell,' he said, 'you and I have always been dear friends.'

'Yes; always.'

'Why should we not be something more than friends?'

To give Captain Dale his due I must declare that his voice was perfectly natural as he asked this question, and that he showed no signs of nervousness, either in face or limbs. He had made up his mind to do it on that occasion, and he did it without any signs of outward disturbance. He asked his question, and then he waited for his answer. In this he was rather hard upon his cousin; for, though the question had certainly been asked in language that could not be mistaken, still the matter had not been put forward with all that fulness which a young lady, under such circumstances, has a right to expect.

They had sat down on the turf close to the ha-ha, and they were so near that Bernard was able to put out his hand with the view of taking that of his cousin within his own. But she contrived to keep her hands locked together, so that he merely held her gently by the wrist.

'I don't quite understand, Bernard,' she said, after a minute's pause.

'Shall we be more than cousins? Shall we be man and wife?'

Now, at least, she could not say that she did not understand. If the question was ever asked plainly, Bernard Dale had asked it plainly. Shall we be man and wife? Few men, I fancy, dare to put it all at once in so abrupt a way, and yet I do not know that the English language affords any better terms for the question.

'Oh, Bernard! you have surprised me.'

'I hope I have not pained you, Bell. I have been long thinking of this, but I am well aware that my own manner, even to you, has not been that of a lover. It is not in me to smile and say soft things, as Crosbie can. But I do not love you the less on that account. I have looked about for a wife, and I have thought that if I could gain you I should be very fortunate.'

He did not then say anything about his uncle, and the eight hundred a year; but he fully intended to do so as soon as an opportunity should serve. He was quite of opinion that eight hundred a year and the good-will of a rich uncle were strong grounds for matrimony, – were grounds even for love; and he did not doubt but his cousin would see the matter in the same light.

'You are very good to me – more than good. Of course I know that. But, oh, Bernard! I did not expect this a bit.'

'But you will answer me, Bell! Or if you would like time to think, or to speak to my aunt, perhaps you will answer me to-morrow?'

'I think I ought to answer you now.'

'Not if it be a refusal, Bell. Think well of it before you do that. I should have told you that our uncle wishes this match, and that he will remove any difficulty there might be about money.'

'I do not care for money.'

'But, as you were saying about Lily, one has to be prudent. Now, in our marriage, everything of that kind would be well arranged. My uncle has promised me that he would at once allow us—'

'Stop, Bernard. You must not be led to suppose that any offer made by my uncle would help to purchase— Indeed, there can be no need for us to talk about money.'

'I wished to let you know the facts of the case, exactly as they are. And as to our uncle, I cannot but think that you would be glad, in such a matter, to have him on your side.'

'Yes, I should be glad to have him on my side; that is, if I were going— But my uncle's wishes could not influence my decision. The fact is, Bernard—'

'Well, dearest, what is the fact?'

'I have always regarded you rather as a brother than as anything else.'

'But that regard may be changed.'

'No; I think not. Bernard, I will go further and speak on at once. It cannot be changed. I know myself well enough to say that with certainty. It cannot be changed.'

'You mean that you cannot love me?'

'Not as you would have me do. I do love you very dearly, –very dearly, indeed. I would go to you in any trouble, exactly as I would go to a brother.'

'And must that be all, Bell?'

'Is not that all the sweetest love that can be felt? But you must not think me ungrateful, or proud. I know well that you are – are proposing to do for me much more than I deserve. Any girl might be proud of such an offer. But, dear Bernard—'

'Bell, before you give me a final answer, sleep upon this and talk it over with your mother. Of course you were unprepared, and I cannot expect that you should promise me so much without a moment's consideration.'

'I was unprepared, and therefore I have not answered you as I should have done. But as it has gone so far, I cannot let you leave me in uncertainty. It is not necessary that I should keep you waiting. In this matter I do know my own mind. Dear Bernard, indeed, indeed it cannot be as you have proposed.'

She spoke in a low voice, and in a tone that had in it something of almost imploring humility; but, nevertheless, it conveyed to her cousin an assurance that she was in earnest; an assurance also that that earnest would not readily be changed. Was she not a Dale? And when did a Dale change his mind? For a while he sat silent by her; and she too, having declared her intention, refrained from further words. For some minutes they thus remained, looking down into the ha-ha. She still kept her old position, holding her hands clasped together over her knees; but he was now lying on his side, supporting his head upon his arm, with his face indeed turned towards her, but with his eyes fixed upon the grass. During this time, however, he was not idle. His cousin's answer, though it had grieved him, had not come upon him as a blow stunning him for a moment, and rendering him unfit for instant thought. He was grieved, more grieved than he had thought he would have been. The thing that he had wanted moderately he now wanted the more in that it was denied to him. But he was able to perceive the exact truth of his position, and to calculate what might be his chances if he went on with his suit, and what his advantage if he at once abandoned it.

'I do not wish to press you unfairly, Bell; but may I ask if any other preference—'

'There is no other preference,' she answered. And then again they were silent for a minute or two.

'My uncle will be much grieved at this,' he said at last.

'If that be all,' said Bell, 'I do not think that we need either of us trouble ourselves. He can have no right to dispose of our hearts.'

'I understand the taunt, Bell.'

'Dear Bernard, there was no taunt. I intended none.'

'I need not speak of my own grief. You cannot but know how deep it must be. Why should I have submitted myself to this mortification had not my heart been concerned? But that I will bear, if I must bear it—' And then he paused, looking up at her.

'It will soon pass away,' she said.

'I will accept it at any rate without complaint. But as to my uncle's feelings, it is open to me to speak, and to you, I should think, to listen without indifference. He has been kind to us both, and loves us two above any other living beings. It's not surprising that he should wish to see us married, and it will not be surprising if your refusal should be a great blow to him.'

'I shall be sorry – very sorry.'

'I also shall be sorry. I am now speaking of him. He has set his heart upon it; and as he has but few wishes, few desires, so is he the more constant in those which he expresses. When he knows this, I fear that we shall find him very stern.'

'Then he will be unjust.'

'No; he will not be unjust. He is always a just man. But he will be unhappy, and will, I fear, make others unhappy. Dear Bell, may not this thing remain for a while unsettled? You will not find that I take advantage of your goodness. I will not intrude it on you again, – say for a fortnight, – or till Crosbie shall be gone.'

'No, no, no,' said Bell.

'Why are you so eager in your noes? There can be no danger in such delay. I will not press you, – and you can let my uncle think that you have at least taken time for consideration.'

'There are things as to which one is bound to answer at once. If I doubted myself, I would let you persuade me. But I do not doubt myself, and I should be wrong to keep you in suspense. Dear, dearest Bernard, it cannot be; and as it cannot be, you, as my brother, would bid me say so clearly. It cannot be.'

As she made this last assurance, they heard the steps of Lily and her lover close to them, and they both felt that it would be well that their intercourse should thus be brought to a close. Neither had known how to get up and leave the place, and yet each had felt that nothing further could be said.

'Did you ever see anything so sweet and affectionate and romantic?' said Lily, standing over them and looking at them. 'And all the while we have been so practical and worldly. Do you know, Bell, that Adolphus seems to think we can't very well keep pigs in London. It makes me so unhappy.'

'It does seem a pity,' said Crosbie, 'for Lily seems to know all about pigs.'

'Of course I do. I haven't lived in the country all my life for nothing. Oh, Bernard, I should so like to see you rolled down into the bottom of the ha-ha. Just remain there, and we'll do it between us.'

Whereupon Bernard got up, as did Bell also, and they all went in to tea.

CHAPTER IX

MRS. DALE'S LITTLE PARTY

THE next day was the day of the party. Not a word more was said on that evening between Bell and her cousin, at least, not a word more of any peculiar note; and when Crosbie suggested to his friend on the following morning that they should both step down and see how the preparations were getting on at the Small House, Bernard declined.

'You forget, my dear fellow, that I'm not in love as you are,' said he.

'But I thought you were,' said Crosbie.

'No; not at all as you are. You are an accepted lover, and will be allowed to do anything, – whip the creams, and tune the piano, if you know how. I'm only a half sort of lover, meditating a mariage de convenance to oblige an uncle, and by no means required by the terms of my agreement to undergo a very rigid amount of drill. Your position is just the reverse.' In saying all which Captain Dale was no doubt very false; but if falseness can be forgiven to a man in any position, it may be forgiven in that which he then filled. So Crosbie went down to the Small House alone.

'Dale wouldn't come,' said he, speaking to the three ladies together, 'I suppose he's keeping himself up for the dance on the lawn.'

'I hope he will be here in the evening,' said Mrs. Dale. But Bell said never a word. She had determined, that under the existing circumstances, it would be only fair to her cousin that his offer and her answer to it should be kept secret. She knew why Bernard did not come across from the Great House with his friend, but she said nothing of her knowledge. Lily looked at her, but looked without speaking; and as for Mrs. Dale, she took no notice of the circumstance.

Thus they passed the afternoon together without further mention of Bernard Dale; and it may be said, at any rate of Lily and Crosbie, that his presence was not missed.

Mrs. Eames, with her son and daughter, were the first to come. 'It is so nice of you to come early,' said Lily, trying on the spur of the moment to say something which should sound pleasant and happy, but in truth using that form of welcome which to my ears sounds always the most ungracious. 'Ten minutes before the time named; and, of course, you must have understood that I meant thirty minutes after it!' That is my interpretation of the words when I am thanked for coming early. But Mrs. Eames was a kind, patient, unexacting woman, who took all civil words as meaning civility. And, indeed, Lily had meant nothing else.

'Yes; we did come early,' said Mrs. Eames, 'because Mary thought she would like to go up into the girls' room and just settle her hair, you know.'

'So she shall,' said Lily, who had taken Mary by the hand.

'And we knew we shouldn't be in the way. Johnny can go out into the garden if there's anything left to be done.'

'He shan't be banished unless he likes it,' said Mrs. Dale. 'If he finds us women too much for his unaided strength—'

John Eames muttered something about being very well as he was, and then got himself into an armchair. He had shaken hands with Lily, trying as he did so to pronounce articulately a little speech which he had prepared for the occasion. 'I have to congratulate you, Lily, and I hope with all my heart that you will be happy.' The words were simple enough, and were not ill-chosen, but the poor young man never got them spoken. The word 'congratulate' did reach Lily's ears, and she understood it all; – both the kindness of the intended speech and the reason why it could not be spoken.

'Thank you, John,' she said; 'I hope I shall see so much of you in London. It will be so nice to have an old Guestwick friend near me.' She had her own voice, and the pulses of her heart better under command than had he; but she also felt that the occasion was trying to her. The man had loved her honestly and truly, – still did love her, paying her the great homage of bitter grief in that he had lost her. Where is the girl who will not sympathize with such love and such grief, if it be shown only because it cannot be concealed, and be declared against the will of him who declares it?

Then came in old Mrs. Hearn, whose cottage was not distant two minutes' walk from the Small House. She always called Mrs. Dale 'my dear,' and petted the girls as though they had been children. When told of Lily's marriage, she had thrown up her hands with surprise, for she had still left in some corner of her drawers remnants of sugar-plums which she had bought for Lily. 'A London man is he? Well, well. I wish he lived in the country. Eight hundred a year, my dear?' she had said to Mrs. Dale. 'That sounds nice down here, because we are all so poor. But I suppose eight hundred a year isn't very much up in London?'

'The squire's coming, I suppose, isn't he?' said Mrs. Hearn, as she seated herself on the sofa close to Mrs. Dale.

'Yes, he'll be here by-and-by; unless he changes his mind, you know. He doesn't stand on ceremony with me.'

'He change his mind! When did you ever know Christopher Dale change his mind?'

'He is pretty constant, Mrs. Hearn.'

'If he promised to give a man a penny, he'd give it. But if he promised to take away a pound, he'd take it, though it cost him years to get it. He's going to turn me out of my cottage, he says.'

'Nonsense, Mrs. Hearn!'

'Jolliffe came and told me' – Jolliffe, I should explain, was the bailiff, – 'that if I didn't like it as it was, I might leave it, and that the squire could get double the rent for it. Now all I asked was that he should do a little painting in the kitchen; and the wood is all as black as his hat.'

'I thought it was understood you were to paint inside.'

'How can I do it, my dear, with a hundred and forty pounds for everything? I must live, you know! And he that has workmen about him every day of the year! And was that a message to send to me, who have lived in the parish for fifty years? Here he is.' And Mrs. Hearn majestically raised herself from her seat as the squire entered the room.

With him entered Mr. and Mrs. Boyce, from the parsonage, with Dick Boyce, the ungrown gentleman, and two girl Boyces, who were fourteen and fifteen years of age. Mrs. Dale, with the amount of good-nature usual on such occasions, asked reproachfully why Jane, and Charles, and Florence, and Bessy, did not come, – Boyce being a man who had his quiver full of them, – and Mrs. Boyce, giving the usual answer, declared that she already felt that they had come as an avalanche.

'But where are the – the – the young men?' asked Lily, assuming a look of mock astonishment.

'They'll be across in two or three hours' time,' said the squire. 'They both dressed for dinner, and, as I thought, made themselves very smart; but for such a grand occasion as this they thought a second dressing necessary. How do you do, Mrs. Hearn? I hope you are quite well. No rheumatism left, eh?' This the squire said very loud into Mrs. Hearn's ear. Mrs. Hearn was perhaps a little hard of hearing; but it was very little, and she hated to be thought deaf. She did not, moreover, like to be thought rheumatic. This the squire knew, and therefore his mode of address was not good-natured.

'You needn't make me jump so, Mr. Dale. I'm pretty well now, thank ye. I did have a twinge in the spring, – that cottage is so badly built for draughts! "I wonder you can live in it," my sister said to me the last time she was over. I suppose I should be better off over with her at Hamersham, only one doesn't like to move, you know, after living fifty years in one parish.'

'You mustn't think of going away from us,' Mrs. Boyce said, speaking by no means loud, but slowly and plainly, hoping thereby to flatter the old woman. But the old woman understood it all. 'She's a sly creature, is Mrs. Boyce,' Mrs. Hearn said to Mrs. Dale, before the evening was out. There are some old people whom it is very hard to flatter, and with whom it is, nevertheless, almost impossible to live unless you do flatter them.

At last the two heroes came in across the lawn at the drawing-room window; and Lily, as they entered, dropped a low curtsey before them, gently swelling down upon the ground with her light muslin dress, till she looked like some wondrous flower that had bloomed upon the carpet, and putting her two hands, with the backs of her fingers pressed together, on the buckle of her girdle, she said, 'We are waiting upon your honours' kind grace, and feel how much we owe to you for favouring our poor abode.' And then she gently rose up again, smiling, oh, so sweetly, on the man she loved, and the puffings and swellings went out of her muslin.

I think there is nothing in the world so pretty as the conscious little tricks of love played off by a girl towards the man she loves, when she has made up her mind boldly that all the world may know that she has given herself away to him.

I am not sure that Crosbie liked it all as much as he should have done. The bold assurance of her love when they two were alone together he did like. What man does not like such assurances on such occasions? But perhaps he would have been better pleased had Lily shown more reticence, – been more secret, as it were, as to her feelings, when others were around them. It was not that he accused her in his thoughts of any want of delicacy. He read her character too well; – was, if not quite aright in his reading of it, at least too nearly so to admit of his making against her any such accusation at that. It was the calf-like feeling that was disagreeable to him. He did not like to be presented, even to the world of Allington, as a victim caught for the sacrifice, and bound with ribbon for the altar. And then there lurked behind it all a feeling that it might be safer that the thing should not be so openly manifested before all the world. Of course, everybody knew that he was engaged to Lily Dale; nor had he, as he said to himself, perhaps too frequently, the slightest idea of breaking from that engagement. But then the marriage might possibly be delayed. He had not discussed that matter yet with Lily, having, indeed, at the first moment of his gratified love, created some little difficulty for himself by pressing for an early day. 'I will refuse you nothing,' she had said to him; 'but do not make it too soon.' He saw, therefore, before him some little embarrassment, and was inclined to wish that Lily would abstain from that manner which seemed to declare to all the world that she was about to be married immediately. 'I must speak to her to-morrow,' he said to himself, as he accepted her salute with a mock gravity equal to her own.

Poor Lily! How little she understood as yet what was passing through his mind. Had she known his wish she would have wrapped up her love carefully in a napkin, so that no one should have seen it, – no one but he, when he might choose to have the treasure uncovered for his sight. And it was all for his sake that she had been thus open in her ways. She had seen girls who were half ashamed of their love; but she would never be ashamed of hers or of him. She had given herself to him; and now all the world might know it, if all the world cared for such knowledge. Why should she be ashamed of that which, to her thinking, was so great an honour to her? She had heard of girls who would not speak of their love, arguing to themselves cannily that there may be many a slip between the cup and the lip. There could be no need of any such caution with her. There could surely be no such slip! Should there be such a fall, – should any such fate, either by falseness or misfortune, come upon her, – no such caution could be of service to save her. The cup would have been so shattered in its fall that no further piecing of its parts would be in any way possible. So much as this she did not exactly say to herself; but she felt it all, and went bravely forward, – bold in her love, and careful to hide it from none who chanced to see it.

They had gone through the ceremony with the cake and teacups, and had decided that, at any rate, the first dance or two should be held upon the lawn when the last of the guests arrived.

'Oh, Adolphus, I am so glad he has come,' said Lily. 'Do try to like him.' Of Dr. Crofts, who was the new comer, she had sometimes spoken to her lover, but she had never coupled her sister's name with that of the doctor, even in speaking to him. Nevertheless, Crosbie had in some way conceived the idea that this Crofts either had been, or was, or was to be, in love with Bell; and as he was prepared to advocate his friend Dale's claims in that quarter, he was not particularly anxious to welcome the doctor as a thoroughly intimate friend of the family. He knew nothing as yet of Dale's offer, or of Bell's refusal, but he was prepared for war, if war should be necessary. Of the squire, at the present moment, he was not very fond; but if his destiny intended to give him a wife out of this family, he should prefer the owner of Allington and nephew of Lord De Guest as a brother-in-law to a village doctor, – as he took upon himself, in his pride, to call Dr. Crofts.

'It is very unfortunate,' said he, 'but I never do like Paragons.'

'But you must like this Paragon. Not that he is a Paragon at all, for he smokes and hunts, and does all manner of wicked things.' And then she went forward to welcome her friend.

Dr. Crofts was a slight, spare man, about five feet nine in height, with very bright dark eyes, a broad forehead, with dark hair that almost curled, but which did not come so forward over his brow as it should have done for purposes of beauty, – with a thin well-cut nose, and a mouth that would have been perfect had the lips been a little fuller. The lower part of his face, when seen alone, had in it somewhat of sternness, which, however, was redeemed by the brightness of his eyes. And yet an artist would have declared that the lower features of his face were by far the more handsome.

Lily went across to him and greeted him heartily, declaring how glad she was to have him there. 'And I must introduce you to Mr. Crosbie,' she said, as though she was determined to carry her point. The two men shook hands with each other, coldly, without saying a word, as young men are apt to do when they are brought together in that way. Then they separated at once, somewhat to the disappointment of Lily. Crosbie stood off by himself, both his eyes turned up towards the ceiling, and looking as though he meant to give himself airs; while Crofts got himself quickly up to the fireplace, making civil little speeches to Mrs. Dale, Mrs. Boyce, and Mrs. Hearn. And then at last he made his way round to Bell.

'I am so glad,' he said, 'to congratulate you on your sister's engagement.'

'Yes,' said Bell; 'we knew that you would be glad to hear of her happiness.'

'Indeed, I am glad, and thoroughly hope that she may be happy. You all like him, do you not?'

'We like him very much.'

'And I am told that he is well off. He is a very fortunate man, – very fortunate, – very fortunate.'

'Of course we think so,' said Bell. 'Not, however, because he is rich.'

'No; not because he is rich. But because, being worthy of such happiness, his circumstances should enable him to marry, and to enjoy it.'

'Yes, exactly,' said Bell. 'That is just it.' Then she sat down, and in sitting down put an end to the conversation. 'That is just it,' she had said. But as soon as the words were spoken she declared to herself that it was not so, and that Crofts was wrong. 'We love him,' she said to herself, 'not because he is rich enough to marry without anxious thought, but because he dares to marry although he is not rich.' And then she told herself that she was angry with the doctor.

After that Dr. Crofts got off towards the door, and stood there by himself, leaning against the wall, with the thumbs of both his hands stuck into the armholes of his waistcoat. People said that he was a shy man. I suppose he was shy, and yet he was a man that was by no means afraid of doing anything that he had to do. He could speak before a multitude without being abashed, whether it was a multitude of men or of women. He could be very fixed too in his own opinion, and eager, if not violent, in the prosecution of his purpose. But he could not stand and say little words, when he had in truth nothing to say. He could not keep his ground when he felt that he was not using the ground upon which he stood. He had not learned the art of assuming himself to be of importance in whatever place he might find himself. It was this art which Crosbie had learned, and by this art that he had flourished. So Crofts retired and leaned against the wall near the door; and Crosbie came forward and shone like an Apollo among all the guests. 'How is it that he does it?' said John Eames to himself, envying the perfect happiness of the London man of fashion.

At last Lily got the dancers out upon the lawn, and then they managed to get through one quadrille. But it was found that it did not answer. The music of the single fiddle which Crosbie had hired from Guestwick was not sufficient for the purpose; and then the grass, though it was perfect for purposes of croquet, was not pleasant to the feet for dancing.

'This is very nice,' said Bernard to his cousin. 'I don't know anything that could be nicer; but perhaps—'

'I know what you mean,' said Lily. 'But I shall stay here. There's no touch of romance about any of you. Look at the moon there at the back of the steeple. I don't mean to go in all night.' Then she walked off by one of the paths, and her lover went after her.

'Don't you like the moon?' she said, as she took his arm, to which she was now so accustomed that she hardly thought of it as she took it.

'Like the moon – well; I fancy I like the sun better. I don't quite believe in moonlight. I think it does best to talk about when one wants to be sentimental.'

'Ah; that is just what I fear. That is what I say to Bell when I tell her that her romance will fade as the roses do. And then I shall have to learn that prose is more serviceable than poetry, and that the mind is better than the heart, and – and that money is better than love. It's all coming, I know; and yet I do like the moonlight.'

'And the poetry, – and the love?'

'Yes. The poetry much, and the love more. To be loved by you is sweeter even than any of my dreams, – is better than all the poetry I have read.'

'Dearest Lily,' and his unchecked arm stole round her waist.

'It is the meaning of the moonlight, and the essence of the poetry,' continued the impassioned girl. 'I did not know then why I liked such things, but now I know. It was because I longed to be loved.'

'And to love.'

'Oh, yes. I would be nothing without that. But that, you know, is your delight, – or should be. The other is mine. And yet it is a delight to love you; to know that I may love you.'

'You mean that this is the realization of your romance.'

'Yes; but it must not be the end of it, Adolphus. You must like the soft twilight, and the long evenings when we shall be alone; and you must read to me the books I love, and you must not teach me to think that the world is hard and dry, and cruel, – not yet. I tell Bell so very often; but you must not say so to me.'

'It shall not be dry and cruel, if I can prevent it.'

'You understand what I mean, dearest. I will not think it dry and cruel, even though sorrow should come upon us, if you— I think you know what I mean.'

'If I am good to you.'

'I am not afraid of that; – I am not the least afraid of that. You do not think that I could ever distrust you? But you must not be ashamed to look at the moonlight, and to read poetry, and to—'

'To talk nonsense, you mean.'

But as he said it, he pressed her closer to his side, and his tone was pleasant to her.

'I suppose I'm talking nonsense now?' she said, pouting. 'You liked me better when I was talking about the pigs; didn't you?'

'No; I like you best now.'

'And why didn't you like me then? Did I say anything to offend you?'

'I like you best now, because—'

They were standing in the narrow pathway of the gate leading from the bridge into the gardens of the Great House, and the shadow of the thick-spreading laurels was around them. But the moonlight still pierced brightly through the little avenue, and she looked up to him, could see the form of his face and the loving softness of his eye.

'Because—,' said he; and then he stooped over her and pressed her closely, while she put up her lips to his, standing on tip-toe that she might reach to his face.

'Oh, my love!' she said. 'My love! my love!'

As Crosbie walked back to the Great House that night, he made a firm resolution that no consideration of worldly welfare should ever induce him to break his engagement with Lily Dale. He went somewhat further also, and determined that he would not put off the marriage for more than six or eight months, or, at the most, ten, if he could possibly get his affairs arranged in that time. To be sure, he must give up everything, – all the aspirations and ambition of his life; but then, as he declared to himself somewhat mournfully, he was prepared to do that. Such were his resolutions, and, as he thought of them in bed, he came to the conclusion that few men were less selfish than he was.

'But what will they say to us for staying away?' said Lily, recovering herself. 'And I ought to be making the people dance, you know. Come along, and do make yourself nice. Do waltz with Mary Eames; – pray, do. If you don't, I won't speak to you all night!'

Acting under which threat, Crosbie did, on his return, solicit the honour of that young lady's hand, thereby elating her into a seventh heaven of happiness. What could the world afford better than a waltz with such a partner as Adolphus Crosbie? And poor Mary Eames could waltz well; though she could not talk much as she danced, and would pant a good deal when she stopped. She put too much of her energy into the motion, and was too anxious to do the mechanical part of the work in a manner that should be satisfactory to her partner. 'Oh! thank you; – it's very nice. I shall be able to go on – again directly.' Her conversation with Crosbie did not get much beyond that, and yet she felt that she had never done better than on this occasion.

Though there were, at most, not above five couples of dancers, and though they who did not dance, such as the squire and Mr. Boyce, and a curate from a neighbouring parish, had, in fact, nothing to amuse them, the affair was kept on very merrily for a considerable number of hours.

Exactly at twelve o'clock there was a little supper, which, no doubt, served to relieve Mrs. Hearn's *ennui*, and at which Mrs. Boyce also seemed to enjoy herself. As to the Mrs. Boyces on such occasions, I profess that I feel no pity. They are generally happy in their children's happiness, or if not, they ought to be. At any rate, they are simply performing a manifest duty, which duty, in their time, was performed on their behalf. But on what account do the Mrs. Hearns betake themselves to such gatherings? Why did that ancient lady sit there hour after hour yawning, longing for her bed, looking every ten minutes at her watch, while her old bones were stiff and sore, and her old ears pained with the noise? It could hardly have been simply for the sake of the supper. After the supper, however, her maid took her across to her cottage, and Mrs. Boyce also then stole away home, and the squire went off with some little parade, suggesting to the young men that they should make no noise in the house as they returned. But the poor curate remained, talking a dull word every now and then to Mrs. Dale, and looking on with tantalized eyes at the joys which the world had prepared for others than him. I must say that I think that public opinion and the bishops together are too hard upon curates in this particular.

In the latter part of the night's delight, when time and practice had made them all happy together, John Eames stood up for the first time to dance with Lily. She had done all she could, short of asking him, to induce him to do her this favour; for she felt that it would be a favour. How great had been the desire on his part to ask her, and, at the same time, how great the repugnance, Lily, perhaps, did not quite understand. And yet she understood much of it. She knew that he was not angry with her. She knew that he was suffering from the injured pride of futile love, almost as much as from the futile love itself. She wished to put him at his ease in this; but she did not quite give him credit for the full sincerity, and the upright, uncontrolled heartiness of his feelings.

At length he did come up to her, and though, in truth, she was engaged, she at once accepted his offer. Then she tripped across the room. 'Adolphus,' she said, 'I can't dance with you, though I said I would. John Eames has asked me, and I haven't stood up with him before. You understand, and you'll be a good boy, won't you?'

Crosbie, not being in the least jealous, was a good boy, and sat himself down to rest, hidden behind a door.

For the first few minutes the conversation between Eames and Lily was of a very matter-of-fact kind. She repeated her wish that she might see him in London, and he said that of course he should come and call. Then there was silence for a little while, and they went through their figure dancing.

'I don't at all know yet when we are to be married,' said Lily, as soon as they were again standing together.

'No; I dare say not,' said Eames.

'But not this year, I suppose. Indeed, I should say, of course not.'

'In the spring, perhaps,' suggested Eames. He had an unconscious desire that it might be postponed to some Greek kalends,* and yet he did not wish to injure Lily.

'The reason I mention it is this, that we should be so very glad if you could be here. We all love you so much, and I should so like to have you here on that day.'

Why is it that girls so constantly do this, – so frequently ask men who have loved them to be present at their marriages with other men? There is no triumph in it. It is done in sheer kindness and affection. They intend to offer something which shall soften and not aggravate the sorrow that they have caused. 'You can't marry me yourself,' the lady seems to say. 'But the next greatest blessing which I can offer you shall be yours, – you shall see me married to somebody else.' I fully appreciate the intention, but in honest truth, I doubt the eligibility of the proffered entertainment.

On the present occasion John Eames seemed to be of this opinion, for he did not at once accept the invitation.

'Will you not oblige me so far as that?' said she softly.

'I would do anything to oblige you,' said he gruffly; 'almost anything.'

'But not that?'

'No; not that. I could not do that.' Then he went off upon

his figure, and when they were next both standing together, they remained silent till their turn for dancing had again come. Why was it, that after that night Lily thought more of John Eames than ever she had thought before; – felt for him, I mean, a higher respect, as for a man who had a will of his own?

And in that quadrille Crofts and Bell had been dancing together, and they also had been talking of Lily's marriage. 'A man may undergo what he likes for himself,' he had said, 'but he has no right to make a woman undergo poverty.'

'Perhaps not,' said Bell.

'That which is no suffering for a man, – which no man should think of for himself, – will make a hell on earth for a woman.'

'I suppose it would, said Bell, answering him without a sign of feeling in her face or voice. But she took in every word that he spoke, and disputed their truth inwardly with all the strength of her heart and mind, and with the very vehemence of her soul. 'As if a woman cannot bear more than a man!' she said to herself, as she walked the length of the room alone, when she had got herself free from the doctor's arm.

CHAPTER X

MRS. LUPEX AND AMELIA ROPER

I SHOULD simply mislead a confiding reader if I were to tell him that Mrs. Lupex was an amiable woman. Perhaps the fact that she was not amiable is the one great fault that should be laid to her charge; but that fault had spread itself so widely, and had cropped forth in so many different places of her life, like a strong rank plant that will show itself all over a garden, that it may almost be said that it made her odious in every branch of life, and detestable alike to those who knew her little and to those who knew her much. If a searcher could have got at the inside spirit of the woman, that searcher would have found that she wished to go right, – that she did make, or at any rate promise to herself that she would make, certain struggles to attain decency and propriety. But it was so natural to her to torment those whose misfortune brought them near to her, and especially that wretched man who in an evil day had taken her to his bosom as his wife, that decency fled from her, and propriety would not live in her quarters.

Mrs. Lupex was, as I have already described her, a woman not without some feminine attraction in the eyes of those who like morning negligence and evening finery, and do not object to a long nose somewhat on one side. She was clever in her way, and could say smart things. She could flatter also, though her very flattery had always in it something that was disagreeable. And she must have had some power of will, as otherwise her husband would have escaped from her before the days of which I am writing. Otherwise, also, she could hardly have obtained her footing and kept it in Mrs. Roper's drawing-room. For though the hundred pounds a year, either paid, or promised to be paid, was matter with Mrs. Roper of vast consideration, nevertheless the first three months of Mrs. Lupex's sojourn in Burton Crescent were not over before the landlady of that house was most anxiously desirous of getting herself quit of her married boarders.

I shall perhaps best describe a little incident that had occurred in Burton Crescent during the absence of our friend Eames, and the manner in which things were going on in that locality, by giving at length two letters which Johnny received by post at Guestwick on the morning after Mrs. Dale's party. One was from his friend Cradell, and the other from the devoted Amelia. In this instance I will give that from the gentleman first, presuming that I shall best consult my reader's wishes by keeping the greater delicacy till the last.

'INCOME-TAX OFFICE, September, 186–,

'My dear Johnny,

'We have had a terrible affair in the Crescent; and I really hardly know how to tell you; and yet I must do it, for I want your advice. You know the sort of standing that I was on with Mrs. Lupex, and perhaps you remember what we were saying on the platform at the station. I have, no doubt, been fond of her society, as I might be of that of any other friend. I knew, of course, that she was a fine woman; and if her husband chose to be jealous, I couldn't help that. But I never intended anything wrong; and, if it was necessary, couldn't I call you as a witness to prove it? I never spoke a word to her out of Mrs. Roper's drawing-room; and Miss Spruce, or Mrs. Roper, or somebody has always been there. You know he drinks horribly sometimes, but I do not think he ever gets downright drunk. Well, he came home last night about nine o'clock after one of these bouts. From what Jemima says [Jemima was Mrs. Roper's parlour-maid], I believe he had been at it down at the theatre for three days. We hadn't seen him since Tuesday. He went straight into the parlour and sent up Jemima to me, to say that he wanted to see me. Mrs. Lupex was in the room and heard the girl summon me, and, jumping up, she declared that if there was going to be blood shed she would leave the house. There was nobody else in the room but Miss Spruce, and she didn't say a word, but took her candle and went upstairs. You must own it looked very uncomfortable. What was I to do with a drunken man down in the parlour? However, she seemed to think I ought to go. "If he comes up here," said she, "I shall be the victim. You little know of what that man is capable when his wrath has been inflamed by wine!" Now, I think you are aware that I am not likely to be very much afraid of any man; but why was I to be got into a row in such a way as this? I hadn't done anything. And then, if there was to be a quarrel, and anything was to come of it, as she seemed to expect, – like bloodshed, I mean, or a fight, or if he were to knock me on the head with the poker, where should I be at my office? A man in a public office, as you and I are, can't quarrel like anybody else. It was this that I felt so much at the moment. "Go down to him," said she, "unless you wish to see me murdered at your feet." Fisher says, that if what I say is true, they must have arranged it all between them. I don't think that; for I do believe that she is really fond of me. And then everybody knows that they never do agree about anything. But she certainly did implore me to go down to him. Well, I went down; and, as I got to the bottom of the stairs, where

I found Jemima, I heard him walking up and down the parlour. "Take care of yourself, Mr. Cradell," said the girl; and I could see by her face that she was in a terrible fright.

'At that moment I happened to see my hat on the hall table, and it occurred to me that I ought to put myself into the hands of a friend. Of course, I was not afraid of that man in the dining-room; but should I have been justified in engaging in a struggle perhaps for dear life, in Mrs. Roper's house? I was bound to think of her interests. So I took up my hat, and deliberately walked out of the front door. "Tell him," said I to Jemima, "that I'm not at home." And so I went away direct to Fisher's, meaning to send him back to Lupex as my friend; but Fisher was at his chess-club.

'As I thought there was no time to be lost on such an occasion as this, I went down to the club and called him out. You know what a cool fellow Fisher is. I don't suppose anything would ever excite him. When I told him the story, he said that he would sleep upon it; and I had to walk up and down before the club while he finished his game. Fisher seemed to think that I might go back to Burton Crescent; but, of course, I knew that that would be out of the question. So it ended in my going home and sleeping on his sofa, and sending for some of my things in the morning. I wanted him to get up and see Lupex before going to the office this morning. But he said it would be better to put it off, and so he will call upon him at the theatre immediately after office hours.

'I want you to write to me at once saying what you know about the matter. I ask you, as I don't want to lug in any of the other people at Roper's. It is very uncomfortable, as I can't exactly leave her at once because of last quarter's money, otherwise I should cut and run; for the house is not the sort of place either for you or me. You may take my word for that, Master Johnny. And I could tell you something, too, about A. R., only I don't want to make mischief. But do you write immediately. And now I think of it, you had better write to Fisher, so that he can show your letter to Lupex, – just saying, that to the best of your belief there had never been anything between her and me but mere friendship; and that, of course, you, as my friend, must have known everything. Whether I shall go back to Roper's to-night will depend on what Fisher says after the interview.

'Good-by, old fellow! I hope you are enjoying yourself, and that L. D. is quite well.

'Your sincere friend,
'Joseph Cradell.'

John Eames read this letter over twice before he opened that from Amelia. He had never yet received a letter from Miss Roper; and felt very little of that ardour for its perusal which young men generally experience on the receipt of a first letter from a young lady. The memory of Amelia was at the present moment distasteful to him; and he would have thrown the letter unopened into the fire, had he not felt it might be dangerous to do so. As regarded his friend Cradell, he could not but feel ashamed of him, – ashamed of him, not for running away from Mr. Lupex, but for excusing his escape on false pretences.

And then, at last, he opened the letter from Amelia. 'Dearest John,' it began; and as he read the words, he crumpled the paper up between his fingers. It was written in a fair female hand, with sharp points instead of curves to the letters, but still very legible, and looking as though there were a decided purport in every word of it.

'Dearest John,

'It feels so strange to me to write to you in such language as this. And yet you are dearest, and have I not a right to call you so? And are you not my own, and am not I yours? [Again he crunched the paper up in his hand, and, as he did so, he muttered words which I need not repeat at length. But still he went on with his letter.] I know that we understand each other perfectly, and when that is the case, heart should be allowed to speak openly to heart. Those are my feelings, and I believe that you will find them reciprocal in your own bosom. Is it not sweet to be loved? I find it so. And, dearest John, let me assure you, with open candour, that there is no room for jealousy in this breast with regard to you. I have too much confidence for that, I can assure you, both in your honour and in my own – I would say charms, only you would call me vain. You must not suppose that I meant what I said about L. D. Of course, you will be glad to see the friends of your childhood; and it would be far from your Amelia's heart to begrudge you such delightful pleasure. Your friends will, I hope, some day be my friends. [Another crunch.] And if there be any one among them, any real L. D. whom you have specially liked, I will receive her to my heart, specially also. [This assurance on the part of his Amelia was too much for him, and he threw the letter from him, thinking whence he might get relief – whether from suicide or from the colonies; but presently he took it up again, and drained the bitter cup to the bottom.] And if I seemed petulant to you before you went away, you must forgive your own Amelia. I had nothing before me but misery for the month of your absence. There is no one here congenial to my feelings, – of course not. And you would not wish me to be happy in your absence, – would you? I can assure you, let your wishes be what they may, I never can be happy again unless you are with me. Write to me one little line, and tell me that you are grateful to me for my devotion.

'And now, I must tell you that we have had a sad affair in the house; and I do not think that your friend Mr. Cradell has behaved at all well. You remember how he has been always going on with Mrs. Lupex. Mother was quite unhappy about it, though she didn't like to say anything. Of course, when a lady's name is concerned, it is particular. But Lupex has become dreadful jealous during the last week; and we all knew that something was coming. She is an artful woman, but I don't think she meant anything bad – only to drive her husband to desperation. He came here yesterday in one of his tantrums, and wanted to see Cradell; but he got frightened, and took his hat and went off. Now, that wasn't quite right. If he was innocent, why didn't he stand his ground and explain the mistake? As mother says, it gives the house such a name. Lupex swore last night that he'd be off to the Income-tax Office this morning, and have Cradell out before all the commissioners, and clerks, and everybody. If he does that, it will get into the papers, and all London will be full of it. She would like it, I know; for all she cares for is to be talked about; but only think what it will be for mother's house. I wish you were here; for your high prudence and courage would set everything right at once, – at least, I think so.

'I shall count the minutes till I get an answer to this, and shall envy the postman who will have your letter before it will reach me. Do write at once. If I do not hear by Monday morning, I shall think that something is the matter. Even though you are among your dear old friends, surely you can find a moment to write to your own Amelia.

'Mother is very unhappy about this affair of the Lupexes. She says that if you were here to advise her she should not mind it so much. It is very hard upon her, for she does strive to make the house respectable and comfortable for everybody. I would send my duty and love to your dear mamma, if I only knew her, as I hope I shall do one day, and to your sister, and to L. D. also, if you like to tell her how we are situated together. So, now, no more from your

'Always affectionate sweetheart,
'Amelia Roper.'

Poor Eames did not feel the least gratified by any part of this fond letter; but the last paragraph of it was the worst. Was it to be endured by him that this woman should send her love to his mother and to his sister, and even to Lily Dale! He felt that there was a pollution in the very mention of Lily's name by such a one as Amelia Roper. And yet Amelia Roper was, as she had assured him, – his own. Much as he disliked her at the present moment, he did believe that he was – her own. He did feel that she had obtained a certain property in him, and that his destiny in life would tie him to her. He had said very few words of love to her at any time – very few, at least, that were themselves of any moment; but among those few there had undoubtedly been one or two in which he had told her that he loved her. And he had written to her that fatal note! Upon the whole, would it not be as well for him to go out to the great reservoir behind Guestwick, by which the Hamersham Canal was fed with its waters, and put an end to his miserable existence?

On that same day he did write a letter to Fisher, and he wrote also to Cradell. As to those letters he felt no difficulty. To Fisher he declared his belief that Cradell was innocent as he was himself as regarded Mrs. Lupex. 'I don't think he is the sort of man to make up to a married woman,' he said, somewhat to Cradell's displeasure, when the letter reached the Income-tax Office; for that gentleman was not averse to the reputation for success in love which the little adventure was, as he thought, calculated to give him among his brother clerks. At the first bursting of the shell, when that desperately jealous man was raging in the parlour, incensed by the fumes both of wine and love, Cradell had felt that the affair was disagreeably painful. But on the morning of the third day – for he had passed two nights on his friend Fisher's sofa – he had begun to be somewhat proud of it, and did not dislike to hear Mrs. Lupex's name in the mouths of the other clerks. When, therefore, Fisher read to him the letter from Guestwick, he hardly was pleased with his friend's tone. 'Ha, ha, ha,' said he, laughing. 'That's just what I wanted him to say. Make up to a married woman, indeed. No; I'm the last man in London to do that sort of thing.'

'Upon my word, Caudle, I think you are,' said Fisher; 'the very last man.'

And then poor Cradell was not happy. On that afternoon he boldly went to Burton Crescent, and ate his dinner there. Neither Mr. or Mrs. Lupex was to be seen, nor were their names mentioned to him by Mrs. Roper. In the course of the evening he did pluck up courage to ask Miss Spruce where they were; but that ancient lady merely shook her head solemnly, and declared that she knew nothing about such goings on – no, not she.

But what was John Eames to do as to that letter from Amelia Roper? He felt that any answer to it would be very dangerous, and yet that he could not safely leave it unanswered. He walked off by himself across Guestwick Common, and through the woods of Guestwick Manor, up by the big avenue of elms in Lord De Guest's park, trying to resolve how he might rescue himself from this scrape. Here, over the same ground, he had wandered scores of times in his earlier years, when he knew nothing beyond the innocency of his country home, thinking of Lily Dale, and swearing to himself that she should be his wife. Here he had strung together his rhymes, and fed his ambition with high hopes, building gorgeous castles in the air, in all of which Lilian reigned as a queen; and though in those days he had known himself to be awkward, poor, uncared for by any in the world except his mother and his sister, yet he had been happy in his hopes – happy in his hopes, even though he had never taught himself really to believe that they would be realized. But now there was nothing in his hopes or thoughts to make him happy. Everything was black, and wretched, and ruinous. What would it matter, after all, even if he should marry Amelia Roper, seeing that Lily was to be given to another? But then the idea of Amelia as he had seen her that night through the chink in the door came upon his memory, and he confessed to himself that life with such a wife as that would be a living death.

At one moment he thought that he would tell his mother everything, and leave her to write an answer to Amelia's letter. Should the worst come to the worst, the Ropers could not absolutely destroy him. That they could bring an action against him, and have him locked up for a term of years, and dismissed from his office, and exposed in all the newspapers, he seemed to know. That might all, however, be endured, if only the gauntlet could be thrown down for him by some one else. The one thing which he felt that he could not do was, to write to a girl whom he had professed to love, and tell her that he did not love her. He knew that he could not himself form such words upon the paper; nor, as he was well aware, could he himself find the courage to tell her to her face that he had changed his mind. He knew that he must become the victim of his Amelia, unless he could find some friendly knight to do battle in his favour; and then again he thought of his mother.

But when he returned home he was as far as ever from any resolve to tell her how he was situated. I may say that his walk had done him no good, and that he had not made up his mind to anything. He had been building those pernicious castles in the air during more than half the time; not castles in the building of which he could make himself happy, as he had done in the old days, but black castles, with cruel dungeons, into which hardly a ray of light could find its way. In all these edifices his imagination pictured to him Lily as the wife of Mr. Crosbie. He accepted that as a fact, and then went to work in his misery, making her as wretched as himself, through the misconduct and harshness of her husband. He tried to think, and to resolve what he would do; but there is no task so hard as that of thinking, when the mind has an objection to the matter brought before it. The mind, under such circumstances, is like a horse that is brought to the water, but refuses to drink. So Johnny returned to his home, still doubting whether or no he would answer Amelia's letter. And if he did not answer it, how would he conduct himself on his return to Burton Crescent?

I need hardly say that Miss Roper, in writing her letter, had been aware of all this, and that Johnny's position had been carefully prepared for him by— his affectionate sweetheart.

CHAPTER XI

SOCIAL LIFE

Mr. and Mrs. Lupex had eaten a sweetbread together in much connubial bliss on that day which had seen Cradell returning to Mrs. Roper's hospitable board. They had together eaten a sweetbread, with some other delicacies of the season, in the neighbourhood of the theatre, and had washed down all unkindness with bitter beer and brandy-and-water. But of this reconciliation Cradell had not heard; and when he saw them come together into the drawing-room, a few minutes after the question he had addressed to Miss Spruce, he was certainly surprised.

Lupex was not an ill-natured man, nor one naturally savage by disposition. He was a man fond of sweetbread and little dinners, and one to whom hot brandy-and-water was too dear. Had the wife of his bosom been a good helpmate to him, he might have gone through the world, if not respectably, at any rate without open disgrace. But she was a woman who left a man no solace except that to be found in brandy-and-water. For eight years they had been man and wife; and sometimes – I grieve to say it – he had been driven almost to hope that she would commit a married woman's last sin, and leave him. In his misery, any mode of escape would have been welcome to him. Had his energy been sufficient he would have taken his scene-painting capabilities off to Australia, – or to the farthest shifting of scenes known on the world's stage. But he was an easy, listless, self-indulgent man; and at any moment, let his misery be as keen as might be, a little dinner, a few soft words, and a glass of brandy-and-water would bring him round. The second glass would make him the fondest husband living; but the third would restore to him the memory of all his wrongs, and give him courage against his wife or all the world, – even to the detriment of the furniture around him, should a stray poker chance to meet his hand. All these peculiarities of his character were not, however, known to Cradell; and when our friend saw him enter the drawing-room with his wife on his arm, he was astonished.

'Mr. Cradell, your hand,' said Lupex, who had advanced as far as the second glass of brandy-and-water, but had not been allowed to go beyond it. 'There has been a misunderstanding between us; let it be forgotten.'

'Mr. Cradell, if I know him,' said the lady, 'is too much the gentleman to bear any anger when a gentleman has offered him his hand.'

'Oh, I'm sure,' said Cradell, 'I'm quite— indeed, I'm delighted to find there's nothing wrong after all.' And then he shook hands with both of them; whereupon Miss Spruce got up, curtseyed low, and also shook hands with the husband and wife.

'You're not a married man, Mr. Cradell,' said Lupex. 'and, therefore, you cannot understand the workings of a husband's heart. There have been moments when my regard for that woman has been too much for me.'

'Now, Lupex, don't,' said she, playfully tapping him with an old parasol which she still held.

'And I do not hesitate to say that my regard for her was too much for me on that night when I sent for you to the dining-room.'

'I'm glad it's all put right now,' said Cradell.

'Very glad, indeed,' said Miss Spruce.

'And, therefore, we need not say any more about it,' said Mrs. Lupex.

'One word,' said Lupex, waving his hand. 'Mr. Cradell, I greatly rejoice that you did not obey my summons on that night. Had you done so, – I confess it now, – had you done so, blood would have been the consequence. I was mistaken. I acknowledge my mistake; – but blood would have been the consequence.'

'Dear, dear, dear,' said Miss Spruce.

'Miss Spruce,' continued Lupex, 'there are moments when the heart becomes too strong for a man.'

'I dare say,' said Miss Spruce.

'Now, Lupex, that will do,' said his wife.

'Yes; that will do. But I think it right to tell Mr. Cradell that I am glad he did not come to me. Your friend, Mr. Cradell, did me the honour of calling on me at the theatre yesterday, at half-past four; but I was in the slings then, and could not very well come down to him. I shall be happy to see you both any day at five, and to bury all unkindness with a chop and glass at the Pot and Poker, in Bow-street.'

'I'm sure you're very kind,' said Cradell.

'And Mrs. Lupex will join us. There's a delightful little snuggery upstairs at the Pot and Poker; and if Miss Spruce will condescend to—'

'Oh, I'm an old woman, sir.'

'No – no – no,' said Lupex, 'I deny that. Come, Cradell, what do you say? – just a snug little dinner for four, you know.'

It was, no doubt, pleasant to see Mr. Lupex in his present mood, – much pleasanter than in that other mood of which blood would have been the consequence; but pleasant as he now was, it was, nevertheless, apparent that he was not quite sober. Cradell, therefore, did not settle the day for the little dinner; but merely remarked that he should be very happy at some future day.

'And now, Lupex, suppose you get off to bed,' said his wife. 'You've had a very trying day, you know.'

'And you. ducky?'

'I shall come presently. Now don't be making a fool of yourself, but get yourself off. Come—' and she stood close up against the open door, waiting for him to pass.

'I rather think I shall remain where I am, and have a glass of something hot,' said he.

'Lupex, do you want to aggravate me again?' said the lady, and she looked at him with a glance of her eye which he thoroughly understood. He was not in a humour for fighting, nor was he at present desirous of blood; so he resolved to go. But as he went he prepared himself for new battles. 'I shall do something desperate, I am sure; I know I shall,' he said, as he pulled off his boots.

'Oh, Mr. Cradell,' said Mrs. Lupex as soon as she had closed the door behind her retreating husband, 'how am I ever to look you in the face again after the events of these last memorable days?' And then she seated herself on the sofa, and hid her face in a cambric* handkerchief.

'As for that,' said Cradell, 'what does it signify, – among friends like us, you know?'

'But that it should be known at your office, – as of course it is, because of the gentleman that went down to him at the theatre! – I don't think I shall ever survive it.'

'You see I was obliged to send somebody, Mrs. Lupex.'

'I'm not finding fault, Mr. Cradell. I know very well that in my melancholy position I have no right to find fault, and I don't pretend to understand gentlemen's feelings towards each other. But to have had my name mentioned up with yours in that way is – Oh! Mr. Cradell, I don't know how I'm ever to look you in the face again.' And she buried hers in her pocket-handkerchief.

'Handsome is as handsome does,' said Miss Spruce; and there was that in her tone of voice which seemed to convey much hidden meaning.

'Exactly so, Miss Spruce,' said Mrs. Lupex; 'and that's my only comfort at the present moment. Mr. Cradell is a gentleman who would scorn to take advantage – I'm quite sure of that.' And then she did contrive to look at him over the edge of the hand which held the handkerchief.

'That I wouldn't, I'm sure,' said Cradell. 'That is to say—' And then he paused. He did not wish to get into a scrape about Mrs. Lupex. He was by no means anxious to encounter her husband in one of his fits of jealousy. But he did like the idea of being talked of as the admirer of a married woman, and he did like the brightness of the lady's eyes. When the unfortunate moth in his semi-blindness whisks himself and his wings within the flame of the candle, and finds himself mutilated and tortured, he even then will not take the lesson, but returns again and again till he is destroyed. Such a moth was poor Cradell. There was no warmth to be got by him from that flame. There was no beauty in the light, – not even the false brilliance of unhallowed love. Injury might come to him, – a pernicious clipping of the wings, which might destroy all power of future flight; injury, and not improbably destruction, if he should persevere. But one may say that no single hour of happiness could accrue to him from his intimacy with Mrs. Lupex. He felt for her no love. He was afraid of her, and, in many respects, disliked her. But to him, in his moth-like weakness, ignorance, and blindness, it seemed to be a great thing that he should be allowed to fly near the candle. Oh! my friends, if you will but think of it, how many of you have been moths, and are now going about ungracefully with wings more or less burnt off, and with bodies sadly scorched!

But before Mr. Cradell could make up his mind whether or no he would take advantage of the present opportunity for another dip into the flame of the candle, – in regard to which proceeding, however, he could not but feel that the presence of Miss Spruce was objectionable, – the door of the room was opened, and Amelia Roper joined the party.

'Oh, indeed; Mrs. Lupex,' she said. 'And Mr. Cradell!'

'And Miss Spruce, my dear,' said Mrs. Lupex, pointing to the ancient lady.

'I'm only an old woman,' said Miss Spruce.

'Oh, yes; I see Miss Spruce,' said Amelia. 'I was not hinting at anything, I can assure you.'

'I should think not, my dear,' said Mrs. Lupex.

'Only I didn't know that you two were quite— That is, when last I heard about it, I fancied— But if the quarrel's made up, there's nobody more rejoiced than I am.'

'The quarrel is made up,' said Cradell.

'If Mr. Lupex is satisfied, I'm sure I am,' said Amelia.

'Mr. Lupex is satisfied,' said Mrs. Lupex; 'and let me tell you, my dear, seeing that you are expecting to get married yourself—'

'Mrs. Lupex, I'm not expecting to get married, – not particularly, by any means.'

'Oh, I thought you were. And let me tell you, that when you've got a husband of your own, you won't find it so easy to keep everything straight. That's the worst of these lodgings, if there is any little thing, everybody knows it. Don't they, Miss Spruce?'

'Lodgings is so much more comfortable than housekeeping,' said Miss Spruce, who lived rather in fear of her relatives, the Ropers.

'Everybody knows it; does he?' said Amelia. 'Why, if a gentleman will come home at night tipsy and threaten to murder another gentleman in the same house; and if a lady—' And then Amelia paused, for she knew that the line-of-battle ship which she was preparing to encounter had within her much power of fighting.

'Well, miss,' said Mrs. Lupex, getting on her feet, 'and what of the lady?'

Now we may say that the battle had begun, and that the two ships were pledged by the general laws of courage and naval warfare to maintain the contest till one of them should be absolutely disabled, if not blown up or sunk. And at this moment it might be difficult for a bystander to say with which of the combatants rested the better chance of permanent success. Mrs. Lupex had doubtless on her side more matured power, a habit of fighting which had given her infinite skill, a courage which deadened her to the feeling of all wounds while the heat of the battle should last, and a recklessness which made her almost indifferent whether she sank or swam. But then Amelia carried the greater guns, and was able to pour in heavier metal than her enemy could use; and she, too, swam in her own waters. Should they absolutely come to grappling and boarding, Amelia would no doubt have the best of it; but Mrs. Lupex would probably be too crafty to permit such a proceeding as that. She was, however, ready for the occasion, and greedy for the fight.

'And what of the lady?' said she, in a tone of voice that admitted of no pacific rejoinder.

'A lady, if she is a lady,' said Amelia, 'will know how to behave herself.'

'And you're going to teach me, are you, Miss Roper? I'm sure I'm ever so much obliged to you. It's Manchester manners, I suppose, that you prefer?'

'I prefer honest manners, Mrs. Lupex, and decent manners, and manners that won't shock a whole house full of people; and I don't care whether they come from Manchester or London.'

'Milliner's manners, I suppose?'

'I don't care whether they are milliner's manners or theatrical, Mrs. Lupex, as long as they're not downright bad manners – as yours are, Mrs. Lupex. And now you've got it. What are you going on for in this way with that young man, till you'll drive your husband into a madhouse with drink and jealousy?'

'Miss Roper! Miss Roper!' said Cradell; 'now really—'

'Don't mind her, Mr. Cradell,' said Mrs. Lupex; 'she's not worthy for you to speak to. And as to that poor fellow Eames, if you've any friendship for him, you'll let him know what she is. My dear, how's Mr. Juniper, of Grogram's house, at Salford? I know all about you, and so shall John Eames, too – poor unfortunate fool of a fellow! Telling me of drink and jealousy, indeed!'

'Yes, telling you! And now you've mentioned Mr. Juniper's name, Mr. Eames, and Mr. Cradell too, may know the whole of it. There's been nothing about Mr. Juniper that I'm ashamed of.'

'It would be difficult to make you ashamed of anything, I believe.'

'But let me tell you this, Mrs. Lupex, you're not going to destroy the respectability of this house by your goings on.'

'It was a bad day for me when I let Lupex bring me into it.'

'Then pay your bill, and walk out of it,' said Amelia, waving her hand towards the door. 'I'll undertake to say there shan't be any notice required. Only you pay mother what you owe, and you're free to go at once.'

'I shall go just when I please, and not one hour before. Who are you, you gipsy, to speak to me in this way?'

'And as for going, go you shall, if we have to call in the police to make you.'

Amelia, as at this period of the fight she stood fronting her foe with her arms akimbo, certainly seemed to have the best of the battle. But the bitterness of Mrs. Lupex's tongue had hardly yet produced its greatest results. I am inclined to think that the married lady would have silenced her who was single, had the fight been allowed to rage, – always presuming that no resort to grappling-irons took place. But at this moment Mrs. Roper entered the room, accompanied by her son, and both the combatants for a moment retreated.

'Amelia, what's all this?' said Mrs. Roper, trying to assume a look of agonized amazement.

'Ask Mrs. Lupex,' said Amelia.

'And Mrs. Lupex will answer,' said that lady. 'Your daughter has come in here, and attacked me – in such language – before Mr. Cradell, too—'

'Why doesn't she pay what she owes, and leave the house?' said Amelia.

'Hold your tongue,' said her brother. 'What she owes is no affair of yours.'

'But it's an affair of mine, when I'm insulted by such a creature as that.'

'Creature!' said Mrs. Lupex. 'I'd like to know which is most like a creature! But I'll tell you what it is, Amelia Roper—' Here, however, her eloquence was stopped, for Amelia had disappeared through the door, having been pushed out of the room by her brother. Whereupon Mrs. Lupex, having found a sofa convenient for the service, betook herself to hysterics. There for the moment we will leave her, hoping that poor Mrs. Roper was not kept late out of her bed.

'What a deuce of a mess Eames will make of it if he marries that girl!' Such was Cradell's reflection as he betook himself to his own room. But of his own part in the night's transactions he was rather proud than otherwise, feeling that the married lady's regard for him had been the cause of the battle which had raged. So, likewise, did Paris derive much gratification from the ten years' siege of Troy.

CHAPTER XII

LILIAN DALE BECOMES A BUTTERFLY

AND now we will go back to Allington. The same morning that brought to John Eames the two letters which were given in the last chapter but one, brought to the Great House, among others, the following epistle for Adolphus Crosbie. It was from a countess, and was written on pink paper, beautifully creamlaid and scented, ornamented with a coronet and certain singularly-entwined initials. Altogether, the letter was very fashionable and attractive, and Adolphus Crosbie was by no means sorry to receive it.

'COURCY CASTLE, September, 186–.

'My dear Mr. Crosbie,

'We have heard of you from the Gazebees, who have come down to us, and who tell us that you are rusticating at a charming little village, in which, among other attractions, there are wood nymphs and water nymphs, to whom much of your time is devoted. As this is just the thing for your taste, I would not for worlds disturb you; but if you should ever tear yourself away from the groves and fountains of Allington, we shall be delighted to welcome you here, though you will find us very unromantic after your late Elysium.

'Lady Dumbello is coming to us, who I know is a favourite

of yours. Or is it the other way, and are you a favourite of hers? I did ask Lady Hartletop, but she cannot get away from the poor marquis, who is, you know, so very infirm. The duke isn't at Gatherum at present, but, of course, I don't mean that that has anything to do with dear Lady Hartletop's not coming to us. I believe we shall have the house full, and shall not want for nymphs either, though I fear they will not be of the wood and water kind. Margaretta and Alexandrina particularly want you to come, as they say you are so clever at making a houseful of people go off well. If you can give us a week before you go back to manage the affairs of the nation, pray do.

'Yours very sincerely,
'Rosina De Courcy.'

The Countess De Courcy was a very old friend of Mr. Crosbie's; that is to say, as old friends go in the world in which he had been living. He had known her for the last six or seven years, and had been in the habit of going to all her London balls, and dancing with her daughters everywhere, in a most good-natured and affable way. He had been intimate, from old family relations, with Mr. Mortimer Gazebee, who, though only an attorney of the more distinguished kind, had married the countess's eldest daughter, and now sat in Parliament for the city of Barchester, near to which Courcy Castle was situated. And, to tell the truth honestly at once, Mr. Crosbie had been on terms of great friendship with Lady De Courcy's daughters, the Ladies Margaretta and Alexandrina – perhaps especially so with the latter, though I would not have my readers suppose by my saying so that anything more tender than friendship had ever existed between them.

Crosbie said nothing about the letter on that morning; but during the day, or, perhaps, as he thought over the matter in bed, he made up his mind that he would accept Lady De Courcy's invitation. It was not only that he would be glad to see the Gazebees, or glad to stay in the same house with that great master in the high art of fashionable life, Lady Dumbello, or glad to renew his friendship with the Ladies Margaretta and Alexandrina. Had he felt that the circumstances of his engagement with Lily made it expedient for him to stay with her till the end of his holidays, he could have thrown over the De Courcys without a struggle. But he told himself that it would be well for him now to tear himself away from Lily; or perhaps he said that it would be well for Lily that he should be torn away. He must not teach her to think that they were to live only in the sunlight of each other's eyes during those months, or perhaps years, which must elapse before their engagement could be carried out. Nor must he allow her to suppose that either he or she were to depend solely upon the other for the amusements and employments of life. In this way he argued the matter very sensibly within his own mind, and resolved, without much difficulty, that he would go to Courcy Castle, and bask for a week in the sunlight of the fashion which would be collected there. The quiet humdrum of his own fireside would come upon him soon enough!

'I think I shall leave you on Wednesday, sir,' Crosbie said to the squire at breakfast on Sunday morning.

'Leave us on Wednesday!' said the squire, who had an old-fashioned idea that people who were engaged to marry each other should remain together as long as circumstances could be made to admit of their doing so. 'Nothing wrong, is there?'

'O dear no! But everything must come to an end some day; and as I must make one or two short visits before I get back to town, I might as well go on Wednesday. Indeed, I have made it as late as I possibly could.'

'Where do you go from here?' asked Bernard.

'Well, as it happens, only into the next county, – to Courcy Castle.' And then there was nothing more said about the matter at that breakfast table.

It had become their habit to meet together on Sunday mornings before church, on the lawn belonging to the Small House, and on this day the three gentlemen walked down together, and found Lily and Bell already waiting for them. They generally had some few minutes to spare on those occasions before Mrs. Dale summoned them to pass through the house to church, and such was the case at present. The squire at these times would stand in the middle of the grass-plot, surveying his grounds, and taking stock of the shrubs, and flowers, and fruit-trees round him; for he never forgot that it was all his own, and would thus use this opportunity, as he seldom came down to see the spot on other days. Mrs. Dale, as she would see him from her own window while she was tying on her bonnet, would feel that she knew what was passing through his mind, and would regret that circumstances had forced her to be beholden to him for such assistance. But, in truth, she did not know all that he thought at such times. 'It is mine,' he would say to himself, as he looked around on the pleasant place. 'But it is well for me that they should enjoy it. She is my brother's widow, and she is welcome; – very welcome.' I think that if those two persons had known more than they did of each other's hearts and minds they might have loved each other better.

And then Crosbie told Lily of his intention. 'On Wednesday!' she said, turning almost pale with emotion as she heard this news. He had told her abruptly, not thinking, probably, that such tidings would affect her so strongly.

'Well, yes. I have written to Lady De Courcy and said Wednesday. It wouldn't do for me exactly to drop everybody, and perhaps—'

'Oh, no! And, Adolphus, you don't suppose I begrudge your going. Only it does seem so sudden; does it not?'

'You see, I've been here over six weeks.'

'Yes; you've been very good. When I think of it, what a six weeks it has been. I wonder whether the difference seems to you as great as it does to me. I've left off being a grub, and begun to be a butterfly.'

'But you mustn't be a butterfly when you're married, Lily.'

'No; not in that sense. But I meant that my real position in the world, – that for which I would fain hope that I was created, – opened to me only when I knew you and knew that you loved me. But mamma is calling us, and we must go through to church. Going on Wednesday! There are only three days more, then!'

'Yes, just three days,' he said, as he took her on his arm and passed through the house on to the road.

'And when are we to see you again?' she asked, as they reached the churchyard.

'Ah, who is to say that yet? We must ask the Chairman of Committees when he will let me go again.' Then there was nothing more said, and they all followed the squire through the little porch and up to the big family-pew in which they all sat. Here the squire took his place in one special corner which he had occupied ever since his father's death, and from which he read the responses loudly and plainly, – so loudly and plainly, that the parish clerk could by no means equal him, though with emulous voice he still made the attempt. 'T' squire 'd like to be squire, and parson, and clerk, and everything; so a would,' the poor clerk would say, when complaining of the ill-usage which he suffered.

If Lily's prayers were interrupted by her new sorrow, I think that her fault in that respect would be forgiven. Of course she had known that Crosbie was not going to remain at Allington much longer. She knew quite as well as he did the exact day on which his leave of absence came to its end, and the hour at which it behoved him to walk into his room at the General Committee Office. She had taught herself to think that he would remain with them up to the end of his vacation, and now she felt as a schoolboy would feel who was told suddenly, a day or two before the time, that the last week of his holidays was to be taken from him. The grievance would have been slight had she known it from the first; but what schoolboy could stand such a shock, when the loss amounted to two-thirds of his remaining wealth? Lily did not blame her lover. She did not even think that he ought to stay. She would not allow herself to suppose that he could propose anything that was unkind. But she felt her loss, and more than once, as she knelt at her prayers, she wiped a hidden tear from her eyes.

Crosbie also was thinking of his departure more than he should have done during Mr. Boyce's sermon. 'It's easy listening to him,' Mrs. Hearn used to say of her husband's successor. 'It don't give one much trouble following him into his arguments.' Mr. Crosbie perhaps found the difficulty greater than did Mrs. Hearn, and would have devoted his mind more perfectly to the discourse had the argument been deeper. It is very hard, that necessity of listening to a man who says nothing. On this occasion Crosbie ignored the necessity altogether, and gave up his mind to the consideration of what it might be expedient that he should say to Lily before he went. He remembered well those few words which he had spoken in the first ardour of his love, pleading that an early day might be fixed for their marriage. And he remembered, also, how prettily Lily had yielded to him. 'Only do not let it be too soon,' she had said. Now he must unsay what he had then said. He must plead against his own pleadings, and explain to her that he desired to postpone the marriage rather than to hasten it – a task which, I presume, must always be an unpleasant one for any man engaged to be married. 'I might as well do it at once,' he said to himself, as he bobbed his head forward into his hands by way of returning thanks for the termination of Mr. Boyce's sermon.

As he had only three days left, it was certainly as well that he should do this at once. Seeing that Lily had no fortune, she could not in justice complain of a prolonged engagement. That was the argument which he used in his own mind. But he as often told himself that she would have very great ground of complaint if she were left for a day unnecessarily in doubt as to this matter. Why had he rashly spoken those hasty words to her in his love, betraying himself into all manner of scrapes, as a schoolboy might do, or such a one as Johnny Eames? What an ass he had been not to have remembered himself and to have been collected, – not to have bethought himself on the occasion of all that might be due to Adolphus Crosbie! And then the idea came upon him whether he had not altogether made himself an ass in this matter. And as he gave his arm to Lily outside the church-door, he shrugged his shoulders while making that reflection. 'It is too late now,' he said to himself; and then turned round and made some sweet little loving speech to her. Adolphus Crosbie was a clever man; and he meant also to be a true man, – if only the temptations to falsehood might not be too great for him.

'Lily,' he said to her, 'will you walk in the fields after lunch?'

Walk in the fields with him! Of course she would. There were only three days left, and would she not give up to him every moment of her time, if he would accept of all her moments? And then they lunched at the Small House, Mrs. Dale having promised to join the dinner-party at the squire's table. The squire did not eat any lunch, excusing himself on the plea that lunch in itself was a bad thing. 'He can eat lunch at his own house,' Mrs. Dale afterwards said to Bell. 'And I've often seen him take a glass of sherry.' While thinking of this, Mrs. Dale made her own dinner. If her brother-in-law would not eat at her board, neither would she eat at his.

And then in a few minutes Lily had on her hat, in place of that decorous, church-going bonnet which Crosbie was wont to abuse with a lover's privilege, feeling well assured that he might say what he liked of the bonnet as long as he would praise the hat. 'Only three days,' she said, as she walked down with him across the lawn at a quick pace. But she said it in a voice which made no complaint, – which seemed to say simply this, – that as the good time was to be so short, they must make the most of it. And what compliment could be paid to a man so sweet as that? What flattery could be more gratifying? All my earthly heaven is with you; and now, for the delight of these immediately present months or so, there are left to me but three days of this heaven! Come, then; I will make the most of what happiness is given to me. Crosbie felt it all as she felt it, and recognized the extent of the debt he owed her. 'I'll come down to them for a day at Christmas, though it be only for a day,' he said to himself. Then he reflected that as such was his intention, it might be well for him to open his present conversation with a promise to that effect.

'Yes, Lily; there are only three days left now. But I wonder whether – I suppose you'll all be at home at Christmas?'

'At home at Christmas? – of course we shall be at home.

You don't mean to say you'll come to us!'

'Well; I think I will, if you'll have me.'

'Oh! that will make such a difference. Let me see. That will only be three months. And to have you here on Christmas Day! I would sooner have you then than on any other day in the year.'

'It will only be for one day, Lily. I shall come to dinner on Christmas Eve, and must go away the day after.'

'But you will come direct to our house!'

'If you can spare me a room.'

'Of course we can. So we could now. Only when you came, you know—' Then she looked up into his face and smiled.

'When I came, I was the squire's friend and your cousin's, rather than yours. But that's all changed now.'

'Yes; you're my friend now, – mine specially. I'm to be now and always your own special, dearest friend; – eh, Adolphus?' And then she exacted from him the repetition of the promise which he had so often given her.

By this time they had passed through the grounds of the Great House and were in the fields. 'Lily,' said he, speaking rather suddenly, and making her feel by his manner that something of importance was to be said; 'I want to say a few words to you about, – business.' And he gave a little laugh as he spoke the last word, making her fully understand that he was not quite at his ease.

'Of course I'll listen. And, Adolphus, pray don't be afraid about me. What I mean is, don't think that I can't bear cares and troubles. I can bear anything as long as you love me. I say that because I'm afraid I seemed to complain about your going. I didn't mean to.'

'I never thought you complained, dearest. Nothing can be better than you are at all times and in every way. A man would be very hard to please if you didn't please him.'

'If I can only please you—'

'You do please me, in everything. Dear Lily, I think I found an angel when I found you. But now about this business. Perhaps I'd better tell you everything.'

'Oh, yes, tell me everything.'

'But then you mustn't misunderstand me. And if I talk

about money, you mustn't suppose that it has anything to do with my love for you.'

'I wish for your sake that I wasn't such a little pauper.'

'What I mean to say is this, that if I seem to be anxious about money, you must not suppose that that anxiety bears any reference whatever to my affection for you. I should love you just the same, and look forward just as much to my happiness in marrying you, whether you were rich or poor. You understand that?'

She did not quite understand him; but she merely pressed his arm, so as to encourage him to go on. She presumed that he intended to tell her something as to their future mode of life – something which he supposed it might not be pleasant for her to hear, and she was determined to show him that she would receive it pleasantly.

'You know,' said he, 'how anxious I have been that our marriage should not be delayed. To me, of course, it must be everything now to call you my own as soon as possible.' In answer to which little declaration of love, she merely pressed his arm again, the subject being one on which she had not herself much to say.

'Of course I must be very anxious, but I find it not so easy as I expected.'

'You know what I said, Adolphus. I said that I thought we had better wait. I'm sure mamma thinks so. And if we can only see you now and then—'

'That will be a matter of course. But, as I was saying— Let me see. Yes, – all that waiting will be intolerable to me. It is such a bore for a man when he has made up his mind on such a matter as marriage, not to make the change at once, especially when he is going to take to himself such a little angel as you are,' and as he spoke these loving words, his arm was again put round her waist; 'but—' and then he stopped. He wanted to make her understand that this change of intention on his part was caused by the unexpected misconduct of her uncle. He desired that she should know exactly how the matter stood; that he had been led to suppose that her uncle would give her some small fortune; that he had been disappointed, and had a right to feel the disappointment keenly; and that in consequence of this blow to his expectations, he must put off his marriage. But he wished her also to understand at the same time that this did not in the least mar his love for her; that he did not join her at all in her uncle's fault. All this he was anxious to convey to her, but he did not know how to get it said in a manner that would not be offensive to her personally, and that should not appear to accuse himself of sordid motives. He had begun by declaring that he would tell her all; but sometimes it is not easy, that task of telling a person everything. There are things which will not get themselves told.

'You mean, dearest,' said she, 'that you cannot afford to marry at once.'

'Yes; that is it. I had expected that I should be able, but—'

Did any man in love ever yet find himself able to tell the lady whom he loved that he was very much disappointed on discovering that she had got no money? If so, his courage, I should say, was greater than his love. Crosbie found himself unable to do it, and thought himself cruelly used because of the difficulty. The delay to which he intended to subject her was occasioned, as he felt, by the squire, and not by himself. He was ready to do his part, if only the squire had been willing to do the part which properly belonged to him. The squire would not; and, therefore, neither could he, – not as yet. Justice demanded that all this should be understood; but when he came to the telling of it, he found that the story would not form itself properly. He must let the thing go, and bear the injustice, consoling himself as best he might by the reflection that he at least was behaving well in the matter.

'It won't make me unhappy, Adolphus.'

'Will it not?' said he. 'As regards myself, I own that I cannot bear the delay with so much indifference.'

'Nay, my love; but you should not misunderstand me,' she said, stopping and facing him on the path in which they were walking. 'I suppose I ought to protest, according to the common rules, that I would rather wait. Young ladies are expected to say so. If you were pressing me to marry at once, I should say so, no doubt. But now, as it is, I will be more honest. I have only one wish in the world, and that is, to be your wife, – to be able to share everything with you. The sooner we can be together the better it will be, – at any rate for me. There; will that satisfy you?'

'My own, own Lily!'

'Yes, your own Lily. You shall have no cause to doubt me, dearest. But I do not expect that I am to have everything exactly as I want it. I say again, that I shall not be unhappy in waiting. How can I be unhappy while I feel certain of your love? I was disappointed just now when you said that you were going so soon; and I am afraid I showed it. But those little things are more unendurable than the big things.'

'Yes; that's very true.'

'But there are three more days, and I mean to enjoy them so much. And then you will write to me: and you will come at Christmas. And next year, when you have your holiday, you will come down to us again; will you not?'

'You may be quite sure of that.'

'And so the time will go by till it suits you to come and take me. I shall not be unhappy.'

'I, at any rate, shall be impatient.'

'Ah, men always are impatient. It is one of their privileges, I suppose. And I don't think that a man ever has the same positive and complete satisfaction in knowing that he is loved, which a girl feels. You are my bird that I have shot at with my own gun; and the assurance of my success is sufficient for my happiness.'

'You have bowled me over, and know that I can't get up again.'

'I don't know about can't. I would let you up quick enough, if you wished it.'

How he made his loving assurance that he did not wish it, never would or could wish it, the reader will readily understand. And then he considered that he might as well leave all those money questions as they now stood. His real object had been to convince her that their joint circumstances did not admit of an immediate marriage; and as to that she completely understood him. Perhaps, during the next three days, some opportunity might arise for explaining the whole matter to Mrs. Dale. At any rate, he had declared his own purpose honestly, and no one could complain of him.

On the following day they all rode over to Guestwick together, – the all consisting of the two girls, with Bernard and Crosbie. Their object was to pay two visits, – one to their very noble and highly exalted ally, the Lady Julia De Guest; and the other to their much humbler and better known friend, Mrs. Eames. As Guestwick Manor lay on their road into the town, they performed the grander ceremony the first. The present Earl De Guest, brother of that Lady Fanny who ran away with Major Dale, was an unmarried nobleman, who devoted himself chiefly to the breeding of cattle. And as he bred very good cattle, taking infinite satisfaction in the employment, devoting all his energies thereto, and abstaining from all prominently evil courses, it should be acknowledged that he was not a bad member of society. He was a thorough-going old Tory, whose proxy was always in the hand of the leader of his party; and who seldom himself went near the metropolis, unless called thither by some occasion of cattle-showing. He was a short, stumpy man, with red cheeks and a round face; who was usually to be seen till dinner-time dressed in a very old shooting coat, with breeches, gaiters, and very thick shoes. He lived generally out of doors, and was almost as great in the preserving of game as in the breeding of oxen. He knew every acre of his own estate, and every tree upon it, as thoroughly as a lady knows the ornaments in her drawing-room. There was no gap in a fence of which he did not remember the exact bearings, no path hither and thither as to which he could not tell the why and the wherefore. He had been in his earlier years a poor man as regarded his income, – very poor, seeing that he was an earl. But he was not at present by any means an impoverished man, having been taught a lesson by the miseries of his father and grandfather, and having learned to live within his means. Now, as he was going down the vale of years, men said that he was becoming rich, and that he had ready money to spend, – a position in which no Lord De Guest had found himself for many generations back. His father and grandfather had been known as spendthrifts; and now men said that this earl was a miser.

There was not much of nobility in his appearance; but they greatly mistook Lord De Guest who conceived that on that account his pride of place was not dear to his soul. His peerage dated back to the time of King John,* and there were but three lords in England whose patents had been conferred before his own. He knew what privileges were due to him on behalf of his blood, and was not disposed to abate one jot of them. He was not loud in demanding them. As he went through the world he sent no trumpeters to the right or left, proclaiming that the Earl de Guest was coming. When he spread his board for his friends, which he did but on rare occasions, he entertained them simply, with a mild, tedious, old-fashioned courtesy. We may say that, if properly treated, the earl never walked over anybody. But he could, if ill-treated, be grandly indignant; and if attacked, could hold his own against all the world. He knew himself to be every inch an earl, pottering about after his oxen with his muddy gaiters and red cheeks, as much as though he were glittering with stars in courtly royal ceremonies among his peers at Westminster; – ay, more an earl than any of those who use their nobility for pageant purposes. Woe be to him who should mistake that old coat for a badge of rural degradation! Now and again some unlucky wight did make such a mistake, and had to do his penance very uncomfortably.

With the earl lived a maiden sister, the Lady Julia. Bernard Dale's father had, in early life, run away with one sister, but no suitor had been fortunate enough to induce the Lady Julia to run with him. Therefore she still lived, in maiden blessedness, as mistress of Guestwick Manor; and as such had no mean opinion of the high position which destiny had called upon her to fill. She was a tedious, dull, virtuous old woman, who gave herself infinite credit for having remained all her days in the home of her youth, probably forgetting, in her present advanced years, that her temptations to leave it had not been strong or numerous. She generally spoke of her sister Fanny with some little contempt, as though that poor lady had degraded herself in marrying a younger brother. She was as proud of her own position as was the earl her brother, but her pride was maintained with more of outward show and less of inward nobility. It was hardly enough for her that the world should know that she was a De Guest, and therefore she had assumed little pompous ways and certain airs of condescension which did not make her popular with her neighbours.

The intercourse between Guestwick Manor and Allington was not very frequent or very cordial. Soon after the running away of the Lady Fanny, the two families had agreed to acknowledge their connection with each other, and to let it be known by the world that they were on friendly terms. Either that course was necessary to them, or the other course, of letting it be known that they were enemies. Friendship was the less troublesome, and therefore the two families called on each other from time to time, and gave each other dinners about once a year. The earl regarded the squire as a man who had deserted his politics, and had thereby forfeited the respect due to him as an hereditary land magnate; and the squire was wont to be-little the earl as one who understood nothing of the outer world. At Guestwick Manor Bernard was to some extent a favourite. He was actually a relative, having in his veins blood of the De Guests, and was not the less a favourite because he was the heir to Allington, and because the blood of the Dales was older even than that of the noble family to which he was allied. When Bernard should come to be the squire, then indeed there might be cordial relations between Guestwick Manor and Allington; unless, indeed, the earl's heir and the squire's heir should have some fresh cause of ill-will between themselves.

They found Lady Julia sitting in her drawing-room alone, and introduced to her Mr. Crosbie in due form. The fact of Lily's engagement was of course known at the manor, and it was quite understood that her intended husband was now brought over that he might be looked at and approved. Lady Julia made a very elaborate curtsey, and expressed a hope that her young friend might be made happy in that sphere of life to which it had pleased God to call her.

'I hope I shall, Lady Julia,' said Lily, with a little laugh; 'at any rate I mean to try.'

'We all try, my dear, but many of us fail to try with sufficient energy of purpose. It is only by doing our duty that we can hope to be happy, whether in single life or in married.'

'Miss Dale means to be a dragon of perfection in the performance of hers,' said Crosbie.

'A dragon!' said Lady Julia. 'No; I hope Miss Lily Dale will never become a dragon.' And then she turned to her nephew. It may be as well to say at once that she never forgave Mr. Crosbie the freedom of the expression which he had used. He had been in the drawing-room of Guestwick Manor for two minutes only, and it did not become him to talk about dragons. 'Bernard,' she said, 'I heard from your mother yesterday. I am afraid she does not seem to be very strong.' And then there was a little conversation, not very interesting in its nature, between the aunt and the nephew as to the general health of Lady Fanny.

'I didn't know my aunt was so unwell,' said Bell.

'She isn't ill,' said Bernard. 'She never is ill; but then she is never well.'

'Your aunt,' said Lady Julia, seeming to put a touch of sarcasm into the tone of her voice as she repeated the word – 'your aunt has never enjoyed good health since she left this house; but that is a long time ago.'

'A very long time,' said Crosbie, who was not accustomed to be left in his chair silent. 'You, Dale, at any rate, can hardly remember it.'

'But I can remember it,' said Lady Julia, gathering herself up. 'I can remember when my sister Fanny was recognized as the beauty of the country. It is a dangerous gift, that of beauty.'

'Very dangerous,' said Crosbie. Then Lily laughed again, and Lady Julia became more angry than ever. What odious man was this whom her neighbours were going to take into their very bosom! But she had heard of Mr. Crosbie before, and Mr. Crosbie also had heard of her.

'By-the-by, Lady Julia,' said he, 'I think I know some very dear friends of yours.'

'Very dear friends is a very strong word. I have not many very dear friends.'

'I mean the Gazebees. I have heard Mortimer Gazebee and Lady Amelia speak of you.'

Whereupon Lady Julia confessed that she did know the Gazebees. Mr. Gazebee, she said, was a man who in early life had wanted many advantages, but still he was a very estimable person. He was now in Parliament, and she understood that he was making himself useful. She had not quite approved of Lady Amelia's marriage at the time, and so she had told her very old friend Lady De Courcy; but— And then Lady Julia said many words in praise of Mr. Gazebee, which seemed to amount to this; that he was an excellent sort of man, with a full conviction of the too great honour done to him by the earl's daughter who had married him, and a complete consciousness that even that marriage had not put him on a par with his wife's relations, or even with his wife. And then it came out that Lady Julia in the course of the next week was going to meet the Gazebees at Courcy Castle.

'I am delighted to think that I shall have the pleasure of seeing you there,' said Crosbie.

'Indeed!' said Lady Julia.

'I am going to Courcy on Wednesday. That, I fear, will be too early to allow of my being of any service to your ladyship.'

Lady Julia drew herself up, and declined the escort which Mr. Crosbie had seemed to offer. It grieved her to find that Lily Dale's future husband was an intimate friend of her friend's, and it especially grieved her to find that he was now going to that friend's house. It was a grief to her, and she showed that it was. It also grieved Crosbie to find that Lady Julia was to be a fellow guest with himself at Courcy Castle; but he did not show it. He expressed nothing but smiles and civil self-congratulation on the matter, pretending that he would have much delight in again meeting Lady Julia; but, in truth, he would have given much could he have invented any manœuvre by which her ladyship might have been kept at home.

'What a horrid old woman she is,' said Lily, as they rode back down the avenue. 'I beg your pardon, Bernard; for, of course, she is your aunt.'

'Yes; she is my aunt; and though I am not very fond of her, I deny that she is a horrid old woman. She never murdered anybody, or robbed anybody, or stole away any other woman's lover.'

'I should think not,' said Lily.

'She says her prayers earnestly, I have no doubt,' continued Bernard, 'and gives away money to the poor, and would sacrifice to-morrow any desire of her own to her brother's wish. I acknowledge that she is ugly, and pompous, and that, being a woman, she ought not to have such a long black beard on her upper lip.'

'I don't care a bit about her beard,' said Lily. 'But why did she tell me to do my duty? I didn't go there to have a sermon preached to me.'

'And why did she talk about beauty being dangerous?' said Bell. 'Of course, we all knew what she meant.'

'I didn't know at all what she meant,' said Lily; 'and I don't know now.'

'I think she's a charming woman, and I shall be especially civil to her at Lady De Courcy's,' said Crosbie.

And in this way, saying hard things of the poor old spinster whom they had left, they made their way into Guestwick, and again dismounted at Mrs. Eames's door.

CHAPTER XIII

A VISIT TO GUESTWICK

As the party from Allington rode up the narrow High-street of Guestwick, and across the market-square towards the small, respectable, but very dull row of new houses in which Mrs. Eames lived, the people of Guestwick were all aware that Miss Lily Dale was escorted by her future husband. The opinion that she had been a very fortunate girl was certainly general among the Guestwickians, though it was not always expressed in open or generous terms. 'It was a great match for her,' some said, but shook their heads at the same time, hinting that Mr. Crosbie's life in London was not all that it should be, and suggesting that she might have been more safe had she been content to bestow herself upon some country neighbour of less dangerous pretensions. Others declared that it was no such great match after all. They knew his income to a penny, and believed that the young people would find it very difficult to keep a house in London unless the old squire intended to assist them. But, nevertheless, Lily was envied as she rode through the town with her handsome lover by her side.

And she was very happy. I will not deny that she had some feeling of triumphant satisfaction in the knowledge that she was envied. Such a feeling on her part was natural, and is natural to all men and women who are conscious that they have done well in the adjustment of their own affairs. As she herself had said, he was her bird, the spoil of her own gun, the product of such capacity as she had in her, on which she was to live, and, if possible, to thrive during the remainder of her life. Lily fully recognized the importance of the thing she was doing, and, in soberest guise, had thought much of this matter of marriage. But the more she thought of it the more satisfied she was that she was doing well. And yet she knew that there was a risk. He who was now everything to her might die; nay, it was possible that he might be other than she thought him to be; that he might neglect her, desert her, or misuse her. But she had resolved to trust in everything, and, having so trusted, she would not provide for herself any possibility of retreat. Her ship should go out into the middle ocean, beyond all ken of the secure port from which it had sailed; her army should fight its battle with no hope of other safety than that which victory gives. All the world might know that she loved him if all the world chose to inquire about the matter. She triumphed in her lover, and did not deny even to herself that she was triumphant.

Mrs. Eames was delighted to see them. It was so good in Mr. Crosbie to come over and call upon such a poor, forlorn woman as her, and so good in Captain Dale; so good also in the dear girls, who, at the present moment, had so much to make them happy at home at Allington! Little things, accounted as bare civilities by others, were esteemed as great favours by Mrs. Eames.

'And dear Mrs. Dale? I hope she was not fatigued when we kept her up the other night so unconscionably late?' Bell and Lily both assured her that their mother was none the worse for what she had gone through; and then Mrs. Eames got up and left the room, with the declared purpose of looking for John and Mary, but bent, in truth, on the production of some cake and sweet wine which she kept under lock and key in the little parlour.

'Don't let's stay here very long,' whispered Crosbie.

'No, not very long,' said Lily. 'But when you come to see my friends you mustn't be in a hurry, Mr. Crosbie.'

'He had his turn with Lady Julia,' said Bell, 'and we must have ours now.'

'At any rate Mrs. Eames won't tell us to do our duty and to beware of being too beautiful,' said Lily.

Mary and John came into the room before their mother returned; then came Mrs. Eames, and a few minutes afterwards, the cake and wine arrived. It certainly was rather dull, as none of the party seemed to be at their ease. The grandeur of Mr. Crosbie was too great for Mrs. Eames and her daughter, and John was almost silenced by the misery of his position. He had not yet answered Miss Roper's letter, nor had he even made up his mind whether he would answer it or no. And then the sight of Lily's happiness did not fill him with all that friendly joy which he should perhaps have felt as the friend of her childhood. To tell the truth, he hated Crosbie, and so he had told himself; and had so told his sister also very frequently since the day of the party.

'I tell you what it is, Molly,' he had said, 'if there was any way of doing it, I'd fight that man.'

'What; and make Lily wretched?'

'She'll never be happy with him. I'm sure she won't. I don't want to do her any harm, but yet I'd like to fight that man, – if I only knew how to manage it.'

And then he bethought himself that if they could both

be slaughtered in such an encounter it would be the only fitting termination to the present state of things. In that way, too, there would be an escape from Amelia, and, at the present moment, he saw none other.

When he entered the room he shook hands with all the party from Allington, but, as he told his sister afterwards, his flesh crept when he touched Crosbie. Crosbie, as he contemplated the Eames family sitting stiff and ill at ease in their own drawing-room chairs, made up his mind that it would be well that his wife should see as little of John Eames as might be when she came to London; – not that he was in any way jealous of her lover. He had learned everything from Lily, – all, at least, that Lily knew, – and regarded the matter rather as a good joke. 'Don't see him too often,' he had said to her, 'for fear he should make an ass of himself.' Lily had told him everything, – all that she could tell; but yet he did not in the least comprehend that Lily had, in truth, a warm affection for the young man whom he despised.

'Thank you, no,' said Crosbie. 'I never do take wine in the middle of the day.'

'But a bit of cake?' And Mrs. Eames by her look implored him to do her so much honour. She implored Captain Dale, also, but they were both inexorable. I do not know that the two girls were at all more inclined to eat and drink than the two men; but they understood that Mrs. Eames would be broken-hearted if no one partook of her delicacies. The little sacrifices of society are all made by women, as are also the great sacrifices of life. A man who is good for anything is always ready for his duty, and so is a good woman always ready for a sacrifice.

'We really must go now,' said Bell, 'because of the horses.' And under this excuse they got away.

'You will come over before you go back to London, John?' said Lily, as he came out with the intention of helping her mount, from which purpose, however, he was forced to recede by the iron will of Mr. Crosbie.

'Yes, I'll come over again – before I go. Good-bye.'

'Good-bye, John,' said Bell. 'Good-bye, Eames,' said Captain Dale. Crosbie, as he seated himself in the saddle, made the very slightest sign of recognition, to which his rival would not condescend to pay any attention. 'I'll manage to have a fight with him in some way,' said Eames to himself as he walked back through the passage of his mother's house. And Crosbie, as he settled his feet in the stirrups, felt that he disliked the young man more and more. It would be monstrous to suppose that there could be aught of jealousy in the feeling; and yet he did dislike him very strongly, and felt almost angry with Lily for asking him to come again to Allington. 'I must put an end to all that,' he said to himself as he rode silently out of town.

'You must not snub my friends, sir,' said Lily, smiling as she spoke, but yet with something of earnestness in her voice. They were out of the town by this time, and Crosbie had hardly uttered a word since they had left Mrs. Eames's door. They were now on the high road, and Bell and Bernard Dale were somewhat in advance of them.

'I never snub anybody,' said Crosbie, petulantly; 'that is, unless they have absolutely deserved snubbing.'

'And have I deserved it? Because I seem to have got it,' said Lily.

'Nonsense, Lily. I never snubbed you yet, and I don't think it likely that I shall begin. But you ought not to accuse me of not being civil to your friends. In the first place I am as civil to them as my nature will allow me to be. And, in the second place—'

'Well; in the second place—?'

'I am not quite sure that you are very wise to encourage that young man's – friendship just at present.'

'That means, I suppose, that I am very wrong to do so?'

'No, dearest, it does not mean that. If I meant so I would tell you so honestly. I mean just what I say. There can, I suppose, be no doubt that he has filled himself with some kind of romantic attachment for you, – a foolish kind of love which I don't suppose he ever expected to gratify, but the idea of which lends a sort of grace to his life. When he meets some young woman fit to be his wife he will forget all about it, but till then he will go about fancying himself a despairing lover. And then such a young man as John Eames is very apt to talk of his fancies.'

'I don't believe for a moment that he would mention my name to any one.'

'But, Lily, perhaps I may know more of young men than you do.'

'Yes, of course you do.'

'And I can assure you that they are generally too well inclined to make free with the names of girls whom they think that they like. You must not be surprised if I am unwilling that any man should make free with your name.'

After this Lily was silent for a minute or two. She felt that an injustice was being done to her and she was not inclined to put up with it, but she could not quite see where the injustice lay. A great deal was owing from her to Crosbie. In very much she was bound to yield to him, and she was anxious to do on his behalf even more than her duty. But yet she had a strong conviction that it would not be well that she should give way to him in everything. She wished to think as he thought as far as possible, but she could not say that she agreed with him when she knew that she differed from him. John Eames was an old friend whom she could not abandon, and so much at the present time she felt herself obliged to say.

'But, Adolphus—'

'Well, dearest?'

'You would not wish me to be unkind to so very old a friend as John Eames? I have known him all my life, and we have all of us had a very great regard for the whole family. His father was my uncle's most particular friend.'

'I think, Lily, you must understand what I mean. I don't want you to quarrel with any of them, or to be what you call unkind. But you need not give special and pressing invitations to this young man to come and see you before he goes back to London, and then to come and see you directly you get to London. You tell me that he has some kind of romantic idea of being in love with you; – of being in despair because you are not in love with him. It's all great nonsense, no doubt, but it seems to me that under such circumstances you'd better – just leave him alone.'

Again Lily was silent. These were her three last days, in which it was her intention to be especially happy, but above all things to make him especially happy. On no account would she say to him sharp words, or encourage in her own heart a feeling of animosity against him, and yet she believed him to be wrong; and so believing could hardly bring herself to bear the injury. Such was her nature, as a Dale. And let it be remembered that very many who can devote themselves for great sacrifices, cannot bring themselves to the endurance of little injuries. Lily could have given up any gratification for her lover, but she could not allow herself to have been in the wrong, believing herself to have been in the right.

'I have asked him now, and he must come,' she said.

'But do not press him to come any more.'

'Certainly not, after what you have said, Adolphus. If he comes over to Allington, he will see me in mamma's house, to which he has always been made welcome by her. Of course I understand perfectly—'

'You understand what, Lily?'

But she had stopped herself, fearing that she might say that which would be offensive to him if she continued.

'What is it you understand, Lily?'

'Do not press me to go on, Adolphus. As far as I can, I will do all that you want me to do.'

'You meant to say that when you find yourself an inmate of my house, as a matter of course you could not ask your own friends to come and see you. Was that gracious?'

'Whatever I may have meant to say, I did not say that. Nor in truth did I mean it. Pray don't go on about it now. These are to be our last days, you know, and we shouldn't waste them by talking of things that are unpleasant. After all poor Johnny Eames is nothing to me; nothing, nothing. How can any one be anything to me when I think of you?'

But even this did not bring Crosbie back at once into a pleasant humour. Had Lily yielded to him and confessed that he was right, he would have made himself at once as pleasant as the sun in May. But this she had not done. She had simply abstained from her argument because she did not choose to be vexed, and had declared her continued purpose of seeing Eames on his promised visit. Crosbie would have had her acknowledge herself wrong, and would have delighted in the privilege of forgiving her. But Lily Dale was one who did not greatly relish forgiveness, or any necessity of being forgiven. So they rode on, if not in silence, without much joy in their conversation. It was now late on the Monday afternoon, and Crosbie was to go early on the Wednesday morning. What if these three last days should come to be marred with such terrible drawbacks as these!

Bernard Dale had not spoken a word to his cousin of his suit, since they had been interrupted by Crosbie and Lily as they were lying on the bank by the ha-ha. He had danced with her again and again at Mrs. Dale's party, and had seemed to revert to his old modes of conversation without difficulty. Bell, therefore, had believed the matter to be over, and was thankful to her cousin, declaring within her own bosom that the whole matter should be treated by her as though it had never happened. To no one, – not even to her mother, would she tell it. To such reticence she bound herself for his sake, feeling that he would be best pleased that it should be so. But now as they rode on together, far in advance of the other couple, he again returned to the subject.

'Bell,' said he, 'am I to have any hope?'

'Any hope as to what, Bernard?'

'I hardly know whether a man is bound to take a single answer on such a subject. But this I know, that if a man's heart is concerned, he is not very willing to do so.'

'When that answer had been given honestly and truly—'

'Oh, no doubt. I don't at all suppose that you were dishonest or false when you refused to allow me to speak to you.'

'But, Bernard, I did not refuse to allow you to speak to me.'

'Something very like it. But, however, I have no doubt you were true enough. But, Bell, why should it be so? If you were in love with any one else I could understand it.'

'I am not in love with any one else.'

'Exactly. And there are so many reasons why you and I should join our fortunes together.'

'It cannot be a question of fortune, Bernard.'

'Do listen to me. Do let me speak, at any rate. I presume I may at least suppose that you do not dislike me.'

'Oh, no.'

'And though you might not be willing to accept any man's hand merely on a question of fortune, surely the fact that our marriage would be in every way suitable as regards money should not set you against it. Of my own love for you I will not speak further, as I do not doubt that you believe what I say; but should you not question your own feelings very closely before you determine to oppose the wishes of all those who are nearest to you?'

'Do you mean mamma, Bernard?'

'Not her especially, though I cannot but think she would like a marriage that would keep all the family together, and would give you an equal claim to the property to that which I have.'

'That would not have a feather's-weight with mamma.'

'Have you asked her?'

'No, I have mentioned the matter to no one.'

'Then you cannot know. And as to my uncle, I have the means of knowing that it is the great desire of his life. I must say that I think some consideration for him should induce you to pause before you give a final answer, even though no consideration for me should have any weight with you.'

'I would do more for you than for him, – much more.'

'Then do this for me. Allow me to think that I have not yet had an answer to my proposal; give me to this day month, to Christmas; till any time that you like to name, so that I may think that it is not yet settled, and may tell uncle Christopher that such is the case.'

'Bernard, it would be useless.'

'It would show him that you are willing to think of it.'

'But I am not willing to think of it; – not in that way. I do know my own mind thoroughly, and I should be very wrong if I were to deceive you.'

'And you wish me to give that as your only answer to my uncle?'

'To tell the truth, Bernard, I do not much care what you may say to my uncle in this matter. He can have no right to interfere in the disposal of my hand, and therefore I need not regard his wishes on the subject. I will explain to you in one word what my feelings are about it. I would accept no man in opposition to mamma's wishes; but not even for her could I accept any man in opposition to my own. But as concerns my uncle, I do not feel myself called on to consult him in any way on such a matter.'

'And yet he is the head of our family.'

'I don't care anything about the family, – not in that way.'

'And he has been very generous to you all.'

'That I deny. He has not been generous to mamma. He is very hard and ungenerous to mamma. He lets her have that house because he is anxious that the Dales should seem to be respectable before the world; and she lives in it, because she thinks it better for us that she should do so. If I had my way, she should leave it to-morrow – or, at any rate, as soon as Lily is married. I would much sooner go into Guestwick, and live as the Eames do.'

'I think you are ungrateful, Bell.'

'No; I am not ungrateful. And as to consulting, Bernard, – I should be much more inclined to consult you than him about my marriage. If you would let me look on you altogether as a brother, I should think little of promising to marry no one whom you did not approve.'

But such an agreement between them would by no means have suited Bernard's views. He had thought, some four or five weeks back, that he was not personally very anxious for this match. He had declared to himself that he liked his cousin well enough; that it would be a good thing for him to settle himself; that his uncle was reasonable in his wishes and sufficiently liberal in his offers; and that, therefore, he would marry. It had hardly occurred to him as probable that his cousin would reject so eligible an offer, and had certainly never occurred to him that he would have to suffer anything from such rejection. He had entertained none of that feeling of which lovers speak when they declare that they are staking their all upon the hazard of a die. It had not seemed to him that he was staking anything, as he gently told his tale of languid love, lying on the turf by the ha-ha. He had not regarded the possibility of disappointment, of sorrow, and of a deeply-vexed mind. He would have felt but little triumph if accepted, and had not thought that he could be humiliated by any rejection. In this frame of mind he had gone to his work; but now he found, to his own surprise, that this girl's answer had made him absolutely unhappy. Having expressed a wish for this thing, the very expression of the wish made him long to possess it. He found, as he rode along silently by her side, that he was capable of more earnestness of desire than he had known himself to possess. He was at this moment unhappy, disappointed, anxious, distrustful of the future, and more intent on one special toy than he had ever been before, even as a boy. He was vexed, and felt himself to be sore at heart. He looked round at her, as she sat silent, quiet, and somewhat sad upon her pony, and declared to himself that she was very beautiful, – that she was a thing to be gained if still there might be the possibility of gaining her. He felt that he really loved her, and yet he was almost angry with himself for so feeling. Why had he subjected himself to this numbing weakness? His love had never given him any pleasure. Indeed he had never hitherto acknowledged it; but now he was driven to do so on finding it to be the source of trouble and pain. I think it is open to us to doubt whether, even yet, Bernard Dale was in love with his cousin; whether he was not rather in love with his own desire. But against himself he found a verdict that he was in love, and was angry with himself and with all the world.

'Ah, Bell,' he said, coming close up to her, 'I wish you could understand how I love you.' And, as he spoke, his cousin unconsciously recognized more of affection in his tone, and less of that spirit of bargaining which had seemed to pervade all his former pleas, than she had ever found before.

'And do I not love you? Have I not offered to be to you in all respects as a sister?'

'That is nothing. Such an offer to me now is simply laughing at me. Bell, I tell you what, – I will not give you up. The fact is, you do not know me yet, – not know me as you must know any man before you choose him for your husband. You and Lily are not alike in this. You are cautious, doubtful of yourself, and perhaps, also, somewhat doubtful of others. My heart is set upon this, and I shall still try to succeed.'

'Ah, Bernard, do not say that! Believe me, when I tell you that it can never be.'

'No; I will not believe you. I will not allow myself to be made utterly wretched. I tell you fairly that I will not believe you. I may surely hope if I choose to hope. No, Bell, I will never give you up, – unless, indeed, I should see you become another man's wife.'

As he said this, they all turned in through the squire's gate, and rode up to the yard in which it was their habit to dismount from their horses.

CHAPTER XIV

JOHN EAMES TAKES A WALK

JOHN EAMES watched the party of cavaliers as they rode away from his mother's door, and then started upon a solitary walk, as soon as the noise of the horses' hoofs had passed away out of the street. He was by no means happy in his mind as he did so. Indeed, he was overwhelmed with care and trouble, and as he went along very gloomy thoughts passed through his mind. Had he not better go to Australia, or Vancouver's Island, or—? I will not name the places which the poor fellow suggested to himself as possible terminations of the long journeys which he might not improbably be called upon to take. That very day, just before the Dales had come in, he had received a second letter from his darling Amelia, written very closely upon the heels of the first. Why had he not answered her? Was he ill? Was he untrue? No; she would not believe that, and therefore fell back upon the probability of his illness. If it was so, she would rush down to see him. Nothing on earth should keep her from the bedside of her betrothed. If she did not get an answer from her beloved John by return of post, she would be down with him at Guestwick by the express train. Here was a position for such a young man as John Eames! And of Amelia Roper we may say that she was a young woman who would not give up her game, as long as the least chance remained of her winning it. 'I must go somewhere,' John said to himself, as he put on his slouched hat and wandered forth through the back streets of Guestwick. What would his mother say when she heard of Amelia Roper? What would she say when she saw her?

He walked away towards the Manor, so that he might roam about the Guestwick woods in solitude. There was a path with a stile, leading off from the high road, about half a mile beyond the lodges through which the Dales had ridden up to the house, and by this path John Eames turned in, and went away till he had left the Manor house behind him, and was in the centre of the Guestwick woods. He knew the whole ground well, having roamed there ever since he was first allowed to go forth upon his walks alone. He had thought of Lily Dale by the hour together, as he had lost himself among the oak-trees; but in those former days he had thought of her with some pleasure. Now he could only think of her as of one gone from him for ever; and then he had also to think of her whom he had taken to himself in Lily's place.

Young men, very young men, – men so young that it may be almost a question whether or no they have as yet reached their manhood, – are more inclined to be earnest and thoughtful when alone than they ever are when with others, even though those others be their elders. I fancy that, as we grow old ourselves, we are apt to forget that it was so with us; and, forgetting it, we do not believe that it is so with our children. We constantly talk of the thoughtlessness of youth. I do not know whether we might not more appropriately speak of its thoughtfulness. It is, however, no doubt, true that thought will not at once produce wisdom. It may almost be a question whether such wisdom as many of us have in our mature years has not come from the dying out of the power of temptation, rather than as the results of thought and resolution. Men, full fledged and at their work, are, for the most part, too busy for much thought; but lads, on whom the work of the world has not yet fallen with all its pressure, – they have time for thinking.

And thus John Eames was thoughtful. They who knew him best accounted him to be a gay, good-hearted, somewhat reckless young man, open to temptation, but also open to good impressions; as to whom no great success could be predicted, but of whom his friends might fairly hope that he might so live as to bring upon them no disgrace and not much trouble. But, above all things, they would have called him thoughtless. In so calling him, they judged him wrong. He was ever thinking, – thinking much of the world as it appeared to him, and of himself as he appeared to the world; and thinking, also, of things beyond the world. What was to be his fate here and hereafter? Lily Dale was gone from him, and Amelia Roper was hanging round his neck like a millstone! What, under such circumstances, was to be his fate here and hereafter?

We may say that the difficulties in his way were not as yet very great. As to Lily, indeed, he had no room for hope; but, then, his love for Lily had, perhaps, been a sentiment rather than a passion. Most young men have to go through that disappointment, and are enabled to bear it without much injury to their prospects or happiness. And in after-life the remembrance of such love is a blessing rather than a curse, enabling the possessor of it to feel that in those early days there was something within him of which he had no cause to be ashamed. I do not pity John Eames much in regard to Lily Dale. And then, as to Amelia Roper, – had he achieved but a tithe of that lady's experience in the world, or possessed a quarter of her audacity, surely such a difficulty as that need not have stood much in his way! What could Amelia do to him if he fairly told her that he was not minded to marry her? In very truth he had never promised to do so. He was in no way bound to her, not even by honour. Honour, indeed, with such as her! But men are cowards before women until they become tyrants; and are easy dupes, till of a sudden they recognize the fact that it is pleasanter to be the victimizer than the victim, – and as easy. There are men, indeed, who never learn the latter lesson.

But, though the cause for fear was so slight, poor John Eames was thoroughly afraid. Little things which, in connection with so deep a sorrow as his, it is almost ridiculous to mention, added to his embarrassments, and made an escape from them seem to him to be impossible. He could not return to London without going to Burton Crescent, because his clothes were there, and because he owed to Mrs. Roper some small sum of money which on his return to London he would not have immediately in his pocket. He must therefore meet Amelia, and he knew that he had not the courage to tell a girl, face to face, that he did not love her, after he had been once induced to say that he did do so. His boldest conception did not go beyond the writing of a letter in which he would renounce her, and removing himself altogether from that quarter of the town in which Burton Crescent was situated. But then about his clothes, and that debt of his? And what if Amelia should in the meantime come down to Guestwick and claim him? Could he in his mother's presence declare that she had no right to make such claim? The difficulties, in truth, were not very great, but they were too heavy for that poor young clerk from the Income-tax Office.

You will declare that he must have been a fool and a coward. Yet he could read and understand Shakespeare. He knew much, – by far too much, – of Byron's poetry by heart. He was a deep critic, often writing down his criticisms in a lengthy journal which he kept. He could write quickly, and with understanding; and I may declare that men at his office had already ascertained that he was no fool. He knew his business, and could do it, – as many men failed to do who were much less foolish before the world. And as to that matter of cowardice, he would have thought it the greatest blessing in the world to be shut up in a room with Crosbie, having permission to fight with him till one of them should have been brought by stress of battle to give up his claim to Lily Dale. Eames was no coward. He feared no man on earth. But he was terribly afraid of Amelia Roper.

He wandered about through the old Manor woods very ill at ease. The post from Guestwick went out at seven, and he must at once make up his mind whether or no he would write to Amelia on that day. He must also make up his mind as to what he would say to her. He felt that he should at least answer her letter, let his answer be what it might. Should he promise to marry her, – say, in ten or twelve years' time? Should he tell her that he was a blighted being, unfit for love, and with humility entreat of her that he might be excused? Or should he write to her mother, telling her that Burton Crescent would not suit him any longer, promising her to send the balance on receipt of his next payment, and asking her to send his clothes in a bundle to the Income-tax Office? Or should he go home to his own mother, and boldly tell it all to her?

He at last resolved that he must write the letter, and as he composed it in his mind he sat himself down beneath an old tree which stood on a spot at which many of the forest tracks met and crossed each other. The letter, as he framed it here, was not a bad letter, if only he could have got it written and posted. Every word of it he chose with precision, and in his mind he emphasized every expression which told his mind clearly and justified his purpose. 'He acknowledged himself to have been wrong in misleading his correspondent, and allowing her to imagine that she possessed his heart. He had not a heart at her disposal. He had been weak not to write to her before, having been deterred from doing so by the fear of giving her pain; but now he felt that he was bound in honour to tell her the truth. Having so told her, he would not return to Burton Crescent, if it would pain her to see him there. He would always have a deep regard for her,' – Oh, Johnny! – 'and would hope anxiously that her welfare in life might be complete.' That was the letter, as he wrote it on the tablets of his mind under the tree; but the getting it put on to paper was a task, as he knew, of greater difficulty. Then, as he repeated it to himself, he fell asleep.

'Young man,' said a voice in his ears as he slept. At first the voice spoke as a voice from his dream without waking him, but when it was repeated, he sat up and saw that a stout gentleman was standing over him. For a moment he did not know where he was, or how he had come there; nor could he recollect, as he saw the trees about him, how long he had been in the wood. But he knew the stout gentleman well enough, though he had not seen him for more than two years. 'Young man,' said the voice, 'if you want to catch rheumatism, that's the way to do it. Why, it's young Eames, isn't it?'

'Yes, my lord,' said Johnny, raising himself up so that he was now sitting, instead of lying, as he looked up into the earl's rosy face.

'I knew your father, and a very good man he was; only he shouldn't have taken to farming. People think they can farm without learning the trade, but that's a very great mistake. I can farm, because I've learned it. Don't you think you'd better get up?' Whereupon Johnny raised himself to his feet. 'Not but what you're very welcome to lie there if you like it. Only, in October, you know—'

'I'm afraid I'm trespassing, my lord,' said Eames. 'I came in off the path, and—'

'You're welcome; you're very welcome. If you'll come up to the house, I'll give you some luncheon.' This hospitable offer, however, Johnny declined, alleging that it was late, and that he was going home to dinner.

'Come along,' said the earl. 'You can't go any shorter way than by the house. Dear, dear, how well I remember your father. He was a much cleverer man than I am, – very much; but he didn't know how to send a beast to market any better than a child. By-the-by, they have put you into a public office, haven't they?'

'Yes, my lord.'

'And a very good thing, too, – a very good thing, indeed. But why were you asleep in the wood? It isn't warm, you know. I call it rather cold.' And the earl stopped, and looked at him, scrutinizing him, as though resolved to inquire into so deep a mystery.

'I was taking a walk, and, thinking of something, I sat down.'

'Leave of absence, I suppose?'

'Yes, my lord.'

'Have you got into trouble? You look as though you were in trouble. Your poor father used to be in trouble.'

'I haven't taken to farming,' said Johnny, with an attempt at a smile.

'Ha, ha, ha, – quite right. No, don't take to farming. Unless you learn it, you know, you might just as well take to shoe-making; – just the same. You haven't got into trouble, then; eh?'

'No, my lord, not particularly.'

'Not particularly! I know very well that young men do get into trouble when they get up to London. If you want any – any advice, or that sort of thing, you may come to me; for I knew your father well. Do you like shooting?'

'I never did shoot anything.'

'Well, perhaps better not. To tell the truth, I'm not very fond of young men who take to shooting without having anything to shoot at. By-the-by, now I think of it, I'll send your mother some game.' It may, however, here be fair to mention that game very often came from Guestwick Manor to Mrs. Eames. 'And look here, cold pheasant for breakfast is the best thing I know of. Pheasants at dinner are rubbish, – mere rubbish. Here we are at the house. Will you come in and have a glass of wine?'

But this John Eames declined, pleasing the earl better by doing so than he would have done by accepting it. Not that the lord was inhospitable or insincere in his offer, but he preferred that such a one as John Eames should receive his proferred familiarity without too much immediate assurance. He felt that Eames was a little in awe of his companion's rank, and he liked him the better for it. He liked him the better for it, and was a man apt to remember his likings. 'If you won't come in, good-bye,' and he gave Johnny his hand.

'Good evening, my lord,' said Johnny.

'And remember this; it is the deuce of a thing to have rheumatism in your loins. I wouldn't go to sleep under a tree, if I were you, – not in October. But you're always welcome to go anywhere about the place.'

'Thank you, my lord.'

'And if you should take to shooting, – but I dare say you won't; and if you come to trouble, and want advice, or that sort of thing, write to me. I knew your father well.' And so they parted, Eames returning on his road towards Guestwick.

For some reason, which he could not define, he felt better after his interview with the earl. There had been something about the fat, good-natured, sensible old man, which had cheered him, in spite of his sorrow. 'Pheasants for dinner are rubbish, – mere rubbish,' he said to himself, over and over again, as he went along the road; and they were the first words which he spoke to his mother, after entering the house.

'I wish we had some of that sort of rubbish,' said she.

'So you will, to-morrow;' and then he described to her his interview.

'The earl was quite right about lying upon the ground. I wonder you can be so foolish. And he is right about your poor father too. But you have got to change your boots; and we shall be ready for dinner almost immediately.'

But Johnny Eames, before he sat down to dinner, did write his letter to Amelia, and did go out to post it with his own hands, – much to his mother's annoyance. But the letter would not get itself written in that strong and appropriate language which had come to him as he was roaming through the woods. It was a bald letter, and somewhat cowardly withal.

Dear Amelia (the letter ran),

'I have received both of yours; and did not answer the first because I felt that there was a difficulty in expressing what I wish to say; and now it will be better that you should allow the subject to stand over till I am back in town. I shall be there in ten days from this. I have been quite well, and am so; but of course am much obliged by your inquiries. I know you will think this very cold; but when I tell you everything, you will agree with me that it is best. If I were to marry, I know that we should be unhappy, because we should have nothing to live on. If I have ever said anything to deceive you, I beg your pardon with all my heart; – but perhaps it will be better to let the subject remain till we shall meet again in London.

'Believe me to be
'Your most sincere friend,
'And I may say admirer, – [Oh, John Eames!]
'John Eames.'

CHAPTER XV

THE LAST DAY

LAST days are wretched days; and so are last moments wretched moments. It is not the fact that the parting is coming which makes these days and moments so wretched, but the feeling that something special is expected from them, which something they always fail to produce. Spasmodic periods of pleasure, of affection, or even of study, seldom fail of disappointment when premeditated. When last days are coming, they should be allowed to come and to glide away without special notice or mention. And as for last moments, there should be none such. Let them ever be ended, even before their presence has been acknowledged.

But Lily Dale had not yet been taught these lessons by her world's experience, and she expected that this sweetest cup of which she had ever drunk should go on being sweet – sweeter and still sweeter – as long as she could press it to her lips. How the dregs had come to mix themselves with the last drops we have already seen; and on that same day – on the Monday evening – the bitter taste still remained; for Crosbie, as they walked about through the garden in the evening, found other subjects on which he thought it necessary to give her sundry hints, intended for her edification, which came to her with much of the savour of a lecture. A girl, when she is thoroughly in love, as surely was the case with Lily, likes to receive hints as to her future life from the man to whom she is devoted; but she would, I think, prefer that such hints should be short, and that the lesson should be implied rather than declared; – that they should, in fact, be hints and not lectures. Crosbie, who was a man of tact, who understood the world and had been dealing with women for many years, no doubt understood all this as well as we do. But he had come to entertain a notion that he was an injured man, that he was giving very much more than was to be given to him, and that therefore he was entitled to take liberties which might not fairly be within the reach of another lover. My reader will say that in all this he was ungenerous. Well; he was ungenerous. I do not know that I have ever said that much generosity was to be expected from him. He had some principles of right and wrong under the guidance of which it may perhaps be hoped that he will not go utterly astray; but his past life had not been of a nature to make him unselfish. He was ungenerous, and Lily felt it, though she would not acknowledge it even to herself. She had been very open with him, – acknowledging the depth of her love for him; telling him that he was now all in all to her; that life without his love would be impossible to her: and in a certain way he took advantage of these strong avowals, treating her as though she were a creature utterly in his power; – as indeed she was.

On that evening he said no more of Johnny Eames, but said much of the difficulty of a man establishing himself with a wife in London, who had nothing but his own moderate income on which to rely. He did not in so many words tell her that if her friends could make up for her two or three thousand pounds, – that being much less than he had expected when he first made his offer, – this terrible difficulty would be removed; but he said enough to make her understand that the world would call him very imprudent in taking a girl who had nothing. And as he spoke of these things, Lily remaining for the most part silent as he did so, it occurred to him that he might talk to her freely of his past life, – more freely than he would have done had he feared that he might lose her by any such disclosures. He had no fear of losing her. Alas! might it not be possible that he had some such hope!

He told her that his past life had been expensive; that, though he was not in debt, he had lived up to every shilling that he had, and that he had contracted habits of expenditure which it would be almost impossible for him to lay aside at a day's notice. Then he spoke of entanglements, meaning, as he did so, to explain more fully what were their nature, – but not daring to do so when he found that Lily was altogether in the dark as to what he meant. No; he was not a generous man, – a very ungenerous man. And yet, during all this time, he thought that he was guided by principle. 'It will be best that I should be honest with her,' he said to himself. And then he told himself, scores of times, that when making his offer he had expected, and had a right to expect, that she would not be penniless. Under those circumstances he had done the best he could for her – offering her his heart honestly, with a quick readiness to make her his own at the earliest day that she might think possible. Had he been more cautious, he need not have fallen into this cruel mistake; but she, at any rate, could not quarrel with him for his imprudence. And still he was determined to stand by his engagement and willing to marry her, although, as he the more thought of it, he felt the more strongly that he would thereby ruin his prospects, and thrust beyond his own reach all those good things which he had hoped to win. As he continued to talk to her he gave himself special credit for his generosity, and felt that he was only doing his duty by her in pointing out to her all the difficulties which lay in the way of their marriage.

At first Lily said some words intended to convey an assurance that she would be the most economical wife that man ever had, but she soon ceased from such promises as these. Her perceptions were keen, and she discovered that the difficulties of which he was afraid were those which he must overcome before his marriage, not any which might be expected to overwhelm him after it. 'A cheap and nasty ménage* would be my aversion,' he said to her. 'It is that which I want to avoid, – chiefly for your sake.' Then she promised him that she would wait patiently for his time – 'even though it should be for seven years,' she said, looking up into his face and trying to find there some sign of approbation. 'That's nonsense,' he said. 'People are not patriarchs nowadays. I suppose we shall have to wait two years. And that's a deuce of a bore, – a terrible bore.' And there was that in the tone of his voice which grated on her feelings, and made her wretched for the moment.

As he parted with her for the night on her own side of the little bridge which led from one garden to the other, he put his arm round her to embrace her and kiss her, as he had often done at that spot. It had become a habit with them to say their evening farewells there, and the secluded little nook amongst the shrubs was inexpressibly dear to Lily. But on the present occasion she made an effort to avoid his caress. She turned from him – very slightly, but it was enough, and he felt it. 'Are you angry with me?' he said. 'Oh, no! Adolphus; how can I be angry with you?' And then she turned to him and gave him her face to kiss almost before he had again asked for it. 'He shall not think that I am unkind to him, – and it will not matter now,' she said to herself, as she walked slowly across the lawn, in the dark, up to her mother's drawing-room window.

'Well, dearest,' said Mrs. Dale, who was there alone; 'did the beards wag merry in the Great Hall this evening?' That was a joke with them, for neither Crosbie nor Bernard Dale used a razor at his toilet.

'Not specially merry. And I think it was my fault, for I have a headache. Mamma, I believe I will go at once to bed.'

'My darling, is there anything wrong?'

'Nothing, mamma. But we had such a long ride; and then Adolphus is going, and of course we have so much to say. To-morrow will be the last day, for I shall only just see him on Wednesday morning; and as I want to be well, if possible, I'll go to bed.' And so she took her candle and went.

When Bell came up, Lily was still awake, but she begged her sister not to disturb her. 'Don't talk to me, Bell,' she said. 'I'm trying to make myself quiet, and I half feel that I should get childish if I went on talking. I have almost more to think of than I know how to manage.' And she strove, not altogether unsuccessfully, to speak with a cheery tone, as though the cares which weighed upon her were not unpleasant in their nature. Then her sister kissed her and left her to her thoughts.

And she had great matter for thinking; so great, that many hours sounded in her ears from the clock on the stairs before she brought her thoughts to a shape that satisfied herself. She did so bring them at last, and then she slept. She did so bring them, toiling over her work with tears that made her pillow wet, with heart-burning and almost with heart-breaking, with much doubting, and many anxious, eager inquiries within her own bosom as to that which she ought to do, and that which she could endure to do. But at last her resolve was taken, and then she slept.

It had been agreed between them that Crosbie should come down to the Small House on the next day after breakfast, and remain there till the time came for riding. But Lily determined to alter this arrangement, and accordingly put on her hat immediately after breakfast, and posted herself at the bridge, so as to intercept her lover as he came. He soon appeared with his friend Dale, and she at once told him her purpose.

'I want to have a talk with you, Adolphus, before you go in to mamma; so come with me into the field.'

'All right,' said he.

'And Bernard can finish his cigar on the lawn. Mamma and Bell will join him there.'

'All right,' said Bernard. So they separated; and Crosbie went away with Lily into the field where they had first learned to know each other in those haymaking days.

She did not say much till they were well away from the house; but answered what words he chose to speak, – not knowing very well of what he spoke. But when she considered that they had reached the proper spot, she began very abruptly.

'Adolphus,' she said, 'I have something to say to you, – something to which you must listen very carefully.' Then he looked at her, and at once knew that she was in earnest.

'This is the last day on which I could say it,' she continued;

'and I am very glad that I have not let the last day go by without saying it. I should not have known how to put it in a letter.'

'What is it, Lily?'

'And I do not know that I can say it properly; but I hope that you will not be hard upon me. Adolphus, if you wish that all this between us should be over, I will consent.'

'Lily!'

'I mean what I say. If you wish it, I will consent; and when I have said so, proposing it myself, you may be quite sure that I shall never blame you, if you take me at my word.'

'Are you tired of me, Lily?'

'No. I shall never be tired of you, – never weary with loving you. I did not wish to say so now; but I will answer your question boldly. Tired of you! I fancy that a girl can never grow tired of her lover. But I would sooner die in the struggle than be the cause of your ruin. It would be better – in every way better.'

'I have said nothing of being ruined.'

'But listen to me. I should not die if you left me, – not be utterly broken-hearted. Nothing on earth can I ever love as I have loved you. But I have a God and a Saviour that will be enough for me. I can turn to them with content, if it be well that you should leave me. I have gone to them, and—' But at this moment she could utter no more words. She had broken down in her effort, losing her voice through the strength of her emotion. As she did not choose that he should see her overcome, she turned from him and walked away across the grass.

Of course he followed her; but he was not so quick after her but that time had been given to her to recover herself. 'It is true,' she said. 'I have the strength of which I tell you. Though I have given myself to you as your wife, I can bear to be divorced from you now, – now. And, my love, though it may sound heartless, I would sooner be so divorced from you, than cling to you as a log that must drag you down under the water, and drown you in trouble and care. I would; – indeed I would. If you go, of course that kind of thing is over for me. But the world has more than that, – much more; and I would make myself happy; – yes, my love, I would be happy. You need not fear that.'

'But, Lily, why is all this said to me here to-day?'

'Because it is my duty to say it. I understand all your position now, though it is only now. It never flashed on me till yesterday. When you proposed to me, you thought that I, – that I had some fortune.'

'Never mind that now, Lily.'

'But you did. I see it all now. I ought perhaps to have told you that it was not so. There has been the mistake, and we are both sufferers. But we need not make the suffering deeper than needs be. My love, you are free, – from this moment. And even my heart shall not blame you for accepting your freedom.'

'And are you afraid of poverty?' he asked her.

'I am afraid of poverty for you. You and I have lived differently. Luxuries, of which I know nothing, have been your daily comforts. I tell you I can bear to part with you, but I cannot bear to become the source of your unhappiness. Yes; I will bear it; and none shall dare in my hearing to speak against you. I have brought you here to say the word; nay, more than that, – to advise you to say it.'

He stood silent for a moment, during which he held her by the hand. She was looking into his face, but he was looking away into the clouds; striving to appear as though he was the master of the occasion. But during those moments his mind was racked with doubt. What if he should take her at her word? Some few would say bitter things against him, but such bitter things had been said against many another man without harming him. Would it not be well for both if he should take her at her word? She would recover and love again, as other girls had done; and as for him, he would thus escape from the ruin at which he had been gazing for the last week past. For it was ruin, – utter ruin. He did love her; so he declared to himself. But was he a man who ought to throw the world away for love? Such men there were; but was he one of them? Could he be happy in that small house, somewhere near the New Road, with five children and horrid misgivings as to the baker's bill? Of all men living, was not he the last that should have allowed himself to fall into such a trap? All this passed through his mind as he turned his face up to the clouds with a look that was intended to be grand and noble.

'Speak to me, Adolphus, and say that it shall be so.'

Then his heart misgave him, and he lacked the courage to extricate himself from his trouble; or, as he afterwards said to himself, he had not the heart to do it. 'If I understand you rightly, Lily, all this comes from no want of love on your own part?'

'Want of love on my part? But you should not ask me that.'

'Until you tell me that there is such a want, I will agree to no parting.' Then he took her hand and put it within his arm. 'No, Lily; whatever may be our cares and troubles, we are bound together, – indissolubly.'

'Are we?' said she; and, as she spoke, her voice trembled, and her hand shook.

'Much too firmly for any such divorce as that. No, Lily, I claim the right to tell you all my troubles; but I shall not let you go.'

'But, Adolphus—' and the hand on his arm was beginning to cling to it again.

'Adolphus,' said he, 'has got nothing more to say on that subject. He exercises the right which he believes to be his own, and chooses to retain the prize which he has won.'

She was now clinging to him in very truth. 'Oh, my love!' she said. 'I do not know how to say it again. It is of you that I am thinking; – of you, of you!'

'I know you are; but you have misunderstood me a little; that's all.'

'Have I? Then listen to me again, once more, my heart's own darling, my love, my husband, my lord! If I cannot be to you at once like Ruth, and never cease from coming after you, my thoughts to you shall be like those of Ruth: * – if aught but death part thee and me, may God do so to me and more also.' Then she fell upon his breast and wept.

He still hardly understood the depth of her character. He

was not himself deep enough to comprehend it all. But yet he was awed by her great love, and exalted to a certain solemnity of feeling which for the time made him rejoice in his late decision. For a few hours he was minded to throw the world behind him, and wear this woman, as such a woman should be worn, – as a comforter to him in all things, and a strong shield against great troubles. 'Lily,' he said, 'my own Lily!'

'Yes, your own, to take when you please, and leave untaken while you please; and as much your own in one way as in the other.' Then she looked up again, and essayed to laugh as she did so. 'You will think I am frantic, but I am so happy. I don't care about your going now; indeed I don't. There; you may go now, this minute, if you like it.' And she withdrew her hand from him. 'I feel so differently from what I have done for the last few days. I am so glad you have spoken to me as you did. Of course I ought to bear all those things with you. But I cannot be unhappy about it now. I wonder if I went to work and made a lot of things, whether that would help?'

'A set of shirts for me, for instance?'

'I could do that, at any rate.'

'It may come to that yet, some of these days.'

'I pray God that it may.' Then again she was serious, and the tears came once more into her eyes. 'I pray God that it may. To be of use to you, – to work for you, – to do something for you that may have in it some sober, earnest purport of usefulness; – that is what I want above all things. I want to be with you at once that I may be of service to you. Would that you and I were alone together, that I might do everything for you. I sometimes think that a very poor man's wife is the happiest, because she does do everything.'

'You shall do everything very soon,' said he; and then they sauntered along pleasantly through the morning hours, and when they again appeared at Mrs. Dale's table, Mrs. Dale and Bell were astonished at Lily's brightness. All her old ways had seemed to return to her, and she made her little saucy speeches to Mr. Crosbie as she had used to do when he was first becoming fascinated by her sweetness. 'You know that you'll be such a swell when you get to that countess's house that you'll forget all about Allington.'

'Of course I shall,' said he.

'And the paper you write upon will be all over coronets, – that is, if ever you do write. Perhaps you will to Bernard some day, just to show that you are staying at a castle.'

'You certainly don't deserve that he should write to you,' said Mrs. Dale.

'I don't expect it for a moment, not till he gets back to London and finds that he has nothing else to do at his office. But I should so like to see how you and Lady Julia get on together. It was quite clear that she regarded you as an ogre; didn't she, Bell?'

'So many people are ogres to Lady Julia,' said Bell.

'I believe Lady Julia to be a very good woman,' said Mrs. Dale, 'and I won't have her abused.'

'Particularly before poor Bernard, who is her pet nephew,' said Lily. 'I dare say Adolphus will become a pet too when she has been a week with him at Courcy Castle. Do try and cut Bernard out.'

From all of which Mrs. Dale learned that some care which had sat heavy on Lily's heart was now lightened, if not altogether removed. She had asked no questions of her daughter, but she had perceived during the past few days that Lily was in trouble, and she knew that such trouble had arisen from her engagement. She had asked no questions, but of course she had been told what was Mr. Crosbie's income, and had been made to understand that it was not to be considered as amply sufficient for all the wants of matrimony. There was little difficulty in guessing what was the source of Lily's care, and as little in now perceiving that something had been said between them by which that care had been relieved.

After that they all rode, and the afternoon went by pleasantly. It was the last day indeed, but Lily had determined that she would not be sad. She had told him that he might go now, and that she would not be discontented at his going. She knew that the morrow would be very blank to her; but she struggled to live up to the spirit of her promise, and she succeeded. They all dined at the Great House, even Mrs. Dale doing so upon this occasion. When they had come in from the garden in the evening, Crosbie talked more to Mrs. Dale than he did even to Lily, while Lily sat a little distant, listening with all her ears, sometimes saying a low-toned word, and happy beyond expression in the feeling that her mother and her lover should understand each other. And it must be understood that Crosbie at this time was fully determined to conquer the difficulties of which he had thought so much, and to fix the earliest day which might be possible for his marriage. The solemnity of that meeting in the field still hung about him, and gave to his present feelings a manliness and a truth of purpose which were too generally wanting to them. If only those feelings would last! But now he talked to Mrs. Dale about her daughter, and about their future prospects, in a tone which he could not have used had not his mind for the time been true to her. He had never spoken so freely to Lily's mother, and at no time had Mrs. Dale felt for him so much of a mother's love. He apologized for the necessity of some delay, arguing that he could not endure to see his young wife without the comfort of a home of her own, and that he was now, as he always had been, afraid of incurring debt. Mrs. Dale disliked waiting engagements, – as do all mothers, – but she could not answer unkindly to such pleadings as this.

'Lily is so very young,' she said, 'that she may well wait for a year or so.'

'For seven years,' said Lily, jumping up and whispering into her mother's ear. 'I shall hardly be six-and-twenty then, which is not at all too old.'

And so the evening passed away very pleasantly.

'God bless you, Adolphus!' Mrs. Dale said to him, as she parted with him at her own door. It was the first time that she had called him by his Christian name. 'I hope you understand how much we are trusting to you.'

'I do, – I do,' said he, as he pressed her hand. Then as he walked back alone, he swore to himself, binding himself to the oath with all his heart, that he would be true to those women, – both to the daughter and to the mother; for the solemnity of the morning was still upon him.

He was to start the next morning before eight, Bernard having undertaken to drive him over to the railway at Guestwick. The breakfast was on the table shortly after seven; and just as the two men had come down, Lily entered the room, with her hat and shawl. 'I said I would be in to pour out your tea,' said she.

It was a silent meal, for people do not know what to say in those last minutes. And Bernard, too, was there; proving how true is the adage which says, that two are company, but that three are not. I think that Lily was wrong to come up on that last morning; but she would not hear of letting him start without seeing him, when her lover had begged her not to put herself to so much trouble. Trouble! Would she not have sat up all night to see even the last of the top of his hat?

Then Bernard, muttering something about the horse, went away. 'I have only one minute to speak to you,' said she, jumping up, 'and I have been thinking all night of what I had to say. It is so easy to think, and so hard to speak.'

'My darling, I understand it all.'

'But you must understand this, that I will never distrust you. I will never ask you to give me up again, or say that I could be happy without you. I could not live without you; that is, without the knowledge that you are mine. But I will never be impatient, never. Pray, pray believe me! Nothing shall make me distrust you.'

'Dearest Lily, I will endeavour to give you no cause.'

'I know you will not; but I specially wanted to tell you that. And you will write, – very soon?'

'Directly I get there.'

'And as often as you can. But I won't bother you; only your letters will make me so happy. I shall be so proud when they come to me. I shall be afraid of writing too much to you, for fear I should tire you.'

'You will never do that.'

'Shall I not? But you must write first, you know. If you could only understand how I shall live upon your letters! And now good-bye. There are the wheels. God bless you, my own, my own!' And she gave herself up into his arms, as she had given herself up into his heart.

She stood at the door as the two men got into the gig, and, as it passed down through the gate, she hurried out upon the terrace, from whence she could see it for a few yards down the lane. Then she ran from the terrace to the gate, and, hurrying through the gate, made her way into the church-yard, from the farther corner of which she could see the heads of the two men till they had made the turn into the main road beyond the parsonage. There she remained till the very sound of the wheels no longer reached her ears, stretching her eyes in the direction they had taken. Then she turned round slowly and made her way out at the church-yard gate, which opened on to the road close to the front door of the Small House.

'I should like to punch his head,' said Hopkins, the gardener, to himself, as he saw the gig driven away and saw Lily trip after it, that she might see the last of him whom it carried. 'And I wouldn't think nothing of doing it; no more I wouldn't,' Hopkins added in his soliloquy. It was generally thought about the place that Miss Lily was Hopkins's favourite; though he showed it chiefly by snubbing her more frequently than he snubbed her sister.

Lily had evidently intended to return home through the front door; but she changed her purpose before she reached the house, and made her way slowly back through the churchyard, and by the gate of the Great House, and by the garden at the back of it, till she crossed the little bridge. But on the bridge she rested awhile, leaning against the railing as she had often leant with him, and thinking of all that had passed since that July day on which she had first met him. On no spot had he so often told her of his love as on this, and nowhere had she so eagerly sworn to him that she would be his own dutiful loving wife.

'And by God's help so I will,' she said to herself, as she walked firmly up to the house. 'He has gone, mamma,' she said, as she entered the breakfast-room. 'And now we'll go back to our workaday ways; it has been all Sunday for me for the last six weeks.'

CHAPTER XVI

MR. CROSBIE MEETS AN OLD CLERGYMAN ON HIS WAY TO COURCY CASTLE

FOR the first mile or two of their journey Crosbie and Bernard Dale sat, for the most part, silent in their gig. Lily, as she ran down to the churchyard corner and stood there looking after them with her loving eyes, had not been seen by them. But the spirit of her devotion was still strong upon them both, and they felt that it would not be well to strike at once into any ordinary topic of conversation. And, moreover, we may presume that Crosbie did feel much at thus parting from such a girl as Lily Dale, with whom he had lived in close intercourse for the last six weeks, and whom he loved with all his heart, – with all the heart that he had for such purposes. In those doubts as to his marriage which had troubled him he had never expressed to himself any disapproval of Lily. He had not taught himself to think that she was other than he would have her be, that he might thus give himself an excuse for parting from her. Nor as yet, at any rate, had he had recourse to that practice, so common with men who wish to free themselves from the bonds with which they have permitted themselves to be bound. Lily had been too sweet to his eyes, to his touch, to all his senses for that. He had enjoyed too keenly the pleasure of being with her, and of hearing her tell him that she loved him, to allow of his being personally tired of her. He had not been so spoilt by his club life but that he had taken exquisite pleasure in all her nice country ways, and soft, kind-hearted, womanly humour. He was by no means tired of Lily. Better than any of his London pleasures was this pleasure of making love in the green fields to Lily Dale. It was the consequences of it that affrighted him. Babies with their belongings would come; and dull evenings, over a dull fire, or else the pining grief of a disappointed woman. He would be driven to be careful as to his clothes, because the ordering of a new coat would entail a serious expenditure. He could go no more among countesses and their daughters, because it would be out of the question that his wife should visit at their houses. All the victories that he had ever won must be given up. He was thinking of this even while the gig was going round the corner near the parsonage house, and while Lily's eyes were still blessed with some view of his departing back; but he was thinking, also, that moment, that there might be other victory in store for him; that it might be possible for him to learn to like that fireside, even though babies should be there, and a woman opposite to him intent on baby care. He was struggling as best he knew how; for the solemnity which Lily had imparted to him had not yet vanished from his spirit.

'I hope that, upon the whole, you feel contented with your visit?' said Bernard to him, at last.

'Contented? Of course I do.'

'That is easily said; and civility to me, perhaps, demands as much. But I know that you have, to some extent, been disappointed.'

'Well; yes. I have been disappointed as regards money. It is of no use denying it.'

'I should not mention it now, only that I want to know that you exonerate me.'

'I have never blamed you; – neither you, nor anybody else; unless, indeed, it has been myself.'

'You mean that you regret what you've done?'

'No; I don't mean that. I am too devotedly attached to that dear girl whom we have just left to feel any regret that I have engaged myself to her. But I do think that had I managed better with your uncle things might have been different.'

'I doubt it. Indeed I know that it is not so; and can assure you that you need not make yourself unhappy on that score. I had thought, as you well know, that he would have done something for Lily; – something, though not as much as he always intended to do for Bell. But you may be sure of this; that he had made up his mind as to what he would do. Nothing that you or I could have said would have changed him.'

'Well; we won't say anything more about it,' said Crosbie.

Then they went on again in silence, and arrived at Guestwick in ample time for the train.

'Let me know as soon as you get to town,' said Crosbie.

'Oh, of course. I'll write to you before that.'

And so they parted. As Dale turned and went, Crosbie felt that he liked him less than he had done before; and Bernard, also, as he was driving him, came to the conclusion that Crosbie would not be so good a fellow as a brother-in-law as he had been as a chance friend. 'He'll give us trouble, in some way; and I'm sorry that I brought him down.' That was Dale's inward conviction in the matter.

Crosbie's way from Guestwick lay, by railway, to Barchester, the cathedral city lying in the next county, from whence he purposed to have himself conveyed over to Courcy. There had, in truth, been no cause for his very early departure, as he was aware that all arrivals at country houses should take place at some hour not much previous to dinner. He had been determined to be so soon upon the road by a feeling that it would be well for him to get over those last hours. Thus he found himself in Barchester at eleven o'clock, with nothing on his hands to do; and having nothing else to do, he went to church. There was a full service at the cathedral, and as the verger marshalled him up to one of the empty stalls, a little spare old man was beginning to chant the Litany. 'I did not mean to fall in for all this,' said Crosbie, to himself, as he settled himself with his arms on the cushion. But the peculiar charm of that old man's voice soon attracted him; – a voice that, though tremulous, was yet strong; and he ceased to regret the saint whose honour and glory had occasioned the length of that day's special service.

'And who is the old gentleman who chanted the Litany?' he asked the verger afterwards, as he allowed himself to be shown round the monuments of the cathedral.

'That's our precentor, sir; Mr. Harding. You must have heard of Mr. Harding.' But Crosbie, with a full apology, confessed his ignorance.

'Well, sir; he's pretty well known too, tho' he is so shy like.

He's father-in-law to our dean, sir; and father-in-law to Archdeacon Grantly also.'

'His daughters have all gone into the profession, then?'

'Why, yes; but Miss Eleanor – for I remember her before she was married at all, – when they lived at the hospital—'

'At the hospital?'

'Hiram's hospital, sir. He was warden,* you know. You should go and see the hospital, sir, if you never was there before. Well, Miss Eleanor, – that was his youngest, – she married Mr. Bold as her first. But now she's the dean's lady.'

'Oh; the dean's lady, is she?'

'Yes, indeed. And what do you think, sir? Mr. Harding might have been dean himself if he'd liked. They did offer it to him.'

'And he refused it?'

'Indeed he did, sir.'

'Nolo decanari.* I never heard of that before. What made him so modest?'

'Just that, sir; because he is modest. He's past his seventy now, – ever so much; but he's just as modest as a young girl. A deal more modest than some of them. To see him and his granddaughter together!'

'And who is his granddaughter?'

'Why, Lady Dumbello, as will be the Marchioness of Hartletop.'

'I know Lady Dumbello,' said Crosbie; not meaning, however, to boast to the verger of his noble acquaintance.

'Oh, do you, sir?' said the man, unconsciously touching his hat at this sign of greatness in the stranger; though in truth he had no love for her ladyship. 'Perhaps you're going to be one of the party at Courcy Castle?'

'Well, I believe I am.'

'You'll find her ladyship there before you. She lunched with her aunt at the deanery as she went through, yesterday; finding it too much trouble to go out to her father's, at Plumstead. Her father is the archdeacon, you know. They do say, – but her ladyship is your friend!'

'No friend at all; only a very slight acquaintance. She's quite as much above my line as she is above her father's.'

'Well, she is above them all. They say she would hardly as much as speak to the old gentleman.'

'What, her father?'

'No, Mr. Harding; he that chanted the Litany just now. There he is, sir, coming out of the deanery.'

They were now standing at the door leading out from one of the transepts, and Mr. Harding passed them as they were speaking together. He was a little, withered, shambling old man, with bent shoulders, dressed in knee-breeches and long black gaiters, which hung rather loosely about his poor old legs, – rubbing his hands one over the other as he went. And yet he walked quickly; not tottering as he walked, but with an uncertain, doubtful step. The verger, as Mr. Harding passed, put his hand to his head, and Crosbie also raised his hat. Whereupon Mr. Harding raised his, and bowed, and turned round as though he were about to speak. Crosbie felt that he had never seen a face on which traits of human kindness were more plainly written. But the old man did not speak. He turned his body half round, and then shambled back, as though ashamed of his intention, and passed on.

'He is of that sort that they make the angels of,' said the verger. 'But they can't make many if they want them all as good as he is. I'm much obliged to you, sir.' And he pocketed the half-crown which Crosbie gave him.

'So that's Lady Dumbello's grandfather,' said Crosbie, to himself, as he walked slowly round the close towards the hospital, by the path which the verger had shown him. He had no great love for Lady Dumbello, who had dared to snub him, – even him. 'They may make an angel of the old gentleman,' he continued to say; 'but they'll never succeed in that way with the granddaughter.'

He sauntered slowly on over a little bridge; and at the gate of the hospital he again came upon Mr. Harding. 'I was going to venture in,' said he, 'to look at the place. But perhaps I shall be intruding?'

'No, no; by no means,' said Mr. Harding. 'Pray come in. I cannot say that I am just at home here. I do not live here, – not now. But I know the ways of the place well, and can make you welcome. That's the warden's house. Perhaps we won't go in so early in the day, as the lady has a very large family. An excellent lady, and a dear friend of mine, – as is her husband.'

'And he is warden, you say?'

'Yes, warden of the hospital. You see the house, sir. Very pretty, isn't it? Very pretty. To my idea it's the prettiest built house I ever saw.'

'I won't go quite so far as that,' said Crosbie.

'But you would if you'd lived there twelve years, as I did. I lived in that house twelve years, and I don't think there's so sweet a spot on the earth's surface. Did you ever see such turf as that?'

'Very nice indeed,' said Crosbie, who began to make a comparison with Mrs. Dale's turf at the Small House, and to determine that the Allington turf was better than that of the hospital.

'I had that turf laid down myself. There were borders there when I first came, with hollyhocks, and those sort of things. The turf was an improvement.'

'There's no doubt of that, I should say.'

'The turf was an improvement, certainly. And I planted those shrubs, too. There isn't such a Portugal laurel as that in the county.'

'Were you warden here, sir?' And Crosbie, as he asked the question, remembered that, in his very young days, he had heard of some newspaper quarrel which had taken place about Hiram's hospital at Barchester.

'Yes, sir. I was warden here for twelve years. Dear, dear, dear! If they had put any gentleman here that was not on friendly terms with me it would have made me very unhappy, – very. But, as it is, I go in and out just as I like; almost as much as I did before they— But they didn't turn me out. There were reasons which made it best that I should resign.'

'And you live at the deanery now, Mr. Harding?'

'Yes; I live at the deanery now. But I am not dean, you know. My son-in-law, Dr. Arabin, is the dean. I have another daughter married in the neighbourhood, and can truly say that my lines have fallen to me in pleasant places.'

Then he took Crosbie in among the old men, into all of whose rooms he went. It was an almshouse for aged men of the city, and before Crosbie had left him Mr. Harding had explained all the circumstances of the hospital, and of the way in which he had left it. 'I didn't like going, you know; I thought it would break my heart. But I could not stay when they said such things as that; – I couldn't stay. And, what is more, I should have been wrong to stay. I see it all now. But when I went out under that arch, Mr. Crosbie, leaning on my daughter's arm, I thought that my heart would have broken.' And the tears even now ran down the old man's cheeks as he spoke.

It was a long story, and it need not be repeated here. And there was no reason why it should have been told to Mr. Crosbie, other than this, – that Mr. Harding was a fond garrulous old man, who loved to indulge his mind in reminiscences of the past. But this was remarked by Crosbie; that, in telling his story, no word was said by Mr. Harding injurious to any one. And yet he had been injured, – injured very deeply. 'It was all for the best,' he said at last; 'especially as the happiness has not been denied to me of making myself at home at the old place. I would take you into the house, which is very comfortable, – very; only it is not always convenient early in the day, where there's a large family.' In hearing which Crosbie was again made to think of his own future home and limited income.

He had told the old clergyman who he was, and that he was on his way to Courcy. 'Where, as I understand, I shall meet a granddaughter of yours.'

'Yes, yes; she is my grandchild. She and I have got into different walks of life now, so that I don't see much of her. They tell me that she does her duty well in that sphere of life to which it has pleased God to call her.'

'That depends,' thought Crosbie, 'on what the duties of a viscountess may be supposed to be.' But he wished his new friend good-bye, without saying anything further as to Lady Dumbello, and, at about six o'clock in the evening, had himself driven up under the portico of Courcy Castle.

CHAPTER XVII

COURCY CASTLE

COURCY CASTLE was very full. In the first place, there was a great gathering there of all the Courcy family. The earl was there, – and the countess, of course. At this period of the year Lady De Courcy was always at home; but the presence of the earl himself had heretofore been by no means so certain. He was a man who had been much given to royal visitings and attendances, to parties in the Highlands, to – no doubt necessary – prolongations of the London season, to sojournings at certain German watering-places, convenient, probably, in order that he might study the ways and ceremonies of German Courts, – and to various other absences from home, occasioned by a close pursuit of his own special aims in life; for the Earl De Courcy had been a great courtier. But of late gout, lumbago, and perhaps also some diminutions in his powers of making himself generally agreeable, had reconciled him to domestic duties, and the earl spent much of his time at home. The countess, in former days, had been heard to complain of her lord's frequent absence. But it is hard to please some women, – and now she would not always be satisfied with his presence.

And all the sons and daughters were there, – excepting Lord Porlock, the eldest, who never met his father. The earl and Lord Porlock were not on terms, and indeed hated each other as only such fathers and such sons can hate. The Honourable George De Courcy was there with his bride, he having lately performed a manifest duty, in having married a young woman with money. Very young she was not, – having reached some years of her life in advance of thirty; but then, neither was the Honourable George very young; and in this respect the two were not ill-sorted. The lady's money had not been very much, – perhaps thirty thousand pounds or so. But then the Honourable George's money had been absolutely none. Now he had an income on which he could live, and therefore his father and mother had forgiven him all his sins, and taken him again to their bosom. And the marriage was matter of great moment, for the elder scion of the house had not yet taken to himself a wife, and the De Courcy family might have to look to this union for an heir. The lady herself was not beautiful, or clever, or of imposing manners – nor was she of high birth. But neither was she ugly, nor unbearably stupid. Her manners were, at any rate, innocent; and as to her birth, – seeing that, from the first, he was not supposed to have had any, – no disappointment was felt. Her father had been a coal-merchant. She was always called Mrs. George, and the effort made respecting her by everybody in and about the family was to treat her as though she were a figure of a woman, a large well-dressed resemblance of a being, whom it was necessary for certain purposes that the De Courcys should carry in their train. Of the Honourable George we may further observe, that, having been a spendthrift all his life, he had now become strictly parsimonious. Having reached the discreet age of forty, he had at last learned that beggary was objectionable; and he, therefore, devoted every energy of his mind to save shillings and pence wherever pence and shillings might be saved. When first this turn came upon him both his father and mother were delighted to observe it; but, although it had hardly yet lasted over twelve months, some evil results were beginning to appear. Though possessed of an income, he would take no steps towards possessing himself of a house. He hung by the paternal mansion, either in town or country; drank the paternal wines, rode the paternal horses, and had even contrived to obtain his wife's dresses from the maternal milliner. In the completion of which little last success, however, some slight family dissent had showed itself.

The Honourable John, the third son, was also at Courcy. He had as yet taken to himself no wife, and as he had not hitherto made himself conspicuously useful in any special walk of life his family were beginning to regard him as a burden. Having no income of his own to save, he had not copied his brother's virtue of parsimony; and, to tell the truth plainly, he had made himself so generally troublesome to his father, that he had been on more than one occasion threatened with expulsion from the family roof.

But it is not easy to expel a son. Human fledglings cannot be driven out of the nest like young birds. An Honourable John turned adrift into absolute poverty will make himself heard of in the world, – if in no other way, by his ugliness as he starves. A thorough-going ne'er-do-well in the upper classes has eminent advantages on his side in the battle which he fights against respectability. He can't be sent to Australia against his will. He can't be sent to the poorhouse without the knowledge of all the world. He can't be kept out of tradesmen's shops; nor, without terrible scandal, can he be kept away from the paternal properties. The earl had threatened, and snarled, and shown his teeth; he was an angry man, and a man who could look very angry; with eyes which could almost become red, and a brow that wrinkled itself in perpendicular wrinkles, sometimes very terrible to behold. But he was an inconsistent man, and the Honourable John had learned to measure his father, and in an accurate balance.

I have mentioned the sons first, because it is to be presumed that they were the elder, seeing that their names were mentioned before those of their sisters in all the peerages. But there were four daughters, – the Ladies Amelia, Rosina, Margaretta, and Alexandrina. They, we may say, were the flowers of the family, having so lived that they had created none of those family feuds which had been so frequent between their father and their brothers. They were discreet, high-bred women, thinking, perhaps, a little too much of their own position in the world, and somewhat apt to put a wrong value on those advantages which they possessed, and on those which they did not possess. The Lady Amelia was already married, having made a substantial if not a brilliant match with Mr. Mortimer Gazebee, a flourishing solicitor, belonging to a firm which had for many years acted as agents to the De Courcy property. Mortimer Gazebee was now member of Parliament for Barchester, partly through the influence of his father-in-law. That this should be so was a matter of great disgust to the Honourable George who thought that the seat should have belonged to him. But as Mr. Gazebee had paid the very heavy expenses of the election out of his own pocket, and as George De Courcy certainly could not have paid them, the justice of his claim may be questionable. Lady Amelia Gazebee was now the happy mother of many babies, whom she was wont to carry with her on her visits to Courcy Castle, and had become an excellent partner to her husband. He would perhaps have liked it better if she had not spoken so frequently to him of her own high position as the daughter of an earl, or so frequently to others of her low position as the wife of an attorney. But, on the whole, they did very well together, and Mr. Gazebee had gotten from his marriage quite as much as he expected when he made it.

The Lady Rosina was very religious; and I do not know that she was conspicuous in any other way, unless it might be that she somewhat resembled her father in her temper. It was of the Lady Rosina that the servants were afraid, especially with reference to that so-called day of rest which, under her dominion, had become to many of them a day of restless torment. It had not always been so with the Lady Rosina; but her eyes had been opened by the wife of a great church dignitary in the neighbourhood, and she had undergone regeneration. How great may be the misery inflicted by an energetic, unmarried, healthy woman in that condition, – a woman with no husband, or children, or duties, to distract her from her work – I pray that my readers may never know.

The Lady Margaretta was her mother's favourite, and she was like her mother in all things, – except that her mother had been a beauty. The world called her proud, disdainful, and even insolent; but the world was not aware that in all that she did she was acting in accordance with a principle which had called for much self-abnegation. She had considered it her duty to be a De Courcy and an earl's daughter at all times; and consequently she had sacrificed to her idea of duty all popularity, adulation, and such admiration as would have been awarded to her as a well-dressed, tall, fashionable, and by no means stupid young woman. To be at all times in something higher than they who were manifestly below her in rank, – that was the effort that she was ever making. But she had been a good daughter, assisting her mother, as best she might, in all family troubles, and never repining at the cold, colourless, unlovely life which had been vouchsafed to her.

Alexandrina was the beauty of the family, and was, in truth, the youngest. But even she was not very young, and was beginning to make her friends uneasy lest she, too, should let the precious season of hay-harvest run by without due use of her summer's sun. She had, perhaps, counted too much on her beauty, which had been beauty according to law rather than beauty according to taste, and had looked, probably, for too bounteous a harvest. That her forehead, and nose, and cheeks, and chin were well formed, no man could deny. Her hair was soft and plentiful. Her teeth were good, and her eyes were long and oval. But the fault of her face was this, – that when you left her you could not remember it. After a first acquaintance you could meet her again and not know her. After many meetings you would fail to carry away with you any portrait of her features. But such as she had been at twenty, such was she now at thirty. Years had not robbed her face of its regularity, or ruffled the smoothness of her too even forehead. Rumour had declared that on more than one, or perhaps more than two occasions, Lady Alexandrina had been already induced to plight her troth in return for proffered love; but we all know that Rumour, when she takes to such topics, exaggerates the truth, and sets down much in malice. The lady was once engaged, the engagement lasting for two years, and the engagement had been broken off, owing to some money difficulties between the gentlemen of the families. Since that she had been somewhat querulous, and was suppose to be uneasy on that subject of her haymaking. Her glass and her maid assured her that her sun shone still as brightly as ever; but her spirit was becoming weary with waiting, and she dreaded lest she should become a terror to all, as was her sister Rosina, or an object of interest to none, as was Margaretta. It was from her especially that this message had been sent to our friend Crosbie; for, during the last spring in London, she and Crosbie had known each other well. Yes, my gentle readers; it is true, as your heart suggests to you.

Under such circumstances Mr. Crosbie should not have gone to Courcy Castle.

Such was the family circle of the De Courcys. Among their present guests I need not enumerate many. First and foremost in all respects was Lady Dumbello, of whose parentage and position a few words were said in the last chapter. She was a lady still very young, having as yet been little more than two years married. But in those two years her triumphs had been many; – so many, that in the great world her standing already equalled that of her celebrated mother-in-law, the Marchioness of Hartletop, who, for twenty years, had owned no great potentate than herself in the realms of fashion. But Lady Dumbello was every inch as great as she; and men said, and women also, that the daughter-in-law would soon be the greater.

'I'll be hanged if I can understand how she does it,' a certain noble peer had once said to Crosbie, standing at the door of Sebright's, during the latter days of the last season. 'She never says anything to any one. She won't speak ten words a whole night through.'

'I don't think she has an idea in her head,' said Crosbie.

'Let me tell you that she must be a very clever woman,' continued the noble peer. 'No fool could do as she does. Remember, she's only a parson's daughter; and as for beauty—'

'I don't admire her for one,' said Crosbie.

'I don't want to run away with her, if you mean that,' said the peer; 'but she is handsome, no doubt. I wonder whether Dumbello likes it.'

Dumbello did like it. It satisfied his ambition to be led about as the senior lacquey in his wife's train. He believed himself to be a great man because the world fought for his wife's presence; and considered himself to be distinguished even among the eldest sons of marquises, by the greatness reflected from the parson's daughter whom he had married. He had now been brought to Courcy Castle, and felt himself proud of his situation because Lady Dumbello had made considerable difficulty in according this week to the Countess De Courcy.

And Lady Julia De Guest was already there, the sister of the other old earl who lived in the next county. She had only arrived on the day before, but had been quick in spreading the news as to Crosbie's engagement. 'Engaged to one of the Dales is he?' said the countess, with a pretty little smile, which showed plainly that the matter was one of no interest to herself. 'Has she got any money?'

'Not a shilling, I should think,' said the Lady Julia.

'Pretty, I suppose?' suggested the countess.

'Why, yes; she is pretty – and a nice girl. I don't know whether her mother and uncle were very wise in encouraging Mr. Crosbie. I don't hear that he has anything special to recommend him, – in the way of money I mean.'

'I dare say it will come to nothing,' said the countess, who liked to hear of girls being engaged and then losing their promised husbands. She did not know that she liked it, but she did; and already had pleasure in anticipating poor Lily's discomfiture. But not the less was she angry with Crosbie, feeling that he was making his way into her house under false pretences.

And Alexandrina also was angry when Lady Julia repeated the same tidings in her hearing. 'I really don't think we care very much about it, Lady Julia,' said she, with a little toss of her head. 'That's three times we've been told of Miss Dale's good fortune.'

'The Dales are related to you, I think?' said Margaretta.

'Not at all,' said Lady Julia, bristling up. 'The lady whom Mr. Crosbie proposes to marry is in no way connected with us. Her cousin, who is the heir to the Allington property, is my nephew by his mother.' And then the subject was dropped.

Crosbie, on his arrival, was shown up into his room, told the hour of dinner, and left to his devices. He had been at the castle before, and knew the ways of the house. So he sat himself down to his table, and began a letter to Lily. But he had not proceeded far, not having as yet indeed made up his mind as to the form in which he would commence it but was sitting idly with the pen in his hand, thinking of Lily, and thinking also how such houses as this in which he now found himself would be soon closed against him, when there came a rap at his door, and before he could answer the Honourable John entered the room.

'Well, old fellow,' said the Honourable John. 'how are you?'

Crosbie had been intimate with John De Courcy, but never felt for him either friendship or liking. Crosbie did not like such men as John de Courcy; but nevertheless, they called each other old fellow, poked each other's ribs, and were very intimate.

'Heard you were here,' continued the Honourable John; 'so I thought I would come up and look after you. Going to be married, ain't you?'

'Not that I know of,' said Crosbie.

'Come, we know better than that. The women have been talking about it for the last three days. I had her name quite pat yesterday, but I've forgot it now. Hasn't got a tanner;* has she?' And the Honourable John had now seated himself upon the table.

'You seem to know a great deal more about it than I do.'

'It is that old woman from Guestwick who told us, then. The women will be at you at once, you'll find. If there's nothing in it, it's what I call a d— shame. Why should they always pull a fellow to pieces in that way? They were going to marry me the other day!'

'Were they indeed, though?'

'To Harriet Twistleton. You know Harriet Twistleton? An uncommon fine girl, you know. But I wasn't going to be caught like that. I'm very fond of Harriet, – in my way, you know; but they don't catch an old bird like me with chaff.'

'I condole with Miss Twistleton for what she has lost.'

'I don't know about condoling. But upon my word that getting married is a very slow thing. Have you seen George's wife?'

Crosbie declared that he had not as yet had that pleasure.

'She's here now, you know. I wouldn't have taken her, not if she had ten times thirty thousand pounds. By Jove, no. But he likes it well enough. Would you believe it now? – he cares for nothing on earth except money. You never saw such a fellow. But I'll tell you what, his nose will be out of joint yet, for Porlock is going to marry. I heard it from Colepepper, who almost lives with Porlock. As soon as Porlock heard that she was in the family way he immediately made up his mind to cut him out.'

'That was a great sign of brotherly love,' said Crosbie.

'I knew he'd do it,' said John; 'and so I told George before he got himself spliced. But he would go on. If he'd remained as he was for four or five years longer there would have been no danger; – for Porlock, you know, is leading the deuce of a life. I shouldn't wonder if he didn't reform now, and take to singing psalms or something of that sort.'

'There's no knowing what a man may come to in this world.'

'By George, no. But I'll tell you what, they'll find no change in me. If I marry it will not be with the intention of giving up life. I say, old fellow, have you got a cigar here?'

'What, to smoke up here, do you mean?'

'Yes; why not? we're ever so far from the women.'

'Not whilst I am occupier of this room. Besides, it's time to dress for dinner.'

'Is it? So it is, by George! But I mean to have a smoke first, I can tell you. So it's all a lie about your being engaged; eh?'

'As far as I know, it is,' said Crosbie. And then his friend left him.

What was he to do at once, now, this very day, as to his engagement? He had felt sure that the report of it would be carried to Courcy by Lady Julia De Guest, but he had not settled down upon any resolution as to what he would do in consequence. It had not occurred to him that he would immediately be charged with the offence, and called upon to plead guilty or not guilty. He had never for a moment meditated any plea of not guilty, but he was aware of an aversion on his part to declare himself as engaged to Lilian Dale. It seemed that by doing so he would cut himself off at once from all pleasure at such houses as Courcy Castle; and, as he argued to himself, why should he not enjoy the little remnant of his bachelor life? As to his denying his engagement to John De Courcy, – that was nothing. Any one would understand that he would be justified in concealing a fact concerning himself from such a one as he. The denial repeated from John's mouth would amount to nothing, – even among John's own sisters. But now it was necessary that Crosbie should make up his mind as to what he would say when questioned by the ladies of the house. If he were to deny the fact to them the denial would be very serious. And, indeed, was it possible that he should make such denial with Lady Julia opposite to him?

Make such a denial! And was it the fact that he could wish to do so, – that he should think of such falsehood, and even meditate on the perpetration of such cowardice? He had held that young girl to his heart on that very morning. He had sworn to her, and had also sworn to himself, that she should have no reason for distrusting him. He had acknowledged most solemnly to himself that, whether for good or for ill, he was bound to her; and could it be that he was already calculating as to the practicability of disowning her? In doing so must he not have told himself that he was a villain? But in truth he made no such calculation. His object was to banish the subject, if it were possible to do so; to think of some answer by which he might create a doubt. It did not occur to him to tell the countess boldly that there was no truth whatever in the report, and that Miss Dale was nothing to him. But might he not skilfully laugh off the subject, even in the presence of Lady Julia? Men who were engaged did so usually, and why should not he? It was generally thought that solicitude for the lady's feelings should prevent a man from talking openly of his own engagement. Then he remembered the easy freedom with which his position had been discussed throughout the whole neighbourhood of Allington, and felt for the first time that the Dale family had been almost indelicate in their want of reticence. 'I suppose it was done to tie me the faster,' he said to himself, as he pulled out the ends of his cravat. 'What a fool I was to come here, or indeed to go anywhere, after settling myself as I have done.' And then he went down into the drawing-room.

It was almost a relief to him when he found that he was

not charged with his sin at once. He himself had been so full of the subject that he had expected to be attacked at the moment of his entrance. He was, however, greeted without any allusion to the matter. The countess, in her own quiet way, shook hands with him as though she had seen him only the day before. The earl, who was seated in his arm-chair, asked some one, out loud, who the stranger was, and then, with two fingers put forth, muttered some apology for a welcome. But Crosbie was quite up to that kind of thing. 'How do, my lord?' he said, turning his face away to some one else as he spoke; and then he took no further notice of the master of the house. 'Not know him, indeed!' Crippled though he was by his matrimonial bond, Crosbie felt that, at any rate as yet, he was the earl's equal in social importance. After that, he found himself in the back part of the drawing-room, away from the elder people, standing with Lady Alexandrina, with Miss Gresham, a cousin of the De Courcys, and sundry other of the younger portion of the assembled community.

'So you have Lady Dumbello here?' said Crosbie.

'Oh, yes; the dear creature!' said Lady Margaretta. 'It was so good of her to come, you know.'

'She positively refused the Duchess of St. Bungay,' said Alexandrina. 'I hope you perceive how good we've been to you in getting you to meet her. People have actually asked to come.'

'I am grateful; but, in truth, my gratitude has more to do with Courcy Castle and its habitual inmates, than with Lady Dumbello. Is he here?'

'Oh, yes! he's in the room somewhere. There he is, standing up by Lady Clandidlem. He always stands in that way before dinner. In the evening he sits down much after the same fashion.'

Crosbie had seen him on first entering the room, and had seen every individual in it. He knew better than to omit the duty of that scrutinizing glance; but it sounded well in his line not to have observed Lord Dumbello.

'And her ladyship is not down?' said he.

'She is generally last,' said Lady Margaretta.

'And yet she has always three women to dress her,' said Alexandrina.

'But when finished, what a success it is!' said Crosbie.

'Indeed it is!' said Margaretta, with energy. Then the door was opened, and Lady Dumbello entered the room.

There was immediately a commotion among them all. Even the gouty old lord shuffled up out of his chair, and tried, with a grin, to look sweet and pleasant. The countess came forward, looking very sweet and pleasant, making little complimentary speeches, to which the viscountess answered simply by a gracious smile. Lady Clandidlem, though she was very fat and heavy, left the viscount, and got up to join the group. Baron Potsneuf, a diplomatic German of great celebrity, crossed his hands upon his breast and made a low bow. The Honourable George, who had stood silent for for the last quarter of an hour, suggested to her ladyship that she must have found the air rather cold; and the Ladies Margaretta and Alexandrina fluttered up with little complimentary speeches to their dear Lady Dumbello, hoping this and beseeching that, as though the 'Woman in White' before them had been the dearest friend of their infancy.

She was a woman in white, being dressed in white silk, with white lace over it, and with no other jewels upon her person than diamonds. Very beautifully she was dressed; doing infinite credit, no doubt, to those three artists who had, between them, succeeded in turning her out of hand. And her face, also, was beautiful, with a certain cold, inexpressive beauty. She walked up the room very slowly, smiling here and smiling there; but still with very faint smiles, and took the place which her hostess indicated to her. One word she said to the countess and two to the earl. Beyond that she did not open her lips. All the homage paid to her she received as though it were clearly her due. She was not in the least embarrassed, nor did she show herself to be in the slightest degree ashamed of her own silence. She did not look like a fool, nor was she even taken for a fool; but she contributed nothing to society but her cold, hard beauty, her gait, and her dress. We may say that she contributed enough, for society acknowledged itself to be deeply indebted to her.

The only person in the room who did not move at Lady Dumbello's entrance was her husband. But he remained unmoved from no want of enthusiasm. A spark of pleasure actually beamed in his eye as he saw the triumphant entrance of his wife. He felt that he had made a match that was becoming to him as a great nobleman, and that the world was acknowledging that he had done his duty. And yet Lady Dumbello had been simply the daughter of a country parson, of a clergyman who had reached no higher rank than that of an archdeacon. 'How wonderfully well that woman has educated her,' the countess said that evening in her dressing-room, to Margaretta. The woman alluded to was Mrs. Grantly, the wife of the parson and mother of Lady Dumbello.

The old earl was very cross because destiny and the table of precedence required him to take out Lady Clandidlem to dinner. He almost insulted her, as she kindly endeavoured to assist him in his infirm step rather than to lean upon him.

'Ugh!' he said, 'it's a bad arrangement that makes two old people like you and me be sent out together to help each other.'

'Speak for yourself,' said her ladyship, with a laugh. 'I, at any rate, can get about without any assistance,' – which, indeed, was true enough.

'It's well for you!' growled the earl, as he got himself into his seat.

And after that he endeavoured to solace his pain by a flirtation with Lady Dumbello on his left. The earl's smiles and the earl's teeth, when he whispered naughty little nothings to pretty young women, were phenomena at which men might marvel. Whatever those naughty nothings were on the present occasion, Lady Dumbello took them all with placidity, smiling graciously, but speaking hardly more than monosyllables.

Lady Alexandrina fell to Crosbie's lot, and he felt gratified that it was so. It might be necessary for him, as a married man, to give up such acquaintances as the De Courcys, but he should like, if possible, to maintain a friendship with Lady Alexandrina. What a friend Lady Alexandrina would be for Lily, if any such friendship were only possible! What an advantage would such an alliance confer upon that dear little girl; – for, after all, though the dear little girl's attractions were very great, he could not but admit to himself that she wanted a something, – a way of holding herself and of speaking, which some people call style. Lily might certainly learn a great deal from Lady Alexandrina; and it was this conviction, no doubt, which made him so sedulous in pleasing that lady on the present occasion.

And she, as it seemed, was well inclined to be pleased. She said no word to him during dinner about Lily; and yet she spoke about the Dales, and about Allington, showing that she knew in what quarters he had been staying, and then she alluded to their last parties in London, – those occasions on which, as Crosbie now remembered, the intercourse between them had almost been tender. It was manifest to him that at any rate she did not wish to quarrel with him. It was manifest, also, that she had some little hesitation in speaking to him about his engagement. He did not for a moment doubt that she was aware of it. And in this way matters went on between them till the ladies left the room.

'So you're going to be married, too,' said the Honourable George, by whose side Crosbie found himself seated when the ladies were gone. Crosbie was employing himself upon a walnut, and did not find it necessary to make any answer.

'It's the best thing a fellow can do,' continued George; 'that is, if he has been careful to look to the main chance, – if he hasn't been caught napping, you know. It doesn't do for a man to go hanging on by nothing till he finds himself an old man.'

'You've feathered your own nest, at any rate.'

'Yes; I've got something in the scramble, and I mean to keep it. Where will John be when the governor goes off the hooks? Porlock wouldn't give him a bit of bread and cheese and a glass of beer to save his life; – that is to say, not if he wanted it.'

'I'm told your elder brother is going to be married.'

'You've heard that from John. He's spreading that about everywhere to take a rise out of me. I don't believe a word of it. Porlock never was a marrying man; – and, what's more, from all I hear, I don't think he'll live long.'

In this way Crosbie escaped from his own difficulty; and when he rose from the dinner-table had not as yet been driven to confess anything to his own discredit.

But the evening was not yet over. When he returned to the drawing-room he endeavoured to avoid any conversation with the countess herself, believing that the attack would more probably come from her than from her daughter. He, therefore, got into conversation first with one and then wtih another of the girls, till at last he found himself again alone with Alexandrina.

'Mr. Crosbie,' she said, in a low voice, as they were standing together over one of the distant tables, with their backs to the rest of the company, 'I want you to tell me something about Miss Lilian Dale.'

'About Miss Lilian Dale!' he said, repeating her words.

'Is she very pretty?'

'Yes; she certainly is pretty.'

'And very nice, and attractive, and clever, – and all that is delightful? Is she perfect?'

'She is very attractive,' said he; 'but I don't think she is perfect.'

'And what are her faults?'

'That question is hardly fair, is it? Supose any one were to ask me what were your faults, do you think I should answer the question?'

'I am quite sure you would, and make a very long list of them, too. But as to Miss Dale, you ought to think her perfect. If a gentleman were engaged to me, I should expect him to swear before all the world that I was the very pink of perfection.'

'But supposing the gentleman were not engaged to you?'

'That would be a different thing.'

'I am not engaged to you,' said Crosbie. 'Such happiness and such honour are, I fear, very far beyond my reach. But, nevertheless, I am prepared to testify as to your perfection anywhere.'

'And what would Miss Dale say?'

'Allow me to assure you that such opinions as I may choose to express of my friends will be my own opinions, and not depend on those of any one else.'

'And you think, then, that you are not bound to be enslaved as yet? How many more months of such freedom are you to enjoy?'

Crosbie remained silent for a minute before he answered, and then he spoke in a serious voice. 'Lady Alexandrina,' said he, 'I would beg from you a great favour.'

'What is the favour, Mr. Crosbie?'

'I am quite in earnest. Will you be good enough, kind enough, enough my friend, not to connect my name again with that of Miss Dale while I am here?'

'Has there been a quarrel?'

'No; there has been no quarrel. I cannot explain to you now why I make this request; but to you I will explain it before I go.'

'Explain it to me!'

'I have regarded you as more than an acquaintance, – as a friend. In days now past there were moments when I was almost rash enough to hope that I might have said even more than that. I confess that I had no warrant for such hopes, but I believe that I may still look on you as a friend?'

'Oh, yes, certainly,' said Alexandrina, in a very low voice, and with a certain amount of tenderness in her tone. 'I have always regarded you as a friend.'

'And therefore I venture to make the request. The subject is not one on which I can speak openly, without regret, at the present moment. But to you, at least, I promise that I will explain it all before I leave Courcy.'

He at any rate succeeded in mystifying Lady Alexandrina. 'I don't believe he is engaged a bit,' she said to Lady Amelia Gazebee that night.

'Nonsense, my dear. Lady Julia wouldn't speak of it in that certain way if she didn't know. Of course he doesn't wish to have it talked about.'

'If ever he has been engaged to her, he has broken it off again,' said Lady Alexandrina.

'I dare say he will, my dear, if you give him encouragement,' said the married sister, with great sisterly good-nature.

CHAPTER XVIII

LILY DALE'S FIRST LOVE-LETTER

CROSBIE was rather proud of himself when he went to bed. He had succeeded in baffling the charge made against him, without saying anything as to which his conscience need condemn him. So, at least, he then told himself. The impression left by what he had said would be that there had been some question of an engagement between him and Lilian Dale, but that nothing at this moment was absolutely fixed. But in the morning his conscience was not quite so clear. What would Lily think and say if she knew it all? Could he dare to tell to her, or to tell any one the real state of his mind?

As he lay in bed, knowing that an hour remained to him before he need encounter the perils of his tub, he felt that he hated Courcy Castle and its inmates. Who was there, among them all, that was comparable to Mrs. Dale and her daughters? He detested both George and John. He loathed the earl. As to the countess herself, he was perfectly indifferent, regarding her as a woman whom it was well to know, but as one only to be known as the mistress of Courcy Castle and a house in London. As to the daughters, he had ridiculed them all from time to time – even Alexandrina, whom he now professed to love. Perhaps in some sort of way he had a weak fondness for her; – but it was a fondness that had never touched his heart. He could measure the whole thing at its worth, – Courcy Castle with its privileges, Lady Dumbello, Lady Clandidlem, and the whole of it. He knew that he had been happier on that lawn at Allington, and more contented with himself, than ever he had been even under Lady Hartletop's splendid roof in Shropshire. Lady Dumbello was satisfied with these things, even in the inmost recesses of her soul; but he was not a male Lady Dumbello. He knew that there was something better, and that that something was within his reach.

But, nevertheless, the air of Courcy was too much for him. In arguing the matter with himself he regarded himself as one infected with a leprosy from which there could be no recovery, and who should, therefore, make his whole life suitable to the circumstances of that leprosy. It was of no use for him to tell himself that the Small House at Allington was better than Courcy Castle. Satan knew that heaven was better than hell; but he found himself to be fitter for the latter place. Crosbie ridiculed Lady Dumbello, even there among her friends, with all the cutting words that his wit could find; but, nevertheless, the privilege of staying in the same house with her was dear to him. It was the line of life into which he had fallen, and he confessed inwardly that the struggle to extricate himself would be too much for him. All that had troubled him while he was yet at Allington, but it overwhelmed him almost with dismay beneath the hangings of Courcy Castle.

Had he not better run from the place at once? He had almost acknowledged to himself that he repented his engagement with Lilian Dale, but he still was resolved that he would fulfil it. He was bound in honour to marry 'that little girl', and he looked sternly up at the drapery over his head, as he assured himself that he was a man of honour. Yes; he would sacrifice himself. As he had been induced to pledge his word, he would not go back from it. He was too much of a man for that!

But had he not been wrong to refuse the result of Lily's wisdom when she told him in the field that it would be better for them to part? He did not tell himself that he had refused her offer merely because he had not the courage to accept it on the spur of the moment. No. 'He had been too good to the poor girl to take her at her word.' It was thus he argued on the matter within his own breast. He had been too true to her; and now the effect would be that they would both be unhappy for life! He could not live in content with a family upon a small income. He was well aware of that.

No one could be harder upon him in that matter than was he himself. But it was too late now to remedy the ill effects of an early education.

It was thus that he debated the matter as he lay in bed, – contradicting one argument by another over and over again; but still in all of them teaching himself to think that this engagement of his was a misfortune. Poor Lily! Her last words to him had conveyed an assurance that she would never distrust him. And she also, as she lay awake in her bed on this the first morning of his absence, thought much of their mutual vows. How true she would be to them! How she would be his wife with all her heart and spirit! It was not only that she would love him; – but in her love she would serve him to her utmost; serve him as regarded this world, and if possible as regarded the next.

'Bell,' she said, 'I wish you were going to be married too.'

'Thank'ye, dear,' said Bell. 'Perhaps I shall some day.'

'Ah; but I'm not joking. It seems such a serious thing. And I can't expect you to talk to me about it now as you would if you were in the same position yourself. Do you think I shall make him happy?'

'Yes, I do, certainly.'

'Happier than he would be with any one else that he might meet? I dare not think that. I think I could give him up to-morrow, if I could see any one that would suit him better.' What would Lily have said had she been made acquainted with all the fascinations of Lady Alexandrina De Courcy?

The countess was very civil to him, saying nothing about his engagement, but still talking to him a good deal about his sojourn at Allington. Crosbie was a pleasant man for ladies in a large house. Though a sportsman, he was not so keen a sportsman as to be always out with the gamekeepers. Though a politician, he did not sacrifice his mornings to the perusal of blue-books* or the preparation of party tactics. Though a reading man, he did not devote himself to study. Though a horseman, he was not often to be found in the stables. He could supply conversation when it was wanted, and could take himself out of the way when his presence among the women was not needed. Between breakfast and lunch on the day following his arrival he talked a good deal to the countess, and made himself very agreeable. She continued to ridicule him gently for his prolonged stay among so primitive and rural a tribe of people as the Dales, and he bore her little sarcasm with the utmost good-humour.

'Six weeks at Allington without a move! Why, Mr. Crosbie, you must have felt yourself to be growing there.'

'So I did – like an ancient tree. Indeed, I was so rooted that I could hardly get away.'

'Was the house full of people all the time?'

'There was nobody there but Bernard Dale, Lady Julia's nephew.'

'Quite a case of Damon and Pythias*. Fancy your going down to the shades of Allington to enjoy the uninterrupted pleasures of friendship for six weeks.'

'Friendship and the partridges.'

'There was nothing else, then?'

'Indeed there was. There was a widow with two very nice daughters, living, not exactly in the same house, but on the same grounds.'

'Oh, indeed. That makes such a difference; doesn't it? You are not a man to bear much privation on the score of partridges, nor a great deal, I imagine, for friendship. But when you talk of pretty girls—'

'It makes a difference, doesn't it?'

'A very great difference. I think I have heard of that Mrs. Dale before. And so her girls are nice?'

'Very nice indeed.'

'Play croquet, I suppose, and eat syllabub* on the lawn? But, really, didn't you get very tired of it?'

'O dear, no. I was happy as the day was long.'

'Going about with a crook, I suppose?'

'Not exactly a live crook; but doing all that kind of thing. I learned a great deal about pigs.'

'Under the guidance of Miss Dale?'

'Yes; under the guidance of Miss Dale.'

'I'm sure one is very much obliged to you for tearing yourself away from such charms, and coming to such unromantic people as we are. But I fancy men always do that sort of thing once or twice in their lives, – and then they talk of their souvenirs. I suppose it won't go beyond a souvenir with you.'

This was a direct question, but still admitted of a fencing answer. 'It has, at any rate, given me one,' said he, 'which will last me my life!'

The countess was quite contented. That Lady Julia's statement was altogether true she had never for a moment doubted. That Crosbie should become engaged to a young lady in the country, whereas he had shown signs of being in love with her daughter in London, was not at all wonderful. Nor, in her eyes, did such practice amount to any great sin. Men did so daily, and girls were prepared for their so doing. A man in her eyes was not to be regarded as safe from attack because he was engaged. Let the young lady who took upon herself to own him have an eye to that. When she looked back on the past careers of her own flock, she had to reckon more than one such disappointment for her own daughters. Others besides Alexandrina had been so treated. Lady De Courcy had had her grand hopes respecting her girls, and after them moderate hopes, and again after them bitter disappointments. Only one had been married, and she was married to an attorney. It was not to be supposed that she would have any very high-toned feelings as to Lily's rights in this matter.

Such a man as Crosbie was certainly no great match for an earl's daughter. Such a marriage, indeed, would, one may say, be but a poor triumph. When the countess, during the last season in town, had observed how matters were going with Alexandrina, she had cautioned her child, taking her to task for her imprudence. But the child had been at this work for fourteen years, and was weary of it. Her sisters had been at the work longer, and had almost given it up in despair. Alexandrina did not tell her parent that her heart was now beyond her control, and that she had devoted herself to Crosbie for ever; but she pouted, saying that she knew very well what she was about, scolding her mother in return, and making Lady De Courcy perceive that the struggle was becoming very weary. And then there were other considerations. Mr. Crosbie had not much certainly in his own possession, but he was a man out of whom something might be made by family influence and his own standing. He was not a hopeless, ponderous man, whom no leaven could raise. He was one of whose position in society the countess and her daughters need not be ashamed. Lady De Courcy had given no expressed consent to the arrangement, but it had come to be understood between her and her daughter that the scheme was to be entertained as admissible.

Then came these tidings of the little girl down at Allington. She felt no anger against Crosbie. To be angry on such a subject would be futile, foolish, and almost indecorous. It was a part of the game which was as natural to her as fielding is to a cricketer. One cannot have it all winnings at any game. Whether Crosbie should eventually become her own son-in-law or not it came to her naturally, as a part of her duty in life, to bowl down the stumps of that young lady at Allington. If Miss Dale knew the game well and could protect her own wicket, let her do so.

She had no doubt as to Crosbie's engagement with Lilian Dale, but she had as little as to his being ashamed of that engagement. Had he really cared for Miss Dale he would not have left her to come to Courcy Castle. Had he been really resolved to marry her, he would not have warded all questions respecting his engagement with fictitious answers. He had amused himself with Lily Dale, and it was to be hoped that the young lady had not thought very seriously about it. That was the most charitable light in which Lady De Courcy was disposed to regard the question.

It behoved Crosbie to write to Lily Dale before dinner. He had promised to do so immediately on his arrival, and he was aware that he would be regarded as being already one day beyond his promise. Lily had told him that she would live upon his letters, and it was absolutely necessary that he should furnish her with her first meal. So he betook himself to his room in sufficient time before dinner, and got out his pen, ink, and paper.

He got out his pen, ink, and paper, and then he found that his difficulties were beginning. I beg that it may be understood that Crosbie was not altogether a villain. He could not sit down and write a letter as coming from his heart, of which as he wrote it he knew the words to be false. He was an ungenerous, worldly, inconstant man, very prone to think well of himself, and to give himself credit for virtues which he did not possess; but he could not be false with premeditated cruelty to a woman he had sworn to love. He could not write an affectionate, warm-hearted letter to Lily, without bringing himself, at any rate for the time, to feel towards her in an affectionate, warm-hearted way. Therefore he now sat himself to work, while his pen yet remained dry in his hand, to remodel his thoughts, which had been turned against Lily and Allington by the craft of Lady De Courcy. It takes some time before a man can do this. He has to struggle with himself in a very uncomfortable way, making efforts which are often unsuccessful. It is sometimes easier to lift a couple of hundredweights than to raise a few thoughts in one's mind which at other moments will come galloping in without a whistle.

He had just written the date of his letter when a little tap came at his door, and it was opened.

'I say, Crosbie,' said the Honourable John, 'didn't you say something yesterday about a cigar before dinner?'

'Not a word,' said Crosbie, in rather an angry tone.

'Then it must have been me,' said John. 'But bring your case with you, and come down to the harness-room, if you won't smoke here. I've had a regular little snuggery fitted up there; and we can go in and see the fellows making up the horses.'

Crosbie wished the Honourable John at the mischief.

'I have letters to write,' said he. 'Besides, I never smoke before dinner.'

'That's nonsense. I've smoked hundreds of cigars with you before dinner. Are you going to turn curmudgeon, too, like George and the rest of them? I don't know what's coming to the world! I suppose the fact is, that little girl at Allington won't let you smoke.'

'The little girl at Allington—' began Crosbie; and then he reflected that it would not be well for him to say anything to his present companion about that little girl. 'I'll tell you what it is,' said he. 'I really have got letters to write which must go by this post. There's my cigar-case on the dressing-table.'

'I hope it will be long before I'm brought to such a state,' said John, taking up the cigars in his hand.

'Let me have the case back,' said Crosbie.

'A present from the little girl, I suppose?' said John. 'All right, old fellow! you shall have it.'

'There would be a nice brother-in-law for a man,' said Crosbie to himself, as the door closed behind the retreating scion of the De Courcy family. And then, again, he took up his pen. The letter must be written, and therefore he threw himself upon the table, resolved that the words should come and the paper be filled.

'COURCY CASTLE, October, 186–

'Dearest Lily, – This is the first letter I ever wrote to you, except those little notes when I sent you my compliments discreetly, – and it sounds so odd. You will think that this does not come as soon as it should; but the truth is that after all I only got in here just before dinner yesterday. I stayed ever so long in Barchester, and came across such a queer character. For you must know I went to church, and afterwards fraternized with the clergyman who did the service; such a gentle old soul, – and, singularly enough, he is the grandfather of Lady Dumbello, who is staying here. I wonder what you'd think of Lady Dumbello, or how you'd like to be shut up in the same house with her for a week?

'But with reference to my staying at Barchester, I must tell you the truth now, though I was a gross impostor the day that I went away. I wanted to avoid a parting on that last morning, and therefore I started much sooner than I need have done. I know you will be very angry with me; but open confession is good for the soul. You frustrated all my little plan by your early rising; and as I saw you standing on the terrace, looking after us as we went, I acknowledged that you had been right, and that I was wrong. When the time came, I was very glad to have you with me at the last moment.

'My own dearest Lily, you cannot think how different this place is from the two houses at Allington, or how much I prefer the sort of life which belongs to the latter. I know that I have been what the world calls worldly, but you will have to cure me of that. I have questioned myself very much since I left you, and I do not think that I am quite beyond the reach of a cure. At any rate, I will put myself trustingly into the doctor's hands. I know it is hard for a man to change his habits; but I can with truth say this for myself, that I was happy at Allington, enjoying every hour of the day, and that here I am ennuyé* by everybody and nearly by everything. One of the girls of the house I do like; but as to other people, I can hardly find a companion among them, let alone a friend. However, it would not have done for me to have broken away from all such alliances too suddenly.

'When I get up to London – and now I really am anxious to get there – I can write to you more at my ease, and more freely than I do here. I know that I am hardly myself among these people, – or rather, I am hardly myself as you know me, and as I hope you always will know me. But, nevertheless, I am not so overcome by the miasma but what I can tell you how truly I love you. Even though my spirit should be here, which it is not, my heart would be on the Allington lawns. That dear lawn and that dear bridge!

'Give my kind love to Bell and your mother. I feel already that I might almost say my mother. And Lily, my darling, write to me at once. I expect your letters to me to be longer, and better, and brighter than mine to you. But I will endeavour to make mine nicer when I get back to town.

'God bless you. Yours, with all my heart,

'A. C.'

As he had waxed warm with his writing he had forced himself to be affectionate, and, as he flattered himself, frank and candid. Nevertheless, he was partly conscious that he was preparing for himself a mode of escape in those allusions of his to his own worldliness; if escape should ultimately be necessary. 'I have tried,' he would say; 'I have struggled honestly, with my best efforts for success; but I am not good enough for such success.' I do not intend to say that he wrote with a premeditated intention of thus using his words; but as he wrote them he could not keep himself from reflecting that they might be used in that way.

He read his letter over, felt satisfied with it, and resolved that he might now free his mind from that consideration for the next forty-eight hours. Whatever might be his sins he had done his duty by Lily! And with this comfortable reflection he deposited his letter in the Courcy Castle letter-box.

CHAPTER XIX

THE SQUIRE MAKES A VISIT TO THE SMALL HOUSE

MRS. DALE acknowledged to herself that she had not much ground for hoping that she should ever find in Crosbie's house much personal happiness for her future life. She did not dislike Mr. Crosbie, nor in any great degree mistrust him; but she had seen enough of him to make her certain that Lily's future home in London could not be a home for her. He was worldly, or, at least, a man of the world. He would be anxious to make the most of his income, and his life would be one long struggle, not perhaps for money, but for those things which money only can give. There are men to whom eight hundred a year is great wealth, and houses to which it brings all the comforts that life requires. But Crosbie was not such a man, nor would his house be such a house. Mrs. Dale hoped that Lily would be happy with him, and satisfied with his modes of life, and she strove to believe that such would be the case; but as regarded herself she was forced to confess that in such a marriage her child would be much divided from her. That pleasant abode to which she had long looked forward that she might have a welcome there in coming years should be among fields and trees, not in some narrow London street. Lily must now become a city lady; but Bell would still be left to her, and it might still be hoped that Bell would find for herself some country home.

Since the day on which Lily had first told her mother of her engagement, Mrs. Dale had found herself talking much more fully and more frequently with Bell than with her younger daughter. As long as Crosbie was at Allington this was natural enough. He and Lily were of course together, while Bell remained with her mother. But the same state of things continued even after Crosbie was gone. It was not that there was any coolness or want of affection between the mother and daughter, but that Lily's heart was full of her lover, and that Mrs. Dale, though she had given her cordial consent to the marriage, felt that she had but few points of sympathy with her future son-in-law. She had never said, even to herself, that she disliked him; nay, she had sometimes declared to herself that she was fond of him. But, in truth, he was not a man after her own heart. He was not one who could ever be to her as her own son and her own child.

But she and Bell would pass hours together talking of Lily's prospects. 'It seems so strange to me,' said Mrs. Dale, 'that she of all girls should have been fancied by such a man as Mr. Crosbie, or that she should have liked him. I cannot imagine Lily living in London.'

'If he is good and affectionate to her she will be happy wherever he is,' said Bell.

'I hope so; – I'm sure I hope so. But it seems as though she will be so far separated from us. It is not the distance, but the manner of life which makes the separation. I hope you'll never be taken so far from me.'

'I don't think I shall allow myself to be taken up to London,' said Bell, laughing. 'But one can never tell. If I do you must follow us, mamma.'

'I do not want another Mr. Crosbie for you, dear.'

'But perhaps I may want one for myself. You need not tremble quite yet, however. Apollos do not come this road every day.'

'Poor Lily! Do you remember when she first called him Apollo? I do, well. I remember his coming here the day after Bernard brought him down, and how you were playing on the lawn, while I was in the other garden. I little thought then what it would come to.'

'But, mamma, you don't regret it?'

'Not if it's to make her happy. If she can be happy with him, of course I shall not regret it; not though he were to take her to the world's end away from us. What else have I to look for but that she and you should both be happy?'

'Men in London are happy with their wives as well as men in the country.'

'Oh, yes; of all women I should be the first to acknowledge that.'

'And as to Adolphus himself, I do not know why we should distrust him.'

'No, my dear; there is no reason. If I did distrust him, I should not have given so ready an assent to the marriage. But, nevertheless—'

'The truth is, you don't like him, mamma.'

'Not so cordially as I hope I may like any man whom you may choose for your husband.'

And Lily, though she said nothing on the subject to Mrs. Dale, felt that her mother was in some degree estranged from her. Crosbie's name was frequently mentioned between them, but in the tone of Mrs. Dale's voice, and in her manner when she spoke of him, there was lacking that enthusiasm and heartiness which real sympathy would have produced. Lily did not analyse her own feelings, or closely make inquiry as to those of her mother, but she perceived that it was not all as she would have wished it to have been. 'I know mamma does not love him,' she said to Bell on the evening of the day on which she received Crosbie's first letter.

'Not as you do, Lily; but she does love him.'

'Not as I do! To say that is nonsense, Bell; of course she does not love him as I do. But the truth is she does not love him at all. Do you think I cannot see it?'

'I'm afraid that you see too much.'

'She never says a word against him; but if she really liked him she would sometimes say a word in his favour. I do not think she would ever mention his name unless you or

I spoke of him before her. If she did not approve of him, why did she not say so sooner?'

'That's hardly fair upon mamma,' said Bell, with some earnestness. 'She does not disapprove of him, and she never did. You know mamma well enough to be sure that she would not interfere with us in such a matter without very strong reason. As regards Mr. Crosbie, she gave her consent without a moment's hesitation.'

'Yes, she did.'

'How can you say, then, that she disapproves of him?'

'I didn't mean to find fault with mamma. Perhaps it will come all right.'

'It will come all right.' But Bell, though she made this very satisfactory promise, was as well aware as either of the others that the family would be divided when Crosbie should have married Lily and taken her off to London.

On the following morning Mrs. Dale and Bell were sitting together. Lily was above in her own room, either writing to her lover, or reading his letter, or thinking of him, or working for him. In some way she was employed on his behalf, and with this object she was alone. It was now the middle of October, and the fire was lit in Mrs. Dale's drawing-room. The window which opened upon the lawn was closed, the heavy curtains had been put back in their places, and it had been acknowledged as an unwelcome fact that the last of the summer was over. This was always a sorrow to Mrs. Dale; but it is one of those sorrows which hardly admit of open expression.

'Bell,' she said, looking up suddenly; 'there's your uncle at the window. Let him in.' For now, since the putting up of the curtains, the window had been bolted as well as closed. So Bell got up, and opened a passage for the squire's entrance. It was not often that he came down in this way, and when he did so it was generally for some purpose which had been expressed before.

'What! fires already?' said he. 'I never have fires at the other house in the morning till the first of November. I like to see a spark in the grate after dinner.'

'I like a fire when I'm cold,' said Mrs. Dale. But this was a subject on which the squire and his sister-in-law had differed before, and as Mr. Dale had some business in hand, he did not now choose to waste his energy in supporting his own views on the question of fires.

'Bell, my dear,' said he, 'I want to speak to your mother for a minute or two on a matter of business. You wouldn't mind leaving us for a little while, would you?' Whereupon Bell collected up her work and went upstairs to her sister. 'Uncle Christopher is below with mamma,' said she, 'talking about business. I suppose it is something to do with your marriage.' But Bell was wrong. The squire's visit had no reference to Lily's marriage.

Mrs. Dale did not move or speak a word when Bell was gone, though it was evident that the squire paused in order that she might ask some question of him. 'Mary,' said he, at last, 'I'll tell you what it is that I have come to say to you.' Whereupon she put the piece of needlework which was in her hands down upon the work-basket before her, and settled herself to listen to him.

'I wish to speak to you about Bell.'

'About Bell?' said Mrs. Dale, as though much surprised that he should have anything to say to her respecting her eldest daughter.

'Yes, about Bell. Here's Lily going to be married, and it will be well that Bell should be married too.'

'I don't see that at all,' said Mrs. Dale. 'I am by no means in a hurry to be rid of her.'

'No, I dare say not. But, of course, you only regard her welfare, and I can truly say that I do the same. There would be no necessity for hurry as to a marriage for her under ordinary circumstances, but there may be circumstances to make such a thing desirable, and I think that there are.' It was evident from the squire's tone and manner that he was very much in earnest; but it was also evident that he found some difficulty in opening out the budget with which he had prepared himself. He hesitated a little in his voice, and seemed to be almost nervous. Mrs. Dale, with some little spice of ill-nature, altogether abstained from assisting him. She was jealous of interference from him about her girls, and though she was of course bound to listen to him, she did so with a prejudice against and almost with a resolve to oppose anything that he might say. When he had finished his little speech about circumstances, the squire paused again; but Mrs. Dale still sat silent, with her eyes fixed upon his face.

'I love your children very dearly,' said he, 'though I believe you hardly give me credit for doing so.'

'I am sure you do,' said Mrs. Dale, 'and they are both well aware of it.'

'And I am very anxious that they should be comfortably established in life. I have no children of my own, and those of my two brothers are everything to me.'

Mrs. Dale had always considered it as a matter of course that Bernard should be the squire's heir, and had never felt that her daughters had any claim on that score. It was a well-understood thing in the family that the senior male Dale should have all the Dale property and all the Dale money. She fully recognized even the propriety of such an arrangement. But it seemed to her that the squire was almost guilty of hypocrisy in naming his nephew and his two nieces together, as though they were the joint heirs of his love. Bernard was his adopted son, and no one had begrudged to the uncle the right of making such adoption. Bernard was everything to him, and as being his heir was bound to obey him in many things. But her daughters were no more to him than any nieces might be to any uncle. He had nothing to do with their disposal in marriage; and the mother's spirit was already up in arms and prepared to do battle for her own independence, and for that of her children. 'If Bernard would marry well,' said she, 'I have no doubt it would be a comfort to you,' – meaning to imply thereby that the squire had no right to trouble himself about any other marriage.

'That's just it,' said the squire. 'It would be a great comfort to me. And if he and Bell could make up their minds together it would, I should think, be a great comfort to you also.'

'Bernard and Bell!' exclaimed Mrs. Dale. No idea of such

a union had ever yet come upon her, and now in her surprise she sat silent. She had always liked Bernard Dale, having felt for him more family affection than for any other of the Dale family beyond her own hearth. He had been very intimate in her house, having made himself almost as a brother to her girls. But she had never thought of him as a husband for either of them.

'Then Bell has not spoken to you about it,' said the squire.

'Never a word.'

'And you had never thought about it?'

'Certainly not.'

'I have thought about it a great deal. For some years I have always been thinking of it. I have set my heart upon it, and shall be very unhappy if it cannot be brought about. They are both very dear to me, – dearer than anybody else. If I could see them man and wife, I should not much care then how soon I left the old place to them.'

There was a purer touch of feeling in this than the squire had ever before shown in his sister-in-law's presence, and more heartiness than she had given him the credit of possessing. And she could not but acknowledge to herself that her own child was included in this unexpected warmth of love, and that she was bound to entertain some gratitude for such kindness.

'It is good of you to think of her,' said the mother; 'very good.'

'I think a great deal about her,' said the squire. 'But that does not much matter now. The fact is, that she has declined Bernard's offer.'

'Has Bernard offered to her?'

'So he tells me; and she has refused him. It may perhaps be natural that she should do so, never having taught herself to look at him in the light of a lover. I don't blame her at all. I am not angry with her.'

'Angry with her! No. You can hardly be angry with her for not being in love with her cousin.'

'I say that I am not angry with her. But I think she might undertake to consider the question. You would like such a match, would you not?'

Mrs. Dale did not at first make any answer, but began to revolve the thing in her mind, and to look at it in various points of view. There was a great deal in such an arrangement which at the first sight recommended it to her very strongly. All the local circumstances were in its favour. As regarded herself it would promise to her all that she had ever desired. It would give her a prospect of seeing very much of Lily; for if Bell were settled at the old family house, Crosbie would naturally be much with his friend. She liked Bernard also; and for a moment or two fancied, as she turned it all over in her mind, that, even yet, if such a marriage were to take place, there might grow up something like true regard between her and the old squire. How happy would be her old age in the Small House, if Bell with her children were living so close to her!

'Well?' said the squire, who was looking very intently into her face.

'I was thinking,' said Mrs. Dale. 'Do you say that she has already refused him?'

'I am afraid she has; but then you know—'

'It must of course be left for her to judge.'

'If you mean that she cannot be made to marry her cousin, of course we all know she can't.'

'I mean rather more than that.'

'What do you mean, then?'

'That the matter must be left altogether to her own decision; that no persuasion must be used by you or me. If he can persuade her, indeed—'

'Yes, exactly. He must persuade her. I quite agree with you that he should have liberty to plead his own cause. But look you here, Mary; – she has always been a very good child to you—'

'Indeed she has.'

'And a word from you would go a long way with her, – as it ought. If she knows that you would like her to marry her cousin, it will make her think it her duty—'

'Ah! but that is just what I cannot try to make her think.'

'Will you let me speak, Mary? You take me up* and scold me before the words are half out of my mouth. Of course I know that in these days a young lady is not to be compelled into marrying anybody; – not but that, as far as I can see, they did better than they do now when they had not quite so much of their own way.'

'I never would take upon myself to ask a child to marry any man.'

'But you may explain to her that it is her duty to give such a proposal much thought before it is absolutely refused. A girl either is in love or she is not. If she is, she is ready to jump down a man's throat; and that was the case with Lily.'

'She never thought of the man till he had proposed to her fully.'

'Well, never mind now. But if a girl is not in love, she thinks she is bound to swear and declare that she never will be so.'

'I don't think Bell ever declared anything of the kind.'

'Yes, she did. She told Bernard that she didn't love him and couldn't love him, – and, in fact, that she wouldn't think anything more about it. Now, Mary, that's what I call being headstrong and positive. I don't want to drive her, and I don't want you to drive her. But here is an arrangement which for her will be a very good one; you must admit that. We all know that she is on excellent terms with Bernard. It isn't as though they had been falling out and hating each other all their lives. She told him that she was very fond of him, and talked nonsense about being his sister, and all that.'

'I don't see that it was nonsense at all.'

'Yes, it was nonsense, – on such an occasion. If a man asks a girl to marry him, he doesn't want her to talk to him about being his sister. I think it is nonsense. If she would only consider about it properly she would soon learn to love him.'

'That lesson, if it be learned at all, must be learned without any tutor.'

'You won't do anything to help me then?'

'I will, at any rate, do nothing to mar you. And, to tell the truth, I must think over the matter fully before I can decide what I had better say to Bell about it. From her not speaking to me—'

'I think she ought to have told you.'

'No, Mr. Dale. Had she accepted him, of course she would have told me. Had she thought of doing so she might probably have consulted me. But if she made up her mind that she must reject him—'

'She oughtn't to have made up her mind.'

'But if she did, it seems natural to me that she should speak of it to no one. She might probably think that Bernard would be as well pleased that it should not be known.'

'Psha, – known! – of course it will be known. As you want time to consider of it, I will say nothing more now. If she were my daughter, I should have no hesitation in telling her what I thought best for her welfare.'

'I have none; though I may have some in making up my mind as to what is best for her welfare. But, Mr. Dale, you may be sure of this; I will speak to her very earnestly of your kindness and love for her. And I wish you would believe that I feel your regard for her very strongly.'

In answer to this he merely shook his head, and hummed and hawed. 'You would be glad to see them married, as regards yourself?' he asked.

'Certainly I would,' said Mrs. Dale. 'I have always liked Bernard, and I believe my girl would be safe with him. But then, you see, it's a question on which my own likings or dislikings should not have any bearing.'

And so they parted, the squire making his way back again through the drawing-room window. He was not above half pleased with his interview; but then he was a man for whom half-pleasure almost sufficed. He rarely indulged any expectation that people would make themselves agreeable to him. Mrs. Dale, since she had come to the Small House, had never been a source of satisfaction to him, but he did not on that account regret that he had brought her there. He was a constant man; urgent in carrying out his own plans, but not sanguine in doing so, and by no means apt to expect that all things would go smooth with him. He had made up his mind that his nephew and his niece should be married, and should he ultimately fail in this, such failure would probably embitter his future life; – but it was not in the nature of the man to be angry in the meantime, or to fume and scold because he met with opposition. He had told Mrs. Dale that he loved Bell dearly. So he did, though he seldom spoke to her with much show of special regard, and never was soft and tender with her. But, on the other hand, he did not now love her the less because she opposed his wishes. He was a constant, undemonstrative man, given rather to brooding than to thinking; harder in his words than in his thoughts, with more of heart than others believed, or than he himself knew; but, above all, he was a man who having once desired a thing would desire it always.

Mrs. Dale, when she was left alone, began to turn over the question in her mind in a much fuller manner than the squire's presence had as yet made possible for her. Would not such a marriage as this be for them all the happiest domestic arrangement which circumstances could afford? Her daughter would have no fortune, but here would be prepared for her all the comforts which fortune can give. She would be received into her uncle's house, not as some penniless, portionless bride whom Bernard might have married and brought home, but as the wife whom of all others Bernard's friends had thought desirable for him. And then, as regarded Mrs. Dale herself, there would be nothing in such a marriage which would not be delightful to her. It would give a realization to all her dreams of future happiness.

But, as she said to herself over and over again, all that must go for nothing. It must be for Bell, and for her only, to answer Bernard's question. In her mind there was something sacred in that idea of love. She would regard her daughter almost as a castaway if she were to marry any man without absolutely loving him, – loving him as Lily loved her lover, with all her heart and all her strength.

With such a convicition as this strong upon her, she felt that she could not say much to Bell that would be of any service.

CHAPTER XX

DR. CROFTS

If there was anything in the world as to which Isabella Dale was quite certain, it was this – that she was not in love with Dr. Crofts. As to being in love wih her cousin Bernard, she had never had occasion to ask herself any question on that head. She liked him very well, but she had never thought of marrying him; and now, when he made his proposal, she could not bring herself to think of it. But as regards Dr. Crofts, she had thought of it, and had made up her mind; – in the manner above described.

It may be said that she could not have been justified in discussing the matter even within her own bosom, unless authorized to do so by Dr. Crofts himself. Let it then be considered that Dr. Crofts had given her some such authority. This may be done in more ways than one; and Miss Dale could not have found herself asking herself questions about him, unless there had been fitting occasion for her to do so.

The profession of a medical man in a small provincial town is not often one which gives to its owner in early life a large income. Perhaps in no career has a man to work harder for what he earns, or to do more work without earning anything. It has sometimes seemed to me as though the young doctors and the old doctors had agreed to divide between them the different results of their profession, – the young doctors doing all the work and the old doctors taking all the money. If this be so it may account for that appearance of premature gravity which is borne by so many of the medical profession. Under such an arrangement a man may be excused for a desire to put away childish things very early in life.

Dr. Crofts had now been practising in Guestwick nearly seven years, having settled himself in that town when he was twenty-three years old, and being at this period about thirty. During those seven years his skill and industry had been so fully admitted that he had succeeded in obtaining the medical care of all the paupers in the union, for which work he was paid at the rate of one hundred pounds a year. He was also assistant-surgeon at a small hospital which was maintained in that town, and held two or three other similar public positions, all of which attested his respectability and general proficiency. They, moreover, thoroughly saved him from any of the dangers of idleness; but, unfortunately, they did not enable him to regard himself as a successful professional man. Whereas old Dr. Gruffen, of whom but few people spoke well, had made a fortune in Guestwick, and even still drew from the ailments of the town a considerable and hardly yet decreasing income. Now this was hard upon Dr. Crofts – unless there was existing some such well-understood arrangement as that above named.

He had been known to the family of the Dales long previous to his settlement at Guestwick, and had been very intimate with them from that time to the present day. Of all the men, young or old, whom Mrs. Dale counted among her intimate friends, he was the one whom she most trusted and admired. And he was a man to be trusted by those who knew him well. He was not bright and always ready, as was Crosbie, nor had he all the practical worldly good sense of Bernard Dale. In mental power I doubt whether he was superior to John Eames; – to John Eames, such as he might become when the period of his hobbledehoyhood should have altogether passed away. But Crofts, compared with the other three, as they all were at present, was a man more to be trusted than any of them. And there was, moreover, about him an occasional dash of humour, without which Mrs. Dale would hardly have regarded him with that thorough liking which she had for him. But it was a quiet humour, apt to show itself when he had but one friend with him, rather than in general society. Crosbie, on the other hand, would be much more bright among a dozen, than he could with a single companion. Bernard Dale was never bright; and as for Johnny Eames—; but in this matter of brightness, Johnny Eames had not yet shown to the world what his character might be.

It was now two years since Crofts had been called upon for medical advice on behalf of his friend Mrs. Dale. She had then been ill for a long period – some two or three months, and Dr. Crofts had been frequent in his visits at Allington. At that time he became very intimate with Mrs. Dale's daughters, and especially so with the eldest. Young unmarried doctors ought perhaps to be excluded from houses in which there are young ladies. I know, at any rate, that many sage matrons hold very strongly to that opinion, thinking, no doubt, that doctors ought to get themselves married before they venture to begin working for a living. Mrs. Dale, perhaps, regarded her own girls as still merely children, for Bell, the elder, was then hardly eighteen; or perhaps she held imprudent and heterodox opinions on this subject; or it may be that she selfishly preferred Dr. Crofts, with all the danger to her children, to Dr. Gruffen, with all the danger to herself. But the result was that the young doctor one day informed himself, as he was riding back to Guestwick, that much of his happiness in this world would depend on his being able to marry Mrs. Dale's eldest daughter. At that time his total income amounted to little more than two hundred a year, and he had resolved within his own mind that Dr. Gruffen was esteemed as much the better doctor by the general public opinion of Guestwick, and that Dr. Gruffen's sandy-haired assistant would even have a better chance of success in the town than himself, should it ever come to pass that the doctor was esteemed too old for personal practice. Crofts had no fortune of his own, and he was aware that Miss Dale had none. Then, under those circumstances, what was he to do?

It is not necessary that we should inquire at any great length into those love passages of the doctor's life which took place three years before the commencement of this narrative. He made no declaration to Bell; but Bell, young as she was, understood well that he would fain have done so, had not his courage failed him, or rather had not his prudence prevented him. To Mrs. Dale he did speak, not openly avowing his love even to her, but hinting at it, and then talking to her of his unsatisfied hopes and professional disappointments. 'It is not that I complain of being poor as I am,' said he; 'or at any rate, not so poor that my poverty must be any source of discomfort to me; but I could hardly marry with such an income as I have at present.'

'But it will increase, will it not?' said Mrs. Dale.

'It may some day, when I am becoming an old man,' he said. 'But of what use will it be to me then?'

Mrs. Dale could not tell him that, as far as her voice in the matter went, he was welcome to woo her daughter and marry her, poor as he was, and doubly poor as they would both be together on such a pittance. He had not even mentioned Bell's name, and had he done so she could only have bade him wait and hope. After that he said nothing further to her upon the subject. To Bell he spoke no word of overt love; but on an autumn day, when Mrs. Dale was already convalescent, and the repetition of his professional visits had become unnecessary, he got her to walk with him through the half hidden shrubbery paths, and then told her things which he should never have told her, if he really wished to bind her heart to his. He repeated that story of his income, and explained to her that his poverty was only grievous to him in that it prevented him from thinking of marriage. 'I suppose it must,' said Bell. 'I should think it wrong to ask any lady to share such an income as mine,' said he. Whereupon Bell had suggested to him that some ladies had incomes of their own, and that he might in that way get over the difficulty. 'I should be afraid of myself in marrying a girl with money,' said he; 'besides, that is altogether out of the question now.' Of course Bell did not ask him why it was out of the question, and for a time they went on walking in silence. 'It is a hard thing to do,' he then said, – not looking at her, but looking at the gravel on which he stood. 'It is a hard thing to do, but I will determine to think of it no further. I believe a man may be as happy single as he may married, – almost.' 'Perhaps more so,' said Bell. Then the doctor left her, and Bell, as I have said before, made up her mind with great firmness that she was not in love with him. I may certainly say that there was nothing in the world as to which she was so certain as she was of this.

And now, in these days, Dr. Crofts did not come over to Allington very often. Had any of the family in the Small

House been ill, he would have been there of course. The squire himself employed the apothecary in the village, or if higher aid was needed, would send for Dr. Gruffen. On the occasion of Mrs. Dale's party, Crofts was there, having been specially invited; but Mrs. Dale's special invitations to her friends were very few, and the doctor was well aware that he must himself make occasion for going there if he desired to see the inmates of the house. But he very rarely made such occasion, perhaps feeling that he was more in his element at the workhouse and the hospital.

Just at this time, however, he made one very great and unexpected step towards success in his profession. He was greatly surprised one morning by being summoned to the Manor House to attend upon Lord De Guest. The family at the Manor had employed Dr. Gruffen for the last thirty years, and Crofts, when he received the earl's message, could hardly believe the words. 'The earl ain't very bad,' said the servant, 'but he would be glad to see you if possible a little before dinner.'

'You're sure he wants to see me?' said Crofts.

'Oh, yes; I'm sure enough of that, sir.'

'It wasn't Dr. Gruffen?'

'No, sir; it wasn't Dr. Gruffen. I believe his lordship's had about enough of Dr. Gruffen. The doctor took to chaffing his lordship one day.'

'Chaffed his lordship; – his hands and feet, and that sort of thing?' suggested the doctor.

'Hands and feet!' said the man. 'Lord bless you, sir, he poked his fun at him, just as though he was nobody. I didn't hear, but Mrs. Connor says that my lord's back was up terribly high.' And so Dr. Crofts got on his horse and rode up to Guestwick Manor.

The earl was alone, Lady Julia having already gone to Courcy Castle. 'How d'ye do, how d'ye do?' said the earl. 'I'm not very ill, but I want to get a little advice from you. It's quite a trifle, but I thought it well to see somebody.' Whereupon Dr. Crofts of course declared that he was happy to wait upon his lordship.

'I know all about you, you know,' said the earl. 'Your

grandmother Stoddard was a very old friend of my aunt's. You don't remember Lady Jemima?'

'No,' said Crofts, 'I never had that honour.'

'An excellent old woman, and knew your grandmother Stoddard well. You see, Gruffen has been attending us for I don't know how many years; but upon my word—' and then the earl stopped himself.

'It's an ill wind that blows nobody any good,' said Crofts, with a slight laugh.

'Perhaps it'll blow me some good, for Gruffen never did me any. The fact is this; I'm very well, you know; – as strong as a horse.'

'You look pretty well.'

'No man could be better, – not of my age. I'm sixty, you know.'

'You don't look as though you were ailing.'

'I'm always out in the open air, and that, I take it, is the best thing for a man.'

'There's nothing like plenty of exercise, certainly.'

'And I'm always taking exercise,' said the earl. 'There isn't a man about the place works much harder than I do. And, let me tell you, sir, when you undertake to keep six or seven hundred acres of land in your own hand, you must look after it, unless you mean to lose money by it.'

'I've always heard that your lordship is a good farmer.'

'Well, yes; wherever the grass may grow about my place, it doesn't grow under my feet. You won't often find me in bed at six o'clock, I can tell you.'

After this Dr. Crofts ventured to ask his lordship as to what special deficiency his own aid was invoked at the present time.

'Ah, I was just coming to that,' said the earl. 'They tell me it's a very dangerous practice to go to sleep after dinner.'

'It's not very uncommon at any rate,' said the doctor.

'I suppose not; but Lady Julia is always at me about it. And, to tell the truth, I think I sleep almost too sound when I get to my arm-chair in the drawing-room. Sometimes my sister really can't wake me; – so, at least, she says.'

'And how's your appetite at dinner?'

'Oh, I'm quite right there. I never eat any luncheon, you know, and enjoy my dinner thoroughly. Then I drink three or four glasses of port wine—'

'And feel sleepy afterwards?'

'That's just it,' said the earl.

It is not perhaps necessary that we should inquire what was the exact nature of the doctor's advice; but it was, at any rate, given in such a way that the earl said he would be glad to see him again.

'And look here, Doctor Crofts, I'm alone just at present. Suppose you come over and dine with me to-morrow; then, if I should go to sleep, you know, you'll be able to let me know whether Lady Julia doesn't exaggerate. Just between ourselves, I don't quite believe all she says about my – my snoring, you know.'

Whether it was that the earl restrained his appetite when at dinner under the doctor's eyes, or whether the mid-day mutton chop which had been ordered for him had the desired effect, or whether the doctor's conversation was more lively than that of the Lady Julia, we will not say; but the earl, on the evening in question, was triumphant. As he sat in his easy-chair after dinner he hardly winked above once or twice; and when he had taken the large bowl of tea, which he usually swallowed in a semi-somnolent condition, he was quite lively.

'Ah, yes,' he said, jumping up and rubbing his eyes; 'I think I do feel lighter. I enjoy a snooze after dinner; I do indeed; I like it; but then, when one comes to go to bed, one does it in such a sneaking sort of way, as though one were in disgrace! And my sister, she thinks it a crime – literally a sin, to go to sleep in a chair. Nobody ever caught her napping! By-the-by, Dr. Croft, did you know that Mr. Crosbie whom Bernard Dale brought down to Allington? Lady Julia and he are staying at the same house now.'

'I met him once at Mrs. Dale's.'

'Going to marry one of the girls, isn't he?'

Whereupon Dr. Crofts explained that Mr. Crosbie was engaged to Lilian Dale.

'Ah, yes; a nice girl, I'm told. You know all those Dales

are connections of ours. My sister Fanny married their uncle Orlando. My brother-in-law doesn't like travelling, and so I don't see very much of him; but of course I'm interested about the family.'

'They're very old friends of mine,' said Crofts.

'Yes, I dare say. There are two girls, are there not?'

'Yes, two.'

'And Miss Lily is the youngest. There's nothing about the elder one getting married, is there?'

'I've not heard anything of it.'

'A very pretty girl she is, too. I remember seeing her at her uncle's last year. I shouldn't wonder if she were to marry her cousin Bernard. He is to have the property, you know; and he's my nephew.'

'I'm not quite sure that it's a good thing for cousins to marry,' said Crofts.

'They do, you know, very often; and it suits some family arrangements. I suppose Dale must provide for them, and that would take one off his hands without any trouble.'

Dr. Crofts didn't exactly see the matter in this light, but he was not anxious to argue it very closely with the earl. 'The younger one,' he said, 'has provided for herself.'

'What; by getting a husband? But I suppose Dale must give her something. They're not married yet, you know, and, from what I hear, that fellow may prove a slippery customer. He'll not marry her unless old Dale gives her something. You'll see if he does. I am told that he has got another string to his bow at Courcy Castle.'

Soon after this, Crofts took his horse and rode home, having promised the earl that he would dine with him again before long.

'It'll be a great convenience to me if you'd come about that time,' said the earl, 'and as you're a bachelor perhaps you won't mind it. You'll come on Thursday at seven, will you? Take care of yourself. It's as dark as pitch. John, go and open the first gates for Dr. Crofts.' And then the earl took himself off to bed.

Crofts, as he rode home, could not keep his mind from thinking of the two girls at Allington. 'He'll not marry her unless old Dale gives her something.' Had it come to that with the world, that a man must be bribed into keeping his engagement with a lady? Was there no romance left among mankind, – no feeling of chivalry? 'He's got another string to his bow at Courcy Castle,' said the earl; and his lordship seemed to be in no degree shocked as he said it. It was in this tone that men spoke of women nowadays, and yet he himself had felt such awe of the girl he loved, and such a fear lest he might injure her in her worldly position, that he had not dared to tell her that he loved her.

CHAPTER XXI

JOHN EAMES ENCOUNTERS TWO ADVENTURES, AND DISPLAYS GREAT COURAGE IN BOTH

LILY thought that her lover's letter was all that it should be. She was not quite aware what might be the course of post between Courcy and Allington, and had not, therefore, felt very grievously disappointed when the letter did not come on the very first day. She had, however, in the course of the morning, walked down to the post-office, in order that she might be sure that it was not remaining there.

'Why, miss, they be all delivered; you know that,' said Mrs. Crump, the post-mistress.

'But one might be left behind, I thought.'

'John Postman went up to the house this very day, with a newspaper for your mamma. I can't make letters for people if folks don't write them.'

'But they are left behind sometimes, Mrs. Crump. He wouldn't come up with one letter if he'd got nothing else for anybody in the street.'

'Indeed but he would then. I wouldn't let him leave a letter here no how, nor yet a paper. It's no good your coming down here for letters, Miss Lily. If he don't write to you, I can't make him do it.' And so poor Lily went home discomfited.

But the letter came on the next morning, and all was right. According to her judgment it lacked nothing, either in fullness or in affection. When he told her how he had planned his early departure in order that he might avoid the pain of parting with her on the last moment, she smiled and pressed the paper, and rejoiced inwardly that she had got the better of him as to that manœuvre. And then she kissed the words which told her that he had been glad to have her with him at the last moment. When he declared that he had been happier at Allington than he was at Courcy, she believed him thoroughly, and rejoiced that it should be so. And when he accused himself of being worldly, she excused him, persuading herself that he was nearly perfect in this respect as in others. Of course a man living in London, and having to earn his bread out in the world, must be more worldly than a country girl; but the fact of his being able to love such a girl, to choose such a one for his wife, – was not that alone sufficient proof that the world had not enslaved him? 'My heart is on the Allington lawns,' he said; and then, as she read the words, she kissed the paper again.

In her eyes, and to her ears, and to her heart, the letter was a beautiful letter. I believe there is no bliss greater than that which a thorough love-letter gives to a girl who knows that in receiving it she commits no fault, – who can open it before her father and mother with nothing more than the slight blush which the consciousness of her position gives her. And of all love-letters the first must be the sweetest! What a value there is in every word! How each expression is scanned and turned to the best account! With what importance are all those little phrases invested, which too soon become mere phrases, used as a matter of course. Crosbie had finished his letter by bidding God bless her; 'And you too,' said Lily, pressing the letter to her bosom.

'Does he say anything particular?' asked Mrs. Dale.

'Yes, mamma; it's all very particular.'

'But there's nothing for the public ear.'

'He sends his love to you and Bell.'

'We are very much obliged to him.'

'So you ought to be. And he says that he went to church going through Barchester, and that the clergyman was the grandfather of that Lady Dumbello. When he got to Courcy Castle Lady Dumbello was there.'

'What a singular coincidence!' said Mrs. Dale.

'I won't tell you a word more about his letter,' said Lily. So she folded it up, and put it in her pocket. But as soon as she found herself alone in her own room, she had it out again, and read it over some half-a-dozen times.

That was the occupation of her morning; – that, and the manufacture of some very intricate piece of work which was intended for the adornment of Mr. Crosbie's person. Her hands, however, were very full of work; – or, rather, she intended that they should be full. She would take with her to her new home, when she was married, all manner of household gear, the produce of her own industry and economy. She had declared that she wanted to do something for her future husband, and she would begin that something at once. And in this matter she did not belie her promises to herself, or allow her good intentions to evaporate unaccomplished. She soon surrounded herself with harder tasks than those embroidered slippers with which she indulged herself immediately after his departure. And Mrs. Dale and Bell, – though in their gentle way they laughed at her, – nevertheless they worked with her, sitting sternly at their long tasks, in order that Crosbie's house might not be empty, when their darling should go to take her place there as his wife.

But it was absolutely necessary that the letter should be answered. It would in her eyes have been a great sin to have let that day's post go without carrying a letter from her to Courcy Castle, – a sin of which she felt no temptation to be guilty. It was an exquisite pleasure to her to seat herself at her little table with her neat desk and small appurtenances for epistle-craft, and to feel that she had a letter to write in which she had truly much to say. Hitherto her correspondence had been uninteresting and almost weak in its nature. From her mother and sister she had hardly yet been parted; and though she had other friends, she had seldom found herself with very much to tell them by post. What could she communicate to Mary Eames at Guestwick, which should be in itself exciting as she wrote it? When she wrote to John Eames, and told 'Dear John' that mamma hoped to have the pleasure of seeing him to tea at such an hour, the work of writing was of little moment to her, though the note when written became one of the choicest treasures of him to whom it was addressed.

But now the matter was very different. When she saw the words 'Dearest Adolphus' on the paper before her, she was startled with their significance. 'And four months ago I had never heard of him,' she said to herself, almost with awe. And now he was more to her, and nearer to her, than even was her sister or her mother! She recollected how she had laughed at him behind his back, and called him a swell on the first day of his coming to the Small House, and how, also, she had striven, in her innocent way, to look her best when called upon to go out and walk with the stranger from London. He was no longer a stranger now, but her own dearest friend.

She had put down her pen that she might think of all this – by no means for the first time – and then resumed it with a sudden start as though fearing that the postman might be in the village before her letter was finished. 'Dearest Adolphus, I need not tell you how delighted I was when your letter was brought to me this morning.' But I will not repeat the whole of her letter here. She had no incident to relate, none even so interesting as that of Mr. Crosbie's encounter with Mr. Harding at Barchester. She had met no Lady Dumbello, and had no counterpart to Lady Alexandrina, of whom, as a friend, she could say a word in praise. John Eames's name she did not mention, knowing that John Eames was not a favourite with Mr. Crosbie; nor had she anything to say of John Eames, that had not been already said. He had, indeed, promised to come over to Allington; but this visit had not been made when Lily wrote her first letter to Crosbie. It was a sweet, good, honest love-letter, full of assurances of unalterable affection and unlimited confidence, indulging in a little quiet fun as to the grandees of Courcy Castle, and ending with a promise that she would he happy and contented if she might receive his letters constantly, and live with the hope of seeing him at Christmas.

'I am in time, Mrs. Crump, am I not?' she said, as she walked into the post-office.

'Of course you be, – for the next half-hour. T' postman – he bain't stirred from t' ale'us yet. Just put it into t' box, wull ye?'

'But you won't leave it there?'

'Leave it there! Did you ever hear the like of that? If you're afeared to put it in, you can take it away; that's all about it, Miss Lily.' And then Mrs. Crump turned away to her avocations at the washing-tub. Mrs. Crump had a bad temper, but perhaps she had some excuse. A separate call was made upon her time with reference to almost every letter brought to her office, and for all this, as she often told her friends in profound disgust, she received a salary no more than 'tuppence farden a day. It don't find me in shoe-leather; no more it don't.' As Mrs. Crump was never seen out of her own house, unless it was in church once a month, this latter assertion about her shoe-leather could hardly have been true.

Lily had received another letter, and had answered it before Eames made his promised visit to Allington. He, as will be remembered, had also had a correspondence. He had answered Miss Roper's letter, and had since that been living in fear of two things; in a lesser fear of some terrible rejoinder from Amelia, and in a greater fear of a more terrible visit from his lady-love. Were she to swoop down in very truth upon his Guestwick home, and declare herself to his mother and sister as his affianced bride, what mode of escape would then be left for him? But that she had not yet done, nor had she even answered his cruel missive.

'What an ass I am to be afraid of her!' he said to himself as he walked along under the elms of Guestwick manor, which overspread the road to Allington. When he first went over to Allington after his return home, he had mounted himself on horseback, and had gone forth brilliant with spurs, and trusting somewhat to the glories of his dress and gloves. But he had then known nothing of Lily's engagement. Now he was contented to walk; and as he had taken up his slouched hat and stick in the passage of his mother's house, he had been very indifferent as to his appearance. He walked quickly along the road, taking for the first three miles the shade of the Guestwick elms, and keeping his feet on the broad greensward which skirts the outside of the earl's palings. 'What an ass I am to be afraid of her!' And as he swung his big stick in his hand, striking a tree here and there, and knocking the stones from his path, he began to question himself in earnest, and to be ashamed of his position in the world. 'Nothing on earth shall make me marry her,' he said; 'not if they bring a dozen actions against me. She knows as well as I do, that I have never intended to marry her. It's a cheat from beginning to end. If she comes down here, I'll tell her so before my mother.' But as the vision of her sudden arrival came before his eyes, he acknowledged to himself that he still held her in great fear. He had told her that he loved her. He had written as much as that. If taxed with so much he must confess his sin.

Then, by degrees, his mind turned away from Amelia Roper to Lily Dale, not giving him a prospect much more replete with enjoyment than that other one. He had said that he would call at Allington before he returned to town, and he was now redeeming his promise. But he did not know why he should go there. He felt that he should sit silent and abashed in Mrs. Dale's drawing-room, confessing by his demeanour that secret which it behoved him now to hide from every one. He could not talk easily before Lily, nor could he speak to her of the only subject which would occupy his thoughts when in her presence. If, indeed, he might find her alone— But perhaps that might be worse for him than any other condition.

When he was shown into the drawing-room there was nobody there. 'They were here a minute ago, all three,' said the servant girl. 'If you'll walk down the garden, Mr. John, you'll be sure to find some of 'em.' So John Eames, with a little hesitation, walked down the garden.

First of all he went the whole way round the walks, meeting nobody. Then he crossed the lawn, returning again to the farther end; and there, emerging from the little path which led from the Great House, he encountered Lily alone. 'Oh, John,' she said, 'how d'ye do? I'm afraid you did not find anybody in the house. Mamma and Bell are with Hopkins, away in the large kitchen-garden.'

'I've just come over,' said Eames, 'because I promised. I said I'd come before I went back to London.'

'And they'll be very glad to see you, and so am I. Shall we go after them into the other grounds? But perhaps you walked over and are tired.'

'I did walk,' said Eames; 'not that I am very tired.' But in truth he did not wish to go after Mrs. Dale, though he was altogether at a loss as to what he would say to Lily while remaining with her. He had fancied that he would like to have some opportunity of speaking to her alone before he went away; – of making some special use of the last interview which he should have with her before she became a married woman. But now the opportunity was there, and he hardly dared to avail himself of it.

'You'll stay and dine with us,' said Lily.

'No, I'll not do that, for I especially told my mother that I would be back.'

'I'm sure it was very good of you to walk so far to see us. If you really are not tired, I think we will go to mamma, as she would be very sorry to miss you.'

This she said, remembering at the moment what had been Crosbie's injunctions to her about John Eames. But John had resolved that he would say those words which he had come to speak, and that, as Lily was there with him, he would avail himself of the chance which fortune had given him.

'I don't think I'll go into the squire's garden,' he said.

'Uncle Christopher is not there. He is about the farm somewhere.'

'If you don't mind, Lily, I think I'll stay here. I suppose they'll be back soon. Of course I should like to see them before I go away to London. But, Lily, I came over now chiefly to see you. It was you who asked me to promise.'

Had Crosbie been right in those remarks of his? Had she been imprudent in her little endeavour to be cordially kind to her old friend? 'Shall we go into the drawing-room?' she said, feeling that she would be in some degree safer there than out among the shrubs and paths of the garden. And I think she was right in this. A man will talk of love out among the lilacs and roses, who would be stricken dumb by the demure propriety of the four walls of a drawing-room. John Eames also had some feeling of this kind, for he determined to remain out in the garden, if he could so manage it.

'I don't want to go in unless you wish it,' he said. 'Indeed, I'd rather stay here. So, Lily, you're going to be married?' And thus he rushed at once into the middle of his discourse.

'Yes,' said she, 'I believe I am.'

'I have not told you yet that I congratulate you.'

'I have known very well that you did so in your heart. I have always been sure that you wished me well.'

'Indeed I have. And if congratulating a person is hoping that she may always be happy, I do congratulate you. But, Lily—' And then he paused, abashed by the beauty, purity, and woman's grace which had forced him to love her.

'I think I understand all that you would say. I do not want ordinary words to tell me that I am to count you among my best friends.'

'No, Lily; you don't understand all that I would say. You have never known how often and how much I have thought of you; how dearly I have loved you.'

'John, you must not talk of that now.'

'I cannot go without telling you. When I came over here, and Mrs. Dale told me that you were to be married to that man—'

'You must not speak of Mr. Crosbie in that way,' she said, turning upon him almost fiercely.

'I did not mean to say anything disrespectful of him to you. I should hate myself if I were to do so. Of course you like him better than anybody else?'

'I love him better than all the world besides.'

'And so do I love you better than all the world besides.' And as he spoke he got up from his seat and stood before her. 'I know how poor I am, and unworthy of you; and only that you are engaged to him, I don't suppose that I should now tell you. Of course you couldn't accept such a one as me. But I have loved you ever since you remember; and now that you are going to be his wife, I cannot but tell you that it is so. You will go and live in London, but as to my seeing you there it will be impossible. I could not go into that man's house.'

'Oh, John.'

'No, never; not if you become his wife. I have loved you as well as he does. When Mrs. Dale told me of it, I thought I should have fallen. I went away without seeing you because I was unable to speak to you. I made a fool of myself, and have been a fool all along. I am foolish now to tell you this, but I cannot help it.'

'You will forget it all when you see some girl that you can really love.'

'And have I not really loved you? Well, never mind. I have said what I came to say, and I will now go. If it ever happens that we are down in the country together, perhaps I may see you again; but never in London. Good-bye, Lily.' And he put out his hand to her.

'And won't you stay for mamma?' she said.

'No. Give her my love, and to Bell. They understand all about it. They will know why I have gone. If ever you should want anybody to do anything for you, remember that I will do it, whatever it is.' And as he paced away from her across the lawn, the special deed in her favour to which his mind was turned, – that one thing which he most longed to do on her behalf, – was an act of corporal chastisement upon Crosbie. If Crosbie would but ill-treat her, – ill-treat her with some antenuptial barbarity, – and if only he could be called in to avenge her wrongs! And as he made his way back along the road towards Guestwick, he built up within his own bosom a castle in the air, for her part in which Lily Dale would by no means have thanked him.

Lily when she was left alone burst into tears. She had certainly said very little to encourage her forlorn suitor, and had so borne herself during the interview that even Crosbie could hardly have been dissatisfied; but now that Eames was gone, her heart became very tender towards him. She felt that she did love him also; – not at all as she loved Crosbie, but still with a love that was tender, soft, and true. If Crosbie could have known all her thoughts at that moment, I doubt whether he would have liked them. She burst into tears, and then hurried away into some nook where she could not be seen by her mother and Bell on their return.

Eames went on his way, walking very quickly, swinging his stick and kicking through the dust, with his heart full of the scene which had just passed. He was angry with himself, thinking that he had played his part badly, accusing himself in that he had been rough to her, and selfish in the expression of his love; and he was angry with her because she had declared to him that she loved Crosbie better than all the world besides. He knew that of course she must do so; – that at any rate it was to be expected that such was the case. Yet he thought, she might have refrained from saying so to him. 'She chooses to scorn me now,' he said to himself; 'but the time may come when she will wish that she had scorned him.' That Crosbie was wicked, bad, and selfish, he believed most fully. He felt sure that the man would ill-use her and make her wretched. He had some slight doubt whether he would marry her, and from this doubt he endeavoured to draw a scrap of comfort. If Crosbie would desert her, and if to him might be accorded the privilege of beating the man to death with his fists because of this desertion, then the world would not be quite blank for him. In all this he was no doubt very cruel to Lily; – but then had not Lily been very cruel to him?

He was still thinking of these things when he came to the first of the Guestwick pastures. The boundary of the earl's property was very plainly marked, for with it commenced also the shady elms along the roadside, and the broad green margin of turf, grateful equally to those who walked and to those who rode. Eames had got himself on to the grass, but, in the fullness of his thoughts, was unconscious of the change in his path, when he was startled by a voice in the next field and the loud bellowing of a bull. Lord De Guest's choice cattle he knew were there, and there was one special bull which was esteemed by his lordship as of great value, and regarded as a high favourite. The people about the place declared that the beast was vicious, but Lord De Guest had often been heard to boast that it was never vicious with him. 'The boys tease him, and the men are almost worse than the boys,' said the earl; 'but he'll never hurt any one that has not hurt him.' Guided by faith in his own teaching the earl had taught himself to look upon his bull as a large, horned, innocent lamb if the flock.

As Eames paused on the road, he fancied that he recognized the earl's voice, and it was the voice of one in distress. Then the bull's roar sounded very plain in his ear, and almost close; upon hearing which he rushed on to the gate, and, without much thinking what he was doing, vaulted over it, and advanced a few steps into the field.

'Halloo!' shouted the earl. 'There's a man. Come on.' And then his continued shoutings hardly formed themselves into intelligible words; but Eames plainly understood that he was invoking assistance under great pressure and stress of circumstances. The bull was making short runs at his owner, as though determined in each run to have a toss at his lordship; and at each run the earl would retreat quickly for a few paces, but he retreated always facing his enemy, and as the animal got near to him, would make digs at his face with the long spud which he carried in his hand. But in thus making good his retreat he had been unable to keep in a direct line to the gate, and there seemed to be great danger lest the bull should succeed in pressing him up against the hedge. 'Come on!' shouted the earl, who was fighting his battle manfully, but was by no means anxious to carry off all the laurels of the victory himself. 'Come on, I say!' Then he stopped in his path, shouted into the bull's face, brandished his spud, and threw about his arms, thinking that he might best dismay the beast by the display of these warlike gestures.

Johnny Eames ran on gallantly to the peer's assistance, as he would have run to that of any peasant in the land. He was one to whom I should be perhaps wrong to attribute at this period of his life the gift of very high courage. He feared many things which no man should fear; but he did not fear personal mishap or injury to his own skin and bones. When Cradell escaped out of the house in Burton Crescent, making his way through the passage into the outer air, he did so because he feared that Lupex would beat him or kick him, or otherwise ill-use him. John Eames would also have desired to escape under similar circumstances; but he would have so desired because he could not endure to be looked upon in his difficulties by the people of the house, and because his imagination would have painted the horrors of a policeman dragging him off with a black eye and a torn coat. There was no one to see him now, and no policeman to take offence. Therefore he rushed to the earl's assistance, brandishing his stick, and roaring in emulation of the bull.

When the animal saw with what unfairness he was treated, and that the number of his foes was doubled, while no assistance had lent itself on his side, he stood for a while, disgusted by the injustice of humanity. He stopped, and throwing his head up to the heavens, bellowed out his complaint. 'Don't come close!' said the earl, who was almost out of breath. 'Keep a little apart. Ugh! ugh! whoop, whoop!' And he threw up his arms manfully, jobbing about with his spud, ever and anon rubbing the perspiration from off his eyebrows with the back of his hand.

As the bull stood pausing, meditating whether under such circumstances flight would not be preferable to gratified passion, Eames made a rush in at him, attempting to hit him on the head. The earl, seeing this, advanced a step also, and got his spud almost up to the animal's eye. But these indignities the beast could not stand. He made a charge, bending his head first towards John Eames, and then, with that weak vacillation which is as disgraceful in a bull as in a general, he changed his purpose, and turned his horns upon his other enemy. The consequence was that his steps carried him in between the two, and that the earl and Eames found themselves for a while behind his tail.

'Now for the gate,' said the earl.

'Slowly does it; slowly does it; don't run!' said Johnny, assuming in the heat of the moment a tone of counsel which would have been very foreign to him under other circumstances.

The earl was not a whit offended. 'All right,' said he, taking with a backward motion the direction of the gate. Then as the bull again faced towards him, he jumped from the ground, labouring painfully with arms and legs, and ever keeping his spud well advanced against the foe. Eames, holding his position a little apart from his friend, stooped low and beat the ground with his stick, and as though defying the creature. The bull felt himself defied, stood still and roared, and then made another vacillating attack.

'Hold on till we reach the gate,' said Eames.

'Ugh! ugh! Whoop! whoop!' shouted the earl. And so gradually they made good their ground.

'Now get over,' said Eames, when they had both reached the corner of the field in which the gate stood.

'And what'll you do?' said the earl.

'I'll go at the hedge to the right.' And Johnny as he spoke dashed his stick about, so as to monopolize, for a moment, the attention of the brute. The earl made a spring at the gate, and got well on to the upper rung. The bull, seeing that his prey was going, made a final rush upon the earl and struck the timber furiously with his head, knocking his lordship down on the other side. Lord De Guest was already over, but not off the rail; and thus, though he fell, he fell in safety on the sward* beyond the gate. He fell in safety, but utterly exhausted. Eames, as he had purposed, made a leap almost sideways at a thick hedge which divided the field from one of the Guestwick copses. There was a fairly broad ditch, and on the other side a quickset hedge, which had, however, been weakened and injured by trespassers at this corner, close to the gate. Eames was young and active and jumped well. He jumped so well that he carried his body full into the middle of the quickset, and then scrambled through to the other side, not without much injury to his clothes, and some damage also to his hands and face.

The beast, recovering from his shock against the wooden bars, looked wistfully at his last retreating enemy, as he still struggled amidst the bushes. He looked at the ditch and at the broken hedge, but he did not understand how weak were the impediments in his way. He had knocked his head against the stout timber, which was strong enough to oppose him, but was dismayed by the brambles which he might have trodden under foot without an effort. How many of us are like the bull, turning away conquered by opposition which should be as nothing to us, and breaking our feet, and worse still, our hearts, against rocks of adamant.* The bull at last made up his mind that he did not dare to face the hedge; so he gave one final roar, and then turning himself round, walked placidly back amidst the herd.

Johnny made his way on to the road by a stile that led out of the copse, and was soon standing over the earl, while the blood ran down his cheeks from the scratches. One of the legs of his trousers had been caught by a stake, and was torn from the hip downward, and his hat was left in the field, the only trophy for the bull. 'I hope you're not hurt, my lord,' he said.

'Oh dear, no; but I'm terribly out of breath. Why, you're bleeding all over. He didn't get at you, did he?'

'It's only the thorns in the hedge,' said Johnny, passing his hand over his face. 'But I've lost my hat.'

'There are plenty more hats,' said the earl.

'I think I'll have a try for it,' said Johnny, with whom the means of getting hats had not been so plentiful as with the earl. 'He looks quiet now.' And he moved towards the gate.

But Lord De Guest jumped upon his feet, and seized the young man by the collar of his coat. 'Go after your hat!' said he. 'You must be a fool to think of it. If you're afraid of catching cold, you shall have mine.'

'I'm not the least afraid of catching cold,' said Johnny. 'Is he often like that, my lord?' And he made a motion with his head towards the bull.

'The gentlest creature alive; he's a lamb generally – just like a lamb. Perhaps he saw my red pocket-handkerchief.' And Lord De Guest showed his friend that he carried such an article. 'But where should I have been if you hadn't come up?'

'You'd have got to the gate, my lord.'

'Yes; with my feet foremost, and four men carrying me. I'm very thirsty. You don't happen to carry a flask, do you?'

'No, my lord, I don't.'

'Then we'll make the best of our way home, and have a glass of wine there.' And on this occasion his lordship intended that his offer should be accepted.

CHAPTER XXII

LORD DE GUEST AT HOME

THE earl and John Eames, after their escape from the bull, walked up to the Manor House together. 'You can write a note to your mother, and I'll send it by one of the boys,' said the earl. This was his lordship's answer when Eames declined to dine at the Manor House, because he would be expected home.

'But I'm so badly off for clothes, my lord,' pleaded Johnny. 'I tore my trousers in the hedge.'

'There will be nobody there beside us two and Dr. Crofts. The doctor will forgive you when he hears the story; and as for me, I didn't care if you hadn't a stitch to your back. You'll have company back to Guestwick, so come along.'

Eames had no further excuse to offer, and therefore did as he was bidden. He was by no means as much at home with the earl now as during those minutes of the combat. He would rather have gone home, being somewhat ashamed of being seen in his present tattered and bare-headed condition by the servants of the house; and moreover, his mind would sometimes revert to the scene which had taken place in the garden at Allington. But he found himself obliged to obey the earl, and so he walked on with him through the woods.

The earl did not say very much, being tired and somewhat thoughtful. In what little he did say he seemed to be specially hurt by the ingratitude of the bull towards himself. 'I never teased him, or annoyed him in any way.'

'I suppose they are dangerous beasts?' said Eames.

'Not a bit of it, if they're properly treated. It must have been my handkerchief, I suppose. I remember that I did blow my nose.'

He hardly said a word in the way of thanks to his assistant. 'Where should I have been if you had not come to me?' he had exclaimed immediately after his deliverance; but having said that he didn't think it necessary to say much more to Eames. But he made himself very pleasant, and by the time he had reached the house his companion was almost glad that he had been forced to dine at the Manor House. 'And now we'll have a drink,' said the earl. 'I don't know how you feel, but I never was so thirsty in my life.'

Two servants immediately showed themselves, and evinced some surprise at Johnny's appearance. 'Has the gentleman hurt hisself, my lord?' asked the butler, looking at the blood upon our friend's face.

'He has hurt his trousers the worst, I believe,' said the earl. 'And if he was to put on any of mine they'd be too short and too big, wouldn't they? I am sorry you should be so uncomfortable, but you mustn't mind it for once.'

'I don't mind it a bit,' said Johnny.

'And I'm sure I don't,' said the earl. 'Mr. Eames is going to dine here, Vickers.'

'Yes, my lord.'

'And his hat is down in the middle of the nineteen acres. Let three or four men go for it.'

'Three or four men, my lord!'

'Yes, – three or four men. There's something wrong with that bull. And you must get a boy with a pony to take a note into Guestwick, to Mrs. Eames. Oh dear, I'm better now,' and he put down the tumbler from which he'd been drinking. 'Write your note here, and then we'll go and see my pet pheasants before dinner.'

Vickers and the footman knew that something had happened of much moment, for the earl was usually very particular about his dinner-table. He expected every guest who sat there to be dressed in such guise as the fashion of the day demanded; and he himself, though his morning costume was by no means brilliant, never dined, even when alone, without having put himself into a suit of black, with a white cravat, and having exchanged the old silver hunting-watch which he carried during the day tied round his neck by a bit of old ribbon, for a small gold watch, with a chain and seals, which in the evening always dangled over his waistcoat. Dr.

Gruffen had once been asked to dinner at Guestwick Manor. 'Just a bachelor's chop,' said the earl; 'for there's nobody at home but myself.' Whereupon Dr. Gruffen had come in coloured trousers, – and had never again been asked to dine at Guestwick Manor. All this Vickers knew well; and now his lordship had brought young Eames home to dine with him with his clothes all hanging about him in a manner which Vickers declared in the servants' hall wasn't more than half-decent. Therefore, they all knew that something very particular must have happened. 'It's some trouble about the bull, I know,' said Vickers; – 'but bless you, the bull couldn't have tore his things in that way!'

Eames wrote his note, in which he told his mother that he had had an adventure with Lord De Guest, and that his lordship had insisted on bringing him home to dinner. 'I have torn my trousers all to pieces,' he added in a postscript, 'and have lost my hat. Everything else is all right.' He was not aware that the earl also sent a short note to Mrs. Eames.

'Dear Madam (ran the earl's note), –

'Your son has, under Providence, probably saved my life. I will leave the story for him to tell. He has been good enough to accompany me home, and will return to Guestwick after dinner with Dr. Crofts, who dines here. I congratulate you on having a son with so much cool courage and good feeling.

'Your very faithful servant,

'De Guest.

'Guestwick Manor, Thursday, October, 186–'

And then they went to see the pheasants. 'Now, I'll tell you what,' said the earl. 'I advise you to take to shooting. It's the amusement of a gentleman when a man chances to have the command of game.'

'But I'm always up in London.'

'No, you're not. You're not up in London now. You always have your holidays. If you choose to try it, I'll see that you have shooting enough while you're here. It's better than going to sleep under the trees. Ha, ha, ha! I wonder what made you lay yourself down there. You hadn't been fighting a bull that day?'

'No, my lord. I hadn't seen the bull then.'

'Well; you think of what I have been saying. When I say a thing, I mean it. You shall have shooting enough if you have a mind to try it.' Then they looked at the pheasants, and pottered about the the place till the earl said it was time to dress for dinner. 'That's hard upon you, isn't it?' said he. 'But, at any rate, you can wash your hands, and get rid of the blood. I'll be down in the little drawing-room five minutes before seven, and I suppose I'll find you there.'

At five minutes before seven Lord De Guest came into the small drawing-room, and saw Johnny seated there, with a book before him. The earl was a little fussy, and showed by his manner that he was not quite at his ease, as some men do when they have any piece of work on hand which is not customary to them. He held something in his hand, and shuffled a little as he made his way up the room. He was dressed, as usual, in black; but his gold chain was not, as usual, dangling over his waistcoat.

'Eames,' he said, 'I want you to accept a little present from me, – just as a memorial of our affair with the bull. It will make you think of it sometimes, when I'm perhaps gone.'

'Oh, my lord—'

'It's my own watch, that I've been wearing for some time; but I've got another; two or three, I believe, somewhere upstairs. You mustn't refuse me. I can't bear being refused. There are two or three little seals, too, which I have worn. I have taken off the one with my arms, because that's of no use to you, and it is to me. It doesn't want a key, but winds up at the handle, in this way,' and the earl proceeded to explain the nature of the toy.

'My lord, you think too much of what happened to-day,' said Eames, stammering.

'No, I don't; I think very little about it. I know what I think of. Put the watch in your pocket before the doctor comes. There; I hear his horse. Why didn't he drive over, and then he could have taken you back?'

'I can walk very well.'

'I'll make that all right. The servant shall ride Croft's horse, and bring back the little phaeton. How d'you do, doctor? You know Eames, I suppose? You needn't look at him in that way. His leg is not broken; it's only his trousers.' And then the earl told the story of the bull.

'Johnny will become quite a hero in town,' said Crofts.

'Yes; I fear he'll get the most of the credit; and yet I was at it twice as long as he was. I'll tell you what, young men, when I got to that gate I didn't think I'd breath enough left in me to get over it. It's all very well jumping into a hedge when you're only two-and-twenty; but when a man comes to be sixty he likes to take his time about such things. Dinner ready, is it? So am I. I quite forgot that mutton-chop of yours to-day, doctor. But I suppose a man may eat a good dinner after a fight with a bull?'

The evening passed by without any very pleasurable excitement, and I regret to say that the earl went fast to sleep in the drawing-room as soon as he had swallowed his cup of coffee. During dinner he had been very courteous to both his guests, but towards Eames he had used a good-humoured and almost affectionate familiarity. He had quizzed him for having been found asleep under the tree, telling Crofts that he had looked very forlorn, – 'So that I haven't a doubt about his being in love,' said the earl. And he had asked Johnny to tell the name of the fair one, bringing up the remnants of his half-forgotten classicalities to bear out the joke. 'If I am to take more of the severe Falernian,'* said he, laying his hand on the decanter of port, 'I must know the lady's name. Whoever she be, I'm well sure you need not blush for her. What! you refuse to tell! Then I'll drink no more.' And so the earl had walked out of the dining-room; but not till he had perceived by his guest's cheeks that the joke had been too true to be pleasant. As he went, however, he leaned with his hand on Eames's shoulder, and the servants looking on saw that the young man was to be a favourite. 'He'll make him his heir,' said Vickers. 'I shouldn't wonder a bit if he don't make him his heir.' But to this the footman objected, endeavouring to prove to Mr. Vickers that, in accordance with the law of the land, his lordship's second cousin, once removed, whom the earl had never seen, but whom he was supposed to hate, must be his heir. 'A hearl can never choose his own heir, like you or me,' said the footman, laying down the law. 'Can't he though really, now? That's very hard on him; isn't it?' said the pretty housemaid. 'Psha,' said Vickers: 'you know nothing about it. My lord could make young Eames his heir to-morrow; that is, the heir of his property. He couldn't make him a hearl, because that must go to the heirs of his body. As to his leaving him the place here, I don't just know how that'd be; and I'm sure Richard don't.'

'But suppose he hasn't got any heirs of his body?' asked the pretty housemaid, who was rather fond of putting down Mr. Vickers.

'He must have heirs of his body,' said the butler. 'Everybody has 'em. If a man don't know 'em himself, the law finds 'em out.' And then Mr. Vickers walked away, avoiding further dispute.

In the meantime, the earl was asleep upstairs, and the two young men from Guestwick did not find that they could amuse themselves with any satisfaction. Each took up a book; but there are times at which a man is quite unable to read, and when a book is only a cover for his idleness or dullness. At last Dr. Crofts suggested, in a whisper, that they might as well begin to think of going home.

'Eh; yes; what?' said the earl: 'I'm not asleep.' In answer to which the doctor said that he thought he'd go home, if his lordship would let him order his horse. But the earl was again fast bound in slumber, and took no further notice of the proposition.

'Perhaps we could get off without waking him,' suggested Eames, in a whisper.

'Eh; what?' said the earl. So they both resumed their books, and submitted themselves to their martyrdom for a further period of fifteen minutes. At the expiration of that time, the footman brought in tea.

'Eh; what? tea!' said the earl. 'Yes, we'll have a little tea. I've heard every word you've been saying.' It was that assertion on the part of the earl which always made Lady

Julia so angry. 'You cannot have heard what I have been saying, Theodore, because I have said nothing,' she would reply. 'But I should have heard it if you had,' the earl would rejoin, snappishly. On the present occasion neither Crofts nor Eames contradicted him, and he took his tea and swallowed it while still three parts asleep.

'If you'll allow me, my lord, I think I'll order my horse,' said the doctor.

'Yes; horse – yes—' said the earl, nodding.

'But what are you to do, Eames, if I ride?' said the doctor.

'I'll walk,' whispered Eames, in his very lowest voice.

'What – what – what?' said the earl, jumping up on his feet. 'Oh, ah, yes; going away, are you? I suppose you might as well, as sit here and see me sleeping. But, doctor – I didn't snore, did I?'

'Only occasionally.'

'Not loud, did I? Come, Eames, did I snore loud?'

'Well, my lord, you did snore rather loud two or three times.'

'Did I?' said the earl, in a voice of great disappointment. 'And yet, do you know, I heard every word you said.'

The small phaeton had been already ordered, and the two young men started back to Guestwick together, a servant from the house riding the doctor's horse behind them. 'Look here, Eames,' said the earl, as they parted on the steps of the hall door. 'You're going back to town the day after to-morrow, you say, so I shan't see you again?'

'No, my lord,' said Johnny.

'Look you here, now. I shall be up for the Cattleshow before Christmas. You must dine with me at my hotel, on the twenty-second of December, Pawkins's, in Jermyn Street; seven o'clock, sharp. Mind you do not forget, now. Put it down in your pocket-book when you get home. Good-bye, doctor; good-bye. I see I must stick to that mutton-chop in the middle of the day.' And then they drove off.

'He'll make him his heir for certain,' said Vickers to himself, as he slowly returned to his own quarters.

'You were returning from Allington, I suppose,' said

Crofts, 'when you came across Lord De Guest and the bull?'

'Yes: I just walked over to say good-bye to them.'

'Did you find them all well?'

'I only saw one. The other two were out.'

'Mrs. Dale, was it?'

'No; it was Lily.'

'Sitting alone, thinking of her fine London lover, of course? I suppose we ought to look upon her as a very lucky girl. I have no doubt she thinks herself so.'

'I'm sure I don't know,' said Johnny.

'I believe he's a very good young man,' said the doctor; 'but I can't say I quite liked his manner.'

'I should think not,' said Johnny.

'But then in all probability he did not like mine a bit better, or perhaps yours either. And if so it's all fair.'

'I don't see that it's a bit fair. He's a snob,' said Eames; 'and I don't believe that I am.' He had taken a glass or two of the earl's 'severe Falernian,' and was disposed to a more generous confidence, and perhaps also to stronger language, than might otherwise have been the case.

'No; I don't think he is a snob,' said Crofts. 'Had he been so, Mrs. Dale would have perceived it.'

'You'll see,' said Johnny, touching up the earl's horse with energy as he spoke. 'You'll see. A man who gives himself airs is a snob; and he gives himself airs. And I don't believe he's a straightforward fellow. It was a bad day for us all when he came among them at Allington.'

'I can't say that I see that.'

'I do. But mind, I haven't spoken a word of this to any one. And I don't mean. What would be the good? I suppose she must marry him now?'

'Of course she must.'

'And be wretched all her life. Oh-h-h-h!' and he muttered a deep groan. 'I'll tell you what it is, Crofts. He is going to take the sweetest girl out of this country that ever was in it, and he don't deserve her.'

'I don't think she can be compared to her sister,' said Crofts slowly.

'What; not Lily?' said Eames, as though the proposition

made by the doctor were one that could not hold water for a minute.

'I have always thought that Bell was the more admired of the two,' said Crofts.

'I'll tell you what,' said Eames. 'I have never yet set my eyes on any human creature whom I thought so beautiful as Lily Dale. And now the beast is going to marry her! I'll tell you what, Crofts; I'll manage to pick a quarrel with him yet.' Whereupon the doctor, seeing the nature of the complaint from which his companion was suffering, said nothing more, either about Lily or about Bell.

Soon after this Eames was at his own door, and was received here by his mother and sister with all the enthusiasm due to a hero. 'He has saved the earl's life!' Mrs. Eames had exclaimed to her daughter on reading Lord De Guest's note. 'Oh, goodness!' and she threw herself back upon the sofa almost in a fainting condition.

'Saved Lord De Guest's life!' said Mary.

'Yes – under Providence,' said Mrs. Eames, as though that latter fact added much to her son's good deed.

'But how did he do it?'

'By cool courage and good feeling – so his lordship says. But I wonder how he really did do it?'

'Whatever way it was, he's torn all his clothes and lost his hat,' said Mary.

'I don't care a bit about that,' said Mrs. Eames. 'I wonder whether the earl has any interest at the Income-tax. What a thing it would be if he could get Johnny a step. It would be seventy pounds a year at once. He was quite right to stay and dine when his lordship asked him. And so Dr. Crofts is there. It couldn't have been anything in the doctoring way, I suppose.'

'No, I should say not; because of what he says of his trousers.' And so the two ladies were obliged to wait for John's return.

'How did you do it, John?' said his mother, embracing him, as soon as the door was opened.

'How did you save the earl's life?' said Mary, who was standing behind her mother.

'Would his lordship really have been killed, if it had not been for you?' asked Mrs. Eames.

'And was he very much hurt?' asked Mary.

'Oh, bother,' said Johnny, on whom the results of the day's work, together with the earl's Falernian, had made some still remaining impression. On ordinary occasions, Mrs. Eames would have felt hurt at being so answered by her son; but at the present moment she regarded him as standing so high in general favour that she took no offence. 'Oh, Johnny, do tell us. Of course we must be very anxious to know it all.'

'There's nothing to tell, except that a bull ran at the earl, as I was going by; so I went into the field and helped him, and then he made me stay and dine with him.'

'But his lordship says that you saved his life,' said Mary.

'Under Providence,' added their mother.

'At any rate, he has given me a gold watch and chain,' said Johnny, drawing the present out of his pocket. 'I wanted a watch badly. All the same, I didn't like taking it.'

'It would have been very wrong to refuse,' said his mother. 'And I am so glad you have been so fortunate. And look here, Johnny: when a friend like that comes in your way, don't turn your back on him.' Then, at last, he thawed beneath their kindness, and told them the whole of the story. I fear that in recounting the earl's efforts with the spud, he hardly spoke of his patron with all that deference which would have been appropriate.

CHAPTER XXIII

MR. PLANTAGENET PALLISER

A WEEK passed over Mr. Crosbie's head at Courcy Castle without much inconvenience to him from the well-known fact of his matrimonial engagement. Both George De Courcy and John De Courcy had in their different ways charged him with his offence, and endeavoured to annoy him by recurring to the subject; but he did not care much for the wit or malice of George or John De Courcy. The countess had hardly alluded to Lily Dale after those few words which she said on the first day of his visit, and seemed perfectly willing to regard his doings at Allington as the occupation natural to a young man in such a position. He had been seduced down to a dull country house, and had, as a matter of course, taken to such amusements as the place afforded. He had shot the partridges and made love to the young lady, taking those little recreations as compensation for the tedium of the squire's society. Perhaps he had gone a little too far with the young lady; but then no one knew better than the countess how difficult it is for a young man to go far enough without going too far. It was not her business to make herself a censor on a young man's conduct. The blame, no doubt, rested quite as much with Miss Dale as with him. She was quite sorry that any young lady should be disappointed; but if girls will be imprudent, and set their caps at men above their mark, they must encounter disappointment. With such language did Lady De Courcy speak of the affair among her daughters, and her daughters altogether agreed with her that it was out of the question that Mr. Crosbie should marry Lily Dale. From Alexandrina he encountered during the week none of that raillery which he had expected. He had promised to explain to her before he left the castle all the circumstances of his acquaintance with Lily, and she at last showed herself determined to demand the fulfilment of this promise; but, previous to that, she said nothing to manifest either offence or a lessened friendship. And I regret to say, that in the intercourse which had taken place between them, that friendship was by no means less tender than it had been in London.

'And when will you tell me what you promised?' she asked him one afternoon, speaking in a low voice, as they were standing together at the window of the billiard-room, in that idle half-hour which always occurs before the necessity for dinner preparation has come. She had been riding and was still in her habit, and he had returned from shooting. She knew that she looked more than ordinarily well in her tall straight hat and riding gear, and was wont to hang about the house, walking skilfully with her upheld drapery, during this period of the day. It was dusk, but not dark, and there was no artificial light in the billiard-room. There had been some pretence of knocking about the balls, but it had been only pretence. 'Even Diana,' she had said, 'could not have played billiards in a habit.' Then she had put down her mace, and they had stood talking together in the recess of a large bow-window.

'And what did I promise?' said Crosbie.

'You know well enough. Not that it is a matter of any special interest to me; only, as you undertook to promise, of course my curiosity has been raised.'

'If it be of no special interest,' said Crosbie, 'you will not object to absolve me from my promise.'

'That is just like you,' she said. 'And how false you men always are! You made up your mind to buy my silence on a distasteful subject by pretending to offer me your future confidence; and now you tell me that you do not mean to confide in me.'

'You begin by telling me that the matter is one that does not in the least interest you.'

'That is so false again! You know very well what I meant. Do you remember what you said to me the day you came? and am I not bound to tell you after that, that your marriage with this or that young lady is not matter of special interest to me? Still, as your friend—'

'Well, as my friend!'

'I shall be glad to know—. But I am not going to beg for your confidence; only I tell you this fairly, that no man is so mean in my eyes, as a man who fights under false colours.'

'And am I fighting under false colours?'

'Yes, you are.' And now, as she spoke, the Lady Alexandrina blushed beneath her hat; and dull as was the remaining light of the evening, Crosbie, looking into her face, saw her heightened colour. 'Yes, you are. A gentleman is fighting under false colours who comes into a house like this, with a public rumour of his being engaged, and then conducts himself as though nothing of the kind existed. Of course, it is not anything to me specially; but that is fighting under false colours. Now, sir, you may redeem the promise you made me when you first came here, – or you may let it alone.'

It must be acknowledged that the lady was fighting her battle with much courage, and also with some skill. In three or four days Crosbie would be gone; and this victory, if it were ever to be gained, must be gained in those three or four days. And if there were to be no victory, then it would be only fair that Crosbie should be punished for his duplicity, and that she should be avenged as far as any revenge might be in her power. Not that she meditated any deep revenge, or was prepared to feel any strong anger. She liked Crosbie as well as she had ever liked any man. She believed that he liked her also. She had no conception of any very strong passion, but conceived that a married life was more pleasant than one of single bliss. She had no doubt that he had promised to make Lily Dale his wife, but so had he previously promised her, or nearly so. It was a fair game, and she would win it if she could. If she failed, she would show her anger; but she would show it in a mild, weak manner, – turning up her nose at Lily before Crosbie's face, and saying little things against himself behind his back. Her wrath would not carry her much beyond that.

'Now, sir, you may redeem the promise you made me when you first came here, – or you may let it alone.' So she spoke and then she turned her face away from him, gazing out into the darkness.

'Alexandrina!' he said.

'Well, sir? But you have no right to speak to me in that style. You know that you have no right to call me by my name in that way!'

'You mean that you insist upon your title?'

'All ladies insist on what you call their title, from gentlemen, except under the privilege of greater intimacy than you have the right to claim. You did not call Miss Dale by her Christian name till you had obtained permission, I suppose?'

'You used to let me call you so.'

'Never! Once or twice, when you have done so, I have not forbidden it, as I should have done. Very well, sir, as you have nothing to tell me, I will leave you. I must confess that I did not think you were such a coward.' And she prepared to go, gathering up the skirts of her habit, and taking up the whip which she had laid on the window-sill.

'Stay a moment, Alexandrina,' he said; 'I am not happy, and you should not say words intended to make me more miserable.'

'And why are you unhappy?'

'Because— I will tell you instantly, if I may believe that I am telling you only, and not the whole household.'

'Of course I shall not talk of it to others. Do you think that I cannot keep a secret?'

'It is because I have promised to marry one woman, and because I love another. I have told you everything now; and if you choose to say again that I am fighting under false colours I will leave the castle before you can see me again.'

'Mr. Crosbie?'

'Now you know it all, and may imagine whether or no I am very happy. I think you said it was time to dress; – suppose we go?' And without further speech the two went off to their separate rooms.

Crosbie, as soon as he was alone in his chamber, sat himself down in his arm-chair, and went to work striving to make up his mind as to his future conduct. It must not be supposed that the declaration just made by him had been produced solely by his difficulty at the moment. The atmosphere of Courcy Castle had been at work upon him for the last week past. And every word that he had heard, and every word that he had spoken, had tended to destroy all that was good and true within him, and to foster all that was selfish and false. He had said to himself a dozen times during that week that he never could be happy with Lily Dale, and that he never could make her happy. And then he had used the old sophistry in his endeavour to teach himself that it was right to do that which he wished to do. Would it not be better for Lily that he should desert her, than marry her against the dictates of his own heart? And if he really did not love her, would he not be committing a greater crime in marrying her than in deserting her? He confessed to himself that he had been very wrong in allowing the outer world to get such a hold upon him that the love of a pure girl like Lily could not suffice for his happiness. But there was the fact, and he found himself unable to contend against it. If by any absolute self-sacrifice he could secure Lily's well-being, he would not hesitate for a moment. But would it be well to sacrifice her as well as himself?

He had discussed the matter in this way within his own breast, till he had almost taught himself to believe that it was his duty to break off his engagement with Lily; and he had almost taught himself to believe that a marriage with a daughter of the house of Courcy would satisfy his ambition and assist him in his battle with the world. That Lady Alexandrina would accept him he felt certain, if he could only induce her to forgive him for his sin in becoming engaged to Miss Dale. How very prone she would be to forgiveness in this matter, he had not divined, having not as yet learned how easily such a woman can forgive such a sin, if the ultimate triumph be accorded to herself.

And there was another reason which operated much with Crosbie, urging him on in his present mood and wishes, though it should have given an exactly opposite impulse to his heart. He had hesitated as to marrying Lily Dale at once, because of the smallness of his income. Now he had a prospect of considerable increase to that income. One of the commissioners at his office had been promoted to some greater commissionership, and it was understood by everybody that the secretary at the General Committee Office would be the new commissioner. As to that there was no doubt. But then the question had arisen as to the place of secretary. Crosbie had received two or three letters on the subject, and it seemed that the likelihood of his obtaining this step in the world was by no means slight. It would increase his official income from seven hundred a year to twelve, and would place him altogether above the world. His friend, the present secretary, had written to him, assuring him that no other probable competitor was spoken of as being in the field against him. If such good fortune awaited him, would it not smooth any present difficulty which lay in the way of his marriage with Lily Dale? But, alas, he had not looked at the matter in that light! Might not the countess help him to this preferment? And if his destiny intended for him the good things of this world, – secretaryships, commissionerships, chairmanships, and such like, would it not be well that he should struggle on in his upward path by such assistance as good connections might give him?

He sat thinking over it all in his own room on that evening. He had written twice to Lily since his arrival at Courcy Castle. His first letter has been given. His second was written much in the same tone; though Lily, as she had read it, had unconsciously felt somewhat less satisfied than she had been with the first. Expressions of love were not wanting, but they were vague and without heartiness. They savoured of insincerity, though there was nothing in the words themselves to convict them. Few liars can lie with the full roundness and self-sufficiency of truth; and Crosbie, as bad as he was, had not yet become bad enough to reach that perfection. He had said nothing to Lily of the hopes of promotion which had been opened to him; but he had again spoken of his own worldliness – acknowledging that he received an unsatisfying satisfaction from the pomps and vanities of Courcy Castle. In fact he was paving the way for that which he had almost resolved that he would do, now he had told Lady Alexandrina that he loved her; and he was obliged to confess to himself that the die was cast.

As he thought of all this, there was not wanting to him some of the satisfaction of an escape. Soon after making that declaration of love at Allington he had begun to feel that in making it he had cut his throat. He had endeavoured to persuade himself that he could live comfortably with his throat cut in that way; and as long as Lily was with him he would believe that he could do so; but as soon as he was again alone he would again accuse himself of suicide. This was his frame of mind even while he was yet at Allington, and his ideas on the subject had become stronger during his sojourn at Courcy. But the self-immolation had not been completed, and he now began to think that he could save himself. I need hardly say that this was not all triumph to him. Even had there been no material difficulty as to his desertion of Lily, – no uncle, cousin, and mother whose anger he must face, – no vision of a pale face, more eloquent of wrong in its silence than even uncle, cousin, and mother, with their indignant storm of words, – he was not altogether heartless. How should he tell all this to the girl who had loved him so well; who had so loved him, that, as he himself felt, her love would fashion all her future life either for weal or for foe? 'I am unworthy of her, and will tell her so,' he said to himself. How many a false hound of a man has endeavoured to salve his own conscience by such mock humility? But he acknowledged at this moment, as he rose from his seat to dress himself, that the die was cast, and that it was open to him now to say what he pleased to Lady Alexandrina. 'Others have gone through the same fire before,' he said to himself, as he walked downstairs, 'and have come out scatheless.' And then he recalled to himself the names of various men of high repute in the world who were supposed to have committed in their younger days some such little mistake as that into which he had been betrayed.

In passing through the hall he overtook Lady Julia De Guest, and was in time to open for her the door of the drawing-room. He then remembered that she had come into the billiard-room at one side, and had gone out at the other, while he was standing with Alexandrina at the window. He had not, however, then thought much of Lady Julia; and as he now stood for her to pass by him through the doorway, he made to her some indifferent remark.

But Lady Julia was on some subjects a stern woman, and not without a certain amount of courage. In the last week she had seen what had been going on, and had become more and more angry. Though she had disowned any family connection with Lily Dale, nevertheless she now felt for her sympathy and almost affection. Nearly every day she had repeated stiffly to the countess some incident of Crosbie's courtship and engagement to Miss Dale, – speaking of it as with absolute knowledge, as a thing settled at all points. This she had done to the countess alone, in the presence of the countess and Alexandrina, and also before all the female guests of the castle. But what she had said was received simply with an incredulous smile. 'Dear me! Lady Julia,' the countess had replied at last, 'I shall begin to think you are in love with Mr. Crosbie yourself; you harp so constantly on this affair of his. One would think that young ladies in your part of the world must find it very difficult to get husbands, seeing that the success of one young lady is trumpeted so loudly.' For the moment, Lady Julia was silenced; but it was not easy to silence her altogether when she had a subject for speech near her heart.

Almost all the Courcy world were assembled in the drawing-room as she now walked into the room with Crosbie at her heels. When she found herself near the crowd she turned round, and addressed him in a voice more audible than that generally required for purposes of drawing-room conversation. 'Mr. Crosbie,' she said, 'have you heard lately from our dear friend, Lily Dale?' And she looked him full in the face, in a manner more significant, probably, than even she had intended it to be. There was, at once, a general hush in the room, and all eyes were turned upon her and upon him.

Crosbie instantly made an effort to bear the attack gallantly, but he felt that he could not quite command his colour, or prevent a sudden drop of perspiration from showing itself upon his brow. 'I had a letter from Allington yesterday,' he said. 'I suppose you have heard of your brother's encounter with the bull?'

'The bull!' said Lady Julia. And it was instantly manifest to all that her attack had been foiled and her flank turned.

'Good gracious! Lady Julia, how very odd you are!' said the countess.

'But what about the bull?' asked the Honourable George.

'It seems that the earl was knocked down in the middle of one of his own fields.'

'Oh, dear!' exclaimed Alexandrina. And sundry other exclamations were made by all the assembled ladies.

'But he wasn't hurt,' said Crosbie. 'A young man named Eames seems to have fallen from the sky and carried off the earl on his back.'

'Ha, ha, ha, ha!' growled the other earl, as he heard of the discomfiture of his brother peer.

Lady Julia, who had received her own letters that day from Guestwick, knew that nothing of importance had happened to her brother; but she felt that she was foiled for that time.

'I hope that there has not really been any accident,' said Mr. Gazebee, with a voice of great solicitude.

'My brother was quite well last night, thank you,' said she. And then the little groups again formed themselves, and Lady Julia was left alone on the corner of a sofa.

'Was that all an invention of yours, sir?' said Alexandrina to Crosbie.

'Not quite. I did get a letter this morning from my friend Bernard Dale, – that old harridan's nephew; and Lord De Guest has been worried by some of his animals. I wish I had told her that his stupid old neck had been broken.'

'Fie, Mr. Crosbie!'

'What business has she to interfere with me?'

'But I mean to ask the same question that she asked, and you won't put me off with a cock-and-bull story like that.' But then, as she was going to ask the question, dinner was announced.

'And is it true that De Guest has been tossed by a bull?' said the earl, as soon as the ladies were gone. He had spoken nothing during dinner except what words he had muttered into the ear of Lady Dumbello. It was seldom that conversation had many charms for him in his own house; but there was a savour of pleasantry in the idea of Lord De Guest having been tossed, by which even he was tickled.

'Only knocked down, I believe,' said Crosbie.

'Ha, ha, ha,' growled the earl; then he filled his glass, and allowed some one else to pass the bottle. Poor man! There was not much left to him now in the world which did amuse him.

'I don't see anything to laugh at,' said Plantagenet Palliser, who was sitting at the earl's right hand, opposite to Lord Dumbello.

'Don't you?' said the earl. 'Ha, ha, ha!'

'I'll be shot if I do. From all I hear De Guest is an uncommon good farmer. And I don't see the joke of tossing a farmer merely because he's a nobleman also. Do you?' and he turned round to Mr. Gazebee, who was sitting on the other side. The earl was an earl, and was also Mr. Gazebee's father-in-law. Mr. Plantagenet Palliser was the heir to a dukedom. Therefore, Mr. Gazebee merely simpered, and did not answer the question put to him. Mr. Palliser said nothing more about it, nor did the earl; and then the joke died away.

Mr. Plantagenet Palliser was the Duke of Omnium's heir – heir to that nobleman's title and to his enormous wealth; and, therefore, was a man of mark in the world. He sat in the House of Commons, of course. He was about five-and-twenty years of age, and was, as yet, unmarried. He did not hunt or shoot or keep a yacht, and had been heard to say that he had never put a foot upon a race-course in his life. He dressed very quietly, never changing the colour or form of his garments; and in society was quiet, reserved, and very often silent. He was tall, slight, and not ill-looking; but more than this cannot be said for his personal appearance – except, indeed, this, that no one could mistake him for other than a gentleman. With his uncle, the duke, he was on good terms – that is to say, they had never quarrelled. A very liberal allowance had been made to the nephew; but the two relatives had no tastes in common, and did not often meet. Once a year Mr. Palliser visited the duke at his great country seat for two or three days, and usually dined with him two or three times during the season in London. Mr. Palliser sat for a borough which was absolutely under the duke's command; but had accepted his seat under the distinct understanding that he was to take whatever part in politics might seem good to himself. Under these well-understood arrangements, the duke and his heir showed to the world quite a pattern of a happy family. 'So different to the earl and Lord Porlock!' the people of West Barsetshire used to say. For the estates, both of the duke and of the earl, were situated in the western division of that county.

Mr. Palliser was chiefly known to the world as a rising politician. We may say that he had everything at his command, in the way of pleasure, that the world could offer him. He had wealth, position, power, and the certainty of attaining the highest rank among, perhaps, the most brilliant nobility in the world. He was courted by all who could get near enough to court him. It is hardly too much to say that he might have selected a bride from all that was most beautiful and best among English women. If he would have bought race-horses, and have expended thousands on the turf, he would have gratified his uncle by doing so. He might have been the master of hounds, or the slaughterer of hecatombs of birds. But to none of these things would he devote himself. He had chosen to be a politician, and in that pursuit he laboured with a zeal and perseverance which would have made his fortune at any profession or in any trade. He was constant in committee-rooms up to the very middle of August. He was rarely absent from any debate of importance, and never from any important division. Though he seldom spoke, he was always ready to speak if his purpose required it. No man gave him credit for any great genius – few even considered that he could become either an orator or a mighty statesman. But the world said that he was a rising man, and old Nestor* of the Cabinet looked on him as one who would be able, at some far future day, to come among them as a younger brother. Hitherto he had declined such inferior offices as had been offered to him, biding his time carefully; and he was as yet tied hand and neck to no party, though known to be Liberal in all his political tendencies. He was a great reader – not taking up a book here, and another there, as chance brought books before him, but working through an enormous course of books, getting up the great subject of the world's history – filling himself full of facts – though perhaps not destined to acquire the power of using those facts otherwise than as precedents. He strove also diligently to become a linguist – not without success, as far as a competent understanding of various languages. He was a thin-minded, plodding, respectable man, willing to devote all his youth to work, in order that in old age he might be allowed to sit among the Councillors of the State.

Hitherto his name had not been coupled by the world

with that of any woman whom he had been supposed to admire; but latterly it had been observed that he had often been seen in the same room with Lady Dumbello. It had hardly amounted to more than this; but when it was remembered how undemonstrative were the two persons concerned – how little disposed was either of them to any strong display of feeling – even this was thought matter to be mentioned. He certainly would speak to her from time to time almost with an air of interest; and Lady Dumbello, when she saw that he was in the room, would be observed to raise her head with some little show of life, and to look round as though there were something there on which it might be worth her while to allow her eyes to rest. When such innuendoes were abroad, no one would probably make more of them than Lady De Courcy. Many, when they heard that Mr. Palliser was to be at the castle, had expressed their surprise at her success in that quarter. Others, when they learned that Lady Dumbello had consented to become her guest, had also wondered greatly. But when it was ascertained that the two were to be there together, her good-natured friends had acknowledged that she was a very clever woman. To have either Mr. Palliser or Lady Dumbello would have been a feather in her cap; but to succeed in getting both, by enabling each to know that the other would be there, was indeed a triumph. As regards Lady Dumbello, however, the bargain was not fairly carried out; for, after all, Mr. Palliser came to Courcy Castle only for two nights and a day, and during the whole of that day he was closeted with sundry large blue-books. As for Lady De Courcy, she did not care how he might be employed. Blue-books and Lady Dumbello were all the same to her. Mr. Palliser had been at Courcy Castle, and neither enemy nor friend could deny the fact.

This was his second evening; and as he had promised to meet his constituents at Silverbridge at one P.M. on the following day, with the view of explaining to them his own conduct and the political position of the world in general; and as he was not to return from Silverbridge to Courcy, Lady Dumbello, if she made any way at all, must take advantage of the short gleam of sunshine which the present hour afforded her. No one, however, could say that she showed any active disposition to monopolize Mr. Palliser's attention. When he sauntered into the drawing-room she was sitting, alone, in a large, low chair, made without arms, so as to admit the full expansion of her dress, but hollowed and round at the back. so as to afford her the support that was necessary to her. She had barely spoken three words since she had left the dining-room, but the time had not passed heavily with her. Lady Julia had again attacked the countess about Lily Dale and Mr. Crosbie, and Alexandrina, driven almost to rage, had stalked off to the farther end of the room, not concealing her special concern in the matter.

'How I do wish they were married and done with,' said the countess; 'and then we should hear no more about them.'

All of which Lady Dumbello heard and understood; and in all of it she took a certain interest. She remembered such things, learning thereby who was who, and regulating her own conduct by what she learned. She was by no means idle at this or at other such times, going through, we may say, a considerable amount of really hard work in her manner of working. There she had sat speechless, unless when acknowledging by a low word of assent some expression of flattery from those around her. Then the door opened, and when Mr. Palliser entered she raised her head, and the faintest possible gleam of satisfaction might have been discerned upon her features. But she made no attempt to speak to him; and when, as he stood at the table, he took up a book and remained thus standing for a quarter of an hour, she neither showed nor felt any impatience. After that Lord Dumbello came in, and he stood at the table without a book. Even then Lady Dumbello felt no impatience.

Plantagenet Palliser skimmed through his little book, and probably learned something. When he put it down he sipped a cup of tea, and remarked to Lady De Courcy that he believed it was only twelve miles to Silverbridge.

'I wish it was a hundred and twelve,' said the countess.

'In that case I should be forced to start to-night,' said Mr. Palliser.

'Then I wish it was a thousand and twelve,' said Lady De Courcy..

'In that case I should not have come at all,' said Mr. Palliser. He did not mean to be uncivil, and had only stated a fact.

'The young men are becoming absolute bears,' said the countess to her daughter Margaretta.

He had been in the room nearly an hour when he did at last find himself standing close to Lady Dumbello: close to her, and without any other very near neighbour.

'I should hardly have expected to find you here,' he said.

'Nor I you,' she answered.

'Though, for the matter of that, we are both near our own homes.'

'I am not near mine.'

'I meant Plumstead; your father's place.'

'Yes; that was my home once.'

'I wish I could show you my uncle's place. The castle is very fine, and he has some good pictures.'

'So I have heard.'

'Do you stay here long?'

'Oh, no. I go to Cheshire the day after to-morrow. Lord Dumbello is always there when the hunting begins.'

'Ah, yes; of course. What a happy fellow he is; never any work to do! His constituents never trouble him, I suppose?'

'I don't think they ever do, much.'

After that Mr. Palliser sauntered away again, and Lady Dumbello passed the rest of the evening in silence. It is to be hoped that they both were rewarded by that ten minutes of sympathetic intercourse for the inconvenience which they had suffered in coming to Courcy Castle.

But that which seems so innocent to us had been looked on in a different light by the stern moralists of that house.

'By Jove!' said the Honourable George to his cousin, Mr. Gresham, 'I wonder how Dumbello likes it.'

'It seems to me that Dumbello takes it very easily.'

'There are some men who will take anything easily,' said George, who, since his own marriage, had learned to have a holy horror of such wicked things.

'She's beginning to come out a little,' said Lady Clandidlem to Lady De Courcy, when the two old women found themselves together over a fire in some back sitting-room. 'Still waters always run deep, you know.'

'I shouldn't at all wonder if she were to go off with him,' said Lady De Courcy.

'He'll never be such a fool as that,' said Lady Clandidlem.

'I believe men will be fools enough for anything,' said Lady De Courcy. 'But, of course, if he did, it would come to nothing afterwards. I know one who would not be sorry. If ever a man was tired of a woman, Lord Dumbello is tired of her.'

But in this, as in almost everything else, the wicked old woman spoke scandal. Lord Dumbello was still proud of his wife, and as fond of her as a man can be of a woman whose fondness depends upon mere pride.

There had not been much that was dangerous in the conversation between Mr. Palliser and Lady Dumbello, but I cannot say the same as to that which was going on at the same moment between Crosbie and Lady Alexandrina. She, as I have said, walked away in almost open dudgeon when Lady Julia recommenced her attack about poor Lily, nor did she return to the general circle during the evening. There were two large drawing-rooms at Courcy Castle, joined together by a narrow link of a room, which might have been called a passage, had it not been lighted by two windows coming down to the floor, carpeted as were the drawing-rooms, and warmed with a separate fireplace. Hither she betook herself, and was soon followed by her married sister Amelia.

'That woman almost drives me mad,' said Alexandrina, as they stood together with their toes upon the fender.

'But, my dear, you of all people should not allow yourself to be driven mad on such a subject.'

'That's all very well, Amelia.'

'The question is this, my dear, – what does Mr. Crosbie mean to do?'

'How should I know?'

'If you don't know, it will be safer to suppose that he is

going to marry this girl, and in that case—'

'Well, what in that case? Are you going to be another Lady Julia? What do I care about the girl?'

'I don't suppose you care much about the girl; and if you care as little about Mr. Crosbie, there's an end of it; only in that case, Alexandrina—'

'Well, what in that case?'

'You know I don't want to preach to you. Can't you tell me at once whether you really like him? You and I have always been good friends.' And the married sister put her arm affectionately round the waist of her who wished to be married.

'I like him well enough.'

'And has he made any declaration to you?'

'In a sort of way he has. Hark, here he is!' And Crosbie, coming in from the larger room, joined the sisters at the fireplace.

'We were driven away by the clack of Lady Julia's tongue,' said the elder.

'I never met such a woman,' said Crosbie.

'There cannot well be many like her,' said Alexandrina. And after that they all stood silent for a minute or two. Lady Amelia Gazebee was considering whether or no she would do well to go and leave the two together. If it were intended that Mr. Crosbie should marry her sister, it would certainly be well to give him an opportunity of expressing such a wish on his own part. But if Alexandrina was simply making a fool of herself, then it would be well for her to stay. 'I suppose she would rather I should go,' said the elder sister to herself; and then, obeying the rule which should guide all our actions from one to another, she went back and joined the crowd.

'Will you come on into the other room?' said Crosbie.

'I think we are very well here,' Alexandrina replied.

'But I wish to speak to you, – particularly,' said he.

'And cannot you speak here?'

'No. They will be passing backwards and forwards.' Lady Alexandrina said nothing further, but led the way into the other large room. That also was lighted, and there were in it four or five persons. Lady Rosina was reading a work on the Millennium, with a light to herself in one corner. Her brother John was asleep in an arm-chair, and a young gentleman and lady were playing chess. There was, however, ample room for Crosbie and Alexandrina to take up a position apart.

'And now, Mr. Crosbie, what have you got to say to me? But, first, I mean to repeat Lady Julia's question, as I told you that I should do. – When did you hear last from Miss Dale?'

'It is cruel in you to ask me such a question, after what I have already told you. You know that I have given to Miss Dale a promise of marriage.'

'Very well, sir. I don't see why you should bring me in here to tell me anything that is so publicly known as that. With such a herald as Lady Julia it is quite unnecessary.'

'If you can only answer me in that tone I will make an end of it at once. When I told you of my engagement, I told you also that another woman possessed my heart. Am I wrong to suppose that you knew to whom I alluded?'

'Indeed, I did not, Mr. Crosbie. I am no conjuror, and I have not scrutinized you so closely as your friend Lady Julia.'

'It is you that I love. I am sure I need hardly say so now.'

'Hardly, indeed, – considering that you are engaged to Miss Dale.'

'As to that I have, of course, to own that I have behaved foolishly; – worse than foolishly, if you choose to say so. You cannot condemn me more absolutely than I condemn myself. But I have made up my mind as to one thing. I will not marry where I do not love.' Oh, if Lily could have heard him as he then spoke! 'It would be impossible for me to speak in terms too high of Miss Dale; but I am quite sure that I could not make her happy as her husband.'

'Why did you not think of that before you asked her?' said Alexandrina. But there was very little of condemnation in her tone.

'I ought to have done so; but it is hardly for you to blame me with severity. Had you, when we were last together in London – had you been less—'

'Less what?'

'Less defiant,' said Crosbie, 'all this might perhaps have been avoided.'

Lady Alexandrina could not remember that she had been defiant; but, however, she let that pass. 'Oh, yes; of course it was my fault.'

'I went down there to Allington with my heart ill at ease, and now I have fallen into this trouble. I tell you all as it has happened. It is impossible that I should marry Miss Dale. It would be wicked in me to do so, seeing that my heart belongs altogether to another. I have told you who is that other; and now may I hope for an answer?'

'An answer to what?'

'Alexandrina, will you be my wife?'

If it had been her object to bring him to a point-blank declaration and proposition of marriage, she had certainly achieved her object now. And she had that trust in her own power of management and in her mother's, that she did not fear that in accepting him she would incur the risk of being served as he was serving Lily Dale. She knew her own position and his too well for that. If she accepted him she would in due course of time become his wife, – let Miss Dale and all her friends say what they might to the contrary. As to that head she had no fear. But nevertheless she did not accept him at once. Though she wished for the prize, her woman's nature hindered her from taking it when it was offered to her.

'How long is it, Mr. Crosbie,' she said, 'since you put the same question to Miss Dale?'

'I have told you everything, Alexandrina, – as I promised that I would do. If you intend to punish me for doing so—'

'And I might ask another question. How long will it be before you put the same question to some other girl?'

He turned round as though to walk away from her in anger; but when he had gone half the distance to the door he returned.

'By heaven!' he said, and he spoke somewhat roughly too, 'I'll have an answer. You at any rate have nothing with which to reproach me. All that I have done wrong, I have done through you, or on your behalf. You have heard my proposal. Do you intend to accept it?'

'I declare you startle me. If you demanded my money or my life, you could not be more imperious.'

'Certainly not more resolute in my determination.'

'And if I decline the honour?'

'I shall think you the most fickle of your sex.'

'And if I were to accept it?'

'I would swear that you were the best, the dearest, and the sweetest of women.'

'I would rather have your good opinion than your bad, certainly,' said Lady Alexandrina. And then it was understood by both of them that that affair was settled. Whenever she was called on in future to speak of Lily, she always called her, 'that poor Miss Dale', but she never again spoke a word of reproach to her future lord about that little adventure. 'I shall tell mamma to-night,' she said to him, as she bade him good-night in some sequestered nook to which they had betaken themselves. Lady Julia's eye was again on them as they came out from the sequestered nook, but Alexandrina no longer cared for Lady Julia.

'George, I cannot quite understand about that Mr. Palliser. Isn't he to be a duke, and oughtn't he to be a lord now?' This question was asked by Mrs. George De Courcy of her husband, when they found themselves together in the seclusion of the nuptial chamber.

'Yes; he'll be Duke of Omnium when the old fellow dies. I think he's one of the slowest fellows I ever came across. He'll take deuced good care of the property, though.'

'But, George, do explain it to me. It is so stupid not to understand, and I am afraid of opening my mouth for fear of blundering.'

'Then keep your mouth shut, my dear. You'll learn all those sort of things in time, and nobody notices it if you don't say anything.'

'Yes, but, George; – I don't like to sit silent all the night. I'd sooner be up here with a novel if I can't speak about anything.'

'Look at Lady Dumbello. She doesn't want to be always talking.'

'Lady Dumbello is very different from me. But do tell me, who is Mr. Palliser?'

'He's the duke's nephew. If he were the duke's son, he would be the Marquis of Silverbridge.'

'And will be plain Mister till his uncle dies?'

'Yes, a very plain Mister.'

'What a pity for him. But, George, – if I have a baby, and if he should be a boy, and if—'

'Oh, nonsense; it will be time enough to talk of that when he comes. I'm going to sleep.'

CHAPTER XXIV

A MOTHER-IN-LAW AND A FATHER-IN-LAW

On the following morning Mr. Plantagenet Palliser was off upon his political mission before breakfast; – either that, or else some private comfort was afforded to him in guise of solitary rolls and coffee. The public breakfast at Courcy Castle was going on at eleven o'clock, and at that hour Mr. Palliser was already closeted with the Mayor of Silverbridge.

'I must get off by the 3.45 train,' said Mr. Palliser. 'Who is there to speak after me?'

'Well, I shall say a few words; and Growdy, – he'll expect them to listen to him. Growdy has always stood very firm by his grace, Mr. Palliser.'

'Mind we are in the room sharp at one. And you can have a fly, for me to get away to the station, ready in the yard. I won't go a moment before I can help. I shall be just an hour and a half myself. No, thank you, I never take any wine in the morning.' And I may here state that Mr. Palliser did get away by the 3.45 train, leaving Mr. Growdy still talking on the platform. Constituents must be treated with respect; but time has become so scarce nowadays that that respect has to be meted out by the quarter of an hour with parsimonious care.

In the meantime there was more leisure at Courcy Castle.

Neither the countess nor Lady Alexandrina came down to breakfast, but their absence gave rise to no special remark. Breakfast at the castle was a morning meal at which people showed themselves, or did not show themselves, as it pleased them. Lady Julia was there looking very glum, and Crosbie was sitting next to his future sister-in-law Margaretta, who already had placed herself on terms of close affection with him. As he finished his tea she whispered into his ear, 'Mr. Crosbie, if you could spare half an hour, mamma would so like to see you in her own room.' Crosbie declared that he would be delighted to wait upon her, and did in truth feel some gratitude in being welcomed as a son-in-law into the house. And yet he felt also that he was being caught, and that in ascending into the private domains of the countess he would be setting the seal upon his own captivity.

Nevertheless, he went with a smiling face and a light step, Lady Margaretta ushering him the way. 'Mamma,' said she; 'I have brought Mr. Crosbie up to you. I did not know that you were here, Alexandrina, or I should have warned him.'

The countess and her youngest daughter had been breakfasting together in the elder lady's sitting-room, and were now seated in a very graceful and well-arranged deshabille.* The tea-cups out of which they had been drinking were made of some elegant porcelain, the teapot and cream-jug were of chased silver and as delicate in their way. The remnant of food consisted of morsels of French roll which had not even been allowed to crumble themselves in a disorderly fashion, and of infinitesimal pats of butter. If the morning meal of the two ladies had been as unsubstantial as the appearance of the fragments indicated, it must be presumed that they intended to lunch early. The countess herself was arrayed in an elaborate morning wrapper of figured silk, but the simple Alexandrina wore a plain white muslin peignoir, fastened with pink ribbon. Her hair, which she usually carried in long rolls, now hung loose over her shoulders, and certainly added something to her stock of female charms. The countess got up as Crosbie entered, and greeted him with an open hand; but Alexandrina kept her seat, and merely nodded at him a little welcome. 'I must run down again,' said Margaretta, 'or I shall have left Amelia with all the cares of the house upon her.'

'Alexandrina has told me all about it,' said the countess, with her sweetest smile; 'and I have given her my approval. I really do think you will suit each other very well.'

'I am very much obliged to you,' said Crosbie. 'I'm sure at any rate of this, – that she will suit me very well.'

'Yes; I think she will. She is a good sensible girl.'

'Psha, mamma; pray don't go on in that Goody Twoshoes sort of way.'

'So you are, my dear. If you were not it would not be well for you to do as you are going to do. If you were giddy and harum-scarum, and devoted to rank and wealth and that sort of thing, it would not be well for you to marry a commoner without fortune. I'm sure Mr. Crosbie will excuse me for saying so much as that.'

'Of course I know,' said Crosbie, 'that I had no right to look so high.'

'Well; we'll say nothing more about it,' said the countess.

'Pray don't,' said Alexandrina. 'It sounds so like a sermon.'

'Sit down, Mr. Crosbie,' said the countess, 'and let us have a little conversation. She shall sit by you, if you like it. Nonsense, Alexandrina, – if he asks it!'

'Don't, mamma; – I mean to remain where I am.'

'Very well, my dear; – then remain where you are. She is a wilful girl, Mr. Crosbie; as you will say when you hear that she has told me all that you told her last night.' Upon hearing this, he changed colour a little, but said nothing. 'She has told me,' continued the countess, 'about that young lady at Allington. Upon my word, I'm afraid you have been very naughty.'

'I have been foolish, Lady De Courcy.'

'Of course; I did not mean anything worse than that. Yes, you have been foolish; – amusing yourself in a thoughtless way, you know, and, perhaps, a little piqued because a certain lady was not to be won so easily as your Royal Highness wished. Well, now, all that must be settled, you know, as quickly as possible. I don't want to ask any indiscreet questions; but if the young lady has really been left with any idea that you meant anything, don't you think you should undeceive her at once?'

'Of course he will, mamma.'

'Of course you will; and it will be a great comfort to Alexandrina to know that the matter is arranged. You hear what Lady Julia is saying almost every hour of her life. Now, of course, Alexandrina does not care what an old maid like Lady Julia may say; but it will be better for all parties that the rumour should be put a stop to. If the earl were to hear it, he might, you know—' And the countess shook her head, thinking that she could thus best indicate what the earl might do, if he were to take it into his head to do anything.

Crosbie could not bring himself to hold any very confidential intercourse with the countess about Lily; but he gave a muttered assurance that he should, as a matter of course, make known the truth to Miss Dale with as little delay as possible. He could not say exactly when he would write, nor whether he would write to her or to her mother; but the thing should be done immediately on his return to town.

'If it will make the matter easier, I will write to Mrs. Dale,' said the countess. But to this scheme Mr. Crosbie objected very strongly.

And then a few words were said about the earl. 'I will tell him this afternoon,' said the countess; 'and then you can see him to-morrow morning. I don't suppose he will say very much, you know; and perhaps he may think, – you won't mind my saying it, I'm sure, – that Alexandrina might have done better. But I don't believe that he'll raise any strong objection. There will be something about settlements, and that sort of thing, of course.' Then the countess went away, and Alexandrina was left with her lover for half an hour. When the half-hour was over, he felt that he would have given all that he had in the world to have back the last four-and-twenty hours of his existence. But he had no hope. To jilt Lily Dale would, no doubt, be within his power, but he knew that he could not jilt Lady Alexandrina De Courcy.

On the next morning at twelve o'clock he had his interview with the father, and a very unpleasant interview it was.

He was ushered into the earl's room, and found the great peer standing on the rug, with his back to the fire, and his hands in his breeches pockets.

'So you mean to marry my daughter?' said he. 'I'm not very well, as you see; I seldom am.'

These last words were spoken in answer to Crosbie's greeting. Crosbie had held out his hand to the earl, and had carried his point so far that the earl had been forced to take one of his own out of his pocket, and give it to his proposed son-in-law.

'If your lordship has no objection. I have, at any rate, her permission to ask for yours.'

'I believe you have not any fortune, have you? She's got none; of course you know that?'

'I have a few thousand pounds, and I believe she has as much.'

'About as much as will buy bread to keep the two of you from starving. It's nothing to me. You can marry her if you like; only, look here, I'll have no nonsense. I've had an old woman in with me this morning, – one of those that are here in the house, – telling me some story about some other girl that you have made a fool of. It's nothing to me how much of that sort of thing you may have done, so that you do none of it here. But, – if you play any prank of that kind with me, you'll find that you've made a mistake.'

Crosbie hardly made any answer to this, but got himself out of the room as quickly as he could.

'You'd better talk to Gazebee about the trifle of money you've got,' said the earl. Then he dismissed the subject from his mind, and no doubt imagined that he had fully done his duty by his daughter.

On the day after this, Crosbie was to go. On the last afternoon, shortly before dinner, he was waylaid by Lady Julia, who had passed the day in preparing traps to catch him.

'Mr. Crosbie,' she said, 'let me have one word with you. Is this true?'

'Lady Julia,' he said, 'I really do not know why you should inquire into my private affairs.'

'Yes, sir, you do know; you know very well. That poor

young lady who has no father and no brother, is my neighbour, and her friends are my friends. She is a friend of my own, and being an old woman, I have a right to speak for her. If this is true, Mr. Crosbie, you are treating her like a villain.'

'Lady Julia, I really must decline to discuss the matter with you.'

'I'll tell everybody what a villain you are; I will, indeed; – a villain and a poor weak silly fool. She was too good for you; that's what she was.' Crosbie, as Lady Julia was addressing to him the last words, hurried upstairs away from her, but her ladyship, standing on a landing-place, spoke up loudly, so that no word should be lost on her retreating enemy.

'We positively must get rid of that woman,' the countess, who heard it all, said to Margaretta. 'She is disturbing the house and disgracing herself every day.'

'She went to papa this morning, mamma.'

'She did not get much by that move,' said the countess.

On the following morning Crosbie returned to town, but just before he left the castle he received a third letter from Lily Dale. 'I have been rather disappointed at not hearing this morning,' said Lily, 'for I thought the postman would have brought me a letter. But I know you'll be a better boy when you get back to London, and I won't scold you. Scold you, indeed! No; I'll never scold you, not though I shouldn't hear for a month.'

He would have given all that he had in the world, three times told, if he could have blotted out that visit to Courcy Castle from the past facts of his existence.

CHAPTER XXV

ADOLPHUS CROSBIE SPENDS AN EVENING AT HIS CLUB

Crosbie, as he was being driven from the castle to the nearest station, in a dog-cart* hired from the hotel, could not keep himself from thinking of that other morning, not yet a fortnight past, on which he had left Allington; and as he thought of it he knew that he was a villain. On this morning Alexandrina had not come out from the house to watch his departure, and catch the last glance of his receding figure. As he had not started very early she had sat with him at the breakfast table; but others also had sat there, and when he got up to go, she did no more than smile softly and give him her hand. It had been already settled that he was to spend his Christmas at Courcy; as it had been also settled that he was to spend it at Allington.

Lady Amelia was, of all the family, the most affectionate to him, and perhaps of them all she was the one whose affection was worth the most. She was not a woman endowed with a very high mind or with very noble feelings. She had begun life trusting to the nobility of her blood for everything, and declaring somewhat loudly among her friends that her father's rank and her mother's birth imposed on her the duty of standing closely by her own order. Nevertheless, at the age of thirty-three she had married her father's man of business, under circumstances which were not altogether creditable to her. But she had done her duty in her new sphere of life with some constancy and a fixed purpose; and now that her sister was going to marry, as she had done, a man much below herself in social standing, she was prepared to do her duty as a sister and a sister-in-law.

'We shall be up in town in November, and of course you'll come to us at once. Albert Villa, you know, in Hamilton Terrace, St. John's Wood. We dine at seven, and on Sundays at two; and you'll always find a place. Mind you come to us, and make yourself quite at home. I do so hope you and Mortimer will get on well together.'

'I'm sure we shall,' said Crosbie. But he had had higher hopes in marrying into this noble family than that of becoming intimate with Mortimer Gazebee. What those hopes were he could hardly define to himself now that he had brought himself so near to the fruition of them. Lady De Courcy had certainly promised to write to her first cousin who was Under-Secretary of State for India, with reference to that secretaryship at the General Committee Office; but Crosbie, when he came to weigh in his mind what good might result to him from this, was disposed to think that his chance of obtaining the promotion would be quite as good without the interest of the Under-Secretary of State for India as with it. Now that he belonged, as we may say, to this noble family, he could hardly discern what were the advantages which he had expected from this alliance. He had said to himself that it would be much to have a countess for a mother-in-law; but now, even already, although the possession to which he had looked was not yet garnered, he was beginning to tell himself that the thing was not worth possessing.

As he sat in the train, with a newspaper in his hand, he went on acknowledging to himself that he was a villain. Lady Julia had spoken the truth to him on the stairs at Courcy, and so he confessed over and over again. But he was chiefly angry with himself for this, – that he had been a villain without gaining anything by his villainy; that he had been a villain, and was to lose so much by his villainy. He made comparison between Lily and Alexandrina, and owned to himself, over and over again, that Lily would make the best wife that a man could take to his bosom. As to Alexandrina, he knew the thinness of her character. She would stick by him, no doubt; and in a circuitous, discontented, unhappy way, would probably be true to her duties as a wife and mother. She would be nearly such another as Lady Amelia Gazebee. But was that a prize sufficiently rich to make him contented with his own prowess and skill in winning it? And was that a prize sufficiently rich to justify him to himself for his terrible villainy? Lily Dale he had loved; and he now declared to himself that he could have continued to love her through his whole life. But what was there for any man to love in Alexandrina De Courcy?

While resolving, during his first four or five days at the castle, that he would throw Lily Dale overboard, he had contrived to quiet his conscience by inward allusions to sundry heroes of romance. He had thought of Lothario, Don Juan, and of Lovelace;* and had told himself that the world had ever been full of such heroes. And the world, too, had treated such heroes well; not punishing them at all as villains, but caressing them rather, and calling them curled darlings. Why should not he be a curled darling as well as another? Ladies had ever been fond of the Don Juan character, and Don Juan had generally been popular with men also. And then he named to himself a dozen modern Lotharios, – men who were holding their heads well above water, although it was known that they had played this lady false, and brought that other one to death's door, or perhaps even to death itself. War and love were alike, and the world was prepared to forgive any guile to militants in either camp.

But now that he had done the deed he found himself forced to look at it from quite another point of view. Suddenly that character of Lothario showed itself to him in a different light, and one in which it did not please him to look at it as belonging to himself. He began to feel that it would be almost impossible for him to write that letter to Lily, which it was absolutely necessary that he should write. He was in a position in which his mind would almost turn itself to thoughts of self-destruction as the only means of escape. A fortnight ago he was a happy man, having everything before him that a man ought to want; and now – now that he was the accepted son-in-law of an earl, and the confident expectant of high promotion – he was the most miserable, degraded wretch in the world!

He changed his clothes at his lodgings in Mount Street and went down to his club to dinner. He could do nothing that night. His letter to Allington must, no doubt, be written at once; but, as he could not send it before the next night's post, he was not forced to set to work upon it that evening. As he walked along Piccadilly on his way to St. James's Square, it occurred to him that it might be well to write a short line to Lily, telling her nothing of the truth, – a note written as though his engagement with her was still unbroken, but yet written with care, saying nothing about that engagement, so as to give him a little time. Then he thought that he would telegraph to Bernard and tell everything to him. Bernard would, of course, be prepared to avenge his cousin in some way, but for such vengeance Crosbie felt that he should care little. Lady Julia had told him that Lily was without father or brother, thereby accusing him of the basest cowardice. 'I wish she had a dozen brothers,' he said to himself. But he hardly knew why he expressed such a wish.

He returned to London on the last day of October, and he found the streets at the West End nearly deserted. He thought, therefore, that he should be quite alone at his club, but as he entered the dinner room he saw one of his oldest and most intimate friends standing before the fire. Fowler Pratt was the man who had first brought him into Sebright's, and had given him almost his earliest start on his successful career in life. Since that time he and his friend Fowler Pratt had lived in close communion, though Pratt had always held a certain ascendancy in their friendship. He was in age a few years senior to Crosbie, and was in truth a man of better parts. But he was less ambitious, less desirous of shining in the world, and much less popular with men in general. He was possessed of a moderate private fortune on which he lived in a quiet, modest manner, and was unmarried, not likely to marry, inoffensive, useless, and prudent. For the first few years of Crosbie's life in London he had lived very much with his friend Pratt, and had been accustomed to depend much on his friend's counsel; but latterly, since he had himself become somewhat noticeable, he had found more pleasure in the society of such men as Dale, who were not his superiors either in age or wisdom. But there had been no coolness between him and Pratt, and now they met with perfect cordiality.

'I thought you were down in Barsetshire,' said Pratt.

'And I thought you were in Switzerland.'

'I have been in Switzerland,' said Pratt.

'And I have been in Barsetshire,' said Crosbie. Then they ordered their dinner together.

'And so you're going to be married?' said Pratt, when the waiter had carried away the cheese.

'Who told you that?'

'Well, but you are? Never mind who told me, if I was told the truth.'

'But if it be not true?'

'I have heard it for the last month,' said Pratt, 'and it has been spoken of as a thing certain; and it is true; is it not?'

'I believe it is,' said Crosbie, slowly.

'Why, what on earth is the matter with you, that you speak of it in that way? Am I to congratulate you, or am I not? The lady, I'm told, is a cousin of Dale's.'

Crosbie had turned his chair from the table round to the fire, and said nothing in answer to this. He sat with his glass of sherry in his hand, looking at the coals, and thinking whether it would not be well that he should tell the whole story to Pratt. No one could give him better advice; and no one, as far as he knew his friend, would be less shocked at the telling of such a story. Pratt had no romance about women, and had never pretended to very high sentiments.

'Come up into the smoking-room and I'll tell you all about it,' said Crosbie. So they went off together, and, as the smoking-room was untenanted, Crosbie was able to tell his story.

He found it very hard to tell; – much harder than he had beforehand fancied. 'I have got into terrible trouble,' he began by saying. Then he told how he had fallen suddenly in love with Lily, how he had been rash and imprudent, how nice she was – 'infinitely too good for such a man as I am,' he said; – how she had accepted him, and then how he had repented. 'I should have told you beforehand,' he then said, 'that I was already half-engaged to Lady Alexandrina De Courcy.' The reader, however, will understand that this half-engagement was a fiction.

'And now you mean that you are altogether engaged to her?'

'Exactly so.'

'And that Miss Dale must be told that, on second thoughts, you have changed your mind?'

'I know that I have behaved very badly,' said Crosbie.

'Indeed you have,' said his friend.

'It is one of those troubles in which a man finds himself involved almost before he knows where he is.'

'Well; I can't look at it exactly in that light. A man may amuse himself with a girl, and I can understand his disappointing her and not offering to marry her, – though even that sort of thing isn't much to my taste. But, by George, to make an offer of marriage to such a girl as that in September, to live for a month in her family as her affianced husband, and then coolly go away to another house in October, and make an offer to another girl of higher rank—'

'You know very well that that has had nothing to do with it.'

'It looks very like it. And how are you going to communicate these tidings to Miss Dale?'

'I don't know,' said Crosbie, who was beginning to be very sore.

'And you have quite made up your mind that you'll stick to the earl's daughter?'

The idea of jilting Alexandrina instead of Lily had never as yet presented itself to Crosbie, and now, as he thought of it, could not perceive that it was feasible.

'Yes,' he said, 'I shall marry Lady Alexandrina; – that is, if I do not cut the whole concern, and my own throat into the bargain.'

'If I were in your shoes I think I should cut the whole concern. I could not stand it. What do you mean to say to Miss Dale's uncle?'

'I don't care a —— for Miss Dale's uncle,' said Crosbie. 'If he were to walk in at that door this moment, I would tell him the whole story, without—'

As he was yet speaking, one of the club servants opened the door of the smoking-room, and seeing Crosbie seated in a lounging-chair near the fire, went up to him with a gentleman's card. Crosbie took the card and read the name. 'Mr. Dale, Allington.'

'The gentleman is in the waiting-room,' said the servant.

Crosbie for the moment was struck dumb. He had declared that very moment that he should feel no personal disinclination to meet Mr. Dale, and now that gentleman was within the walls of the club, waiting to see him!

'Who's that?' asked Pratt. And then Crosbie handed him the card. 'Whew-w-w-hew,' whistled Pratt.

'Did you tell the gentleman I was here?' asked Crosbie.

'I said I thought you were upstairs, sir.'

'That will do,' said Pratt. 'The gentleman will no doubt wait for a minute.' And then the servant went out of the room. 'Now, Crosbie, you must make up your mind. By one of these women and all her friends you will ever be regarded as a rascal, and they of course will look out to punish you with such punishment as may come to their hands. You must now choose which shall be the sufferer.'

The man was a coward at heart. The reflection that he might, even now, at this moment, meet the old squire on pleasant terms, – or at any rate not on terms of defiance, pleaded more strongly in Lily's favour than had any other argument since Crosbie had first made up his mind to abandon her. He did not fear personal ill-usage; – he was not afraid lest he should be kicked or beaten; but he did not dare to face the just anger of the angry man.

'If I were you,' said Pratt, 'I would not go down to that man at the present moment for a trifle.'

'But what can I do?'

'Shirk away out of the club. Only if you do that it seems to me that you'll have to go on shirking for the rest of your life.'

'Pratt, I must say that I expected something more like friendship from you.'

'What can I do for you? There are positions in which it is impossible to help a man. I tell you plainly that you have behaved very badly. I do not see that I can help you.'

'Would you see him?'

'Certainly not, if I am to be expected to take your part.'

'Take any part you like, – only tell him the truth.'

'And what is the truth?'

'I was part engaged to that other girl before; and then, when I came to think of it, I knew that I was not fit to marry Miss Dale. I know I have behaved badly; but, Pratt, thousands have done the same thing before.'

'I can only say that I have not been so unfortunate as to reckon any of those thousands among my friends.'

'You mean to tell me, then, that you are going to turn your back on me?' said Crosbie.

'I haven't said anything of the kind. I certainly won't undertake to defend you, for I don't see that your conduct admits of defence. I will see this gentleman if you wish it, and tell him anything that you may desire me to tell him.'

At this moment the servant returned with a note for Crosbie. Mr. Dale had called for paper and envelope, and sent up to him the following missive: – 'Do you intend to come down to me? I know that you are in the house.' 'For heaven's sake go to him,' said Crosbie. 'He is well aware that I was deceived about his niece, – that I thought he was to give her some fortune. He knows all about that, and that when I learned from him that she was to have nothing—'

'Upon my word, Crosbie, I wish you could find another messenger.'

'Ah! you do not understand,' said Crosbie in his agony. 'You think that I am inventing this plea about her fortune now. It isn't so. He will understand. We have talked all this over before, and he knew how terribly I was disappointed. Shall I wait for you here, or will you come to my lodgings? Or I will go down to the Beaufort, and will wait for you there.' And it was finally arranged that he should get himself out of this club and wait at the other for Pratt's report of the interview.

'Do you go down first,' said Crosbie.

'Yes: I had better,' said Pratt. 'Otherwise you may be seen. Mr. Dale would have his eye upon you, and there would be a row in the house.' There was a smile of sarcasm on Pratt's face as he spoke which angered Crosbie even in his misery, and made him long to tell his friend that he would not trouble him with this mission, – that he would manage his own affairs himself; but he was weakened and mentally humiliated by the sense of his own rascality, and had already lost the power of asserting himself, and of maintaining his ascendancy. He was beginning to recognize the fact that he had done that for which he must endure to be kicked, to be kicked morally if not materially; and that it was no longer possible for him to hold his head up without shame.

Pratt took Mr. Dale's note in his hand and went down into the stranger's room. There he found the squire standing, so that he could see through the open door of the room to the foot of the stairs down which Crosbie must descend before he could leave the club. As a measure of first precaution the ambassador closed the door; then he bowed to Mr. Dale, and asked him if he would take a chair.

'I wanted to see Mr. Crosbie,' said the squire.

'I have your note to that gentleman in my hand,' said he. 'He has thought it better that you should have this interview with me; – and under all the circumstances perhaps it is better.'

'Is he such a coward that he dare not see me?'

'There are some actions, Mr. Dale, that will make a coward of any man. My friend Crosbie is, I take it, brave enough in the ordinary sense of the word, but he has injured you.'

'It is all true, then?'

'Yes, Mr. Dale; I fear it is all true.'

'And you call that man your friend! Mr. ——; I don't know what your name is.'

'Pratt; – Fowler Pratt. I have known Crosbie for fourteen years, – ever since he was a boy; and it is not my way, Mr. Dale, to throw over an old friend under any circumstances.'

'Not if he committed a murder?'

'No; not though he committed a murder.'

'If what I hear is true, this man is worse than a murderer.'

'Of course, Mr. Dale, I cannot know what you have heard. I believe that Mr. Crosbie has behaved very badly to your niece, Miss Dale; I believe that he was engaged to marry her, or, at any rate, that some such proposition had been made.'

'Proposition! Why, sir, it was a thing so completely understood that everybody knew it in the county. It was so positively fixed that there was no secret about it. Upon my honour, Mr. Pratt, I can't as yet understand it. If I remember right, it's not a fortnight since he left my house at Allington, – not a fortnight. And that poor girl was with him on the morning of his going as his betrothed bride. Not a fortnight since! And now I've had a letter from an old family friend telling me that he is going to marry one of Lord De Courcy's daughters! I went instantly off to Courcy, and found that he had started for London. Now, I have followed him here; and you tell me it's all true.'

'I am afraid it is, Mr. Dale; too true.'

'I don't understand it; I don't, indeed. I cannot bring myself to believe that the man who was sitting the other day at my table should be so great a scoundrel. Did he mean it all the time that he was there?'

'No; certainly not. Lady Alexandrina De Courcy was, I believe, an old friend of his; – with whom, perhaps, he had had some lover's quarrel. On his going to Courcy they made it up; and this is the result.'

'And that is to be sufficient for my poor girl?'

'You will, of course, understand that I am not defending Mr. Crosbie. The whole affair is very sad, – very sad, indeed. I can only say, in his excuse, that he is not the first man who has behaved badly to a lady.'

'And that is his message to me, is it? And that is what I am to tell my niece? "You have been deceived by a scoundrel. But what then? You are not the first!" Mr. Pratt, I give you my word as a gentleman, I do not understand it. I have lived a good deal out of the world, and am, therefore, perhaps, more astonished than I ought to be.'

'Mr. Dale, I feel for you—'

'Feel for me! What is to become of my girl? And do you suppose that I will let this other marriage go on; that I will not tell the De Courcys, and all the world at large, what sort of a man this is; – that I will not get at him to punish him? Does he think that I will put up with this?'

'I do not know what he thinks; I must only beg that you will not mix me up in the matter – as though I were a participator in his offence.'

'Will you tell him from me that I desire to see him?'

'I do not think that that would do any good.'

'Never mind, sir; you have brought me this message; will you have the goodness now to take back mine to him?'

'Do you mean at once – this evening, – now?'

'Yes, at once – this evening, – now; – this minute.'

'Ah; he has left the club; he is not here now, he went when I came to you.'

'Then he is a coward as well as a scoundrel.' In answer to which assertion, Mr. Fowler Pratt merely shrugged his shoulders.

'He is a coward as well as a scoundrel. Will you have the kindness to tell your friend from me that he is a coward and a scoundrel, – and a liar, sir.'

'If it be so, Miss Dale is well quit of her engagement.'

'That is your consolation, is it? That may be all very well nowadays; but when I was a young man, I would sooner have burnt out my tongue than have spoken in such a way on such a subject. I would, indeed. Good-night, Mr. Pratt. Pray make your friend understand that he has not yet seen the last of the Dales; although, as you hint, the ladies of that family will no doubt have learned that he is not fit to associate with them.' Then, taking up his hat, the squire made his way out of the club.

'I would not have done it,' said Pratt to himself, 'for all the beauty, and all the wealth, and all the rank that ever were owned by a woman.'

CHAPTER XXVI

LORD DE COURCY IN THE BOSOM OF HIS FAMILY

LADY JULIA DE GUEST had not during her life written many letters to Mr. Dale of Allington, nor had she ever been very fond of him. But when she felt certain how things were going at Courcy, or rather, as we may say, how they had already gone, she took pen in hand, and sat herself to work, doing, as she conceived, her duty by her neighbour.

'My dear Mr. Dale (she said),

'I believe I need make no secret of having known that your niece Lilian is engaged to Mr. Crosbie, of London. I think it proper to warn you that if this be true Mr. Crosbie is behaving himself in a very improper manner here. I am not a person who concerns myself much in the affairs of other people; and under ordinary circumstances, the conduct of Mr. Crosbie would be nothing to me, – or, indeed, less than nothing; but I do to you as I would wish that others should do unto me. I believe it is only too true that Mr. Crosbie has proposed to Lady Alexandrina De Courcy, and been accepted by her. I think you will believe that I would not say this without warrant, and if there be anything in it, it may be well, for the poor young lady's sake, that you should put yourself in the way of learning the truth.

'Believe me to be yours sincerely,
'Julia De Guest.

'Courcy Castle, Thursday.'

The squire had never been very fond of any of the De Guest family, and had, perhaps, liked Lady Julia the least of them all. He was wont to call her a meddling old woman, – remembering her bitterness and pride in those now long bygone days in which the gallant major had run off with Lady Fanny. When he first received this letter, he did not, on the first reading of it, believe a word of its contents. 'Cross-grained old harridan,' he said out loud to his nephew. 'Look what that aunt of yours has written to me.' Bernard read the letter twice, and as he did so his face became hard and angry.

'You don't mean to say you believe it?' said the squire.

'I don't think it will be safe to disregard it.'

'What! you think it possible that your friend is doing as she says.'

'It is certainly possible. He was angry when he found that Lily had no fortune.'

'Heavens, Bernard! And you can speak of it in that way?'

'I don't say that it is true; but I think we should look to it. I will go to Courcy Castle and learn the truth.'

The squire at last decided that he would go. He went to Courcy Castle, and found that Crosbie had started two hours before his arrival. He asked for Lady Julia, and learned from her that Crosbie had actually left the house as the betrothed husband of Lady Alexandrina.

'The countess, I am sure, will not contradict it, if you will see her,' said Lady Julia. But this the squire was unwilling to do. He would not proclaim the wretched condition of his niece more loudly than was necessary, and therefore he started on his pursuit of Crosbie. What was his success on that evening we have already learned.

Both Lady Alexandrina and her mother heard of Mr. Dale's arrival at the castle, but nothing was said between them on the subject. Lady Amelia Gazebee heard of it also, and she ventured to discuss the matter with her sister.

'You don't know exactly how far it went, do you?'

'No; yes; – not exactly, that is,' said Alexandrina.

'I suppose he did say something about marriage to the girl?'

'Yes, I'm afraid he did.'

'Dear, dear! It's very unfortunate. What sort of people are those Dales? I suppose he talked to you about them.'

'No, he didn't; not very much. I dare say she is an artful, sly thing! It's a great pity men should go on in such a way.'

'Yes, it is,' said Lady Amelia. 'And I do suppose that in this case the blame has been more with him than with her. It's only right I should tell you that.'

'But what can I do?'

'I don't say you can do anything; but it's as well you should know.'

'But I don't know, and you don't know; and I can't see that there is any use talking about it now. I knew him a long while before she did, and if she has allowed him to make a fool of her, it isn't my fault.'

'Nobody says it is, my dear.'

'But you seem to preach to me about it. What can I do for the girl? The fact is, he don't care for her a bit, and never did.'

'Then he shouldn't have told her that he did.'

'That's all very well, Amelia; but people don't always do exactly all that they ought to do. I suppose Mr. Crosbie isn't the first man that has proposed to two ladies. I dare say it was wrong, but I can't help it. As to Mr. Dale coming here with a tale of his niece's wrongs, I think it very absurd, – very absurd indeed. It makes it look as though there had been a scheme to catch Mr. Crosbie, and it's my belief that there was such a scheme.'

'I only hope that there'll be no quarrel.'

'Men don't fight duels nowadays, Amelia.'

'But do you remember what Frank Gresham did to Mr.

Moffat when he behaved so badly to poor Augusta?'

'Mr. Crosbie isn't afraid of that kind of thing. And I always thought that Frank was very wrong, – very wrong indeed. What's the good of two men beating each other in the street?'

'Well; I'm sure I hope there'll be no quarrel. But I own I don't like the look of it. You see the uncle must have known all about it, and have consented to the marriage, or he would not have come here.'

'I don't see that it can make any difference to me, Amelia.'

'No, my dear, I don't see that it can. We shall be up in town soon, and I will see as much as possible of Mr. Crosbie. The marriage, I hope, will take place soon.'

'He talks of February.'

'Don't put it off, Alley, whatever you do. There are so many slips, you know, in these things.'

'I'm not a bit afraid of that,' said Alexandrina, sticking up her head.

'I dare say not; and you may be sure that we will keep an eye on him. Mortimer will get him up to dine with us as often as possible, and as his leave of absence is all over, he can't get out of town. He is to be here at Christmas, isn't he?'

'Of course he is.'

'Mind you keep him to that. And as to these Dales, I would be very careful, if I were you, not to say anything unkind of them to any one. It sounds badly in your position.' And with this last piece of advice Lady Amelia Gazebee allowed the subject to drop.

On that day Lady Julia returned to her own home. Her adieux to the whole family at Courcy Castle were very cold, but about Mr. Crosbie and his lady-love at Allington she said no further word to any of them. Alexandrina did not show herself at all on the occasion, and indeed had not spoken to her enemy since that evening on which she had felt herself constrained to retreat from the drawing-room.

'Good-bye,' said the countess. 'You have been so good to come, and we have enjoyed it so much.'

'I thank you very much. Good morning,' said Lady Julia, with a stately courtesy.

'Pray remember me to your brother. I wish we could have seen him; I hope he has not been hurt by the – the bull.' And then Lady Julia went her way.

'What a fool I have been to have that woman in the house,' said the countess, before the door was closed behind her guest's back.

'Inded you have,' said Lady Julia, screaming back through the passage. Then there was a long silence, then a suppressed titter, and after that a loud laugh.

'Oh, mamma, what shall we do?' said Lady Amelia.

'Do!' said Margaretta; 'why should we do anything? She has heard the truth for once in her life.'

'Dear Lady Dumbello, what will you think of us?' said the countess, turning round to another guest, who was also just about to depart. 'Did any one ever know such a woman before?'

'I think she's very nice,' said Lady Dumbello, smiling.

'I can't quite agree with you there,' said Lady Clandidlem. 'But I do believe she means to do her best. She is very charitable, and all that sort of thing.'

'I'm sure I don't know,' said Rosina. 'I asked her for a subscription to the mission for putting down the Papists in the west of Ireland, and she refused me point-blank.'

'Now, my dear, if you're quite ready,' said Lord Dumbello, coming into the room. Then there was another departure; but on this occasion the countess waited till the doors were shut, and the retreating footsteps were no longer heard. 'Have you observed,' said she to Lady Clandidlem, 'that she has not held her head up since Mr. Palliser went away?'

'Indeed I have,' said Lady Clandidlem. 'As for poor Dumbello, he's the blindest creature I ever saw in my life.'

'We shall hear of something before next May,' said Lady De Courcy, shaking her head; 'but for all that she'll never be Duchess of Omnium.'

'I wonder what your mamma will say of me when I go away to-morrow,' said Lady Clandidlem to Margaretta, as they walked across the hall together.

'She won't say that you are going to run away with any gentleman,' said Margaretta.

'At any rate, not with the earl,' said Lady Clandidlem. 'Ha! ha! ha! Well, we are all very good-natured, are we not? The best is that it means nothing.'

Thus by degrees all the guests went, and the family of the De Courcys was left to the bliss of their own domestic circle. This, we may presume, was not without its charms, seeing that there were so many feelings in common between the mother and her children. There were drawbacks to it, no doubt, arising, perhaps chiefly from the earl's bodily infirmities. 'When your father speaks to me,' said Mrs. George to her husband, 'he puts me in such a shiver that I cannot open my mouth to answer him.'

'You should stand up to him,' said George. 'He can't hurt you, you know. Your money's your own; and if I'm ever to be the heir, it won't be by his doing.'

'But he gnashes his teeth at me.'

'You shouldn't care for that, if he don't bite. He used to gnash them at me; and when I had to ask him for money I didn't like it; but now I don't mind him a bit. He threw the peerage at me one day, but it didn't go within a yard of my head.'

'If he throws anything at me, George, I shall drop upon the spot.'

But the countess had a worse time with the earl than any of her children. It was necessary that she should see him daily, and necessary also that she should say much that he did not like to hear, and make many petitions that caused him to gnash his teeth. The earl was one of those men who could not endure to live otherwise than expensively, and yet was made miserable by every recurring expense. He ought to have known by this time that butchers, and bakers, and corn-chandlers,* and coal-merchants will not supply their goods for nothing; and yet it always seemed as though he had expected that at this special period they would do so. He was an embarrassed man, no doubt, and had not been fortunate in his speculations at Newmarket or Homburg; * but, nevertheless, he had still the means of living without daily torment; and it must be supposed that his self-imposed sufferings, with regard to money, rose rather from his disposition than his necessities. His wife never knew whether he were really ruined, or simply pretending it. She had now become so used to her position in this respect, that she did not allow fiscal considerations to mar her happiness. Food and clothing had always come to her, – including velvet gowns, new trinkets, and a man-cook, – and she presumed that they would continue to come. But that daily conference with her husband was almost too much or her. She struggled to avoid it; and, as far as the ways and means were concerned, would have allowed them to arrange themselves, if he would only have permitted it. But he insisted on seeing her daily in his own sitting-room; and she had acknowledged to her favourite daughter, Margaretta, that those half-hours would soon be the death of her. 'I sometimes feel,' she said, 'that I am going mad before I can get out.' And she reproached herself, probably without reason, in that she had brought much of this upon herself. In former days the earl had been constantly away from home, and the countess had complained. Like many other women she had not known when she was well off. She had complained, urging upon her lord that he should devote more of his time to his own hearth. It is probable that her ladyship's remonstrances had been less efficacious than the state of his own health in producing that domestic constancy which he now practised; but it is certain that she looked back with bitter regret to the happy days when she was deserted, jealous, and querulous. 'Don't you wish we could get Sir Omicron to order him to the German Spas?' she had said to Margaretta. Now Sir Omicron was the great London physician, and might, no doubt, do much in that way.

But no such happy order had as yet been given; and, as far as the family could foresee, paterfamilias intended to pass the winter with them at Courcy. The guests, as I have said, were all gone and none but the family were in the house when her ladyship waited upon her lord one morning at twelve o'clock, a few days after Mr. Dale's visit to the castle. He always breakfasted alone, and after breakfast found in a French novel and a cigar what solace those innocent recreations were still able to afford him. When the novel no longer excited him and when he was saturated with smoke, he would send for his wife. After that his valet would dress him. 'She gets it worse than I do,' the man declared in the servants' hall; 'and minds it a deal more. I can give warning, and she can't.'

'Better? No, I ain't better,' the husband said, in answer to his wife's inquiries. 'I never shall be better while you keep that cook in the kitchen.'

'But where are we to get another if we send him away?'

'It's not my business to find cooks. I don't know where you're to get one. It's my belief you won't have a cook at all before long. It seems you have got two extra men into the house without telling me.'

'We must have servants, you know, when there is company. It wouldn't do to have Lady Dumbello here, and no one to wait on her.'

'Who asked Lady Dumbello? I didn't.'

'I'm sure, my dear, you liked having her here.'

'D—— Lady Dumbello!' and then there was a pause. The countess had no objection whatsoever to the above proposition, and was rejoiced that that question of the servants was allowed to slip aside, through the aid of her ladyship.

'Look at that letter from Porlock,' said the earl; and he pushed over to the unhappy mother a letter from her eldest son. Of all her children he was the one she loved the best; but him she was never allowed to see under her own roof. 'I sometimes think that he is the greatest rascal with whom I ever had occasion to concern myself,' said the earl.

She took the letter and read it. The epistle was certainly not one which a father could receive with pleasure from his son; but the disagreeable nature of its contents was the fault rather of the parent than of the child. The writer intimated that certain money due to him had not been paid with necessary punctuality, and that unless he received it, he should instruct his lawyer to take some authorized legal proceedings. Lord De Courcy had raised certain moneys on the family property, which he could not have raised without the co-operation of his heir, and had bound himself, in return for that co-operation, to pay a certain fixed income to his eldest son. This he regarded as an allowance from himself; but Lord Porlock regarded it as his own, by lawful claim. The son had not worded his letter with any affectionate phraseology. 'Lord Porlock begs to inform Lord De Courcy—' Such had been the commencement.

'I suppose he must have his money; else how can he live?' said the countess, trembling.

'Live!' shouted the earl. 'And so you think it proper that he should write such a letter as that to his father!'

'It is all very unfortunate,' she replied.

'I don' know where the money's to come from. As for him, if he were starving, it would serve him right. He's a disgrace to the name and the family. From all I hear, he won't live long.'

'Oh, De Courcy, don't talk of it in that way!'

'What way am I to talk of it? If I say that he's my greatest comfort, and living as becomes a nobleman, and as a fine, healthy man of his age, with a good wife and a lot of legitimate children, will that make you believe it? Women are such fools. Nothing that I say will make him worse than he is.'

'But he may reform.'

'Reform! He's over forty, and when I last saw him he looked nearly sixty. There; – you may answer his letter; I won't.'

'And about the money?'

'Why doesn't he write to Gazebee about his dirty money? Why does he trouble me? I haven't got his money. Ask Gazebee about his money. I won't trouble myself about it.' Then there was another pause, during which the countess folded the letter, and put it in her pocket.

'How long is George going to remain here with that woman?' he asked.

'I'm sure she is very harmless,' pleaded the countess.

'I always think when I see her that I'm sitting down to dinner with my own housemaid. I never saw such a woman. How he can put up with it! But I don't suppose he cares for anything.'

'It has made him very steady.'

'Steady!'

'And as she will be confined before long it may be as well that she should remain here. If Porlock doesn't marry, you know—'

'And so he means to live here altogether, does he? I'll tell you what it is, – I won't have it. He's better able to keep a house over his own head and his wife's than I am to do it for them, and so you may tell them. I won't have it. D'ye hear?' Then there was another short pause. 'D'ye hear?' he shouted at her.

'Yes; of course I hear. I was only thinking you wouldn't wish me to turn them out, just as her confinement is coming on.'

'I know what that means. Then they'd never go. I won't have it; and if you don't tell them I will.' In answer to this Lady De Courcy promised that she would tell them, thinking perhaps that the earl's mode of telling might not be beneficial in that particular epoch which was now coming in the life of Mrs. George.

'Did you know,' said he, breaking out on a new subject, 'that a man had been here named Dale, calling on somebody in this house?' In answer to which the countess acknowledged that she had known it.

'Then why did you keep it from me?' And that gnashing of the teeth took place which was so specially objectionable to Mrs. George.

'It was a matter of no moment. He came to see Lady Julia De Guest.'

'Yes; but he came about that man Crosbie.'

'I suppose he did.'

'Why have you let that girl be such a fool? You'll find he'll play her some knave's trick.'

'Oh dear, no.'

'And why should she want to marry such a man as that?'

'He's quite a gentleman, you know, and very much thought of in the world. It won't be at all bad for her, poor thing. It is so very hard for a girl to get married nowadays without money.'

'And so they're to take up with anybody. As far as I can

see, this is a worse affair than that of Amelia.'

'Amelia has done very well, my dear.'

'Oh, if you call it doing well for your girls, I don't. I call it doing uncommon badly; about as bad as they well can do. But it's your affair. I have never meddled with them, and don't intend to do it now.'

'I really think she'll be happy, and she is devotedly attached to the young man.'

'Devotedly attached to the young man!' The tone and manner in which the earl repeated these words were such as to warrant an opinion that his lordship might have done very well on the stage had his attention been called to that profession. 'It makes me sick to hear people talk in that way. She wants to get married, and she's a fool for her pains; – I can't help that; only remember that I'll have no nonsense here about that other girl. If he gives me trouble of that sort, by ——, I'll be the death of him. When is the marriage to be?'

'They talk of February.'

'I won't have any tomfoolery and expense. If she chooses to marry a clerk in an office, she shall marry him as clerks are married.'

'He'll be the secretary before that, De Courcy.'

'What difference does that make? Secretary, indeed! What sort of men do you suppose secretaries are? A beggar that came from nobody knows where! I won't have any tomfoolery; – d'ye hear?' Whereupon the countess said that she did hear, and soon afterwards managed to escape. The valet then took his turn; and repeated, after his hour of service, that 'Old Nick' in his tantrums had been more like the Prince of Darkness than ever.

CHAPTER XXVII

'ON MY HONOUR, I DO NOT UNDERSTAND IT'

In the meantime Lady Alexandrina endeavoured to realize to herself all the advantages and disadvantages of her own position. She was not possessed of strong affections, nor of depth of character, nor of high purpose; but she was no fool, nor was she devoid of principle. She had asked herself many times whether her present life was so happy as to make her think that a permanent continuance in it would suffice for her desires, and she had always replied to herself that she would fain change to some other life if it were possible. She had also questioned herself as to her rank, of which she was quite sufficiently proud, and had told herself that she could not degrade herself in the world without a heavy pang. But she had at last taught herself to believe that she had more to gain by becoming the wife of such a man as Crosbie than by remaining as an unmarried daughter of her father's house. There was much in her sister Amelia's position which she did not envy, but there was less to envy in that of her sister Rosina. The Gazebee house in St. John's Wood Road was not so magnificent as Courcy Castle; but then it was less dull, less embittered by torment, and was moreover her sister's own.

'Very many do marry commoners,' she had said to Margaretta.

'Oh, yes, of course. It makes a difference, you know, when a man has a fortune.'

Of course it did make a difference. Crosbie had no fortune, was not even so rich as Mr. Gazebee, could keep no carriage, and would have no country house. But then he was a man of fashion, was more thought of in the world than Mr. Gazebee, might probably rise in his own profession, – and was at any rate thoroughly presentable. She would have preferred a gentleman with 5,000*l*. a year; but then as no gentleman with 5,000*l*. a year came that way, would she not be happier with Mr. Crosbie than she would be with no husband at all? She was not very much in love with Mr. Crosbie, but she thought that she could live with him comfortably, and that on the whole it would be a good thing to be married.

And she made certain resolves as to the manner in which she would do her duty by her husband. Her sister Amelia was paramount in her own house, ruling indeed with a moderate, endurable dominion, and ruling much to her husband's advantage. Alexandrina feared that she would not be allowed to rule, but she could at any rate try. She would do all in her power to make him comfortable, and would be specially careful not to irritate him by any insistence on her own higher rank. She would be very meek in this respect; and if children should come she would be as painstaking about them as though her own father had been merely a clergyman or a lawyer. She thought also much about poor Lilian Dale, asking herself sundry questions, with an idea of being high-principled as to her duty in that respect. Was she wrong in taking Mr. Crosbie away from Lilian Dale? In answer to these questions she was able to assure herself comfortably that she was not wrong. Mr. Crosbie would not, under any circumstances, marry Lilian Dale. He had told her so more than once, and that in a solemn way. She could therefore be doing no harm to Lilian Dale. If she entertained any inner feeling that Crosbie's fault in jilting Lilian Dale was less than it would have been had she herself not been an earl's daughter, – that her own rank did in some degree extenuate her lover's falseness, – she did not express it in words even to herself.

She did not get very much sympathy from her own family. 'I'm afraid he does not think much of his religious duties. I'm told that young men of that sort seldom do,' said Rosina. 'I don't say you're wrong,' said Margaretta. 'By no means. Indeed I think less of it now than I did when Amelia did the same thing. I shouldn't do it myself, that's all.' Her father told her that he supposed she knew her own mind. Her mother, who endeavoured to comfort and in some sort to congratulate her, nevertheless harped constantly on the fact that she was marrying a man without rank and without a fortune. Her congratulations were apologetic, and her comfortings took the guise of consolation. 'Of course you won't be rich, my dear; but I really think you'll do very well. Mr. Crosbie may be received anywhere, and you never need be ashamed of him.' By which the countess implied that her elder married daughter was occasionally called on to be ashamed of her husband. 'I wish he could keep a carriage for you, but perhaps that will come some day.' Upon the whole

Alexandrina did not repent, and stoutly told her father that she did know her own mind.

During all this time Lily Dale was as yet perfect in her happiness. That delay of a day or two in the receipt of the expected letter from her lover had not disquieted her. She had promised him that she would not distrust him, and she was firmly minded to keep her promises. Indeed no idea of breaking it came to her at this time. She was disappointed when the postman would come and bring no letter for her, – disappointed, as is the husbandman when the longed-for rain does not come to refresh the parched earth; but she was in no degree angry. 'He will explain it,' she said to herself. And she assured Bell that men never recognized the hunger and thirst after letters which women feel when away from those whom they love.

Then they heard at the Small House that the squire had gone away from Allington. During the last few days Bernard had not been much with them, and now they heard the news, not through their cousin, but from Hopkins. 'I really can't undertake to say, Miss Bell, where the master's gone to. It's not likely the master'd tell me where he was going to; not unless it was about seeds, or the likes of that.'

'He has gone very suddenly,' said Bell.

'Well, miss, I've nothing to say to that. And why shouldn't he go suddenly if he likes? I only know he had his gig, and went to the station. If you was to bury me alive I couldn't tell you more.'

'I should like to try,' said Lily as they walked away. 'He is such a cross old thing. I wonder whether Bernard has gone with my uncle.' And then they thought no more about it.

On the day after that Bernard came down to the Small House, but he said nothing by way of accounting for the squire's absence. 'He is in London, I know,' said Bernard.

'I hope he'll call on Mr. Crosbie,' said Lily. But on this subject Bernard said not a word. He did ask Lily whether she had heard from Adolphus, in answer to which she replied, with as indifferent a voice as she could assume, that she had not had a letter that morning.

'I shall be angry with him if he's not a good correspondent,' said Mrs. Dale, when she and Lily were alone together.

'No, mamma, you mustn't be angry with him. I won't let you be angry with him. Please to remember he's my lover and not yours.'

'But I can see you when you watch for the postman.'

'I won't watch for the postman any more if it makes you have bad thoughts about him. Yes, they are bad thoughts. I won't have you think that he doesn't do everything that is right.'

On the next morning the postman brought a letter, or rather a note, and Lily at once saw that it was from Crosbie. She had contrived to intercept it near the back door, at which the postman called, so that her mother should not watch her watchings, nor see her disappointment if none should come. 'Thank you, Jane,' she said, very calmly, when the eager, kindly girl ran to her with the little missive; and she walked off to some solitude, trying to hide her impatience. The note had seemed so small that it amazed her; but when she opened it the contents amazed her more. There was neither beginning nor end. There was no appellation of love, and no signature. It contained but two lines. 'I will write to you at length to-morrow. This is my first day in London, and I have been so driven about that I cannot write.' That was all, and it was scrawled on half a sheet of note-paper. Why, at any rate, had he not called her his dearest Lily? Why had he not assured her that he was ever her own? Such expressions, meaning so much, may be conveyed in a glance of the pen. 'Ah,' she said, 'if he knew how I hunger and thirst after his love!'

She had but a moment left to her before she must join her mother and sister, and she used that moment in remembering her promise. 'I know it is all right,' she said to herself. 'He does not think of these things as I do. He had to write at the last moment, – as he was leaving his office.' And then, with a quiet, smiling face, she walked into the breakfast-parlour.

'What does he say, Lily?' asked Bell.

'What would you give to know?' said Lily.

'I wouldn't give twopence for the whole of it,' said Bell. 'When you get anybody to write to you letters, I wonder whether you'll show them to everybody?'

'But if there's any special London news, I suppose we might hear it,' said Mrs. Dale.

'But suppose there's no special London news, mamma. The poor man had only been in town one day, you know: and there never is any news at this time of the year.'

'Had he seen uncle Christopher?'

'I don't think he had; but he doesn't say. We shall get all the news from him when he comes. He cares much more about London news than Adolphus does.' And then there was no more said about the letter.

But Lily had read her two former letters over and over again at the breakfast-table; and though she had not read them aloud, she had repeated many words out of them, and had so annotated upon them that her mother, who had heard her, could have almost rewritten them. Now, she did not even show the paper; and then her absence, during which she had read the letter, had hardly exceeded a minute or two. All this Mrs. Dale observed, and she knew that her daughter had been again disappointed.

In fact that day Lily was very serious, but she did not appear to be unhappy. Early after breakfast Bell went over to the parsonage, and Mrs. Dale and her youngest daughter sat together over their work. 'Mamma,' she said, 'I hope you and I are not to be divided when I go to live in London.'

'We shall never be divided in heart, my love.'

'Ah, but that will not be enough for happiness, though perhaps enough to prevent absolute unhappiness. I shall want to see you, touch you, and pet you as I do now.' And she came and knelt on the cushion at her mother's feet.

'You will have some one else to caress and pet, – perhaps many others.'

'Do you mean to say that you are going to throw me off, mamma?'

'God forbid, my darling. It is not mothers that throw off their children. What shall I have left when you and Bell are gone from me?'

'But we will never be gone. That's what I mean. We are to be just the same to you always, even though we are married. I must have my right to be here as much as I have it now; and, in return, you shall have your right to be there. His house must be a home to you, – not a cold place which you may visit now and again, with your best clothes on. You know what I mean, when I say that we must not be divided.'

'But, Lily—'

'Well, mamma?'

'I have no doubt we shall be happy together, – you and I.'

'But you were going to say more than that.'

'Only this, – that your house will be his house, and will be full without me. A daughter's marriage is always a painful parting.'

'Is it, mamma?'

'Not that I would have it otherwise than it is. Do not think that I would wish to keep you at home with me. Of course you will both marry and leave me. I hope that he to whom you are going to devote yourself may be spared to love you and protect you.' Then the widow's heart became too full, and she put away her child from her that she might hide her face.

'Mamma, mamma, I wish I was not going from you.'

'No, Lily; do not say that. I should not be contented with life if I did not see both my girls married. I think that it is the only lot which can give to a woman perfect content and satisfaction. I would have you both married. I should be the most selfish being alive if I wished otherwise.'

'Bell will settle herself near you, and then you will see more of her and love her better than you do me.'

'I shall not love her better.'

'I wish she would marry some London man, and then you would come with us, and be near to us. Do you know, mamma, I sometimes think you don't like this place here.'

'Your uncle has been very kind to give it to us.'

'I know he has; and we have been very happy here. But if Bell should leave you—'

'Then should I go also. Your uncle has been very kind, but I sometimes feel that his kindness is a burden which I should not be strong enough to bear solely on my own shoulders. And what should keep me here, then?' Mrs. Dale as she said this felt that the 'here' of which she spoke extended beyond the limits of the home which she held through the charity of her brother-in-law. Might not all the world, as far as she was concerned in it, be contained in that 'here'? How was she to live if both her children should be taken away from her? She had already realized the fact that Crosbie's house could never be a home to her, – never even a temporary home. Her visits there must be of that full-dressed nature to which Lily had alluded. It was impossible that she could explain this to Lily. She would not prophesy that the hero of her girl's heart would be inhospitable to his wife's mother; but such had been her reading of Crosbie's character. Alas, alas, as matters were to go, his hospitality or inhospitality would be matter of small moment to them.

Again in the afternoon the two sisters were together, and Lily was still more serious than her wont. It might almost have been gathered from her manner that this marriage of hers was about to take place at once, and that she was preparing to leave her home. 'Bell,' she said, 'I wonder why Dr. Crofts never comes to see us now?'

'It isn't a month since he was here, at our party.'

'A month! But there was a time when he made some pretext for being here every other day.'

'Yes, when mamma was ill.'

'Ay, and since mamma was well, too. But I suppose I must not break the promise you made me give you. He's not to be talked about even yet, is he?'

'I didn't say he was not to be talked about. You know what I meant, Lily; and what I meant then, I mean now.'

'And how long will it be before you mean something else? I do hope it will come some day, – I do indeed.'

'It never will, Lily. I once fancied that I cared for Dr. Crofts, but it was only fancy. I know it, because—' She was going to explain that her knowledge on that point was assured to her, because since that day she had felt that she might have learned to love another man. But that other man had been Mr. Crosbie, and so she stopped herself.

'I wish he would come and ask you himself.'

'He will never do so. He would never ask such a question without encouragement, and I shall give him none. Nor will he ever think of marrying till he can do so without, – without what he thinks to be imprudence as regards money. He has courage enough to be poor himself without unhappiness, but he has not courage to endure poverty with a wife. I know well what his feelings are.'

'Well, we shall see,' said Lily. 'I shouldn't wonder if you were married first now, Bell. For my part I'm quite prepared to wait for three years.'

Late on that evening the squire returned to Allington, Bernard having driven over to meet him at the station. He had telegraphed to his nephew that he would be back by a late train, and no more than this had been heard from him since he went. On that day Bernard had seen none of the ladies at the Small House. With Bell at the present moment it was impossible that he should be on easy terms. He could not meet her alone without recurring to the one special subject of interest between them, and as to that he did not choose to speak without much forethought. He had not known himself, when he had gone about his wooing so lightly, thinking it a slight thing, whether or no he might be accepted. Now it was no longer a slight thing to him. I do not know that it was love that made him so eager; not good, honest, downright love. But he had set his heart upon the object, and with the wilfulness of a Dale was determined that it should be his. He had no remotest idea of giving up his cousin, but he had at last persuaded himself that she was not to be won without some toil, and perhaps also some delay.

Nor had he been in a humour to talk either to Mrs. Dale or to Lily. He feared that Lady Julia's news was true, – that there might be in it something of truth; and while thus in doubt he could not go down to the Small House. So he hung about the place by himself, with a cigar in his mouth, fearing that something evil was going to happen, and when the message came for him, almost shuddered as he seated himself in the gig. What would it become him to do in this emergency if Crosbie had truly been guilty of the villainy with which Lady Julia had charged him? Thirty years ago he would have called the man out, and shot at him till one of them was hit. Nowadays it was hardly possible for a man to do that; and yet what would the world say of him if he allowed such an injury as this to pass without vengeance?

His uncle, as he came forth from the station with his travelling-bag in his hand, was stern, gloomy, and silent. He came out and took his place in the gig almost without speaking. There were strangers about, and therefore his nephew at first could ask no question, but as the gig turned the corner out of the station-house yard he demanded the news.

'What have you heard?' he said.

But even then the squire did not answer at once. He shook his head, and turned away his face, as though he did not choose to be interrogated.

'Have you seen him, sir?' asked Bernard.

'No, he has not dared to see me.'

'Then it is true?'

'True? – yes, it is all true. Why did you bring the scoundrel here? It has been your fault.'

'No, sir; I must contradict that. I did not know him for a scoundrel.'

'But it was your duty to have known him before you brought him here among them. Poor girl! how is she to be told?'

'Then she does not know it?'

'I fear not. Have you seen them?'

'I saw them yesterday, and she did not know it then; she may have heard it to-day.'

'I don't think so. I believe he has been too great a coward to write to her. A coward indeed! How can any man find the courage to write such a letter as that?'

By degrees the squire told his tale. How he had gone to Lady Julia, had made his way to London, had tracked Crosbie to his club, and had there learned the whole truth from Crosbie's friend, Fowler Pratt, we already know. 'The coward escaped me while I was talking to the man he sent down,' said the squire. 'It was a concerted plan, and I think he was right. I should have brained him in the hall of the club.' On the following morning Pratt had called upon him at his inn with Crosbie's apology. 'His apology!' said the squire. 'I have it in my pocket. Poor reptile; wretched worm of a man! I cannot understand it. On my honour, Bernard, I do not understand it. I think men are changed since I knew much of them. It would have been impossible for me to write such a letter as that.' He went on telling how Pratt had brought him this letter, and had stated that Crosbie declined an interview. 'The gentleman had the goodness to assure me that no good could come from such a meeting. "You mean," I answered, "that I cannot touch pitch and not be defiled!" He acknowledged that the man was pitch. Indeed, he could not say a word for his friend.'

'I know Pratt. He is a gentleman. I am sure he would not excuse him.'

'Excuse him! How could any one excuse him? Words would not be found to excuse him.' And then he sat silent for some half mile. 'On my honour, Bernard, I can hardly yet bring myself to believe it. It is so new to me. It makes me feel that the world is changed, and that it is no longer worth a man's while to live in it.'

'And he is engaged to this other girl?'

'Oh, yes; with the full consent of the family. It is all arranged, and the settlements, no doubt, in the lawyer's hands by this time. He must have gone away from here determined to throw her over. Indeed, I don't suppose he ever meant to marry her. He was just passing away his time here in the country.'

'He meant it up to the time of his leaving.'

'I don't think it. Had he found me able and willing to give her a fortune, he might, perhaps, have married her. But I don't think he meant it for a moment after I told him that she would have nothing. Well, here we are. I may say that I never before came back to my own house with so sore a heart.'

They sat silently over their supper, the squire showing more open sorrow than might have been expected from his character. 'What am I to say to them in the morning?' he repeated over and over again. 'How am I to do it? And if I tell the mother, how is she to tell her child?'

'Do you think that he has given no intimation of his purpose?'

'As far as I can tell, none. That man Pratt knew that he had not done so yesterday afternoon. I asked him what were the intentions of his blackguard friend, and he said that he did not know – that Crosbie would probably have written to me. Then he brought me this letter. There it is,' and the squire threw the letter over the table; 'read it and let me have it back. He thinks probably that the trouble is now over as far as he is concerned.'

It was a vile letter to have written – not because the language was bad, or the mode of expression unfeeling, or the facts falsely stated – but because the thing to be told was in itself so vile. There are deeds which will not bear a gloss – sins as to which the perpetrator cannot speak otherwise than as a reptile; circumstances which change a man and put upon him the worthlessness of vermin. Crosbie had struggled hard to write it, going home to do it after his last interview on that night with Pratt. But he had sat moodily in his chair at his lodgings, unable to take the pen in hand. Pratt was to come to him at his office on the following morning, and he went to bed resolving that he would write it at his desk. On the next day Pratt was there before a word of it had been written.

'I can't stand this kind of thing,' said Pratt. 'If you mean me to take it, you must write it at once.' Then, with inward groaning, Crosbie sat himself at his table, and the words at last were forthcoming. Such words as they were! 'I know that I can have no excuse to make to you – or to her. But, circumstanced as I now am, the truth is the best. I feel that I should not make Miss Dale happy; and, therefore, as an honest man, I think I best do my duty by relinquishing the honour which she and you had proposed for me.' There was more of it, but we all know of what words such letters are composed, and how men write when they feel themselves constrained to write as reptiles.

'As an honest man!' repeated the squire. 'On my honour,

Bernard, as a gentleman, I do not understand it. I cannot believe it possible that the man who wrote that letter was sitting the other day as a guest at my table.'

'What are we to do to him?' said Bernard, after a while.

'Treat him as you would a rat. Throw your stick at him, if he comes under your feet; but beware, above all things, that he does not get into your house. That is too late for us now.'

'There must be more than that, uncle.'

'I don't know what more. There are deeds for committing which a man is doubly damned, because he has screened himself from overt punishment by the nature of his own villainy. We have to remember Lily's name, and do what may best tend to her comfort. Poor girl! poor girl!'

Then they were silent, till the squire rose and took his bed candle. 'Bernard,' he said, 'let my sister-in-law know early to-morrow that I will see her here, if she will be good enough to come to me after breakfast. Do not have anything else said at the Small House. It may be that he has written to-day.'

Then the squire went to bed, and Bernard sat over the dining-room fire, meditating on it all. How would the world expect that he should behave to Crosbie? and what should he do when he met Crosbie at the club?

CHAPTER XXVIII

THE BOARD

Crosbie, as we already know, went to his office in Whitehall on the morning after his escape from Sebright's, at which establishment he left the Squire of Allington in conference with Fowler Pratt. He had seen Fowler Pratt again that same night, and the course of the story will have shown what took place at that interview.

He went early to his office, knowing that he had before him the work of writing two letters, neither of which would run very glibly from his pen. One was to be his missive to the squire, to be delivered by his friend; the other, that fatal epistle to poor Lily, which, as the day passed away, he found himself utterly unable to accomplish. The letter to the squire he did write, under certain threats; and, as we have seen, was considered to have degraded himself to the vermin rank of humanity by the meanness of his production.

But on reaching his office he found that other cares awaited him, – cares which he would have taken much delight in bearing, had the state of his mind enabled him to take delight in anything. On entering the lobby of his office, at ten o'clock, he became aware that he was received by the messengers assembled there with almost more than their usual deference. He was always a great man at the General Committee Office; but there are shades of greatness and shades of difference, which, though quite beyond the powers of definition, nevertheless manifest themselves clearly to the experienced ear and eye. He walked through to his own apartment, and there found two official letters addressed to him lying on his table. The first which came to hand, though official, was small, and marked private, and it was addressed in the handwriting of his old friend, Butterwell, the outgoing secretary. 'I shall see you in the morning, nearly as soon as you get this,' said the semi-official note; 'but I must be the first to congratulate you on the acquisition of my old shoes. They will be very easy in the wearing to you, though they pinched my corns a little at first. I dare say they want new soling, and perhaps they are a little down at heels; but you will find some excellent cobbler to make them all right, and will give them a grace in the wearing which they have sadly lacked since they came into my possession. I wish you much joy with them,' &c., &c. He then opened the larger official letter, but that had now but little interest for him. He could have made a copy of the contents without seeing them. The Board of Commissioners had had great pleasure in promoting him to the office of secretary, vacated by the promotion of Mr. Butterwell to a seat at their own Board; and then the letter was signed by Mr. Butterwell himself.

How delightful to him would have been this welcome on his return to his office had his heart in other respects been free from care! And as he thought of this, he remembered all Lily's charms. He told himself how much she excelled the noble scion of the De Courcy stock, with whom he was now destined to mate himself; how the bride he had rejected excelled the one he had chosen in grace, beauty, faith, freshness, and all feminine virtues. If he could only wipe out the last fortnight from the facts of his existence! But fortnights such as those are not to be wiped out, – not even with many sorrowful years of tedious scrubbing.

And at this moment it seemed to him as though all those impediments which had frightened him when he had thought of marrying Lily Dale were withdrawn. That which would have been terrible with seven or eight hundred a year, would have been made delightful with twelve or thirteen. Why had his fate been so unkind to him? Why had not this promotion come to him but one fortnight earlier? Why had it not been declared before he had made his visit to that terrible castle? He even said to himself that if he had positively known the fact before Pratt had seen Mr. Dale, he would have sent a different message to the squire, and would have braved the anger of all the race of the De Courcys. But in that he lied to himself, and he knew that he did so. An earl, in his imagination, was hedged by so strong a divinity, that his treason towards Alexandrina could do no more than peep at what it would. It had been considered but little by him, when the project first offered itself to his mind, to jilt the niece of a small rural squire; but it was not in him to jilt the daughter of a countess.

That house full of babies in St. John's Wood appeared to him now under a very different guise from that which it wore as he sat in his room at Courcy Castle on the evening of his arrival there. Then such an establishment had to him the flavour of a graveyard. It was as though he were going to bury himself alive. Now that it was out of his reach, he thought of it as a paradise upon earth. And then he considered what sort of a paradise Lady Alexandrina would make for him. It was astonishing how ugly was the Lady Alexandrina, how old, how graceless, how destitute of all pleasant charm, seen through the spectacles which he wore at the present moment.

During his first hour at the office he did nothing. One or

two of the younger clerks came in and congratulated him with much heartiness. He was popular at his office, and they had got a step by his promotion. Then he met one or two of the elder clerks, and was congratulated with much less heartiness. 'I suppose it's all right,' said one bluff old gentleman. 'My time is gone by, I know. I married too early to be able to wear a good coat when I was young, and I never was acquainted with any lords or lords' families.' The sting of this was the sharper because Crosbie had begun to feel how absolutely useless to him had been all that high interest and noble connection which he had formed. He had really been promoted because he knew more about his work than any of the other men, and Lady De Courcy's influential relation at the India Board had not yet even had time to write a note upon the subject.

At eleven Mr. Butterwell came into Crosbie's room, and the new secretary was forced to clothe himself in smiles. Mr. Butterwell was a pleasant, handsome man of about fifty, who had never yet set the Thames on fire, and had never attempted to do so. He was perhaps a little more civil to great men and a little more patronizing to those below him than he would have been had he been perfect. But there was something frank and English even in his mode of bowing before the mighty ones, and to those who were not mighty he was rather too civil than either stern or supercilious. He knew that he was not very clever, but he knew also how to use those who were clever. He seldom made any mistake, and was very scrupulous not to tread on men's corns. Though he had no enemies, yet he had a friend or two; and we may therefore say of Mr. Butterwell that he had walked his path in life discreetly. At the age of thirty-five he had married a lady with some little fortune, and now he lived a pleasant, easy, smiling life in a villa at Putney. When Mr. Butterwell heard, as he often did hear, of the difficulty which an English gentleman has of earning his bread in his own country, he was wont to look back on his own career with some complacency. He knew that he had not given the world much; yet he had received largely, and no one had begrudged it to him. 'Tact,' Mr. Butterwell used to say to himself, as he walked along the paths of his Putney villa. 'Tact. Tact. Tact.'

'Crosbie,' he said, as he entered the room cheerily, 'I congratulate you with all my heart. I do, indeed. You have got the step early in life, and you deserve it thoroughly; – much better than I did when I was appointed to the same office.'

'Oh, no,' said Crosbie, gloomily.

'But I say, Oh, yes. We are deuced lucky to have such a man, and so I told the commissioners.'

'I'm sure I'm very much obliged to you.'

'I've known it all along, – before you left even. Sir Raffle Buffle had told me he was to go to the Income-tax Office. The chair is two thousand there, you know; and I had been promised the first seat at the Board.'

'Ah; – I wish I'd known,' said Crosbie.

'You are much better as you are,' said Butterwell. 'There's no pleasure like a surprise! Besides, one knows a thing of that kind, and yet doesn't know it. I don't mind saying now that I knew it, – swearing that I knew it, – but I wouldn't have said so to a living being the day before yesterday. There are such slips between the cups and the lips. Suppose Sir Raffle had not gone to the Income-tax!'

'Exactly so,' said Crosbie.

'But it's all right now. Indeed I sat at the Board yesterday, though I signed the letter afterwards. I'm not sure that I don't lose more than I gain.'

'What! with three hundred a year more and less work?'

'Ah, but look at the interest of the thing. The secretary sees everything and knows everything. But I'm getting old, and, as you say, the lighter work will suit me. By the by, will you come down to Putney to-morrow? Mrs. Butterwell will be delighted to see the new secretary. There's nobody in town now, so you can have no ground for refusing.'

But Mr. Crosbie did find some ground for refusing. It would have been impossible for him to have sat and smiled at Mrs. Butterwell's table in his present frame of mind. In a mysterious, half-explanatory manner, he let Mr. Butterwell know that private affairs of importance made it absolutely necessary that he should remain that evening in town. 'And indeed,' as he said, 'he was not his own master just at present.'

'By the by, – of course not. I had quite forgotten to congratulate you on that head. So you're going to be married? Well; I'm very glad, and hope you'll be as lucky as I have been.'

'Thank you,' said Crosbie, again rather gloomily.

'A young lady from near Guestwick, isn't it; or somewhere in those parts?'

'N – no,' stammered Crosbie. 'The lady comes from Barsetshire.'

'Why, I heard the name. Isn't she a Bell, or Tait, or Ball, or some such name as that?'

'No,' said Crosbie, assuming what boldness he could command. 'Her name is De Courcy.'

'One of the earl's daughters?'

'Yes,' said Crosbie.

'Oh, I beg your pardon. I'd heard wrong. You're going to be allied to a very noble family, and I am heartily glad to hear of your success in life.' Then Butterwell shook him very cordially by the hand, – having offered him no such special testimony of approval when under the belief that he was going to marry a Bell, a Tait, or a Ball. All the same, Mr. Butterwell began to think that there was something wrong. He had heard from an indubitable source that Crosbie had engaged himself to a niece of a squire with whom he had been staying near Guestwick, – a girl without any money; and Mr. Butterwell, in his wisdom, had thought his friend Crosbie to be rather a fool for his pains. But now he was going to marry one of the De Courcys! Mr. Butterwell was rather at his wits' ends.

'Well; we shall be sitting at two, you know, and of course you'll come to us. If you're at leisure before that I'll make over what papers I have to you. I've not been a Lord Eldon* in my office, and they won't break your back.'

Immediately after that Fowler Pratt had been shown into Crosbie's room, and Crosbie had written the letter to the squire under Pratt's eye.

He could take no joy in his promotion. When Pratt left him he tried to lighten his heart. He endeavoured to throw

Lily and her wrongs behind him, and fix his thoughts on his advancing successes in life; but he could not do it. A self-imposed trouble will not allow itself to be banished. If a man lose a thousand pounds by a friend's fault, or by a turn in the wheel of fortune, he can, if he be a man, put his grief down and trample it under foot; he can exorcise the spirit of his grievance, and bid the evil one depart from out of his house. But such exorcism is not to be used when the sorrow has come from a man's own folly and sin; – especially not if it has come from his own selfishness. Such are the cases which make men drink; which drive them on to the avoidance of all thought; which create gamblers and reckless prodigals; which are the promoters of suicide. How could he avoid writing this letter to Lily? He might blow his brains out, and so let there be an end of it all. It was to such reflections that he came, when he sat himself down endeavouring to reap satisfaction from his promotion.

But Crosbie was not a man to commit suicide. In giving him his due I must protest that he was too good for that. He knew too well that a pistol-bullet could not be the be-all and the end-all here, and there was too much manliness in him for so cowardly an escape. The burden must be borne. But how was he to bear it? There he sat till it was two o'clock, neglecting Mr. Butterwell and his office papers, and not stirring from his seat till a messenger summoned him before the Board. The Board, as he entered the room, was not such a Board as the public may, perhaps, imagine such Boards to be. There was a round table, with a few pens lying about, and a comfortable leathern arm-chair at the side of it, farthest from the door. Sir Raffle Buffle was leaving his late colleagues, and was standing with his back to the fire-place, talking very loudly. Sir Raffle was a great bully, and the Board was uncommonly glad to be rid of him; but as this was to be his last appearance at the Committee Office, they submitted to his voice meekly. Mr. Butterwell was standing close to him, essaying to laugh mildly at Sir Raffle's jokes. A little man, hardly more than five feet high, with small but honest-looking eyes, and close-cut hair, was standing behind the arm-chair, rubbing his hands together, and longing for the departure of Sir Raffle, in order that he might sit down. This was Mr. Optimist, the new chairman, in praise of whose appointment the Daily Jupiter* had been so loud, declaring that the present Minister was showing himself superior to all Ministers who had ever gone before him, in giving promotion solely on the score of merit. The Daily Jupiter, a fortnight since, had published a very eloquent article, strongly advocating the claims of Mr. Optimist, and was naturally pleased to find that its advice had been taken. Has not an obedient Minister a right to the praise of those powers which he obeys?

Mr. Optimist was, in truth, an industrious little gentleman, very well connected, who had served the public all his life, and who was, at any rate, honest in his dealings. Nor was he a bully, such as his predecessor. It might, however, be a question whether he carried guns enough for the command in which he was now to be employed. There was but one other member of the Board, Major Fiasco by name, a discontented, broken-hearted, silent man, who had been sent to the General Committee Office some few years before because he was not wanted anywhere else. He was a man who had intended to do great things when he entered public life, and had possessed the talent and energy for things moderately great. He had also possessed to a certain extent the ear of those high in office; but, in some way, matters had not gone well with him, and in running his course he had gone on the wrong side of the post. He was still in the prime of life, and yet all men knew that Major Fiasco had nothing further to expect from the public or from the Government. Indeed, there were not wanting those who said that Major Fiasco was already in receipt of a liberal income, for which he gave no work in return; that he merely filled a chair for four hours a day four or five days a week, signing his name to certain forms and documents, reading, or pretending to read, certain papers, but, in truth, doing no good. Major Fiasco, on the other hand, considered himself to be a deeply injured individual, and he spent his life in brooding over his wrongs. He believed now in nothing and in nobody. He had begun public life striving to be honest, and he now regarded all around him as dishonest. He had no satisfaction in any man other than that which he found when some event would show to him that this or that other compeer of his own had proved himself to be self-interested, false, or fraudulent. 'Don't tell me, Butterwell,' he would say – for with Mr. Butterwell he maintained some semi-official intimacy, and he would take that gentleman by the button-hole, holding him close. 'Don't tell me. I know what men are. I've seen the world. I've been looking at things with my eyes open. I knew what he was doing.' And then he would tell of the sly deed of some official known well to them both, not denouncing it by any means, but affecting to take it for granted that the man in question was a rogue. Butterwell would shrug his shoulders, and laugh gently, and say that, upon his word, he didn't think the world so bad as Fiasco made it out to be.

Nor did he; for Butterwell believed in many things. He believed in his Putney villa on this earth, and he believed also that he might achieve some sort of Putney villa in the world beyond without undergoing present martyrdom. His Putney villa first, with all its attendant comforts, and then his duty to the public afterwards. It was thus that Mr. Butterwell regulated his conduct; and as he was solicitous that the villa should be as comfortable a home to his wife as to himself, and that it should be specially comfortable to his friends, I do not think that we need quarrel with his creed.

Mr. Optimist believed in everything, but especially he believed in the Prime Minister, in the Daily Jupiter, in the General Committee Office, and in himself. He had long thought that everything was nearly right; but now that he himself was chairman at the General Committee Office, he was quite sure that everything must be right. In Sir Raffle Buffle, indeed, he had never believed; and now it was, perhaps, the greatest joy of his life that he should never again be called upon to hear the tones of that terrible knight's hated voice.

Seeing who were the components of the new Board, it may be presumed that Crosbie would look forward to enjoying a not uninfluential position in his office. There were, indeed, some among the clerks who did not hesitate to say that the new secretary would have it pretty nearly all his own way. As for 'Old Opt,' there would be, they said, no difficulty about him. Only tell him that such and such a decision was his own, and he would be sure to believe the teller. Butterwell was not fond of work, and had been accustomed to lean upon Crosbie for many years. As for Fiasco, he would be cynical in words, but wholly indifferent in deed. If the whole office were made to go to the mischief, Fiasco, in his own grim way, would enjoy the confusion.

'Wish you joy, Crosbie,' said Sir Raffle, standing up on the rug, waiting for the new secretary to go up to him and shake hands. But Sir Raffle was going, and the new secretary did not indulge him.

'Thank ye, Sir Raffle,' said Crosbie, without going near the rug.

'Mr. Crosbie, I congratulate you most sincerely,' said Mr. Optimist. 'Your promotion has been the result altogether of your own merit. You have been selected for the high office which you are now called upon to fill solely because it has been thought that you are the most fit man to perform the onerous duties attached to it. Hum – h-m – ha. As regards my share in the recommendation which we found ourselves bound to submit to the Treasury, I must say that I never felt less hesitation in my life, and I believe I may declare as much as regards the other members of the Board.' And Mr. Optimist looked around him for approving words. He had come forward from his standing ground behind his chair to welcome Crosbie, and had shaken his hand cordially. Fiasco also had risen from his seat, and had assured Crosbie in a whisper that he had feathered his nest uncommon well. Then he had sat down again.

'Indeed you may, as far as I am concerned,' said Butterwell.

'I told the Chancellor of the Exchequer,' said Sir Raffle, speaking very loud and with much authority, 'that unless he had some first-rate man to send from elsewhere I could name a fitting candidate. "Sir Raffle," he said, "I mean to keep it in the office, and therefore shall be glad of your opinion." "In that case, Mr. Chancellor," said I, "Mr. Crosbie must be the man." "Mr. Crosbie shall be the man," said the Chancellor. And Mr. Crosbie is the man.'

'Your friend Sark spoke to Lord Brock about it,' said Fiasco. Now the Earl of Sark was a young nobleman of much influence at the present moment, and Lord Brock was the Prime Minister. 'You should thank Lord Sark.'

'Had as much to do with it as if my footman had spoken,' said Sir Raffle.

'I am very much obliged to the Board for their good opinion,' said Crosbie, gravely. 'I am obliged to Lord Sark as well, – and also to your footman, Sir Raffle, if, as you seem to say, he has interested himself in my favour.'

'I didn't say anything of the kind,' said Sir Raffle. 'I thought it right to make you understand that it was my opinion, given, of course, officially, which prevailed with the Chancellor of the Exchequer. Well, gentlemen, as I shall be wanted in the city, I will say good morning to you. Is my carriage ready, Boggs?' Upon which the attendant messenger opened the door, and the great Sir Raffle Buffle took his final departure from the scene of his former labours.

'As to the duties of your new office' – and Mr. Optimist continued his speech, taking no other notice of the departure of his enemy than what was indicated by an increased brightness of his eye and a more satisfactory tone of voice – 'you will find yourself quite familiar with them.'

'Indeed he will,' said Butterwell.

'And I am quite sure that you will perform them with equal credit to yourself, satisfaction to the department, and advantage to the public. We shall always be glad to have your opinion on any subject of importance that may come before us; and as regards the internal discipline of the office, we feel that we may leave it safely in your hands. In any matter of importance you will, of course, consult us, and I feel very confident that we shall go on together with great comfort and with mutual confidence.' Then Mr. Optimist looked at his brother commissioners, sat down in his armchair, and taking in his hands some papers before him, began the routine business of the day.

It was nearly five o'clock when, on this special occasion,

the secretary returned from the board-room to his own office. Not for a moment had the weight been off his shoulders while Sir Raffle had been bragging or Mr. Optimist making his speech. He had been thinking, not of them, but of Lily Dale; and though they had not discovered his thoughts, they had perceived that he was hardly like himself.

'I never saw a man so little elated by good fortune in my life,' said Mr. Optimist.

'Ah, he's got something on his mind,' said Butterwell. 'He's going to be married, I believe.'

'If that's the case, it's no wonder he shouldn't be elated,' said Major Fiasco, who was himself a bachelor.

When in his own room again, Crosbie at once seized on a sheet of note-paper, as though by hurrying himself on with it he could get that letter to Allington written. But though the paper was before him, and the pen in his hand, the letter did not, would not, get itself written. With what words was he to begin it? To whom should it be written? How was he to declare himself the villain which he had made himself? The letters from his office were taken away every night shortly after six, and at six o'clock he had not written a word. 'I will do it at home to-night,' he said to himself, and then, tearing off a scrap of paper, he scratched those few lines which Lily received, and which she had declined to communicate to her mother or sister. Crosbie, as he wrote them, conceived that they would in some way prepare the poor girl for the coming blow, – that they would make her know that all was not right; but in so supposing he had not counted on the constancy of her nature, nor had he thought of the promise which she had given him that nothing should make her doubt him. He wrote the scrap, and then taking his hat walked off through the gloom of the November evening up Charing Cross and St. Martin's Lane, towards the Seven Dials and Bloomsbury into regions of the town with which he had no business, and which he never frequented. He hardly knew where he went or wherefore. How was he to escape from the weight of the burden which was now crushing him? It seemed to him as though he would change his position with thankfulness for that of the junior clerk in his office, if only that junior clerk had upon his mind no such betrayal of trust as that of which he was guilty.

At half-past seven he found himself at Sebright's, and there he dined. A man will dine, even though his heart be breaking. Then he got into a cab, and had himself taken home to Mount Street. During his walk he had sworn to himself that he would not go to bed that night till the letter was written and posted. It was twelve before the first words were marked on the paper, and yet he kept his oath. Between two and three, in the cold moonlight, he crawled out and deposited his letter in the nearest post-office.

CHAPTER XXIX

JOHN EAMES RETURNS TO BURTON CRESCENT

JOHN EAMES and Crosbie returned to town on the same day. It will be remembered how Eames had assisted Lord De Guest in the matter of the bull, and how great had been the earl's gratitude on the occasion. The memory of this, and the strong encouragement which he received from his mother and sister for having made such a friend by his gallantry, lent some slight satisfaction to his last hours at home. But his two misfortunes were too serious to allow of anything like real happiness. He was leaving Lily behind him, engaged to be married to a man whom he hated, and he was returning to Burton Crescent, where he would have to face Amelia Roper, – Amelia either in her rage or in her love. The prospect of Amelia in her rage was very terrible to him; but his greatest fear was of Amelia in her love. He had in his letter declined matrimony; but what if she talked down all his objections, and carried him off to church in spite of himself!

When he reached London and got into a cab with his portmanteau, he could hardly fetch up courage to bid the man drive him to Burton Crescent. 'I might as well go to an hotel for the night,' he said to himself, 'and then I can learn how things are going on from Cradell at the office.' Nevertheless, he did give the direction to Burton Crescent, and when it was once given felt ashamed to change it. But, as he was driven up to the well-known door, his heart was so low within him that he might almost be said to have lost it. When the cabman demanded whether he should knock, he could not answer; and when the maid-servant at the door greeted him, he almost ran away.

'Who's at home?' said he, asking the question in a very low voice.

'There's missus,' said the girl, 'and Miss Spruce, and Mrs. Lupex. He's away somewhere, in his tantrums again; and there's Mr.—'

'Is Miss Roper here?' he said, still whispering.

'Oh, yes! Miss Mealyer's here,' said the girl, speaking in a cruelly loud voice. 'She was in the dining-room just now, putting out the table. Miss Mealyer!' And the girl, as she called out the name, opened the dining-room door. Johnny Eames felt that his knees were too weak to support him.

But Miss Mealyer was not in the dining-room. She had perceived the advancing cab of her sworn adorer, and had thought it expedient to retreat from her domestic duties and fortify herself among her brushes and ribbons. Had it been possible that she should know how very weak and cowardly was the enemy against whom she was called upon to put herself in action, she might probably have fought her battle somewhat differently, and have achieved a speedy victory, at the cost of an energetic shot or two. But she did not know. She thought it probable that she might obtain power over him and manage him; but it did not occur to her that his legs were so weak beneath him that she might almost blow him over with a breath. None but the worst and most heartless of women know the extent of their own power over men; – as none but the worst and most heartless of men know the extent of their power over women. Amelia Roper was not a good specimen of the female sex, but there were worse women than she was.

'She ain't there, Mr. Eames; but you'll see her in the drawen-room,' said the girl. 'And it's she'll be glad to see you back again, Mr. Eames.' But he scrupulously passed the door of the upstairs sitting-room, not even looking within it, and contrived to get himself into his own chamber without having encountered anybody. 'Here's yer 'ot water, Mr. Eames,' said the girl, coming up to him after an interval of half-an-hour; 'and dinner'll be on the table in ten minutes. Mr. Cradell is come in, and so is missus's son.'

It was still open to him to go out and dine at some eating-house in the Strand. He could start out, leaving word that he was engaged, and so postpone the evil hour. He had almost made up his mind to do so, and certainly would have done it, had not the sitting-room door opened as he was on the landing-place. The door opened, and he found himself confronting the assembled company. First came Cradell, and leaning on his arm, I regret to say, was Mrs. Lupex, – *Egyptia conjux!** Then there came Miss Spruce with young Roper; Amelia and her mother brought up the rear together. There was no longer question of flight now; and poor Eames, before he knew what he was doing, was carried down into the dining-room with the rest of the company. They were all glad to see him, and welcomed him back warmly, but he was so much beside himself that he could not ascertain whether Amelia's voice was joined with the others. He was already seated at table, and had before him a plate of soup, before he recognized the fact that he was sitting between Mrs. Roper and Mrs. Lupex. The latter lady had separated herself from Mr. Cradell as she entered the room. 'Under all the circumstances perhaps it will be better for us to be apart,' she said. 'A lady can't make herself too safe; can she, Mrs. Roper? There's no danger between you and me, is there, Mr. Eames, – specially when Miss Amelia is opposite?' The last words, however, were intended to be whispered into his ear.

But Johnny made no answer to her; contenting himself for the moment with wiping the perspiration from his brow. There was Amelia opposite to him, looking at him – the very Amelia to whom he had written, declining the honour of marrying her. Of what her mood towards him might be, he could form no judgment from her looks. Her face was simply stern and impassive, and she seemed inclined to eat her dinner in silence. A slight smile of derision had passed across her face as she heard Mrs. Lupex whisper, and it might have been discerned that her nose, at the same time, became somewhat elevated; but she said not a word.

'I hope you've enjoyed yourself, Mr. Eames, among the vernal beauties of the country,' said Mrs. Lupex.

'Very much, thank you,' he replied.

'There's nothing like the country at this autumnal season of the year. As for myself, I've never been accustomed to remain in London after the breaking up of the *beau monde*. We've usually been to Broadstairs, which is a very charming place, with most elegant society, but now—' and she shook her head, by which all the company knew that she intended to allude to the sins of Mr. Lupex.

'I'd never wish to sleep out of London for my part,' said Mrs. Roper. 'When a woman's got a house over her head, I don't think her mind's ever easy out of it.'

She had not intended any reflection on Mrs. Lupex for not having a house of her own, but that lady immediately bristled up. 'That's just what the snails say, Mrs. Roper. And as for having a house of one's own, it's a very good thing, no doubt, sometimes; but that's according to circumstances. It has suited me lately to live in lodgings, but there's no knowing whether I mayn't fall lower than that yet, and have—' but here she stopped herself, and looking over at Mr. Cradell nodded her head.

'And have to let them,' said Mrs. Roper. 'I hope you'll be more lucky with your lodgers than I have been with some of mine. Jemima, hand the potatoes to Miss Spruce. Miss Spruce, do let me send you a little more gravy? There's plenty here, really.' Mrs. Roper was probably thinking of Mr. Todgers.*

'I hope I shall,' said Mrs. Lupex. 'But, as I was saying, Broadstairs is delightful. Were you ever at Broadstairs, Mr. Cradell?'

'Never, Mrs. Lupex. I generally go abroad in my leave. One sees more of the world, you know. I was at Dieppe last June, and found that very delightful – though rather lonely.

I shall go to Ostend this year; only December is so late for Ostend. It was a deuced shame my getting December, wasn't it, Johnny?'

'Yes, it was,' said Eames. 'I managed better.'

'And what have you been doing, Mr. Eames?' said Mrs. Lupex, with one of her sweetest smiles. 'Whatever it may have been, you've not been false to the cause of beauty, I'm sure.' And she looked over to Amelia with a knowing smile. But Amelia was engaged upon her plate, and went on with her dinner without turning her eyes either on Mrs. Lupex or on John Eames.

'I haven't done anything particular,' said Eames. 'I've just been staying with my mother.'

'We've been very social here, haven't we, Miss Amelia?' continued Mrs. Lupex. 'Only now and then a cloud comes across the heavens, and the lights at the banquet are darkened.' Then she put her handkerchief up to her eyes, sobbing deeply, and they all knew that she was again alluding to the sins of her husband.

As soon as dinner was over the ladies with young Mr. Roper retired, and Eames and Cradell were left to take their wine over the dining-room fire, – or their glass of gin and water, as it might be. 'Well, Caudle, old fellow,' said one. 'Well, Johnny, my boy,' said the other. 'What's the news at the office?' said Eames.

'Muggeridge has been playing the very mischief.' Muggeridge was the second clerk in Cradell's room. 'We're going to put him into Coventry and not speak to him except officially. But to tell you the truth, my hands have been so full here at home, that I haven't thought much about the office. What am I to do about that woman?'

'Do about her? How do about her?'

'Yes; what am I to do about her? How am I to manage with her? There's Lupex off again in one of his fits of jealousy.'

'But it's not your fault, I suppose?'

'Well; I can't just say. I'm fond of her, and that's the long and the short of it; deuced fond of her.'

'But, my dear Caudle, you know she's that man's wife.'

'Oh, yes, I know all about it. I'm not going to defend myself. It's wrong, I know, – pleasant, but wrong. But what's a fellow to do? I suppose in strict morality I ought to leave the lodgings. But, by George, I don't see why a man's to be turned out in that way. And then I couldn't make a clean score with old mother Roper. But I say, old fellow, who gave you the gold chain?'

'Well; it was an old family friend at Guestwick; or rather, I should say, a man who said he knew my father.'

'And he gave you that because he knew your governor! Is there a watch to it?'

'Yes, there's a watch. It wasn't exactly that. There was some trouble about a bull. To tell the truth, it was Lord De Guest; the queerest fellow, Caudle, you ever met in your life; but such a trump. I've got to go and dine with him at Christmas.' And then the old story of the bull was told.

'I wish I could find a lord in a field with a bull,' said Cradell. We may, however, be permitted to doubt whether Mr. Cradell would have earned a watch even if he had had his wish.

'You see,' continued Cradell, reverting to the subject on which he most delighted to talk, 'I'm not responsible for that man's ill-conduct.'

'Does anybody say you are?'

'No; nobody says so. But people seem to think so. When he is by I hardly speak to her. She is thoughtless and giddy, as women are, and takes my arm, and that kind of thing, you know. It makes him mad with rage, but upon my honour I don't think she means any harm.'

'I don't suppose she does,' said Eames.

'Well; she may or she mayn't. I hope with all my heart she doesn't.'

'And where is he now?'

'This is between ourselves, you know; but she went to find him this afternoon. Unless he gives her money she can't stay here, nor, for the matter of that, will she be able to go away. If I mention something to you, you won't tell any one?'

'Of course I won't.'

'I wouldn't have it known to any one for the world. I've

lent her seven pounds ten. It's that which makes me so short with mother Roper.'

'Then I think you're a fool for your pains.'

'Ah, that's so like you. I always said you'd no feeling of real romance. If I cared for a woman I'd give her the coat off my back.'

'I'd do better than that,' said Johnny. 'I'd give her the heart out of my body. I'd be chopped up alive for a girl I loved; but it shouldn't be for another man's wife.'

'That's a matter of taste. But she's been to Lupex to-day at that house he goes to in Drury Lane. She had a terrible scene there. He was going to commit suicide in the middle of the street, and she declares that it all comes from jealousy. Think what a time I have of it – standing always, as one may say, on gunpowder. He may turn up here any moment, you know. But, upon my word, for the life of me I cannot desert her. If I were to turn my back on her she wouldn't have a friend in the world. And how's L. D.? I'll tell you what it is – you'll have some trouble with the divine Amelia.'

'Shall I?'

'By Jove, you will. But how's L. D. all this time?'

'L. D. is engaged to be married to a man named Adolphus Crosbie,' said poor Johnny, slowly. 'If you please, we will not say any more about her.'

'Whew – w – w! That's what makes you so down in the mouth! L. D. going to marry Crosbie! Why, that's the man who is to be the new secretary at the General Committee Office. Old Huffle Scuffle, who was their chair, has come to us, you know. There's been a general move at the G. C., and this Crosbie has got to be secretary. He's a lucky chap, isn't he?'

'I don't know anything about his luck. He's one of those fellows that make me hate them the first time I look at them. I've a sort of feeling that I shall live to kick him some day.'

'That's the time, is it? Then I suppose Amelia will have it all her own way now.'

'I'll tell you what, Caudle. I'd sooner get up through the trap-door, and throw myself off the roof into the area, than marry Amelia Roper.'

'Have you and she had any conversation since you came back?'

'Not a word.'

'Then I tell you fairly you've got trouble before you. Amelia and Maria – Mrs. Lupex, I mean – are as thick as thieves just at present, and they have been talking you over. Maria – that is, Mrs. Lupex – lets it all out to me. You'll have to mind where you are, old fellow.'

Eames was not inclined to discuss the matter any further, so he finished his toddy in silence. Cradell, however, who felt that there was something in his affairs of which he had reason to be proud, soon returned to the story of his own very extraordinary position. 'By Jove, I don't know that a man was ever so circumstanced,' he said. 'She looks to me to protect her, and yet what can I do?'

At last Cradell got up, and declared that he must go to the ladies. 'She's so nervous, that unless she has some one to countenance her she becomes unwell.'

Eames declared his purpose of going to the divan, or to the theatre, or to take a walk in the streets. The smiles of beauty had no longer charms for him in Burton Crescent.

'They'll expect you to take a cup of tea the first night,' said Cradell; but Eames declared that they might expect it.

'I'm in no humour for it,' said he. 'I'll tell you what, Cradell, I shall leave this place, and take rooms for myself somewhere. I'll never go into a lodging-house again.'

As he so spoke, he was standing at the dining-room door; but he was not allowed to escape in this easy way. Jemima, as he went out into the passage, was there with a three-cornered note in her hand. 'From Miss Mealyer,' she said. 'Miss Mealyer is in the back parlour all by herself.'

Poor Johnny took the note, and read it by the lamp over the front door.

'Are you not going to speak to me on the day of your return? It cannot be that you will leave the house without seeing me for a moment. I am in the back parlour.'

When he had read these words, he paused in the passage, with his hat on. Jemima, who could not understand why any young man should hesitate as to seeing his lady-love in the back parlour alone, whispered to him again, in her audible way, 'Miss Mealyer is there, sir; and all the rest on 'em's upstairs!' So compelled, Eames put down his hat, and walked with slow steps into the back parlour.

How was it to be with the enemy? Was he to encounter Amelia in anger, or Amelia in love? She had seemed to be stern and defiant when he had ventured to steal a look at her across the dining-table, and now he expected that she would turn upon him with loud threatenings and protestations as to her wrongs. But it was not so. When he entered the room she was standing with her back to him, leaning on the mantel-piece, and at the first moment she did not essay to speak. He walked into the middle of the room and stood there, waiting for her to begin.

'Shut the door!' she said, looking over her shoulder. 'I suppose you don't want the girl to hear all you've got to say to me!'

Then he shut the door; but still Amelia stood with her back to him, leaning upon the mantel-piece.

It did not seem that he had much to say, for he remained perfectly silent.

'Well!' said Amelia, after a long pause, and she then again looked over her shoulder. 'Well, Mr. Eames!'

'Jemima gave me your note, and so I've come,' said he.

'And is this the way we meet!' she exclaimed, turning suddenly upon him, and throwing her long black hair back over her shoulders. There certainly was some beauty about her. Her eyes were large and bright, and her shoulders were well turned. She might have done as an artist's model for a Judith, but I doubt whether any man, looking well into her face, could think that she would do well as a wife. 'Oh, John, is it to be thus, after love such as ours?' And she clasped her hands together, and stood before him.

'I don't know what you mean,' said Eames.

'If you are engaged to marry L. D., tell me so at once. Be a man, and speak out, sir.'

'No,' said Eames; 'I am not engaged to marry the lady to whom you allude.'

'On your honour?'

'I won't have her spoken about. I'm not going to marry her, and that's enough.'

'Do you think that I wish to speak of her? What can L. D. be to me as long as she is nothing to you? Oh, Johnny, why did you write me that heartless letter?' Then she leaned upon his shoulder – or attempted to do so.

I cannot say that Eames shook her off, seeing that he lacked the courage to do so; but he shuffled his shoulder about so that the support was uneasy to her, and she was driven to stand erect again. 'Why did you write that cruel letter?' she said again.

'Because I thought it best, Amelia. What's a man to do with ninety pounds a year, you know?'

'But your mother allows you twenty.'

'And what's a man to do with a hundred and ten?'

'Rising five pounds every year,' said the well-informed Amelia. 'Of course we should live here, with mamma, and you would just go on paying her as you do now. If your heart was right, Johnny, you wouldn't think so much about money. If you loved me – as you said you did—' Then a little sob came, and the words were stopped. The words were stopped, but she was again upon his shoulder. What was he to do? In truth, his only wish was to escape, and yet his arm, quite in opposition to his own desires, found its way round her waist. In such a combat a woman has so many points in her favour! 'Oh, Johnny,' she said again, as soon as she felt the pressure of his arm. 'Gracious, what a beautiful watch you've got,' and she took the trinket out of his pocket. 'Did you buy that?'

'No; it was given to me.'

'John Eames, did L. D. give it you?'

'No, no, no,' he shouted, stamping on the floor as he spoke.

'Oh, I beg your pardon,' said Amelia, quelled for the moment by his energy. 'Perhaps it was your mother.'

'No; it was a man. Never mind about the watch now.'

'I wouldn't mind anything, Johnny, if you would tell me that you loved me again. Perhaps I oughtn't to ask you, and it isn't becoming in a lady; but how can I help it, when you know you've got my heart. Come upstairs and have tea with us now, won't you?'

What was he to do? He said that he would go up and have tea; and as he led her to the door he put down his face and kissed her. Oh, Johnny Eames! But then a woman in such a contest has so many points in her favour.

CHAPTER XXX

'IS IT FROM HIM?'

I HAVE already declared that Crosbie wrote and posted the fatal letter to Allington, and we must now follow it down to that place. On the morning following the squire's return to his own house, Mrs. Crump, the post-mistress at Allington, received a parcel by post directed to herself. She opened it, and found an enclosure addressed to Mrs. Dale, with a written request that she would herself deliver it into that lady's own hand at once. This was Crosbie's letter.

'It's from Miss Lily's gentleman,' said Mrs. Crump, looking at the handwriting. 'There's something up, or he wouldn't be writing to her mamma in this way.' But Mrs. Crump lost no time in putting on her bonnet, and trudging up with the letter to the Small House. 'I must see the missus herself,' said Mrs. Crump. Whereupon Mrs. Dale was called downstairs into the hall, and there received the packet. Lily was in the breakfast-parlour, and had seen the post-mistress arrive; – had seen also that she carried a letter in her hand. For a moment she had thought that it was for her, and imagined that the old woman had brought it herself from simple good-nature. But Lily, when she heard her mother mentioned, instantly withdrew and shut the parlour door. Her heart misgave her that something was wrong, but she hardly tried to think what it might be. After all, the regular postman might bring the letter she herself expected. Bell was not yet downstairs, and she stood alone over the tea-cups on the breakfast-table, feeling that there was something for her to fear. Her mother did not come at once into the room, but, after a pause of a moment or two, went again upstairs. So she remained, either standing against the table, or at the window, or seated in one of the two arm-chairs, for a space of ten minutes, when Bell entered the room.

'Isn't mamma down yet?' said Bell.

'Bell,' said Lily, 'something has happened. Mamma has got a letter.'

'Happened! What has happened? Is anybody ill? Who is the letter from?' And Bell was going to return through the door in search of her mother.

'Stop, Bell,' said Lily. 'Do not go to her yet. I think it's from – Adolphus.'

'Oh, Lily, what do you mean?'

'I don't know, dear. We'll wait a little longer. Don't look like that, Bell.' And Lily strove to appear calm, and strove almost successfully.

'You have frightened me so,' said Bell.

'I am frightened myself. He only sent me one line yesterday, and now he has sent nothing. If some misfortune should have happened to him! Mrs. Crump brought down the letter herself to mamma, and that is odd, you know.'

'Are you sure it was from him?'

'No; I have not spoken to her. I will go up to her now. Don't you come, Bell. Oh! Bell, do not look so unhappy.' She then went over and kissed her sister, and after that, with very gentle steps, made her way up to her mother's room. 'Mamma, may I come in?' she said.

'Oh! my child!'

'I know it is from him, mamma. Tell me all at once.'

Mrs. Dale had read the letter. With quick, glancing eyes, she had made herself mistress of its whole contents, and was already aware of the nature and extent of the sorrow which had come upon them. It was a sorrow that admitted of no hope. The man who had written that letter could never return again; nor if he should return could he be welcomed back to them. The blow had fallen, and it was to be borne. Inside the letter to herself had been a very small note addressed to Lily. 'Give her the enclosed,' Crosbie had said in his letter, 'if you do not now think it wrong to do so. I have left it open, that you may read it.' Mrs. Dale, however, had not yet read it, and she now concealed it beneath her handkerchief.

I will not repeat at length Crosbie's letter to Mrs. Dale. It covered four sides of letter-paper, and was such a letter that any man who wrote it must have felt himself to be a rascal. We saw that he had difficulty in writing it, but the miracle was, that any man could have found it possible to write it. 'I know that you will curse me,' said he; 'and I deserve to be cursed. I know that I shall be punished for this, and I must bear my punishment. My worst punishment will be this, – that I never more shall hold up my head again.' And then, again, he said: – 'my only excuse is my conviction that I should never make her happy. She has been brought up as an angel, with pure thoughts, with holy hopes, with belief in all that is good, and high, and noble. I have been surrounded through my whole life by things low, and mean, and ignoble. How could I live with her, or she with me? I know now that this is so; but my fault has been that I did not know it when I was there with her. I choose to tell you all,' he continued, towards the end of the letter, 'and therefore I let you know that I have engaged myself to marry another woman. Ah! I can foresee how bitter will be your feelings when you read this: but they will not be so bitter as mine while I write it. Yes; I am already engaged to one who will suit me, and whom I may suit. You will not expect me to speak ill of her who is to be near and dear to me. But she is one with whom I may mate myself without an inward conviction that I shall destroy all her happiness by doing so. Lilian,' he said, 'shall always have my prayers; and I trust that she may soon forget, in the love of an honest man, that she ever knew one so dishonest as – Adolphus Crosbie.'

Of what like must have been his countenance as he sat writing such words of himself under the ghastly light of his own small, solitary lamp? Had he written his letter at his office, in the day-time, with men coming in and out of his room, he could hardly have written of himself so plainly. He would have bethought himself that the written words might remain, and be read hereafter by other eyes than those for which they were intended. But, as he sat alone, during the small hours of the night, almost repenting of his sin with true repentance, he declared to himself that he did not care who might read them. They should, at any rate, be true. Now they had been read by her to whom they had been addressed, and the daughter was standing before the mother to hear her doom.

'Tell me at once,' Lily had said; but in what words was her mother to tell her?

'Lily,' she said, rising from her seat, and leaving the two letters on the couch; that addressed to the daughter was hidden beneath a handkerchief, but that which she had read she left open and in sight. She took both the girl's hands in hers as she looked into her face, and spoke to her. 'Lily, my child!' Then she burst into sobs, and was unable to tell her tale.

'Is it from him, mamma? May I read it? He cannot be—'

'It is from Mr. Crosbie.'

'Is he ill, mamma? Tell me at once. If he is ill I will go to him.'

'No, my darling, he is not ill. Not yet; – do not read it yet. Oh, Lily! It brings bad news; very bad news.'

'Mamma, if he is not in danger, I can read it. Is it bad to him, or only bad to me?'

At this moment the servant knocked, and not waiting for an answer, half opened the door.

'If you please, ma'am, Mr. Bernard is below, and wants to speak to you.'

'Mr. Bernard! ask Miss Bell to see him.'

'Miss Bell is with him, ma'am, but he says that he specially wants to speak to you.'

Mrs. Dale felt that she could not leave Lily alone. She could not take the letter away, nor could she leave her child with the letter open.

'I cannot see him,' said Mrs. Dale. 'Ask him what it is. Tell him I cannot come down just at present.' And then the servant went, and Bernard left his message with Bell.

'Bernard,' she had said, 'do you know of anything?? Is there anything wrong about Mr. Crosbie?' Then, in a few words, he told her all, and understanding why his aunt had not come down to him, he went back to the Great House. Bell, almost stupefied by the tidings, seated herself at the table unconsciously, leaning upon her elbows.

'It will kill her,' she said to herself. 'My Lily, my darling Lily! It will surely kill her!'

But the mother was still with the daughter, and the story was still untold.

'Mamma,' said Lily, 'whatever it is, I must, of course, be made to know it. I begin to guess the truth. It will pain you to say it. Shall I read the letter?'

Mrs. Dale was astonished at her calmness. It could not be that she had guessed the truth, or she would not stand like that, with tearless eyes and unquelled courage before her.

'You shall read it, but I ought to tell you first. Oh, my child, my own one!' Lily was now leaning against the bed, and her mother was standing over her, caressing her.

'Then, tell me,' said she. 'But I know what it is. He has thought it all over while away from me, and he finds that it must not be as we have supposed. Before he went I offered to release him, and now he knows that he had better accept my offer. Is it so, mamma?' In answer to this Mrs. Dale did not speak, but Lily understood from her signs that it was so.

'He might have written it to me, myself,' said Lily, very proudly. 'Mamma, we will go down to breakfast. He has sent nothing to me, then?'

'There is a note. He bids me read it, but I have not opened it. It is here.'

'Give it me,' said Lily, almost sternly. 'Let me have his last words to me;' and she took the note from her mother's hands.

'Lily,' said the note, 'your mother will have told you all. Before you read these few words you will know that you have trusted one who was quite untrustworthy. I know that you will hate me. – I cannot even ask you to forgive me. You will let me pray that you may yet be happy. – A. C.'

She read these few words, still leaning against the bed. Then she got up, and walking to a chair, seated herself with her back to her mother. Mrs. Dale moving silently after her stood over the back of the chair, not daring to speak to her.

So she sat for some five minutes, with her eyes fixed upon the open window, and with Crosbie's note in her hand.

'I will not hate him, and I do forgive him,' she said at last, struggling to command her voice, and hardly showing that she could not altogether succeed in her attempt. 'I may not write to him again, but you shall write and tell him so. Now we will go down to breakfast.' And so saying, she got up from her chair.

Mrs. Dale almost feared to speak to her, her composure was so complete, and her manner so stern and fixed. She hardly knew how to offer pity and sympathy, seeing that pity seemed to be so little necessary, and that even sympathy was not demanded. And she could not understand all that Lily had said. What had she meant by the offer to release him? Had there, then, been some quarrel between them before he went? Crosbie had made no such allusion in his letter. But Mrs. Dale did not dare to ask any questions.

'You frighten me, Lily,' she said. 'Your very calmness frightens me.'

'Dear mamma!' and the poor girl absolutely smiled as she embraced her mother. 'You need not be frightened by my calmness. I know the truth well. I have been very unfortunate; – very. The brightest hopes of my life are all gone; – and I shall never again see him whom I love beyond all the world!' Then at last she broke down, and wept in her mother's arms.

There was not a word of anger spoken then against him who had done all this. Mrs. Dale felt that she did not dare to speak in anger against him, and words of anger were not likely to come from poor Lily. She, indeed, hitherto did not know the whole of his offence, for she had not read his letter.

'Give it me, mamma,' she said at last. 'It has to be done sooner or later.'

'Not now, Lily. I have told you all, – all that you need know at present.'

'Yes; now, mamma,' and again that sweet silvery voice became stern. 'I will read it now, and there shall be an end.' Whereupon Mrs. Dale gave her the letter and she read it in silence. Her mother, though standing somewhat behind her, watched her narrowly as she did so. She was now lying over upon the bed, and the letter was on the pillow, as she propped herself upon her arm. Her tears were running, and ever and again she would stop to dry her eyes. Her sobs, too, were very audible, but she went on steadily with her reading till she came to the line on which Crosbie told that he had already engaged himself to another woman. Then her mother could see that she paused suddenly, and that a shudder slightly convulsed all her limbs.

'He has been very quick,' she said, almost in a whisper; and then she finished the letter. 'Tell him, mamma,' she said, 'that I do forgive him, and I will not hate him. You will tell him that, – from me; will you not?' And then she raised herself from the bed.

Mrs. Dale would give her no such assurance. In her present mood her feelings against Crosbie were of a nature which she herself hardly could understand or analyse. She felt that if he were present she could almost fly at him as would a tigress. She had never hated before as she now hated this man. He was to her a murderer, and worse than a murderer. He had made his way like a wolf into her little fold, and torn her ewe-lamb and left her maimed and mutilated for life. How could a mother forgive such an offence as that, or consent to be the medium through which forgiveness should be expressed?

'You must, mamma; or, if you do not, I shall do so. Remember that I love him. You know what it is to have loved one single man. He has made me very unhappy; I hardly know yet how unhappy. But I have loved him and do love him. I believe, in my heart, that he still loves me. Where this has been there must not be hatred and unforgiveness.'

'I will pray that I may become able to forgive him,' said Mrs. Dale.

'But you must write to him those words. Indeed you must, mamma! "She bids me tell you that she has forgiven you, and will not hate you." Promise me that!'

'I can make no promise now, Lily. I will think about it, and endeavour to do my duty.'

Lily was now seated, and was holding the skirt of her mother's dress.

'Mamma,' she said, looking up into her mother's face, 'you must be very good to me now; and I must be very good to you. We shall be always together now. I must be your friend and counsellor; and be everything to you, more than ever. I must fall in love with you now;' and she smiled again, and the tears were almost dry upon her cheeks.

At last they went down to the breakfast room, from which Bell had not moved. Mrs. Dale entered the room first, and Lily followed, hiding herself for a moment behind her mother. Then she came forward boldly, and taking Bell in her arms, clasped her close to her bosom.

'Bell,' she said, 'he has gone.'

'Lily! Lily! Lily!' said Bell, weeping.

'He has gone! We shall talk it over in a few days, and shall know how to do so without losing ourselves in misery. To-day we will say no more about it. I am so thirsty, Bell; do give me my tea;' and she sat herself down at the breakfast-table.

Lily's tea was given to her, and she drank it. Beyond that I cannot say that any of them partook with much heartiness of the meal. They sat there, as they would have sat if no terrible thunderbolt had fallen among them, and no word further was spoken about Crosbie and his conduct. Immediately after breakfast they went into the other room, and Lily, as was her wont, sat herself immediately down to her drawing. Her mother looked at her with wistful eyes, longing to bid her spare herself, but she shrank from interfering with her. For a quarter of an hour Lily sat over her board, with her brush or pencil in her hand, and then she rose up and put it away.

'It is no good pretending,' she said. 'I am only spoiling the things; but I will be better to-morrow. I'll go away and lie down by myself, mamma.' And so she went.

Soon after this Mrs. Dale took her bonnet and went up to the Great House, having received her brother-in-law's message from Bell.

'I know what he has to tell me,' she said; 'but I might as well go. It will be necessary that we should speak to each other about it.' So she walked across the lawn, and up into the hall of the Great House. 'Is my brother in the book-room?' she said to one of the maids; and then knocking at the door went in unannounced.

The squire rose from his air-chair, and came forward to meet her.

'Mary,' he said, 'I believe you know it all.'

'Yes,' she said. 'You can read that,' and she handed him Crosbies letter. 'How was one to know that any man could be so wicked as that?'

'And she has heard it?' asked the squire. 'Is she able to bear it?'

'Wonderfully! She has amazed me by her strength. It frightens me; for I know that a relapse must come. She has never sunk for a moment beneath it. For myself, I feel as though it were her strength that enables me to bear my share of it.' And then she described to the squire all that had taken place that morning.

'Poor child!' said the squire. 'Poor child! What can we do for her? Would it be good for her to go away for a time? She is a sweet, good, lovely girl, and has deserved better than that. Sorrow and disappointment come to us all; but they are doubly heavy when they come so early.'

Mrs. Dale was almost surprised at the amount of sympathy which he showed.

'And what is to be his punishment?' she asked.

'The scorn which men and women will feel for him; those, at least, whose esteem or scorn are matters of concern to any one. I know no other punishment. You would not have Lily's name brought before a tribunal of law?'

'Certainly not that.'

'And I will not have Bernard calling him out. Indeed, it would be for nothing; for in these days a man is not expected to fight duels.'

'You cannot think that I would wish that.'

'What punishment is there, then? I know of none. There are evils which a man may do, and no one can punish him. I know of nothing. I went up to London after him, but he contrived to crawl out of my way. What can you do to a rat but keep clear of him?'

Mrs. Dale had felt in her heart that it would be well if Crosbie could be beaten till all his bones were sore. I hardly know whether such should have been a woman's thought, but it was hers. She had no wish that he should be made to fight a duel. In that there would have been much that was wicked, and in her estimation nothing that was just. But she felt that if Bernard would thrash the coward for his cowardice she would love her nephew better than ever she had loved him. Bernard also had considered it probable that he might be expected to horsewhip the man who had jilted his cousin, and, as regarded the absolute bodily risk, he would not have felt any insuperable objection to undertake the task. But such a piece of work was disagreeable to him in many ways. He hated the idea of a row at his club. He was most desirous that his cousin's name should not be made public. He wished to avoid anything that might be impolitic. A wicked thing had been done, and he was quite ready to hate Crosbie as Crosbie ought to be hated; but as regarded himself, it made him unhappy to think that the world might probably expect him to punish the man who had so lately been his friend. And then he did not know where to catch him, or how to thrash him when caught. He was very sorry for his cousin, and felt strongly that Crosbie should not be allowed to escape. But what was he to do?

'Would she like to go anywhere?' said the squire again, anxious, if he could, to afford solace by some act of generosity. At this moment he would have settled a hundred a year for life upon his niece if by so doing he could have done her any good.

'She will be better at home,' said Mrs. Dale. 'Poor thing. For a while she will wish to avoid going out.'

'I suppose so;' and then there was a pause. 'I'll tell you what, Mary; I don't understand it. On my honour I don't understand it. It is to me as wonderful as though I had caught the man picking my pence out of my pocket. I don't think any man in the position of a gentleman would have done such a thing when I was young. I don't think any man would have dared to do it. But now it seems that a man may act in that way and no harm come to him. He had a friend in London who came to me and talked about it as though it were some ordinary, everyday transaction of life. Yes; you may come in, Bernard. The poor child knows it all now.'

Bernard offered to his aunt what of solace and sympathy he had to offer, and made some sort of half-expressed apology for having introduced this wolf into their flock. 'We always thought very much of him at his club,' said Bernard.

'I don't know much about your London clubs nowadays,' said his uncle, 'nor do I wish to do so if the society of that man can be endured after what he has now done.'

'I don't suppose half-a-dozen men will ever know anything about it,' said Bernard.

'Umph!' ejaculated the squire. He could not say that he wished Crosbie's villainy to be widely discussed, seeing that Lily's name was so closely connected with it. But yet he could not support the idea that Crosbie should not be punished by the frown of the world at large. It seemed to him that from this time forward any man speaking to Crosbie should be held to have disgraced himself by so doing.

'Give her my best love,' he said, as Mrs. Dale got up to take her leave; 'my very best love. If her old uncle can do anything for her she has only to let me know. She met the man in my house, and I feel that I owe her much. Bid her come and see me. It will be better for her than moping at home. And Mary' – this he said to her, whispering into her ear – 'think of what I said to you about Bell.'

Mrs. Dale, as she walked back to her own house, acknowledged to herself that her brother-in-law's manner was different to her from anything that she had hitherto known of him.

During the whole of that day Crosbie's name was not mentioned at the Small House. Neither of the girls stirred out, and Bell spent the greater part of the afternoon sitting, with her arm round her sister's waist, upon the sofa. Each of them had a book; but though there was little spoken, there was as little read. Who can describe the thoughts that were passing through Lily's mind as she remembered the hours which she had passed with Crosbie, of his warm assurances of love, of his accepted caresses, of her uncontrolled and acknowledged joy in his affection? It had all been holy to her then; and now those things which were then sacred had been made almost disgraceful by his fault. And yet as she thought of this she declared to herself over and over again that she would forgive him; – nay, that she had forgiven him. 'And he shall know it, too,' she said, speaking almost out loud.

'Lily, dear Lily,' said Bell, 'turn your thoughts away from it for a while, if you can.'

'They won't go away,' said Lily. And that was all that was said between them on the subject.

Everybody would know it! I doubt whether that must not be one of the bitterest drops in the cup which a girl in such circumstances is made to drain. Lily perceived early in the day that the parlour-maid well knew that she had been jilted. The girl's manner was intended to convey sympathy; but it did convey pity; and Lily for a moment felt angry. But she remembered that it must be so, and smiled upon the girl, and spoke kindly to her. What mattered it? All the world would know it in a day or two.

On the following day she went up, by her mother's advice, to see her uncle.

'My child,' said he, 'I am sorry for you. My heart bleeds for you.'

'Uncle,' she said, 'do not mind it. Only do this for me, – do not talk about it, – I mean to me.'

'No, no; I will not. That there should ever have been in my house so great a rascal—'

'Uncle! uncle! I will not have that! I will not listen to a word against him from any human being, – not a word! Remember that!' And her eyes flashed as she spoke.

He did not answer her, but took her hand and pressed it, and then she left him. 'The Dales were ever constant!' he said to himself, as he walked up and down the terrace before his house. 'Ever constant!'

CHAPTER XXXI

THE WOUNDED FAWN

NEARLY two months passed away, and it was now Christmas time at Allington. It may be presumed that there was no intention at either house that the mirth should be very loud. Such a wound as that received by Lily Dale was one from which recovery could not be quick, and it was felt by all the family that a weight was upon them which made gaiety impracticable. As for Lily herself it may be said that she bore her misfortune with all a woman's courage. For the first week she stood up as a tree that stands against the wind, which is soon to be shivered to pieces because it will not bend. During that week her mother and sister were frightened by her calmness and endurance. She would perform her daily task. She would go out through the village, and appear at her place in church on the first Sunday. She would sit over her book of an evening, keeping back her tears; and would chide her mother and sister when she found that they were regarding her with earnest anxiety.

'Mamma, let it all be as though it had never been,' she said.

'Ah, dear! if that were but possible!'

'God forbid that it should be possible inwardly,' Lily replied, 'but it is possible outwardly. I feel that you are more tender to me than you used to be, and that upsets me. If you would only scold me because I am idle, I should soon be better.' But her mother could not speak to her as she perhaps might have spoken had no grief fallen upon her pet. She could not cease from those anxious tender glances which made Lily know that she was looked on as a fawn wounded almost to death.

At the end of the first week she gave way. 'I won't get up, Bell,' she said one morning, almost petulantly. 'I am ill; – I had better lie here out of the way. Don't make a fuss about it. I'm stupid and foolish, and that makes me ill.'

Thereupon Mrs. Dale and Bell were frightened, and

looked into each other's blank faces, remembering stories of poor broken-hearted girls who had died because their loves had been unfortunate, – as small wax tapers whose lights are quenched if a breath of wind blows upon them too strongly. But then Lily was in truth no such slight taper as that. Nor was she the stem that must be broken because it will not bend. She bent herself to the blast during that week of illness, and then arose with her form still straight and graceful, and with her bright light unquenched.

After that she would talk more openly to her mother about her loss, – openly and with a true appreciation of the misfortune which had befallen her; but with an assurance of strength which seemed to ridicule the idea of a broken heart. 'I know that I can bear it,' she said, 'and that I can bear it without lasting unhappiness. Of course I shall always love him, and must feel almost as you felt when you lost my father.'

In answer to this Mrs. Dale could say nothing. She could not speak out her thoughts about Crosbie, and explain to Lily that he was unworthy of her love. Love does not follow worth, and is not given to excellence; – nor is it destroyed by ill-usage, nor killed by blows and mutilation. When Lily declared that she still loved the man who had so ill-used her, Mrs. Dale would be silent. Each perfectly understood the other, but on that matter even they could not interchange their thoughts with freedom.

'You must promise never to be tired of me, mamma,' said Lily.

'Mothers do not often get tired of their children, whatever the children may do of their mothers.'

'I'm not so sure of that when the children turn out old maids. And I mean to have a will of my own, too, mamma; and a way also, if it be possible. When Bell is married I shall consider it a partnership, and I shan't do what I'm told any longer.'

'Forewarned will be forearmed.'

'Exactly; – and I don't want to take you by surprise. For a year or two longer, till Bell is gone, I mean to be dutiful; but it would be very stupid for a girl to be dutiful all her life.'

All of which Mrs. Dale understood thoroughly. It amounted to an assertion on Lily's part that she had loved once and could never love again; that she had played her game, hoping, as other girls hope, that she might win the prize of a husband; but that, having lost, she could never play the game again. It was that inward conviction on Lily's part which made her say such words to her mother. But Mrs. Dale would by no means allow herself to share this conviction. She declared to herself that time would cure Lily's wound, and that her child might yet be crowned by the bliss of a happy marriage. She would not in her heart consent to that plan in accordance with which Lily's destiny in life was to be regarded as already fixed. She had never really liked Crosbie as a suitor, and would herself have preferred John Eames, with all the faults of his hobbledehoyhood on his head. It might yet come to pass that John Eames' love might be made happy.

But in the meantime Lily, as I have said, had become strong in her courage. and recommenced the work of living with no lackadaisical self-assurance that because she had been made more unhappy than others, therefore she should allow herself to be more idle. Morning and night she prayed for him, and daily, almost hour by hour, she assured herself that it was still her duty to love him. It was hard, this duty of loving, without any power of expressing such love. But still she would do her duty.

'Tell me at once, mamma,' she said one morning, 'when you hear that the day is fixed for his marriage. Pray don't keep me in the dark.'

'It is to be in February,' said Mrs. Dale.

'But let me know the day. It must not be to me like ordinary days. But do not look unhappy, mamma; I am not going to make a fool of myself. I shan't steal off and appear in the church like a ghost.' And then, having uttered her little joke, a sob came, and she hid her face on her mother's bosom. In a moment she raised it again. 'Believe me, mamma, that I am not unhappy,' she said.

After the expiration of that second week Mrs. Dale did write a letter to Crosbie:

'I suppose (she said) it is right that I should acknowledge the receipt of your letter. I do not know that I have aught else to say to you. It would not become me as a woman to say what I think of your conduct, but I believe that your conscience will tell you the same things. If it did not, you must, indeed, be hardened. I have promised my child that I will send to you a message from her. She bids me tell you that she has forgiven you, and that she does not hate you. May God also forgive you, and may you recover his love.

'Mary Dale.

'I beg that no rejoinder may be made to this letter, either to myself or to any of my family.'

The squire wrote no answer to the letter which he had received, nor did he take any steps towards the immediate punishment of Crosbie. Indeed he had declared that no such steps could be taken, explaining to his nephew that such a man could be served only as one serves a rat.

'I shall never see him,' he said once again; 'if I did, I should not scruple to hit him on the head with my stick; but I should think ill of myself to go after him with such an object.'

And yet it was a terrible sorrow to the old man that the scoundrel who had so injured him and his should escape scot-free. He had not forgiven Crosbie. No idea of forgiveness had ever crossed his mind. He would have hated himself had he thought it possible that he could be induced to forgive such an injury. 'There is an amount of rascality in it, – of low meanness, which I do not understand,' he would say over and over again to his nephew. And then as he would walk alone on the terrace he would speculate within his own mind whether Bernard would take any steps towards avenging his cousin's injury. 'He is right,' he would say to himself; 'Bernard is quite right. But when I was young I could not have stood it. In those days a gentleman might have a fellow out who had treated him as he has treated us. A man was satisfied in feeling that he had done something. I suppose the world is different nowadays.' The world is different; but the squire by no means acknowledged in his heart that there had been any improvement.

Bernard also was greatly troubled in his mind. He would have had no objection to fight a duel with Crosbie, had duels in these days been possible. But he believed them to be no longer possible, – at any rate without ridicule. And if he could not fight the man, in what other way was he to punish him? Was it not the fact that for such a fault the world afforded no punishment? Was it not in the power of a man like Crosbie to amuse himself for a week or two at the expense of a girl's happiness for life, and then to escape absolutely without any ill effects to himself? 'I shall be barred out of my club lest I should meet him,' Bernard said to himself, 'but he will not be barred out.' Moreover, there was a feeling within him that the matter would be one of triumph to Crosbie rather than otherwise. In having secured for himself the pleasure of his courtship with such a girl as Lily Dale, without encountering the penalty usually consequent upon such amusement, he would be held by many as having merited much admiration. He had sinned against all the Dales, and yet the suffering arising from his sin was to fall upon the Dales exclusively. Such was Bernard's reasoning, as he speculated on the whole affair, sadly enough, – wishing to be avenged, but not knowing where to look for vengeance. For myself I believe him to have been altogether wrong as to the light in which he supposed that Crosbie's falsehood would be regarded by Crosbie's friends. Men will still talk of such things lightly, professing that all is fair in love as it is in war, and speaking almost with envy of the good fortunes of a practised deceiver. But I have never come across the man who thought in this way with reference to an individual case. Crosbie's own judgment as to the consequences to himself of what he had done was more correct than that formed by Bernard Dale. He had regarded the act as venial as long as it was still to do, – while it was still within his power to leave it undone; but from the moment of its accomplishment it had forced itself upon his own view in its proper light. He knew that he had been a scoundrel, and he knew that other men would so think of him. His friend Fowler

Pratt, who had the reputation of looking at women simply as toys, had so regarded him. Instead of boasting of what he had done, he was afraid of alluding to any matter connected with his marriage as a man is of talking of the articles which he has stolen. He had already felt that men at his club looked askance at him; and, though he was no coward as regarded his own skin and bones, he had an undefined fear lest some day he might encounter Bernard Dale purposely armed with a stick. The squire and his nephew were wrong in supposing that Crosbie was unpunished.

And as the winter came on he felt that he was closely watched by the noble family of De Courcy. Some of that noble family he had already learned to hate cordially. The Honourable John came up to town in November, and persecuted him vilely; – insisted upon having dinners given to him at Sebright's, upon smoking throughout the whole afternoon in his future brother-in-law's rooms, and upon borrowing his future brother-in-law's possessions; till at last Crosbie determined that it would be wise to quarrel with the Honourable John, – and he quarrelled with him accordingly, turning him out of his rooms, and telling him in so many words that he would have no more to do with him.

'You'll have to do it, as I did,' Mortimer Gazebee had said to him; 'I didn't like it because of the family, but Lady Amelia told me that it must be so.' Whereupon Crosbie took the advice of Mortimer Gazebee.

But the hospitality of the Gazebees was perhaps more distressing to him than even the importunities of the Honourable John. It seemed as though his future sister-in-law was determined not to leave him alone. Mortimer was sent to fetch him up for the Sunday afternoons, and he found that he was constrained to go to the villa in St. John's Wood, even in opposition to his own most strenuous will. He could not quite analyse the circumstances of his own position, but he felt as though he were a cock with his spurs cut off, – as a dog with his teeth drawn. He found himself becoming humble and meek. He had to acknowledge to himself that he was afraid of Lady Amelia, and almost even afraid of Mortimer Gazebee. He was aware that they watched him, and knew all his goings out and comings in. They called him Adolphus, and made him tame. That coming evil day in February was dinned into his ears. Lady Amelia would go and look at furniture for him, and talked by the hour about bedding and sheets. 'You had better get your kitchen things at Tomkins'. They're all good, and he'll give you ten per cent off if you pay him ready money, – which of course you will, you know!' Was it for this that he had sacrificed Lily Dale? – for this that he had allied himself with the noble house of De Courcy?

Mortimer had been at him about the settlements from the very first moment of his return to London, and had already bound him up hand and foot. His life was insured, and the policy was in Mortimer's hands. His own little bit of money had been already handed over to be tied up with Lady Alexandrina's little bit. It seemed to him that in all the arrangements made the intention was that he should die off speedily, and that Lady Alexandrina should be provided with a decent little income, sufficient for St. John's Wood. Things were to be so settled that he could not even spend the proceeds of his own money, or of hers. They were to go, under the fostering hand of Mortimer Gazebee, in paying insurances. If he would only die the day after his marriage, there would really be a very nice sum of money for Alexandrina, almost worthy of the acceptance of an earl's daughter. Six months ago he would have considered himself able to turn Mortimer Gazebee round his finger on any subject that could be introduced between them. When they chanced to meet Gazebee had been quite humble to him, treating him almost as a superior being. He had looked down on Gazebee from a very great height. But now it seemed as though he were powerless in this man's hands.

But perhaps the countess had become his greatest aversion. She was perpetually writing to him little notes in which she gave him multitudes of commissions, sending him about as though he had been her servant. And she pestered him with advice which was even worse than her commissions, telling him of the style of life in which Alexandrina would expect to live, and warning him very frequently that such a one as he could not expect to be admitted within the bosom of so noble a family without paying very dearly for that inestimable privilege. Her letters had become odious to him, and he would chuck them on one side, leaving them for the whole day unopened. He had already made up his mind that he would quarrel with the countess also, very shortly after his marriage; indeed, that he would separate himself from the whole family if it were possible. And yet he had entered into this engagement mainly with the view of reaping those advantages which would accrue to him from being allied to the De Courcys! The squire and his nephew were wretched in thinking that this man was escaping without punishment, but they might have spared themselves that misery.

It had been understood from the first that he was to spend his Christmas at Courcy Castle. From this undertaking it was quite out of his power to enfranchise himself; but he resolved that his visit should be as short as possible. Christmas Day unfortunately came on a Monday, and it was known to the De Courcy world that Saturday was almost a *dies non** at the General Committee Office. As to those three days there was no escape for him; but he made Alexandrina understand that the three Commissioners were men of iron as to any extension of those three days. 'I must be absent again in February, of course,' he said, almost making his wail audible in the words he used, 'and therefore it is quite impossible that I should stay now beyond the Monday.' Had there been attractions for him at Courcy Castle I think he might have arranged with Mr. Optimist for a week or ten days. 'We shall be all alone,' the countess wrote to him, 'and I hope you will have an opportunity of learning more of our ways than you have ever really been able to do as yet.' This was bitter as gall to him. But in this world all valuable commodities have their price; and when men such as Crosbie aspire to obtain for themselves an alliance with noble families, they must pay the market price for the article which they purchase.

'You'll all come up and dine with us on Monday,' the squire said to Mrs. Dale, about the middle of the previous week.

'Well, I think not,' said Mrs. Dale; 'we are better, perhaps, as we are.'

At this moment the squire and his sister-in-law were on much more friendly terms than had been usual with them, and he took her reply in good part, understanding her feeling. Therefore, he pressed his request, and succeeded.

'I think you're wrong,' he said; 'I don't suppose that we shall have a very merry Christmas. You and the girls will hardly have that whether you eat your pudding here or at the Great House. But it will be better for us all to make the attempt. It's the right thing to do. That's the way I look at it.'

'I'll ask Lily,' said Mrs. Dale.

'Do, do. Give her my love, and tell her from me that, in spite of all that has come and gone, Christmas Day should still be to her a day of rejoicing. We'll dine about three, so that the servants can have the afternoon.'

'Of course we'll go,' said Lily; 'why not? We always do. And we'll have blind-man's-buff with all the Boyces, as we had last year, if uncle will ask them up.' But the Boyces were not asked up for that occasion.

But Lily, though she put on it all so brave a face, had much to suffer, and did in truth suffer greatly. If you, my reader, ever chanced to slip into the gutter on a wet day, did you not find that the sympathy of the bystanders was by far the severest part of your misfortune? Did you not declare to yourself that all might yet be well, if the people would only walk on and not look at you? And yet you cannot blame those who stood and pitied you; or, perhaps, essayed to rub you down, and assist you in the recovery of your bedaubed hat. You, yourself, if you see a man fall, cannot walk by as though nothing uncommon had happened to him. It was so with Lily. The people of Allington could not regard her with their ordinary eyes. They would look at her tenderly, knowing that she was a wounded fawn, and thus they aggravated the soreness of her wound. Old Mrs. Hearn condoled with her, telling her that very likely she would be better off as she was. Lily would not lie about it in any way. 'Mrs. Hearn,' she said, 'the subject is painful to me.' Mrs.

Hearn said no more about it, but on every meeting between them she looked the things she did not say. 'Miss Lily!' said Hopkins, one day, 'Miss Lily!' – and as he looked up into her face a tear had almost formed itself in his old eye – 'I knew what he was from the start. Oh, dear! oh, dear! If I could have had him killed!' 'Hopkins, how dare you?' said Lily. 'If you speak to me again in such a way, I will tell my uncle.' She turned away from him; but immediately turned back again, and put out her little hand to him. 'I beg your pardon,' she said. 'I know how kind you are, and I love you for it.' And then she went away. 'I'll go after him yet, and break the dirty neck of him,' said Hopkins to himself as he walked down the path.

Shortly before Christmas Day she called, with her sister, at the vicarage. Bell, in the course of the visit, left the room with one of the Boyce girls, to look at the last chrysanthemums of the year. Then Mrs. Boyce took advantage of the occasion to make her little speech. 'My dear Lily,' she said, 'you will think me cold if I do not say one word to you.' 'No, I shall not,' said Lily, almost sharply, shrinking from the finger that threatened to touch her sore. 'There are things which should never be talked about.' 'Well, well; perhaps so,' said Mrs. Boyce. But for a minute or two she was unable to fall back upon any other topic, and sat looking at Lily with painful tenderness. I need hardly say what were Lily's sufferings under such a gaze; but she bore it, acknowledging to herself in her misery that the fault did not lie with Mrs. Boyce. How could Mrs. Boyce have looked at her otherwise than tenderly?

It was settled, then, that Lily was to dine up at the Great House on Christmas Day, and thus show to the Allington world that she was not to be regarded as a person shut out from the world by the depth of her misfortune. That she was right there can, I think, be no doubt; but as she walked across the little bridge, with her mother and sister, after returning from church, she would have given much to be able to have turned round, and have gone to bed instead of to her uncle's dinner.

CHAPTER XXXII

PAWKINS'S IN JERMYN STREET

THE show of fat beasts in London* took place this year on the twentieth day of December, and I have always understood that a certain bullock exhibited by Lord De Guest was declared by the metropolitan butchers to have realized all the possible excellences of breeding, feeding, and condition. No doubt the butchers of the next half-century will have learned much better, and the Guestwick beast, could it be embalmed and then produced, would excite only ridicule at the agricultural ignorance of the present age; but Lord De Guest took the praise that was offered to him, and found himself in a seventh heaven of delight. He was never so happy as when surrounded by butchers, graziers, and salesmen who were able to appreciate the work of his life, and who regarded him as a model nobleman. 'Look at that fellow,' he said to Eames, pointing to the prize bullock. Eames had joined his patron at the show after his office hours, looking on upon the living beef by gaslight. 'Isn't he like his sire? He was got by Lambkin, you know.'

'Lambkin,' said Johnny, who had not as yet been able to learn much about the Guestwick stock.

'Yes, Lambkin. The bull that we had the trouble with. He has just got his sire's back and fore-quarters. Don't you see?'

'I daresay,' said Johnny, who looked very hard, but could not see.

'It's very odd,' exclaimed the earl, 'but do you know, that bull has been as quiet since that day, – as quiet as – as anything. I think it must have been my pocket-handkerchief.'

'I daresay it was,' said Johnny; – 'or perhaps the flies.'

'Flies!' said the earl, angrily. 'Do you suppose he isn't used to flies? Come away. I ordered dinner at seven, and it's past six now. My brother-in-law, Colonel Dale, is up in town, and he dines with us.' So he took Johnny's arm, and led him off through the show, calling his attention as he went to several beasts which were inferior to his own.

And then they walked down through Portman Square

and Grosvenor Square, and across Piccadilly to Jermyn Street. John Eames acknowledged to himself that it was odd that he should have an earl leaning on his arm as he passed along through the streets. At home, in his own life, his daily companions were Cradell and Amelia Roper, Mrs. Lupex and Mrs. Roper. The difference was very great, and yet he found it quite as easy to talk to the earl as to Mrs. Lupex.

'You know the Dales down at Allington, of course,' said the earl.

'Oh, yes, I know them.'

'But, perhaps, you never met the colonel.'

'I don't think I ever did.'

'He's a queer sort of fellow; – very well in his way, but he never does anything. He and my sister live at Torquay, and as far as I can find out, they neither of them have any occupation of any sort. He's come up to town now because we both had to meet our family lawyers and sign some papers, but he looks on the journey as a great hardship. As for me, I'm a year older than he is, but I wouldn't mind going up and down from Guestwick every day.'

'It's looking after the bull that does it,' said Eames.

'By George! you're right, Master Johnny. My sister and Crofts may tell me what they like, but when a man's out in the open air for eight or nine hours every day, it doesn't much matter where he goes to sleep after that. This is Pawkins's, – capital good house, but not so good as it used to be while old Pawkins was alive. Show Mr. Eames up into a bedroom to wash his hands.'

Colonel Dale was much like his brother in face, but was taller, even thinner, and apparently older. When Eames went into the sitting-room, the colonel was there alone, and had to take upon himself the trouble of introducing himself. He did not get up from his arm-chair, but nodded gently at the young man. 'Mr. Eames, I believe? I knew your father at Guestwick, a great many years ago;' then he turned his face back towards the fire and sighed.

'It's very cold this afternoon,' said Johnny, trying to make conversation.

'It's always cold in London,' said the colonel.

'If you had to be here in August you wouldn't say so.'

'God forbid,' said the colonel, and he sighed again, with his eyes fixed upon the fire. Eames had heard of the very gallant way in which Orlando Dale had persisted in running away with Lord De Guest's sister, in opposition to very terrible obstacles, and as he now looked at the intrepid lover, he thought that there must have been a great change since those days. After that nothing more was said till the earl came down.

Pawkins's house was thoroughly old-fashioned in all things, and the Pawkins of that day himself stood behind the earl's elbow when the dinner began, and himself removed the cover from the soup tureen. Lord De Guest did not require much personal attention, but he would have felt annoyed if this hadn't been done. As it was he had a civil word to say to Pawkins about the fat cattle, thereby showing that he did not mistake Pawkins for one of the waiters. Pawkins then took his lordship's orders about the wine and retired.

'He keeps up the old house pretty well,' said the earl to his brother-in-law. 'It isn't like what it was thirty years ago, but then everything of that sort has got worse and worse.'

'I suppose it has,' said the colonel.

'I remember when old Pawkins had as good a glass of port as I've got at home, – or nearly. They can't get it now, you know.'

'I never drink port,' said the colonel. 'I seldom take anything after dinner, except a little negus.'*

His brother-in-law said nothing, but made a most eloquent grimace as he turned his face towards his soup-plate. Eames saw it, and could hardly refrain from laughing. When, at half-past nine o'clock, the colonel retired from the room, the earl, as the door was closed, threw up his hands, and uttered the one word 'negus!' Then Eames took heart of grace and had his laughter out.

The dinner was very dull, and before the colonel went to bed Johnny regretted that he had been induced to dine at Pawkins's. It might be a very fine thing to be asked to dinner with an earl, and John Eames had perhaps received at his office some little accession of dignity from the circumstance, of which he had been not unpleasantly aware; but, as he sat at the table, on which there were four or five apples and a plate of dried nuts, looking at the earl, as he endeavoured to keep his eyes open, and at the colonel, to whom it seemed absolutely a matter of indifference whether his companions were asleep or awake, he confessed to himself that the price he was paying was almost too dear. Mrs. Roper's tea-table was not pleasant to him, but even that would have been preferable to the black dinginess of Pawkins's mahogany, with the company of two tired old men, with whom he seemed to have no mutual subject of conversation. Once or twice he tried a word with the colonel, for the colonel sat with his eyes open looking at the fire. But he was answered with monosyllables, and it was evident to him that the colonel did not wish to talk. To sit still, with his hands closed over each other on his lap, was work enough for Colonel Dale during his after-dinner hours.

But the earl knew what was going on. During that terrible conflict between him and his slumber, in which the drowsy god fairly vanquished him for some twenty minutes, his conscience was always accusing him of treating his guests badly. He was very angry with himself, and tried to arouse himself and talk. But his brother-in-law would not help him in his efforts; and even Eames was not bright in rendering him assistance. Then for twenty minutes he slept soundly, and at the end of that he woke himself with one of his own snorts. 'By George!' he said, jumping up and standing on the rug, 'we'll have some coffee;' and after that he did not sleep any more.

'Dale,' said he, 'won't you take some more wine?'

'Nothing more, said the colonel, still looking at the fire, and shaking his head very slowly.

'Come, Johnny, fill your glass.' He had already got into the way of calling his young friend Johnny, having found that Mrs. Eames generally spoke of her son by that name.

'I have been filling my glass all the time,' said Eames, taking the decanter again in his hand as he spoke.

'I'm glad you've found something to amuse you, for it has seemed to me that you and Dale haven't had much to say to each other. I've been listening all the time.'

'You've been asleep,' said the colonel.

'Then there's been some excuse for my holding my tongue,' said the earl. 'By-the-by, Dale, what do you think of that fellow Crosbie?'

Eames' ears were instantly on the alert, and the spirit of dullness vanished from him.

'Think of him?' said the colonel.

'He ought to have every bone in his skin broken,' said the earl.

'So he ought,' said Eames, getting up from his chair in his eagerness, and speaking in a tone somewhat louder than was perhaps becoming in the presence of his seniors. 'So he ought, my lord. He is the most abominable rascal that ever I met in my life. I wish I was Lily Dale's brother.' Then he sat down again, remembering that he was speaking in the presence of Lily's uncle, and of the father of Bernard Dale, who might be supposed to occupy the place of Lily's brother.

The colonel turned his head round, and looked at the young man with surprise. 'I beg your pardon, sir,' said Eames, 'but I have known Mrs. Dale and your nieces all my life.'

'Oh, have you?' said the colonel. 'Nevertheless it is, perhaps, as well not to make too free with a young lady's name. Not that I blame you in the least, Mr. Eames.'

'I should think not,' said the earl. 'I honour him for his feeling. Johnny, my boy, if ever I am unfortunate enough to meet that man, I shall tell him my mind, and I believe you will do the same.' On hearing this John Eames winked at the earl, and made a motion with his head towards the colonel, whose back was turned to him. And then the earl winked back at Eames.

'De Guest,' said the colonel, 'I think I'll go upstairs; I always have a little arrowroot* in my own room.'

'I'll ring the bell for a candle,' said the host. Then the colonel went, and as the door was closed behind him, the earl raised his two hands and uttered that single word, 'negus!' Whereupon Johnny burst out laughing, and coming round to the fire, sat himself down in the arm-chair which the colonel had left.

'I've no doubt it's all right.' said the earl; 'but I shouldn't like to drink negus myself, nor yet to have arrowroot up in my bedroom.'

'I don't suppose there's any harm in it.'

'Oh dear, no; I wonder what Pawkins says about him. But I suppose they have them of all sorts in an hotel.'

'The waiter didn't seem to think much of it when he brought it.'

'No, no. If he'd asked for senna and salts,* the waiter wouldn't have showed any surprise. By-the-by, you touched him up about that poor girl.'

'Did I, my lord? I didn't mean it.'

'You see he's Bernard Dale's father, and the question is, whether Bernard shouldn't punish the fellow for what he has done. Somebody ought to do it. It isn't right that he should escape. Somebody ought to let Mr. Crosbie know what a scoundrel he has made himself.'

'I'd do it to-morrow, only I'm afraid—'

'No, no, no,' said the earl; 'you are not the right person at all. What have you got to do with it? You've merely known them as family friends, but that's not enough.'

'No, I suppose not,' said Eames, sadly.

'Perhaps it's best as it is,' said the earl. 'I don't know that any good would be got by knocking him over the head. And if we are to be Christians, I suppose we ought to be Christians.'

'What sort of a Christian has he been?'

'That's true enough; and if I was Bernard, I should be very apt to forget my Bible lessons about meekness.'

'Do you know, my lord, I should think it the most Christian thing in the world to pitch into him; I should, indeed. There are some things for which a man ought to be beaten black and blue.'

'So that he shouldn't do them again?'

'Exactly. You might say it isn't Christian to hang a man.'

'I'd always hang a murderer. It wasn't right to hang men for stealing a sheep.'

'Much better hang such a fellow as Crosbie,' said Eames.

'Well, I believe so. If any fellow wanted now to curry favour with the young lady, what an opportunity he'd have.'

Johnny remained silent for a moment or two before he answered. 'I'm not so sure of that,' he said, mournfully, as though grieving at the thought that there was no chance of currying favour with Lily by thrashing her late lover.

'I don't pretend to know much about girls,' said Lord De Guest; 'but I should think it would be so. I should fancy that nothing would please her so much as hearing that he had caught it, and that all the world knew that he'd caught it.' The earl had declared that he didn't know much about girls, and in so saying, he was no doubt right.

'If I thought so,' said Eames, 'I'd find him out to-morrow.'

'Why so? what difference does it make to you?' Then there was another pause, during which Johnny looked very sheepish.

'You don't mean to say that you're in love with Miss Lily Dale?'

'I don't know much about being in love with her,' said Johnny, turning very red as he spoke. And then he made up his mind, in a wild sort of way, to tell all the truth to his friend. Pawkins's port wine may, perhaps, have had something to do with the resolution. 'But I'd go through fire and water for her, my lord. I knew her years before he had ever seen her, and have loved her a great deal better than he will ever love any one. When I heard that she had accepted him, I had half a mind to cut my own throat, – or else his.'

'Highty tighty,' said the earl.

'It's very ridiculous, I know,' said Johnny, 'and of course she would never have accepted me.'

'I don't see that at all.'

'I haven't a shilling in the world.'

'Girls don't care much for that.'

'And then a clerk in the Income-tax Office! It's such a poor thing.'

'The other fellow was only a clerk in another office.'

The earl living down at Guestwick did not understand that the Income-tax Office in the city, and the General Committee Office at Whitehall, were as far apart as Dives and Lazarus,* and separated by as impassable a gulf.

'Oh, yes,' said Johnny; 'but his office is another kind of thing, and then he was a swell himself.'

'By George, I don't see it,' said the earl.

'I don't wonder a bit at her accepting a fellow like that. I hated him the first moment I saw him; but that's no reason she should hate him. He had that sort of manner, you know. He was a swell, and girls like that kind of thing. I never felt angry with her, but I could have eaten him.' As he spoke he looked as though he would have made some such attempt had Crosbie been present.

'Did you ever ask her to have you?' said the earl.

'No; how could I ask her, when I hadn't bread to give her?'

'And you never told her – that you were in love with her, I mean, and all that kind of thing.'

'She knows it now,' said Johnny; 'I went to say good-bye to her the other day, – when I thought she was going to be married. I could not help telling her then.'

'But it seems to me, my dear fellow, that you ought to be very much obliged to Crosbie; – that is to say, if you've a mind to—'

'I know what you mean, my lord. I am not a bit obliged to him. It's my belief that all this will about kill her. As to myself, if I thought she'd ever have me—'

Then he was again silent, and the earl could see that the tears were in his eyes.

'I think I begin to understand it,' said the earl, 'and I'll give you a bit of advice. You come down and spend your Christmas with me at Guestwick.'

'Oh, my lord!'

'Never mind my-lording me, but do as I tell you. Lady Julia sent you a message, though I forgot all about it till now. She wants to thank you herself for what you did in the field.'

'That's all nonsense, my lord.'

'Very well; you can tell her so. You may take my word for this, too, – my sister hates Crosbie quite as much as you do. I think she'd "pitch into him," as you call it, herself, if she knew how. You come down to Guestwick for the Christmas, and then go over to Allington and tell them all plainly what you mean.'

'I couldn't say a word to her now.'

'Say it to the squire, then. Go to him, and tell him what you mean, – holding your head up like a man. Don't talk to me about swells. The man who means honestly is the best swell I know. He's the only swell I recognise. Go to old Dale, and say you come from me, – from Guestwick Manor. Tell him that if he'll put a little stick under the pot to make it boil, I'll put a bigger one. He'll understand what that means.'

'Oh, no, my lord.'

'But I say, oh yes;' and the earl, who was now standing on the rug before the fire, dug his hands deep down into his trousers' pockets. 'I'm very fond of that girl, and would do much for her. You ask Lady Julia if I didn't say so to her before I ever knew of your casting a sheep's-eye that way. And I've a sneaking kindness for you too, Master Johnny. Lord bless you, I knew your father as well as I ever knew any man; and to tell the truth, I believe I helped to ruin him. He held land of me, you know, and there can't be any doubt that he did ruin himself. He knew no more about a beast when he'd done, than – than – than that waiter. If he'd gone on to this day he wouldn't have been any wiser.'

Johnny sat silent, with his eyes full of tears. What was he to say to his friend?

'You come down with me,' continued the earl, 'and you'll find we'll make it all straight. I daresay you're right about not speaking to the girl just at present. But tell everything to the uncle, and then to the mother. And, above all things, never think that you're not good enough yourself. A man should never think that. My belief is that in life people will take you very much at your own reckoning. If you are made of dirt, like that fellow Crosbie, you'll be found out at last, no doubt. But then I don't think you are made of dirt.'

'I hope not.'

'And so do I. You can come down, I suppose, with me the day after to-morrow?'

'I'm afraid not. I have had all my leave.'

'Shall I write to old Buffle, and ask it as a favour?'

'No,' said Johnny; 'I shouldn't like that. But I'll see to-morrow, and then I'll let you know. I can go down by the mail train on Saturday, at any rate.'

'That won't be comfortable. See and come with me if you can. Now, good-night, my dear fellow, and remember this, – when I say a thing I mean it. I think I may boast that I never yet went back from my word.'

The earl as he spoke gave his left hand to his guest, and looking somewhat grandly up over the young man's head, he tapped his own breast thrice with his right hand. As he went through the little scene, John Eames felt that he was every inch an earl.

'I don't know what to say to you, my lord.'

'Say nothing, – not a word more to me. But say to yourself that faint heart never won fair lady. Good-night, my dear boy, good-night. I dine out to-morrow, but you can call and let me know at about six.'

Eames then left the room without another word, and walked out into the cold air of Jermyn Street. The moon was clear and bright, and the pavement in the shining light seemed to be as clean as a lady's hand. All the world was altered to him since he had entered Pawkins's Hotel. Was it then possible that Lily Dale might even yet become his wife? Could it be true that he, even now, was in a position to go boldly to the Squire of Allington, and tell him what were his views with reference to Lily? And how far would he be justified in taking the earl at his word? Some incredible amount of wealth would be required before he could marry Lily Dale. Two or three hundred pounds a year at the very least! The earl could not mean him to understand that any such sum as that would be made up with such an object! Nevertheless he resolved as he walked home to Burton Crescent that he would go down to Guestwick, and that he would obey the earl's behest. As regarded Lily herself he felt that nothing could be said to her for many a long day as yet.

'Oh, John, how late you are!' said Amelia, slipping out from the back parlour, as he let himself in with his latch-key.

'Yes, I am; – very late,' said John, taking his candle, and passing her by on the stairs without another word.

CHAPTER XXXIII

'THE TIME WILL COME'

'DID you hear that young Eames is staying at Guestwick Manor?'

As these were the first words which the squire spoke to Mrs. Dale as they walked together up to the Great House, after church, on Christmas Day, it was clear enough that the tidings of Johnny's visit, when told to him, had made some impression.

'At Guestwick Manor!' said Mrs. Dale. 'Dear me! Do you hear that, Bell? There's promotion for Master Johnny!'

'Don't you remember, mamma,' said Bell, 'that he helped his lordship in his trouble with the bull?'

Lily, who remembered accurately all the passages of her last interview with John Eames, said nothing, but felt, in some sort, sore at the idea that he should be so near her at such a time. In some unconscious way she had liked him for coming to her and saying all that he did say. She valued him more highly after that scene than she did before. But now, she would feel herself injured and hurt if he ever made his way into her presence under circumstances as they existed.

'I should not have thought that Lord De Guest was the man to show so much gratitude for so slight a favour,' said the squire. 'However, I am going to dine there to-morrow.'

'To meet young Eames?' said Mrs. Dale.

'Yes, – especially to meet young Eames. At least, I've been very specially asked to come, and I've been told that he is to be there.'

'And is Bernard going?'

'Indeed I'm not,' said Bernard. 'I shall come over and dine with you.'

A half-formed idea flitted across Lily's mind, teaching her to imagine for a moment that she might possibly be concerned in this arrangement. But the thought vanished as quickly as it came, merely leaving some soreness behind it. There are certain maladies which make the whole body sore. The patient, let him be touched on any point, – let him even be nearly touched, – will roar with agony as though his whole body had been bruised. So it is also with maladies of the mind. Sorrows such as that of poor Lily's leave the heart sore at every point, and compel the sufferer to be ever in fear of new wounds. Lily bore her cross bravely and well; but not the less did it weigh heavily upon her at every turn because she had the strength to walk as though she did not bear it. Nothing happened to her, or in her presence, that did not in some way connect itself with her misery. Her uncle was going over to meet John Eames at Lord De Guest's. Of course the men there would talk about her, and all such talking was an injury to her.

The afternoon of that day did not pass away brightly. As long as the servants were in the room the dinner went on much as other dinners. At such times a certain amount of hypocrisy must always be practised in closely domestic circles. At mixed dinner-parties people can talk before Richard and William the same words that they would use if Richard and William were not there. People so mixed do not talk together their inward home thoughts. But when close friends are together, a little conscious reticence is practised till the door is tiled. At such a meeting as this that conscious reticence was of service, and created an effort which was salutary. When the door was tiled, and when the servants were gone, how could they be merry together? By what mirth should the beards be made to wag on that Christmas Day?

'My father has been up in town,' said Bernard. 'He was with Lord De Guest at Pawkins's.'

'Why didn't you go and see him?' asked Mrs. Dale.

'Well, I don't know. He did not seem to wish it. I shall go down to Torquay in February. I must be up in London, you know, in a fortnight, for good.' Then they were all silent again for a few minutes. If Bernard could have owned the truth, he would have acknowledged that he had not gone up to London, because he did not yet know how to treat Crosbie when he should meet him. His thoughts on this matter threw some sort of shadow across poor Lily's mind, making her feel that her wound was again opened.

'I want him to give up his profession altogether,' said the squire, speaking firmly and slowly. 'It would be better, I think, for both of us that he should do so.'

'Would it be wise at his time of life,' said Mrs. Dale, 'and when he has been doing so well?'

'I think it would be wise. If he were my son it would be thought better that he should live here upon the property, among the people who are to become his tenants, than remain up in London, or perhaps be sent to India. He has one profession as the heir of this place, and that, I think, should be enough.'

'I should have but an idle life of it down here,' said Bernard.

'That would be your own fault. But if you did as I would have you, your life would not be idle.' In this he was alluding to Bernard's proposed marriage, but as to that nothing further could be said in Bell's presence. Bell understood it all, and sat quite silent, with demure countenance; – perhaps even with something of sternness in her face.

'But the fact is,' said Mrs. Dale, speaking in a low tone, and having well considered what she was about to say, 'that Bernard is not exactly the same as your son.'

'Why not?' said the squire. 'I have even offered to settle the property on him if he will leave the service.'

'You do not owe him so much as you would owe your son; and, therefore, he does not owe you as much as he would owe his father.'

'If you mean that I cannot constrain him, I know that well enough. As regards money, I have offered to do for him quite as much as any father would feel called upon to do for an only son.'

'I hope you don't think me ungrateful,' said Bernard.

'No, I do not; but I think you unmindful. I have nothing more to say about it, however; – not about that. If you should marry—' And then he stopped himself, feeling that he could not go on in Bell's presence.

'If he should marry,' said Mrs. Dale, 'it may well be that his wife would like a house of her own.'

'Wouldn't she have this house?' said the squire, angrily. 'Isn't it big enough? I only want one room for myself, and I'd give up that if it were necessary.'

'That's nonsense,' said Mrs. Dale.

'It isn't nonsense,' said the squire.

'You'll be squire of Allington for the next twenty years,' said Mrs. Dale. 'And as long as you are the squire, you'll be master of this house; at least, I hope so. I don't approve of monarchs abdicating in favour of young people.'

'I don't think uncle Christopher would look at all well like Charles the Fifth,' said Lily.

'I would always keep a cell for you, my darling, if I did,' said the squire, regarding her with that painful, special tenderness. Lily, who was sitting next to Mrs. Dale, put her hand out secretly and got hold of her mother's, thereby indicating that she did not intend to occupy the cell offered to her by her uncle; or to look to him as the companion of her monastic seclusion. After that there was nothing more then said as to Bernard's prospects.

'Mrs. Hearn is dining at the vicarage, I suppose?' asked the squire.

'Yes; she went in after church,' said Bell. 'I saw her go with Mrs. Boyce.'

'She told me she never would dine with them again after dark in winter,' said Mrs. Dale. 'The last time she was there, the boy let the lamp blow out as she was going home, and she lost her way. The truth was, she was angry because Mr. Boyce didn't go with her.'

'She's always angry,' said the squire. 'She hardly speaks to me now. When she paid her rent the other day to Jolliffe, she said she hoped it would do me much good; as though she thought me a brute for taking it.'

'So she does,' said Bernard.

'She's very old, you know,' said Bell.

'I'd give her the house for nothing, if I were you, uncle,' said Lily.

'No, my dear; if you were me you would not. I should be very wrong to do so. Why should Mrs. Hearn have her house for nothing, any more than her meat or her clothes? It would be much more reasonable were I to give her so much money into her hand yearly; but it would be wrong in me to do so, seeing that she is not an object of charity; – and it would be wrong in her to take it.'

'And she wouldn't take it,' said Mrs. Dale.

'I don't think she would. But if she did, I'm sure she would grumble because it wasn't double the amount. And if Mr. Boyce had gone home with her, she would have grumbled because he walked too fast.'

'She is very old,' said Bell again.

'But, nevertheless, she ought to know better than to speak disparagingly of me to my servants. She should have more respect for herself.' And the squire showed by the tone of his voice that he thought very much about it.

It was very long and very dull that Christmas evening, making Bernard feel strongly that he would be very foolish to give up his profession, and tie himself down to a life at Allington. Women are more accustomed than men to long, dull, unemployed hours; and, therefore, Mrs. Dale and her daughters bore the tedium courageously. While he yawned, stretched himself, and went in and out of the room, they sat demurely, listening as the squire laid down the law on small matters, and contradicting him occasionally when the spirit of either of them prompted her specially to do so. 'Of course you know much better than I do,' he would say. 'Not at all,' Mrs. Dale would answer. 'I don't pretend to know anything about it. But—' So the evening wore itself away; and when the squire was left alone at half-past nine, he did not feel that the day had passed badly with him. That was his style of life, and he expected no more from it than he got. He did not look to find things very pleasant, and, if not happy, he was, at any rate, contented.

'Only think of Johnny Eames being at Guestwick Manor!' said Bell, as they were going home.

'I don't see why he shouldn't be there,' said Lily. 'I would rather it should be he than I, because Lady Julia is so grumpy.'

'But asking your uncle Christopher especially to meet him!' said Mrs. Dale. 'There must be some reason for it.' Then Lily felt the soreness come upon her again, and spoke no further upon the subject.

We all know that there was a special reason, and that Lily's soreness was not false in its mysterious forebodings. Eames, on the evening after his dinner at Pawkins's, had seen the earl, and explained to him that he could not leave town till the Saturday evening; but that he could remain over the Tuesday. He must be at his office by twelve on Wednesday, and could manage to do that by an early train from Guestwick.

'Very well, Johnny,' said the earl, talking to his young friend with the bedroom candle in his hand, as he was going up to dress. 'Then I'll tell you what; I've been thinking of it. I'll ask Dale to come over to dinner on Tuesday; and if he'll come, I'll explain the whole matter to him myself. He's a man of business, and he'll understand. If he won't come, why then you must go over to Allington, and find him, if you can, on the Tuesday morning; or I'll go to him myself, which will be better. You mustn't keep me now, as I am ever so much too late.'

Eames did not attempt to keep him, but went away feeling that the whole matter was being arranged for him in a very wonderful way. And when he got to Allington he found that the squire had accepted the earl's invitation. Then he declared to himself that there was no longer any possibility of retractation for him. Of course he did not wish to retract. The one great longing of his life was to call Lily Dale his own. But he felt afraid of the squire, – that the squire would despise him and snub him, and that the earl would perceive that he had made a mistake when he saw how his client was scorned and snubbed. It was arranged that the earl was to take the squire into his own room for a few minutes before dinner, and Johnny felt that he would be hardly able to stand his ground in the drawing-room when the two old men should make their appearance together.

He got on very well with Lady Julia, who gave herself no airs, and made herself very civil. Her brother had told her the whole story, and she felt as anxious as he did to provide Lily with another husband in place of that horrible man Crosbie. 'She has been very fortunate in her escape,' she said to her brother; 'very fortunate.' The earl agreed with this, saying that in his opinion his own favourite Johnny would make much the nicer lover of the two. Lady Julia had her doubts as to Lily's acquiescence. 'But, Theodore, he must not speak to Miss Lilian Dale herself about it yet a while.'

'No,' said the earl; 'not for a month or so.'

'He will have a better chance if he can remain silent for six months,' said Lady Julia.

'Bless my soul! somebody else will have picked her up before that,' said the earl.

In answer to this Lady Julia merely shook her head.

Johnny went over to his mother on Christmas Day after church, and was received by her and by his sister with great honour. And she gave him many injunctions as to his behaviour at the earl's table, even descending to small details about his boots and linen. But Johnny had already begun to feel at the Manor that, after all, people are not so very different in their ways of life as they are supposed to be. Lady Julia's manners were certainly not quite those of Mrs. Roper; but she made the tea very much in the same way in which it was made at Burton Crescent, and Eames found that he could eat his egg, at any rate on the second morning, without any tremor in his hand, in spite of the coronet on the silver egg-cup. He did feel himself to be rather out of his place in the Manor pew on the Sunday, conceiving that all the congregation was looking at him; but he got over this on Christmas Day, and sat quite comfortably in his soft corner during the sermon, almost going to sleep. And when he walked with the earl after church to the gate over which the noble peer had climbed in his agony, and inspected the hedge through which he had thrown himself, he was quite at home with his little jokes, bantering his august companion as to the mode of his somersault. But be it always remembered that there are two modes in which a young man may be free and easy with his elder and superior, – the mode pleasant and the mode offensive. Had it been in Johnny's nature to try the latter, the earl's back would soon have been up, and the play would have been over. But it was not in Johnny's nature to do so, and therefore it was that the earl liked him.

At last came the hour of dinner on Tuesday, or at least the hour at which the squire had been asked to show himself at the Manor House. Eames, as by agreement with his patron, did not come down so as to show himself till after the interview. Lady Julia, who had been present at their discussions, had agreed to receive the squire; and then a servant was to ask him to step into the earl's own room. It was pretty to see the way in which the three conspired together, planning and plotting with an eagerness that was beautifully green and fresh.

'He can be as cross as an old stick when he likes it,' said the earl, speaking of the squire; 'and we must take care not to rub him the wrong way.'

'I shan't know what to say to him when I come down,' said Johnny.

'Just shake hands with him and don't say anything,' said Lady Julia.

'I'll give him some port wine that ought to soften his heart,' said the earl, 'and then we'll see how he is in the evening.'

Eames heard the wheels of the squire's little open carriage and trembled. The squire, unconscious of all schemes, soon found himself with Lady Julia, and within two minutes of his entrance was walked off to the earl's private room. 'Certainly,' he said, 'certainly;' and followed the man-servant. The earl, as he entered, was standing in the middle of the room, and his round rosy face was a picture of good-humour.

'I'm very glad you've come, Dale,' said he. 'I've something I want to say to you.'

Mr. Dale, who neither in heart nor in manner was so light a man as the earl, took the proffered hand of his host, and bowed his head slightly, signifying that he was willing to listen to anything.

'I think I told you,' continued the earl, 'that young John Eames is down here; but he goes back to-morrow, as they can't spare him at his office. He's a very good fellow, – as far as I am able to judge, an uncommonly good young man. I've taken a great fancy to him myself.'

In answer to this Mr. Dale did not say much. He sat down, and in some general terms expressed his good-will towards all the Eames family.

'As you know, Dale, I'm a very bad hand at talking, and therefore I won't beat about the bush in what I've got to say at present. Of course we've all heard of that scoundrel Crosbie, and the way he has treated your niece Lilian.'

'He is a scoundrel, – an unmixed scoundrel. But the less we say about that the better. It is ill mentioning a girl's name in such a matter as that.'

'But, my dear Dale, I must mention it at the present moment. Dear young child, I would do anything to comfort her! And I hope that something may be done to comfort her. Do you know that that young man was in love with her long before Crosbie ever saw her?'

'What; – John Eames!'

'Yes, John Eames. And I wish heartily for his sake that he had won her regard before she had met that rascal whom you had to stay down at your house.'

'A man cannot help these things, De Guest,' said the squire.

'No, no, no! There are such men about the world, and it is impossible to know them at a glance. He was my nephew's friend, and I am not going to say that my nephew was in fault. But I wish, – I only say that I wish, – she had first known what are this young man's feelings towards her.'

'But she might not have thought of him as you do.'

'He is an uncommonly good-looking young fellow; straight made, broad in the chest, with a good, honest eye, and a young man's proper courage. He has never been taught to give himself airs like a dancing monkey; but I think he's all the better for that.'

'But it's too late now, De Guest.'

'No, no; that's just where it is. It mustn't be too late! That child is not to lose her whole life because a villain has played her false. Of course she'll suffer. Just at present it wouldn't do, I suppose, to talk to her about a new sweetheart. But, Dale, the time will come; the time will come; – the time always does come.'

'It has never come to you and me,' said the squire, with the slightest possible smile on his dry cheeks. The story of their lives had been so far the same; each had loved, and each had been disappointed, and then each had remained single through life.

'Yes, it has,' said the earl, with no slight touch of feeling and even of romance in what he said. 'We have retricked our beams* in our own ways, and our lives have not been desolate. But for her, – you and her mother will look forward to see her married some day.'

'I have not thought about it.'

'But I want you to think about it. I want to interest you in this fellow's favour; and in doing so, I mean to be very open with you. I suppose you'll give her something?'

'I don't know, I'm sure,' said the squire, almost offended at an inquiry of such a nature.

'Well, then, whether you do or not, I'll give him something,' said the earl. 'I shouldn't have ventured to meddle in the matter had I not intended to put myself in such a position with reference to him as would justify me in asking the question.' And the peer as he spoke drew himself up to his full height. 'If such a match can be made, it shall not be a bad marriage for your niece in a pecuniary point of view. I shall have pleasure in giving to him; but I shall have more pleasure if she can share what I give.'

'She ought to be very much obliged to you,' said the squire.

'I think she would be if she knew young Eames. I hope the day may come when she will be so. I hope that you and I may see them happy together, and that you too may thank me for having assisted in making them so. Shall we go in to

Lady Julia now?' The earl had felt that he had not quite succeeded; that his offer had been accepted somewhat coldly, and had not much hope that further good could be done on that day, even with the help of his best port wine.

'Half a moment,' said the squire. 'There are matters as to which I never find myself able to speak quickly, and this certainly seems to be one of them. If you will allow me I will think over what you have said, and then see you again.'

'Certainly, certainly.'

'But for your own part in the matter, for your great generosity and kind heart, I beg to offer you my warmest thanks.' Then the squire bowed low, and preceded the earl out of the room.

Lord De Guest still felt that he had not succeeded. We may probably say, looking at the squire's character and peculiarities, that no marked success was probable at the first opening-out of such a subject. He had said of himself that he was never able to speak quickly in matters of moment; but he would more correctly have described his own character had he declared that he could not think of them quickly. As it was, the earl was disappointed; but had he been able to read the squire's mind, his disappointment would have been less strong. Mr Dale knew well enough that he was being treated well, and that the effort being made was intended with kindness to those belonging to him; but it was not in his nature to be demonstrative and quick at expressions of gratitude. So he entered the drawing-room with a cold, placid face, leading Eames, and Lady Julia also, to suppose that no good had been done.

'How do you do, sir?' said Johnny, walking up to him in a wild sort of manner, – going through a premeditated lesson, but doing it without any presence of mind.

'How do you do, Eames?' said the squire, speaking with a very cold voice. And then there was nothing further said till the dinner was announced.

'Dale, I know you drink port,' said the earl when Lady Julia left them. 'If you say you don't like that, I shall say you know nothing about it.'

'Ah! that's the '20,' said the squire, tasting it.

'I should rather think it is,' said the earl. 'I was lucky enough to get it early, and it hasn't been moved for thirty years. I like to give it to a man who knows it, as you do, at the first glance. Now there's my friend Johnny, there; it's thrown away upon him.'

'No, my lord, it is not. I think it's uncommonly nice.'

'Uncommonly nice! So is champagne, or ginger-beer, or lollipops, – for those who like them. Do you mean to tell me you can taste wine with half a pickled orange in your mouth?'

'It'll come to him soon enough,' said the squire.

'Twenty port won't come to him when he is as old as we are,' said the earl, forgetting that by that time sixty port will be as wonderful to the then living seniors of the age as was his own pet vintage to him.

The good wine did in some sort soften the squire; but, as a matter of course, nothing further was said as to the new matrimonial scheme. The earl did observe, however, that Mr. Dale was civil, and even kind, to his own young friend, asking a question here and there as to his life in London, and saying something about the work at the Income-tax Office.

'It is hard work,' said Eames. 'If you're under the line, they make a great row about it, send for you, and look at you as though you'd been robbing the bank; but they think nothing of keeping you till five.'

'But how long do you have for lunch and reading the papers?' said the earl.

'Not ten minutes. We take a paper among twenty of us for half the day. That's exactly nine minutes to each; and as for lunch, we only have a biscuit dipped in ink.'

'Dipped in ink!' said the squire.

'It comes to that, for you have to be writing while you munch it.'

'I hear all about you,' said the earl; 'Sir Raffle Buffle is a crony of mine.'

'I don't suppose he ever heard my name as yet,' said Johnny. 'But do you really know him well, Lord De Guest?'

'Haven't see him these thirty years; but I did know him.'

'We call him old Huffle Scuffle.'

'Huffle Scuffle! Ha, ha, ha! He always was Huffle Scuffle; a noisy, pretentious, empty-headed fellow. But I oughtn't to say so before you, young man. Come, we'll go into the drawing-room.'

'And what did he say?' asked Lady Julia as soon as the squire was gone.

There was no attempt at concealment, and the question was asked in Johnny's presence.

'Well, he did not say much. And coming from him, that ought to be taken as a good sign. He is to think of it, and let me see him again. You hold your head up, Johnny, and remember that you shan't want a friend on your side. Faint heart never won fair lady.'

At seven o'clock on the following morning Eames started on his return journey, and was at his desk at twelve o'clock, – as per agreement with his taskmaster at the Income-tax Office.

CHAPTER XXXIV

THE COMBAT

I HAVE said that John Eames was at his office punctually at twelve; but an incident had happened before his arrival there very important in the annals which are now being told, – so important that it is essentially necessary that it should be described with some minuteness of detail.

Lord De Guest, in the various conversations which he had had with Eames as to Lily Dale and her present position, had always spoken of Crosbie with the most vehement abhorrence. 'He is a damned blackguard,' said the earl, and the fire had come out of his round eyes as he spoke. Now the earl was by no means given to cursing and swearing, in the sense which is ordinarily applied to these words. When he made use of such a phrase as that quoted above, it was to be presumed that he in some sort meant what he said; and so he did, and had intended to signify that Crosbie by his conduct had merited all such condemnation as was the fitting punishment for blackguardism of the worst description.

'He ought to have his neck broken,' said Johnny.

'I don't know about that,' said the earl. 'The present times have become so pretty behaved that corporal punishment seems to have gone out of fashion. I shouldn't care so much about that, if any other punishment had taken its place. But it seems to me that a blackguard such as Crosbie can escape now altogether unscathed.'

'He hasn't escaped yet,' said Johnny.

'Don't you go and put your finger in the pie and make a fool of yourself,' said the earl. If it had behoved any one to resent in any violent fashion the evil done by Crosbie, Bernard Dale, the earl's nephew, should have been the avenger. This the earl felt, but under these circumstances he was disposed to think that there should be no such violent vengeance. 'Things were different when I was young,' he said to himself. But Eames gathered from the earl's tone that the earl's words were not strictly in accordance with his thoughts, and he declared to himself over and over again that Crosbie had not yet escaped.

He got into the train at Guestwick, taking a first-class ticket, because the earl's groom in livery was in attendance upon him. Had he been alone he would have gone in a cheaper carriage. Very weak in him, was it not? little also, and mean? My friend, can you say that you would not have done the same at his age? Are you quite sure that you would not do the same now that you are double his age? Be that as it may, Johnny Eames did that foolish thing, and gave the groom in livery half-a-crown into the bargain.

'We shall have you down again soon, Mr. John,' said the groom, who seemed to understand that Mr. Eames was to be made quite at home at the manor.

He went fast to sleep in the carriage, and did not awake till the train was stopped at the Barchester Junction.

'Waiting for the up-train from Barchester, sir,' said the guard. 'They're always late.' Then he went to sleep again, and was aroused in a few minutes by some one entering the carriage in a great hurry. The branch train had come in, just as the guardians of the line then present had made up their minds that the passengers on the main line should not be kept waiting any longer. The transfer of men, women, and luggage was therefore made in great haste, and they who were now taking their new seats had hardly time to look about them. An old gentleman, very red about the gills, first came into Johnny's carriage, which up to that moment he had shared with an old lady. The old gentleman was abusing everybody, because he was hurried, and would not take himself well into the compartment, but stuck in the doorway, standing on the step.

'Now, sir, when you're quite at leisure,' said a voice behind the old man, which instantly made Eames start up in his seat.

'I am not at all at leisure,' said the old man; 'and I'm not going to break my legs if I know it.'

'Take your time, sir,' said the guard.

'So I mean,' said the old man, seating himself in the corner nearest to the open door, opposite to the old lady. Then Eames saw plainly that it was Crosbie who had first spoken, and that he was getting into the carriage.

Crosbie at the first glance saw no one but the old gentleman and the old lady, and he immediately made for the unoccupied corner seat. He was busy with his umbrella and his dressing-bag, and a little flustered by the pushing and hurrying. The carriage was actually in motion before he perceived that John Eames was opposite to him: Eames had, instinctively, drawn up his legs so as not to touch him. He felt that he had become very red in the face, and to tell the truth, the perspiration had broken out upon his brow. It was a great occasion, – great in its imminent trouble, and great in its opportunity for action. How was he to carry himself at the first moment of his recognition by his enemy, and what was he to do afterwards?

It need hardly be explained that Crosbie had also been spending his Christmas with a certain earl of his acquaintance, and that he too was returning to his office. In one respect he had been much more fortunate than poor Eames, for he had been made happy with the smiles of his lady love. Alexandrina and the countess had fluttered about him softly, treating him as a tame chattel, now belonging to the noble house of De Courcy, and in this way he had been initiated into the inner domesticities of that illustrious family. The two extra men-servants, hired to wait upon Lady Dumbello, had vanished. The champagne had ceased to flow in a perennial stream. Lady Rosina had come out from her solitude and had preached at him constantly. Lady Margaretta had given him some lessons in economy. The Honourable John, in spite of a late quarrel, had borrowed five pounds from him. The Honourable George had engaged to come and stay with his sister during the next May. The earl had used a father-in-law's privilege, and had called him a fool. Lady Alexandrina had told him more than once, in rather a tart voice, that this must be done, and that that must be done; and the countess had given him her orders as though it was his duty, in the course of nature, to obey every word that fell from her. Such had been his Christmas delights; and now, as he returned back from the enjoyment of them, he found himself confronted in the railway carriage with Johnny Eames!

The eyes of the two met, and Crosbie made a slight inclination of his head. To this Eames gave no acknowledgment whatever, but looked straight into the other's face. Crosbie immediately saw that they were not to know each other, and was well contented that it should be so. Among all his many troubles, the enmity of John Eames did not go for much. He showed no appearance of being disconcerted, though our friend had shown much. He opened his bag, and taking out a book was soon deeply engaged in it, pursuing his studies as though the man opposite was quite unknown to him. I will not say that his mind did not run away from his book, for indeed there were many things of which he found it impossible not to think; but it did not revert to John Eames. Indeed, when the carriages reached Paddington, he had in truth all but forgotten him; and as he stepped out of the carriage, with his bag in his hand, was quite free from any remotest trouble on his account.

But it had not been so with Eames himself. Every moment of the journey had for him been crowded with thought as to what he would do now that chance had brought his enemy within his reach. He had been made quite wretched by the intensity of his thinking; and yet, when the carriages stopped, he had not made up his mind. His face had been covered with perspiration ever since Crosbie had come across him, and his limbs had hardly been under his own command. Here had come to him a great opportunity, and he felt so little confidence in himself that he almost knew that he would not use it properly. Twice and thrice he had almost flown at Crosbie's throat in the carriage, but he was restrained by an idea that the world and the police would be against him if he did such a thing in the presence of that old lady.

But when Crosbie turned his back upon him, and walked out, it was absolutely necessary that he should do something. He was not going to let the man escape, after all that he had said as to the expediency of thrashing him. Any other disgrace would be preferable to that. Fearing, therefore, lest his enemy should be too quick for him, he hurried out after him, and only just gave Crosbie time to turn round and face the carriages before he was upon him. 'You confounded scoundrel!' he screamed out. 'You confounded scoundrel!' and seized him by the throat, throwing himself upon him, and almost devouring him by the fury of his eyes.

The crowd upon the platform was not very dense, but there were quite enough of people to make a very respectable audience for this little play. Crosbie, in his dismay, retreated a step or two, and his retreat was much accelerated by the weight of Eames's attack. He endeavoured to free his throat from his foe's grasp; but in that he failed entirely. For the minute, however, he did manage to escape any positive blow, owing his safety in that respect rather to Eames's awkwardness than to his own efforts. Something about the police he was just able to utter and there was, as a matter of course, an immediate call for a supply of those functionaries. In about three minutes three policemen, assisted by six porters, had captured our poor friend Johnny; but this had not been done quick enough for Crosbie's purposes. The bystanders, taken by surprise, had allowed the combatants to fall back upon Mr. Smith's book-stall,* and there Eames laid his foe prostrate among the newspapers, falling himself into the yellow shilling-novel depôt by the over fury of his own energy; but as he fell, he contrived to lodge one blow with his fist in Crosbie's right eye, – one telling blow; and Crosbie had, to all intents and purposes, been thrashed.

'Con – founded scoundrel, rascal, blackguard!' shouted Johnny, with what remnants of voice were left to him, as the police dragged him off. 'If you only knew – what he's – done.' But in the meantime the policemen held him fast.

As a matter of course the first burst of public sympathy went with Crosbie. He had been assaulted, and the assault had come from Eames. In the British bosom there is so firm a love of well-constituted order, that these facts alone were sufficient to bring twenty knights to the assistance of the three policemen and the six porters; so that for Eames, even had he desired it, there was no possible chance of escape. But he did not desire it. One only sorrow consumed him at present. He had, as he felt, attacked Crosbie, but had attacked him in vain. He had had his opportunity, and had misused it. He was perfectly unconscious of that happy blow, and was in absolute ignorance of the great fact that his enemy's eye was already swollen and closed, and that in another hour it would be as black as his hat.

'He is a con – founded rascal!' ejaculated Eames, as the policemen and porters hauled him about. 'You don't know what he's done.'

'No, we don't,' said the senior constable; 'but we know what you have done. I say, Bushers, where's that gentleman? He'd better come along with us.'

Crosbie had been picked up from among the newspapers by another policeman and two or three other porters, and was attended also by the guard of the train, who knew him, and knew that he had come up from Courcy Castle. Three or four hangers-on were standing also around him, together with a benevolent medical man who was proposing to him an immediate application of leeches. If he could have done as he wished, he would have gone his way quietly, allowing Eames to do the same. A great evil had befallen him, but he could in no way mitigate that evil by taking the law of the man who had attacked him. To have the thing as little talked about as possible should be his endeavour. What though he should have Eames locked up and fined, and scolded by a police magistrate? That would not in any degree lessen his calamity. If he could have parried the attack, and got the better of his foe; if he could have administered the black eye instead of receiving it, then indeed he could have laughed the matter off at his club, and his original crime would have been somewhat glozed over by his success in arms. But such good fortune had not been his. He was forced, however, on the moment to decide as to what he would do.

'We've got him here in custody, sir,' said Bushers, touching his hat. It had become known from the guard that Crosbie was somewhat of a big man, a frequent guest at Courcy Castle, and of repute and station in the higher regions of the Metropolitan world. 'The magistrates will be sitting at Paddington, now, sir – or will be by the time we get there.'

By this time some mighty railway authority had come upon the scene and made himself cognizant of the facts of the row, – a stern official who seemed to carry the weight of many engines on his brow; one at the very sight of whom smokers would drop their cigars, and porters close their fists against sixpences; a great man with an erect chin, a quick step, and a well-brushed hat powerful with an elaborately upturned brim. This was the platform-superintendent, dominant even over the policemen.

'Step into my room, Mr. Crosbie,' he said. 'Stubbs, bring that man in with you.' And then, before Crosbie had been able to make up his mind as to any other line of conduct, he found himself in the superintendent's room, accompanied by the guard, and by the two policemen who conducted Johnny Eames between them.

'What's all this?' said the superintendent, still keeping on his hat, for he was aware how much of the excellence of his personal dignity was owing to the arrangement of that article; and as he spoke he frowned upon the culprit with his utmost severity. 'Mr. Crosbie, I am very sorry that you should have been exposed to such brutality on our platform.'

'You don't know what he has done,' said Johnny. 'He is

the most confounded scoundrel living. He has broken—' But then he stopped himself. He was going to tell the superintendent that the confounded scoundrel had broken a beautiful young lady's heart; but he bethought himself that he would not allude more specially to Lily Dale in that hearing.

'Do you know who he is, Mr. Crosbie?' said the superintendent.

'Oh, yes,' said Crosbie, whose eye was already becoming blue. 'He is a clerk in the Income-tax Office, and his name is Eames. I believe you had better leave him to me.'

But the superintendent at once wrote down the words 'Income-tax Office – Eames,' on his tablet. 'We can't allow a row like that to take place on our platform and not notice it. I shall bring it before the directors. It's a most disgraceful affair, Mr. Eames – most disgraceful.'

But Johnny by this time had perceived that Crosbie's eye was in a state which proved satisfactorily that his morning's work had not been thrown away, and his spirits were rising accordingly. He did not care two straws for the superintendent or even for the policemen, if only the story could be made to tell well for himself hereafter. It was his object to have thrashed Crosbie, and now, as he looked at his enemy's face, he acknowledged that Providence had been good to him.

'That's your opinion,' said Johnny.

'Yes, sir, it is,' said the superintendent; 'and I shall know how to represent the matter to your superiors, young man.'

'You don't know all about it,' said Eames; 'and I don't suppose you ever will. I had made up my mind what I'd do the first time I saw that scoundrel there; and now I've done it. He'd have got much worse in the railway carriage, only there was a lady there.'

'Mr. Crosbie, I really think we had better take him before the magistrates.'

To this, however, Crosbie objected. He assured the superintendent that he would himself know how to deal with the matter – which, however, was exactly what he did not know. Would the superintendent allow one of the railway servants to get a cab for him, and to find his luggage? He was very anxious to get home without being subjected to any more of Mr. Eames's insolence.

'You haven't done with Mr. Eames's insolence yet, I can tell you. All London shall hear of it, and shall know why. If you have any shame in you, you shall be ashamed to show your face.'

Unfortunate man! Who can say that punishment – adequate punishment – had not overtaken him? For the present, he had to sneak home with a black eye, with the knowledge inside him that he had been whipped by a clerk in the Income-tax Office; and for the future – he was bound over to marry Lady Alexandrina De Courcy!

He got himself smuggled off in a cab, without being forced to go again upon the platform – his luggage being brought to him by two assiduous porters. But in all this there was very little balm for his hurt pride. As he ordered the cabman to drive to Mount Street, he felt that he had ruined himself by that step in life which he had taken at Courcy Castle. Whichever way he looked he had no comfort. 'D— the fellow!' he said, almost out loud in the cab; but though he did with his outward voice allude to Eames, the curse in his inner thoughts was uttered against himself.

Johnny was allowed to make his way down to the platform, and there find his own carpet-bag. One young porter, however, came up and fraternized with him.

'You guve it him tidy just at that last moment, sir. But, laws, sir, you should have let out at him at fust. What's the use of clawing a man's neck-collar?'

It was then a quarter past eleven, but, nevertheless, Eames appeared at his office precisely at twelve.

CHAPTER XXXV

VÆ VICTIS*

CROSBIE had two engagements for that day; one being his natural engagement to do his work at his office, and the other an engagement, which was now very often becoming as natural, to dine at St. John's Wood with Lady Amelia

Gazebee. It was manifest to him when he looked at himself in the glass that he could keep neither of these engagements. 'Oh, laws, Mr. Crosbie,' the woman of the house exclaimed when she saw him.

'Yes, I know,' said he. 'I've had an accident and got a black eye. What's a good thing for it?'

'Oh! an accident!' said the woman, who knew well that that mark had been made by another man's fist. 'They do say that a bit of raw beef is about the best thing. But then it must be held on constant all the morning.'

Anything would be better than leeches, which tell long-enduring tales, and therefore Crosbie sat through the greater part of the morning holding the raw beef to his eye. But it was necessary that he should write two notes as he held it, one to Mr. Butterwell at his office, and the other to his future sister-in-law. He felt that it would hardly be wise to attempt any entire concealment of the nature of his catastrophe, as some of the circumstances would assuredly become known. If he said that he had fallen over the coal-scuttle, or on to the fender, thereby cutting his face, people would learn that he had fibbed, and would learn also that he had had some reason for fibbing. Therefore he constructed his notes with a phraseology that bound him to no details. To Butterwell he said that he had had an accident – or rather a row – and that he had come out of it with considerable damage to his frontispiece. He intended to be at the office on the next day, whether able to appear decently there or not. But for the sake of decency he thought it well to give himself that one half-day's chance. Then to the Lady Amelia he also said that he had had an accident, and had been a little hurt. 'It is nothing at all serious, and affects only my appearance, so that I had better remain in for a day. I shall certainly be with you on Sunday. Don't let Gazebee trouble himself to come to me, as I shan't be at home after to-day.' Gazebee did trouble himself to come to Mount Street so often, and South Audley Street, in which was Mr. Gazebee's office, was so disagreeably near to Mount Street, that Crosbie inserted this in order to protect himself if possible. Then he gave special orders that he was to be at home to no one, fearing that Gazebee would call for him after the hours of business – to make him safe and carry him off bodily to St. John's Wood.

The beefsteak and the dose of physic and the cold-water application which was kept upon it all night was not efficacious in dispelling that horrid, black-blue colour by ten o'clock on the following morning.

'It certainly have gone down, Mr. Crosbie; it certainly have,' said the mistress of the lodgings, touching the part affected with her finger. 'But the black won't go out of them all in a minute; it won't indeed. Couldn't you just stay in one more day?'

'But will one day do it, Mrs. Phillips?'

Mrs. Phillips couldn't take upon herself to say that it would. 'They mostly come with little red streaks across the black before they goes away,' said Mrs. Phillips, who would seem to have been the wife of a prize-fighter, so well was she acquainted with black eyes.

'And that won't be till to-morrow,' said Crosbie, affecting to be mirthful in his agony.

'Not till the third day; – and then they wears themselves out gradual. I never knew leeches do any good.'

He stayed at home the second day, and then resolved that he would go to his office, black eye and all. In that morning's newspaper he saw an account of the whole transaction, saying how Mr. C— of the office of General Committees, who was soon about to lead to the hymeneal altar the beautiful daughter of the Earl De C—, had been made the subject of a brutal personal attack on the platform of the Great Western Railway Station, and how he was confined to his room from the injuries which he had received. The paragraph went on to state that the delinquent had, as it was believed, dared to raise his eyes to the same lady, and that his audacity had been treated with scorn by every member of the noble family in question. 'It was, however, satisfactory to know,' so said the newspaper, 'that Mr. C— had amply avenged himself, and had so flogged the young man in question, that he had been unable to stir from his bed since the occurrence.'

On reading this Crosbie felt that it would be better that

he should show himself at once, and tell as much of the truth as the world would like to ascertain at last without his telling. So on that third morning he put on his hat and gloves, and had himself taken to his office, though the red-streaky period of his misfortune had hardly even yet come upon him. The task of walking along the office passage, through the messengers' lobby, and into his room, was very disagreeable. Of course everybody looked at him, and of course he failed in his attempt to appear as though he did not mind it. 'Boggs,' he said to one of the men as he passed by, 'just see if Mr. Butterwell is in his room,' and then, as he expected, Mr. Butterwell came to him after the expiration of a few minutes.

'Upon my word, that is serious,' said Mr. Butterwell, looking into the secretary's damaged face. 'I don't think I would have come out if I had been you.'

'Of course it's disagreeable,' said Crosbie; 'but it's better to put up with it. Fellows do tell such horrid lies if a man isn't seen for a day or two. I believe it's best to put a good face upon it.'

'That's more than you can do just at present, eh, Crosbie?' And then Mr. Butterwell tittered. 'But how on earth did it happen? The paper says that you pretty well killed the fellow who did it.'

'The paper lies, as papers always do. I didn't touch him at all.'

'Didn't you, though? I should like to have had a poke at him after getting such a tap in the face as that.'

'The policemen came, and all that sort of thing. One isn't allowed to fight it out in a row of that kind as one would have to do on Salisbury Heath. Not that I mean to say that I could lick the fellow. How's a man to know whether he can or not?'

'How, indeed, unless he gets a licking, – or gives it? But who was he, and what's this about his having been scorned by the noble family?'

'Trash and lies, of course. He had never seen any of the De Courcy people.'

'I suppose the truth is, it was about that other – eh, Crosbie? I knew you'd find yourself in some trouble before you'd done.'

'I don't know what it was about, or why he should have made such a brute of himself. You have heard about those people at Allington?'

'Oh, yes; I have heard about them.'

'God knows, I didn't mean to say anything against them. They knew nothing about it.'

'But the young fellow knew them? Ah, yes, I see all about it. He wants to step into your shoes. I can't say that he sets about it in a bad way. But what do you mean to do?'

'Nothing.'

'Nothing! Won't that look queer? I think I should have him before the magistrates.'

'You see, Butterwell, I am bound to spare that girl's name. I know I have behaved badly.'

'Well, yes; I fear you have.'

Mr. Butterwell said this with some considerable amount of decision in his voice, as though he did not intend to mince matters, or in any way to hide his opinion. Crosbie had got into a way of condemning himself in this matter of his marriage, but was very anxious that others, on hearing such condemnation from him, should say something in the way of palliating his fault. It would be so easy for a friend to remark that such little peccadilloes were not altogether uncommon, and that it would sometimes happen in life that people did not know their own minds. He had hoped for some such benevolence from Fowler Pratt, but had hoped in vain. Butterwell was a good-natured, easy man, anxious to stand well with all about him, never pretending to any very high tone of feeling or of morals; and yet Butterwell would say no word of comfort to him. He could get no one to slur over his sin for him, as though it were no sin, – only an unfortunate mistake; no one but the De Courcys, who had, as it were, taken possession of him and swallowed him alive.

'It can't be helped now,' said Crosbie. 'But as for that fellow who made such a brutal attack on me the other morning, he knows that he is safe behind her petticoats. I can do nothing which would not make some mention of her name necessary.'

'Ah, yes; I see,' said Butterwell. 'It's very unfortunate; very. I don't know that I can do anything for you. Will you come before the Board to-day?'

'Yes; of course I shall,' said Crosbie, who was becoming very sore. His sharp ear had told him that all Butterwell's respect and cordiality were gone, – at any rate for the time. Butterwell, though holding the higher official rank, had always been accustomed to treat him as though he, the inferior, were to be courted. He had possessed, and had known himself to possess, in his office as well as in the outside world, a sort of rank much higher than that which from his position he could claim legitimately. Now he was being deposed. There could be no better touchstone in such a matter than Butterwell. He would go as the world went, but he would perceive almost intuitively how the world intended to go. 'Tact, tact, tact,' as he was in the habit of saying to himself when walking along the paths of his Putney villa. Crosbie was now secretary, whereas a few months before he had been simply a clerk; but, nevertheless, Mr. Butterwell's instinct told him that Crosbie had fallen. Therefore he declined to offer any sympathy to the man in his misfortune, and felt aware, as he left the secretary's room, that it might probably be some time before he visited it again.

Crosbie resolved in his soreness that henceforth he would brazen it out. He would go to the Board, with as much indifference as to his black eye as he was able to assume, and if any one said aught to him he would be ready with his answer. He would go to his club, and let him who intended to show him any slight beware of his wrath. He could not turn upon John Eames, but he could turn upon others if it were necessary. He had not gained for himself a position before the world, and held it now for some years, to allow himself to be crushed at once because he had made a mistake. If the world, his world, chose to go to war with him, he would be ready for the fight. As for Butterwell, – Butterwell the incompetent, Butterwell the vapid, – for Butterwell, who in every little official difficulty had for years past come to him, he would let Butterwell know what it was to be thus disloyal to one who had condescended to be his friend. He would show them all at the Board that he scorned them, and could be their master. Then, too, as he was making some other resolves as to his future conduct, he made one or two resolutions respecting the De Courcy people. He would make it known to them that he was not going to be their very humble servant. He would speak out his mind with considerable plainness; and if upon that they should choose to break off this 'alliance,' they might do so; he would not break his heart. And as he leaned back in his arm-chair, thinking of all this, an idea made its way into his brain, – a floating castle in the air, rather than the image of a thing that might by possibility be realized; and in this castle in the air he saw himself kneeling again at Lily's feet, asking her pardon, and begging that he might once more be taken to her heart.

'Mr. Crosbie is here to-day,' said Mr. Butterwell to Mr. Optimist.

'Oh, indeed,' said Mr. Optimist, very gravely; for he had heard all about the row at the railway station.

'They've made a monstrous show of him.'

'I am very sorry to hear it. It's so – so – so— If it were one of the younger clerks, you know, we should tell him that it was discreditable to the department.'

'If a man gets a blow in the eye, he can't help it, you know. He didn't do it himself, I suppose,' said Major Fiasco.

'I am well aware that he didn't do it himself,' continued Mr. Optimist; 'but I really think that, in his position, he should have kept himself out of any such encounter.'

'He would have done so if he could, with all his heart,' said the major. 'I don't suppose he liked being thrashed any better than I should.'

'Nobody gives me a black eye,' said Mr. Optimist.

'Nobody has as yet,' said the major.

'I hope they never will,' said Mr. Butterwell. Then, the hour for their meeting having come round, Mr. Crosbie came into the Board-room.

'We have been very sorry to hear of this misfortune,' said Mr. Optimist, very gravely.

'Not half so sorry as I have been,' said Crosbie, with a laugh. 'It's an uncommon nuisance to have a black eye, and to go about looking like a prize-fighter.'

'And like a prize-fighter that didn't win his battle too,' said Fiasco.

'I don't know that there's much difference as to that,' said Crosbie. 'But the whole thing is a nuisance, and if you please, we won't say anything more about it.'

Mr. Optimist almost entertained an opinion that it was his duty to say something more about it. Was not he the chief Commissioner, and was not Mr. Crosbie secretary to the Board? Ought he, looking at their respective positions, to pass over without a word of notice such a manifest impropriety as this? Would not Sir Raffle Buffle have said something had Mr. Butterwell, when secretary, come to the office with a black eye? He wished to exercise all the full rights of a chairman; but, nevertheless, as he looked at the secretary he felt embarrassed, and was unable to find the proper words. 'H – m, ha, well; we'll go to business now, if you please,' he said, as though reserving to himself the right of returning to the secretary's black eye when the more usual business of the Board should be completed. But when the more usual business of the Board had been completed, the secretary left the room without any further reference to his eye.

Crosbie, when he got back to his own apartment, found Mortimer Gazebee waiting there for him.

'My dear fellow,' said Gazebee, 'this is a very nasty affair.'

'Uncommonly nasty,' said Crosbie; 'so nasty that I don't mean to talk about it to anybody.'

'Lady Amelia is quite unhappy.' He always called her Lady Amelia, even when speaking of her to his own brothers and sisters. He was too well behaved to take the liberty of calling an earl's daughter by her plain Christian name, even though that earl's daughter was his own wife. 'She fears that you have been a good deal hurt.'

'Not at all hurt; but disfigured, as you see.'

'And so you beat the fellow well that did it?'

'No, I didn't,' said Crosbie, very angrily. 'I didn't beat him at all. You don't believe everything you read in the newspapers, do you?'

'No, I don't believe everything. Of course I didn't believe about his having aspired to an alliance with the Lady Alexandrina. That was untrue, of course.' Mr. Gazebee showed by the tone of his voice that imprudence so unparalleled as that was quite incredible.

'You shouldn't believe anything; except this – that I have got a black eye.'

'You certainly have got that. Lady Amelia thinks you would be more comfortable if you would come up to us this evening. You can't go out, of course; but Lady Amelia said, very good-naturedly, that you need not mind with her.'

'Thank you, no; I'll come on Sunday.'

'Of course Lady Alexandrina will be very anxious to hear from her sister; and Lady Amelia begged me very particularly to press you to come.'

'Thank you, no; not to-day.'

'Why not?'

'Oh, simply because I shall be better at home.'

'How can you be better at home? You can have anything that you want. Lady Amelia won't mind, you know.'

Another beefsteak to his eye, as he sat in the drawing-room, a cold-water bandage, or any little medical appliance of that sort; – these were the things which Lady Amelia would, in her domestic good nature, condescend not to mind!

'I won't trouble her this evening,' said Crosbie.

'Well, upon my word, I think you're wrong. All manner of stories will get down to Courcy Castle, and to the countess's ears; and you don't know what harm may come of it. Lady Amelia thinks she had better write and explain it; but she can't do so till she has heard something about it from you.'

'Look here, Gazebee. I don't care one straw what story finds its way down to Courcy Castle.'

'But if the earl were to hear anything and be offended?'

'He may recover from his offence as he best likes.'

'My dear fellow; that's talking wildly, you know.'

'What on earth do you suppose the earl can do to me? Do you think I'm going to live in fear of Lord De Courcy all my life, because I'm going to marry his daughter? I shall write to Alexandrina myself to-day, and you can tell her sister so. I'll be up to dinner on Sunday, unless my face makes it altogether out of the question.'

'And you won't come in time for church?'

'Would you have me go to church with such a face as this ?'

Then Mr. Mortimer Gazebee went, and when he got home he told his wife that Crosbie was taking things with a high hand. 'The fact is, my dear, that he's ashamed of himself, and therefore tries to put a bold face upon it.'

'It was very foolish of him throwing himself in the way of that young man, – very; and so I shall tell him on Sunday. If he chooses to give himself airs to me, I shall make him understand that he is very wrong. He should remember now that the way in which he conducts himself is a matter of moment to all our family.'

'Of course he should,' said Mr. Gazebee.

When the Sunday came the red-streaky period had arrived, but had by no means as yet passed away. The men at the office had almost become used to it; but Crosbie, in spite of his determination to go down to the club, had not yet shown himself elsewhere. Of course he did not go to church, but at five he made his appearance at the house in St. John's Wood. They always dined at five on Sundays, having some idea that by doing so they kept the Sabbath better than they would have done had they dined at seven. If keeping the Sabbath consists in going to bed early, or is in any way assisted by such a practice, they were right. To the cook that semi-early dinner might perhaps be convenient, as it gave her an excuse for not going to church in the afternoon, as the servants' and children's dinner gave her a similar excuse in the morning. Such little attempts at goodness, – proceeding half the way, or perhaps, as in this instance, one quarter of the way, on the disagreeable path towards goodness, – are very common with respectable people, such as Lady Amelia.

If she would have dined at one o'clock, and have eaten cold meat, one perhaps might have felt that she was entitled to some praise.

'Dear, dear, dear; this is very sad, isn't it, Adolphus?' she said on first seeing him.

'Well, it is sad, Amelia,' he said. He always called her Amelia, because she called him Adolphus; but Gazebee himself was never quite pleased when he heard it. Lady Amelia was older than Crosbie, and entitled to call him anything she liked; but he should have remembered the great difference in their rank. 'It is sad, Amelia,' he said. 'But will you oblige me in one thing?'

'What thing, Adolphus?'

'Not to say a word more about it. The black eye is a bad thing, no doubt, and has troubled me much; but the sympathy of my friends has troubled me a great deal more. I had all the family commiseration from Gazebee on Friday, and if it is repeated again, I shall lie down and die.'

'Shall 'oo die, uncle Dolphus, 'cause 'oo've got a bad eye?' asked De Courcy Gazebee, the eldest hope of the family, looking up into his face.

'No, my hero,' said Crosbie, taking the boy up into his arms, 'not because I've got a black eye. There isn't very much harm in that, and you'll have a great many before you leave school. But because the people will go on talking about it.'

'But aunt Dina 'on't like 'oo, if 'oo've got an ugly bad eye.'

'But, Adolphus,' said Lady Amelia, settling herself for an argument, 'that's all very well, you know – and I'm sure I'm very sorry to cause you any annoyance, – but really one doesn't know how to pass over such a thing without speaking of it. I have had a letter from mamma.'

'I hope Lady De Courcy is quite well.'

'Quite well, thank you. But as a matter of course she is very anxious about this affair. She had read what has been said in the newspapers, and it may be necessary that Mortimer should take it up, as the family solicitor.'

'Quite out of the question,' said Adolphus.

'I don't think I should advise any such step as that,' said Gazebee.

'Perhaps not; very likely not. But you cannot be surprised, Mortimer, that my mother under such circumstances should wish to know what are the facts of the case.'

'Not at all surprised,' said Gazebee.

'Then once for all, I'll tell you the facts. As I got out of the train a man I'd seen once before in my life made an attack upon me, and before the police came up, I got a blow in the face. Now you know all about it.'

At that moment dinner was announced. 'Will you give Lady Amelia your arm?' said the husband.

'It's a very sad occurrence,' said Lady Amelia with a slight toss of her head, 'and, I'm afraid, will cost my sister a great deal of vexation.'

'You agree with De Courcy, do you, that aunt Dina won't like me with an ugly black eye?'

'I really don't think it's a joking matter,' said the Lady Amelia. And then there was nothing more said about it during the dinner.

There was nothing more said about it during the dinner, but it was plain enough from Lady Amelia's countenance that she was not very well pleased with her future brother-in-law's conduct. She was very hospitable to him, pressing him to eat; but even in doing that she made repeated little references to his present unfortunate state. She told him that she did not think fried plum-pudding would be bad for him, but that she would recommend him not to drink port-wine after dinner. 'By-the-by, Mortimer, you'd better have some claret up,' she remarked. 'Adolphus shouldn't take anything that is heating.'

'Thank you,' said Crosbie. 'I'll have some brandy-and-water, if Gazebee will give it me.'

'Brandy-and-water!' said Lady Amelia. Crosbie in truth was not given to the drinking of brandy-and-water; but he was prepared to call for raw gin, if he were driven much further by Lady Amelia's solicitude.

At these Sunday dinners the mistress of the house never went away into the drawing-room, and the tea was always brought in to them at the table on which they had dined. It was another little step towards keeping holy the first day of the week. When Lady Rosina was there, she was indulged with the sight of six or seven solid good books which were laid upon the mahogany as soon as the bottles were taken off it. At her first prolonged visit she had obtained for herself the privilege of reading a sermon; but as on such occasions both Lady Amelia and Mr. Gazebee would go to sleep, – and as the footman had also once shown a tendency that way, – the sermon had been abandoned. But the master of the house, on these evenings, when his sister-in-law was present, was doomed to sit in idleness, or else to find solace in one of the solid good books. But Lady Rosina just now was in the country, and therefore the table was left unfurnished.

'And what am I to say to my mother?' said Lady Amelia, when they were alone.

'Give her my kindest regards,' said Crosbie. It was quite clear, both to the husband and to the wife, that he was preparing himself for rebellion against authority.

For some ten minutes there was nothing said. Crosbie amused himself by playing with the boy whom he called Dicksey, by way of a nickname for De Courcy.

'Mamma, he calls me Dicksey. Am I Dicksey? I'll call 'oo old Cross, and then aunt Dina 'on't like 'oo.'

'I wish you would not call the child nicknames, Adolphus. It seems as though you would wish to cast a slur upon the one which he bears.'

'I should hardly think that he would feel disposed to do that,' said Mr. Gazebee.

'Hardly, indeed,' said Crosbie.

'It has never yet been disgraced in the annals of our country by being made into a nickname,' said the proud daughter of the house. She was probably unaware that among many of his associates her father had been called Lord De Curse'ye, from the occasional energy of his language. 'And any such attempt is painful in my ears. I think something of my family, I can assure you, Adolphus, and so does my husband.'

'A very great deal,' said Mr. Gazebee.

'So do I of mine,' said Crosbie. 'That's natural to all of us.

One of my ancestors came over with William the Conqueror. I think he was one of the assistant cooks in the king's tent.'

'A cook!' said young De Courcy.

'Yes, my boy, a cook. That was the way most of our old families were made noble. They were cooks, or butlers to the kings – or sometimes something worse.'

'But your family isn't noble?'

'No – I'll tell you how that was. The king wanted this cook to poison half-a-dozen of his officers who wished to have a way of their own; but the cook said, "No, my Lord King; I am a cook, not an executioner." So they sent him into the scullery, and when they called all the other servants barons and lords, they only called him Cookey. They've changed the name to Crosbie since that, by degrees.'

Mr. Gazebee was awestruck, and the face of the Lady Amelia became very dark. Was it not evident that this snake, when taken into their innermost bosoms that they might there warm him, was becoming an adder, and preparing to sting them? There was very little more conversation that evening, and soon after the story of the cook, Crosbie got up and went away to his own home.

CHAPTER XXXVI

'SEE, THE CONQUERING HERO COMES'

JOHN EAMES had reached his office precisely at twelve o'clock, but when he did so he hardly knew whether he was standing on his heels or his head. The whole morning had been to him one of intense excitement, and latterly, to a certain extent, one of triumph. But he did not at all know what might be the results. Would he be taken before a magistrate and locked up? Would there be a row at the office? Would Crosbie call him out, and, if so, would it be incumbent on him to fight a duel with pistols? What would Lord De Guest say – Lord De Guest, who had specially warned him not to take upon himself the duty of avenging Lily's wrongs? What would all the Dale family say of his conduct? And, above all, what would Lily say and think? Nevertheless, the feeling of triumph was predominant; and now, at this interval of time, he was beginning to remember with pleasure the sensation of his fist as it went into Crosbie's eye.

During his first day at the office he heard nothing about the affair, nor did he say a word of it to any one. It was known in his room that he had gone down to spend his Christmas holiday with Lord De Guest, and he was treated with some increased consideration accordingly. And, moreover, I must explain, in order that I may give Johnny Eames his due, he was gradually acquiring for himself a good footing among the Income-tax officials. He knew his work, and did it with some manly confidence in his own powers, and also with some manly indifference to the occasional frowns of the mighty men of the department. He was, moreover, popular – being somewhat of a Radical in his official demeanour, and holding by his own rights, even though mighty men should frown. In truth, he was emerging from his hobbledehoyhood and entering upon his young manhood, having probably to go through much folly and some false sentiment in that period of his existence, but still with fair promise of true manliness beyond, to those who were able to read the signs of his character.

Many questions on that first day were asked him about the glories of his Christmas, but he had very little to say on the subject. Indeed nothing could have been much more commonplace than his Christmas visit, had it not been for the one great object which had taken him down to that part of the country, and for the circumstance with which his holiday had been ended. On neither of these subjects was he disposed to speak openly; but as he walked home to Burton Crescent with Cradell, he did tell him of the affair with Crosbie.

'And you went in at him on the station?' asked Cradell, with admiring doubt.

'Yes, I did. If I didn't do it there, where was I to do it? I'd said I would, and therefore when I saw him I did it.' Then the whole affair was told as to the black eye, the police, and the superintendent. 'And what's to come next?' asked our hero.

'Well, he'll put it in the hands of a friend, of course: as I did with Fisher in that affair with Lupex. And, upon my word, Johnny, I shall have to do something of the kind again. His conduct last night was outrageous; would you believe it—'

'Oh, he's a fool.'

'He's a fool you wouldn't like to meet when he's in one of his mad fits, I can tell you that. I absolutely had to sit up in my own bedroom all last night. Mother Roper told me that if I remained in the drawing-room she would feel herself obliged to have a policeman in the house. What could I do, you know? I made her have a fire for me, of course.'

'And then you went to bed.'

'I waited ever so long, because I thought that Maria would want to see me. At last she sent me a note. Maria is so imprudent, you know. If he had found anything in her writing, it would have been terrible, you know, – quite terrible. And who can say whether Jemima mayn't tell?'

'And what did she say?'

'Come; that's tellings, Master Johnny. I took very good care to take it with me to the office this morning, for fear of accidents.'

But Eames was not so widely awake to the importance of his friend's adventures as he might have been had he not been weighted with adventures of his own.

'I shouldn't care so much,' said he, 'about that fellow Crosbie going to a friend, as I should about his going to a police magistrate.'

'He'll put it in a friend's hands, of course,' said Cradell, with the air of a man who from experience was well up in such matters. 'And I suppose you'll naturally come to me. It's a deuced bore to a man in a public office, and all that kind of thing, of course. But I'm not the man to desert my friend. I'll stand by you, Johnny, my boy.'

'Oh, thank you,' said Eames, 'I don't think that I shall want that.'

'You must be ready with a friend, you know.'

'I should write down to a man I know in the country, and ask his advice,' said Eames; 'an older sort of friend, you know.'

'By Jove, old fellow, take care what you're about. Don't let them say of you that you show the white feather.* Upon my honour, I'd sooner have anything said of me than that. I would, indeed, – anything.'

'I'm not afraid of that,' said Eames, with a touch of scorn in his voice. 'There isn't much thought about white feathers nowadays, – not in the way of fighting duels.'

After that Cradell managed to carry back the conversation to Mrs. Lupex and his own peculiar position, and as Eames did not care to ask from his companion further advice in his own matters, he listened nearly in silence till they reached Burton Crescent.

'I hope you found the noble earl well,' said Mrs. Roper to him, as soon as they were all seated at dinner.

'I found the noble earl pretty well, thank you,' said Johnny.

It had become plainly understood by all the Roperites that Eames's position was quite altered since he had been honoured with the friendship of Lord De Guest. Mrs. Lupex, next to whom he always sat at dinner, with a view to protecting her as it were from the dangerous neighbourhood of Cradell, treated him with a marked courtesy. Miss Spruce always called him 'sir.' Mrs. Roper helped him the first of the gentlemen, and was mindful about his fat and gravy, and Amelia felt less able than she was before to insist upon the possession of his heart and affections. It must not be supposed that Amelia intended to abandon the fight, and allow the enemy to walk off with his forces; but she felt herself constrained to treat him with a deference that was hardly compatible with the perfect equality which should attend any union of hearts.

'It is such a privilege to be on visiting terms with the nobility,' said Mrs. Lupex. 'When I was a girl, I used to be very intimate—'

'You ain't a girl any longer, and so you'd better not talk about it,' said Lupex. Mr. Lupex had been at that little shop in Drury Lane after he came down from his scene-painting.

'My dear, you needn't be a brute to me before all Mrs. Roper's company. If, led away by feelings which I will not now describe, I left my proper circles in marrying you, you need not before all the world teach me how much I have to regret.' And Mrs. Lupex, putting down her knife and fork, applied her handkerchief to her eyes.

'That's pleasant for a man over his meals, isn't it?' said Lupex, appealing to Miss Spruce. 'I have plenty of that kind of thing, and you can't think how I like it.'

'Them whom God has joined together, let no man put asunder,' said Miss Spruce. 'As for me myself, I'm only an old woman.'

This little ebullition threw a gloom over the dinner-table, and nothing more was said on the occasion as to the glories of Eames's career. But, in the course of the evening, Amelia heard of the encounter which had taken place at the railway station, and at once perceived that she might use the occasion for her own purposes.

'John,' she whispered to her victim, finding an opportunity for coming upon him when almost alone, 'what is this I hear? I insist upon knowing. Are you going to fight a duel?'

'Nonsense,' said Johnny.

'But it is not nonsense. You don't know what my feelings will be, if I think that such a thing is going to happen. But then you are so hard-hearted!'

'I ain't hard-hearted a bit, and I'm not going to fight a duel.'

'But is it true that you beat Mr. Crosbie at the station?'

'It is true. I did beat him.'

'Oh, John! not that I mean to say you were wrong, and indeed I honour you for the feeling. There can be nothing so dreadful as a young man's deceiving a young woman and leaving her after he has won her heart – particularly when she has had his promise in plain words, or, perhaps, even in black and white.' John thought of that horrid, foolish, wretched note which he had written. 'And a poor girl, if she can't right herself by a breach of promise, doesn't know what to do. Does she, John?'

'A girl who'd right herself that way wouldn't be worth having.'

'I don't know about that. When a poor girl is in such a position, she has to be aided by her friends. I suppose, then, Miss Lily Dale won't bring a breach of promise against him.'

This mention of Lily's name in such a place was sacrilege in the ears of poor Eames. 'I cannot tell,' said he, 'what may be the intention of the lady of whom you speak. But from what I know of her friends, I should not think that she will be disgraced by such a proceeding.'

'That may be all very well for Miss Lily Dale—' Amelia said, and then she hesitated. It would not be well, she thought, absolutely to threaten him as yet, – not as long as there was any possibility that he might be won without a threat. 'Of course I know all about it,' she continued. 'She was your L. D., you know. Not that I was ever jealous of her. To you she was no more than one of childhood's friends. Was she, Johnny?'

He stamped his foot upon the floor, and then jumped up from his seat. 'I hate all that sort of twaddle about childhood's friends, and you know I do. You'll make me swear that I'll never come into this room again.'

'Johnny!'

'So I will. The whole thing makes me sick. And as for that Mrs. Lupex—'

'If this is what you learn, John, by going to a lord's house, I think you had better stay at home with your own friends.'

'Of course I had; – much better stay at home with my own friends. Here's Mrs. Lupex, and at any rate I can't stand her.' So he went off, and walked round the Crescent, and down to the New Road, and almost into the Regent's Park, thinking of Lily Dale and of his own cowardice with Amelia Roper.

On the following morning he received a message, at about one o'clock, by the mouth of the Boardroom messenger, informing him that his presence was required in the Boardroom. 'Sir Raffle Buffle has desired your presence, Mr. Eames.'

'My presence, Tupper! what for?' said Johnny, turning upon the messenger almost with dismay.

'Indeed I can't say, Mr. Eames; but Sir Raffle Buffle has desired your presence in the Boardroom.'

Such a message as that in official life always strikes awe into the heart of a young man. And yet, young men generally come forth from such interviews without having received any serious damage, and generally talk about the old gentlemen whom they have encountered with a good deal of light-spirited sarcasm, – or chaff, as it is called in the slang phraseology of the day. It is that same 'majesty which doth hedge a king' that does it. The turkey-cock in his own farmyard is master of the occasion, and the thought of him creates fear. A bishop in his lawn, a judge on the bench, a chairman in the big room at the end of a long table, or a policeman with his bull's-eye lamp upon his beat, can all make themselves terrible by means of those appanages of majesty which have been vouchsafed to them. But how mean is the policeman in his own home, and how few thought much of Sir Raffle Buffle as he sat asleep after dinner in his old slippers! How well can I remember the terror created within me by the air of outraged dignity with which a certain fine old gentleman, now long since gone, could rub his hands slowly, one on the other, and look up to the ceiling, slightly shaking his head, as though lost in the contemplation of my iniquities! I would become sick in my stomach, and feel as though my ankles had been broken. That upward turn of the eye unmanned me so completely that I was speechless as regarded any defence. I think that that old man could hardly have known the extent of his own power.

Once upon a time a careless lad, having the charge of a bundle of letters addressed to the King, – petitions and such like, which in the course of business would not get beyond the hands of some lord-in-waiting's deputy assistant, – sent the bag which contained them to the wrong place; to Windsor, perhaps, if the Court were in London; or to St. James's, if it were at Windsor. He was summoned; and the great man of the occasion contented himself with holding his hands up to the heavens as he stood up from his chair, and exclaiming twice, 'Mis-sent the Monarch's pouch! Mis-sent the Monarch's pouch!' That young man never knew how he escaped from the Board-room; but for a time he was deprived of all power of exertion, and could not resume his work till he had had six months' leave of absence, and been brought round upon rum and asses' milk. In that instance the peculiar use of the word Monarch had a power which the official magnate had never contemplated. The story is traditional; but I believe that the circumstance happened as lately as in the days of George the Third.

John Eames could laugh at the present chairman of the Income-tax Office with great freedom, and call him old Huffle Scuffle, and the like; but now that he was sent for, he also, in spite of his Radical propensities, felt a little weak about his ankle joints. He knew, from the first hearing of the message, that he was wanted with reference to that affair at the railway station. Perhaps there might be a rule that any clerk should be dismissed who used his fists in any public place. There were many rules entailing the punishment of dismissal for many offences, – and he began to think that he did remember something of such a regulation. However, he got up, looked once around him upon his friends, and then followed Tupper into the Board-room.

'There's Johnny been sent for by old Scuffles,' said one clerk.

'That's about his row with Crosbie,' said another. 'The Board can't do anything to him for that.'

'Can't it?' said the first. 'Didn't young Outonites have to resign because of that row at the Cider Cellars, though his cousin, Sir Constant Outonites, did all that he could for him?'

'But he was regularly up the spout with accommodation bills.'

'I tell you that I wouldn't be in Eames's shoes for a trifle. Crosbie is secretary at the Committee Office where Scuffles was chairman before he came here; and of course they're as thick as thieves. I should't wonder if they didn't make him go down and apologize.'

'Johnny won't do that,' said the other.

In the meantime John Eames was standing in the august presence. Sir Raffle Buffle was throned in his great oak armchair at the head of a long table in a very large room; and by him, at the corner of the table, was seated one of the assistant secretaries of the office. Another member of the Board was also at work upon the long table; but he was reading and signing papers at some distance from Sir Raffle, and paid no heed whatever to the scene. The assistant secretary, looking on, could see that Sir Raffle was annoyed by this want of attention on the part of his colleague, but all this was lost upon Eames.

'Mr. Eames?' said Sir Raffle, speaking with a peculiarly harsh voice, and looking at the culprit through a pair of gold-rimmed glasses, which he perched for the occasion upon his big nose. 'Isn't that Mr. Eames?'

'Yes,' said the assistant secretary, 'this is Eames.'

'Ah!' – and then there was a pause. 'Come a little nearer, Mr. Eames, will you?' and Johnny drew nearer, advancing noiselessly over the Turkey carpet.

'Let me see; in the second class, isn't he? Ah! Do you know, Mr. Eames, that I have received a letter from the secretary to the Directors of the Great Western Railway Company, detailing circumstances which, – if truly stated in that letter, – redound very much to your discredit?'

'I did get into a row there yesterday, sir.'

'Got into a row! It seems to me that you have got into a very serious row, and that I must tell the Directors of the Great Western Railway Company that the law must be allowed to take its course.'

'I shan't mind that, sir, in the least,' said Eames, brightening up a little under this view of the case.

'Not mind that, sir!' said Sir Raffle – or rather, he shouted out the words at the offender before him. I am inclined to think that he overdid it, missing the effect which a milder tone might have attained. Perhaps there was lacking to him some of that majesty of demeanour and dramatic propriety of voice which had been so efficacious in the little story as to the King's bag of letters. As it was, Johnny gave a slight jump, but after his jump he felt better than he had been before. 'Not mind, sir, being dragged before the criminal tribunals of your country, and being punished as a felon, – or rather for a misdemeanour, – for an outrage committed on a public platform! Not mind it! What do you mean, sir?'

'I mean, that I don't think the magistrate would say very much about it, sir. And I don't think Mr. Crosbie would come forward.'

'But Mr. Crosbie must come forward, young man. Do you suppose that an outrage against the peace of the Metropolis is to go unpunished because he may not wish to pursue the matter? I'm afraid you must be very ignorant, young man.'

'Perhaps I am,' said Johnny.

'Very ignorant indeed, – very ignorant indeed. And are you aware, sir, that it would become a question with the Commissioners of this Board whether you could be retained in the service of this department if you were publicly punished by a police magistrate for such a disgraceful outrage as that?'

Johnny looked round at the other Commissioner, but that gentleman did not raise his face from his papers.

'Mr. Eames is a very good clerk,' whispered the assistant secretary, but in a voice which made his words audible to Eames; 'one of the best young men we have,' he added, in a voice which was not audible.

'Oh, – ah; very well. Now, I'll tell you what, Mr. Eames, I hope this will be a lesson to you, – a very serious lesson.'

The assistant secretary, leaning back in his chair so as to be a little behind the head of Sir Raffle, did manage to catch the eye of the other Commissioner. The other Commissioner, barely looking round, smiled a little, and then the assistant secretary smiled also. Eames saw this and he smiled too.

'Whether any ulterior consequences may still await the breach of the peace of which you have been guilty, I am not yet prepared to say,' continued Sir Raffle. 'You may go now.'

And Johnny returned to his own place, with no increased reverence for the dignity of the chairman.

On the following morning one of his colleagues showed him with great glee the passage in the newspaper which informed the world that he had been so desperately beaten by Crosbie that he was obliged to keep his bed at the present time in consequence of the flogging that he had received. Then his anger was aroused, and he bounced about the big room of the Income-tax Office, regardless of assistant secretaries, head clerks, and all other official grandees whatsoever, denouncing the iniquities of the public press, and declaring his opinion that it would be better to live in Russia than in a country which allowed such audacious falsehoods to be propagated.

'He never touched me, Fisher; I don't think he ever tried; but, upon my honour, he never touched me.'

'But, Johnny, it was bold in you to make up to Lord De Courcy's daughter,' said Fisher.

'I never saw one of them in my life.'

'He's going it altogether among the aristocracy, now,' said another; 'I suppose you wouldn't look at anybody under a viscount?'

'Can I help what that thief of an editor puts into his paper? Flogged! Huffle Scuffle told me I was a felon, but that wasn't half so bad as this fellow;' and Johnny kicked the newspaper across the room.

'Indict him for a libel,' said Fisher.

'Particularly for saying you wanted to marry a countess's daughter,' said another clerk.

'I never heard such a scandal in my life,' declared a third; 'and then to say that the girl wouldn't look at you.'

But not the less was it felt by all in the office that Johnny Eames was becoming a leading man among them, and that he was one with whom each of them would be pleased to be intimate. And even among the grandees this affair of the railway station did him no real harm. It was known that Crosbie had deserved to be thrashed, and known that Eames had thrashed him. It was all very well for Sir Raffle Buffle to talk of police magistrates and misdemeanours, but all the world at the Income-tax Office knew very well that Eames had come out from that affair with his head upright, and his right foot foremost.

'Never mind about the newspaper,' a thoughtful old senior

clerk said to him. 'As he did get the licking and you didn't, you can afford to laugh at the newspaper.'

'And you wouldn't write to the editor?'

'No, no; certainly not. No one thinks of defending himself to a newspaper except an ass; – unless it be some fellow who wants to have his name puffed. You may write what's as true as the gospel, but they'll know how to make fun of it.'

Johnny therefore gave up his idea of an indignant letter to the editor, but he felt that he was bound to give some explanation of the whole matter to Lord De Guest. The affair had happened as he was coming from the earl's house, and all his own concerns had now been made so much a matter of interest to his kind friend, that he thought that he could not with propriety leave the earl to learn from the newspapers either the facts or the falsehoods. And, therefore, before he left his office he wrote the following letter: –

'INCOME-TAX OFFICE, December 29, 186–.

'My Lord,—'

He thought a good deal about the style in which he ought to address the peer, never having hitherto written to him. He began, 'My dear Lord,' on one sheet of paper, and then put it aside, thinking that it looked over-bold.

'My Lord,—

'As you have been so very kind to me, I feel that I ought to tell you what happened the other morning at the railway station, as I was coming back from Guestwick. That scoundrel Crosbie got into the same carriage with me at the Barchester Junction, and sat opposite to me all the way up to London. I did not speak a word to him, or he to me; but when he got out at the Paddington Station, I thought I ought not to let him go away, so I— I can't say that I thrashed him as I wished to do; but I made an attempt, and I did give him a black eye. A whole quantity of policemen got round us, and I hadn't a fair chance. I know you will think that I was wrong, and perhaps I was; but what could I do when he sat opposite to me there for two hours, looking as though he thought himself the finest fellow in all London?

'They've put a horrible paragraph into one of the newspapers, saying that I got so "flogged" that I haven't been able to stir since. It is an atrocious falsehood, as is all the rest of the newspaper account. I was not touched. He was not nearly so bad a customer as the bull, and seemed to take it all very quietly. I must acknowledge, though, that he didn't get such a beating as he deserved.

'Your friend Sir R. B. sent for me this morning, and told me I was a felon. I didn't seem to care much for that, for he might as well have called me a murderer or a burglar; but I shall care very much indeed if I have made you angry with me. But what I most fear is the anger of some one else, – at Allington.

'Believe me to be, my Lord,

'Yours very much obliged and most sincerely,

'JOHN EAMES.'

'I knew he'd do it if ever he got the opportunity,' said the earl when he had read his letter; and he walked about his room striking his hands together, and then thrusting his thumbs into his waistcoat pockets. 'I knew he was made of the right stuff,' and the earl rejoiced greatly in the prowess of his favourite. 'I'd have done it myself if I'd seen him. I do believe I would.' Then he went back to the breakfast-room and told Lady Julia. 'What do you think?' said he; 'Johnny Eames has come across Crosbie, and given him a desperate beating.'

'No!' said Lady Julia, putting down her newspaper and spectacles, and expressing by the light of her eyes anything but Christian horror at the wickedness of the deed.

'But he has, though. I knew he would if he saw him.'

'Beaten him! Actually beaten him!'

'Sent him home to Lady Alexandrina with two black eyes.'

'Two black eyes! What a young pickle! But did he get hurt himself?'

'Not a scratch, he says.'

'And what'll they do to him?'

'Nothing. Crosbie won't be fool enough to do anything. A

man becomes an outlaw when he plays such a game as he has played. Anybody's hand may be raised against him with impunity. He can't show his face, you know. He can't come forward and answer questions as to what he has done. There are offences which the law can't touch, but which outrage public feeling so strongly that any one may take upon himself the duty of punishing them. He has been thrashed, and that will stick to him till he dies.'

'Do tell Johnny from me that I hope he didn't get hurt,' said Lady Julia. The old lady could not absolutely congratulate him on his feat of arms, but she did the next thing to it.

But the earl did congratulate him, with a full open assurance of his approval.

'I hope,' he said, 'I should have done the same at your age, under similar circumstances, and I'm very glad that he proved less difficult than the bull. I'm quite sure you didn't want any one to help you with Master Crosbie. As for that other person at Allington, if I understand such matters at all, I think she will forgive you.' It may, however, be a question whether the earl did understand such matters at all. And then he added, in a postscript: 'When you write to me again, – and don't be long first, begin your letter, "My dear Lord De Guest," – that is the proper way.'

CHAPTER XXXVII

AN OLD MAN'S COMPLAINT

'HAVE you been thinking again of what I was saying to you, Bell?' Bernard said to his cousin one morning.

'Thinking of it, Bernard? Why should I think more of it? I had hoped that you had forgotten it yourself.'

'No,' he said; 'I am not so easy-hearted as that. I cannot look on such a thing as I would the purchase of a horse, which I could give up without sorrow if I found that the animal was too costly for my purse. I did not tell you that I loved you till I was sure of myself, and having made myself sure I cannot change at all.'

'And yet you would have me change.'

'Yes, of course I would. If your heart be free now, it must of course be changed before you come to love any man. Such change as that is to be looked for. But when you have loved, then it will not be easy to change you.'

'But I have not.'

'Then I have a right to hope. I have been hanging on here, Bell, longer than I ought to have done, because I could not bring myself to leave you without speaking of this again. I did not wish to seem to you to be importunate—'

'If you could only believe me in what I say.'

'It is not that I do not believe. I am not a puppy or a fool, to flatter myself that you must be in love with me. I believe you well enough. But still it is possible that your mind may alter.'

'It is impossible.'

'I do not know whether my uncle or your mother have spoken to you about this.'

'Such speaking would have no effect.'

In fact, her mother had spoken to her, but she truly said that such speaking would have no effect. If her cousin could not win the battle by his own skill, he might have been quite sure, looking at her character as it was known to him, that he would not be able to win it by the skill of others.

'We have all been made very unhappy,' he went on to say, 'by this calamity which has fallen on poor Lily.'

'And because she has been deceived by the man she did love, I am to make matters square by marrying a man I—' and then she paused. 'Dear Bernard, you should not drive me to say words which will sound harsh to you.'

'No words can be harsher than those which you have already spoken. But, Bell, at any rate, you may listen to me.'

Then he told her how desirable it was with reference to all the concerns of the Dale family that she should endeavour to look favourably on his proposition. It would be good for them all, he said, especially for Lily, as to whom, at the present moment, their uncle felt so kindly. He, as Bernard pleaded, was so anxious at heart for this marriage, that he would do anything that was asked of him if he were gratified. But if he were not gratified in this, he would feel that he had ground for displeasure.

Bell, as she had been desired to listen, did listen very patiently. But when her cousin had finished, her answer was very short. 'Nothing that my uncle can say, or think, or do, can make any difference in this,' said she.

'You will think nothing, then, of the happiness of others.'

'I would not marry a man I did not love, to ensure any amount of happiness to others; – at least I know I ought not to do so. But I do not believe I should ensure any one's happiness by this marriage. Certainly not yours.'

After this Bernard had acknowledged to himself that the difficulties in his way were great. 'I will go away till next autumn,' he said to his uncle.

'If you would give up your profession and remain here, she would not be so perverse.'

'I cannot do that, sir. I cannot risk the well-being of my life on such a chance.' Then his uncle had been angry with him, as well as with his niece. In his anger he determined that he would go again to his sister-in-law, and, after some unreasonable fashion, he resolved that it would become him to be very angry with her also, if she declined to assist him with all her influence as a mother.

'Why should they not both marry?' he said to himself. Lord De Guest's offer as to young Eames had been very generous. As he had then declared, he had not been able to express his own opinion at once; but on thinking over what the earl had said, he had found himself very willing to heal the family wound in the manner proposed, if any such healing might be possible. That, however, could not be done quite as yet. When the time should come, and he thought it might come soon, – perhaps in the spring when the days should be fine and the evenings again long, – he would be willing to take his share with the earl in establishing that new household. To Crosbie he had refused to give anything, and there was upon his conscience a shade of remorse in that he had so refused. But if Lily could be brought to love this other man, he would be more open-handed. She should have her share as though she was in fact his daughter. But then, if he intended to do so much for them at the Small House, should not they in return do something also for him? So thinking, he went again to his sister-in-law, determined to explain his views, even though it might be at the risk of some hard words between them. As regarded himself, he did not much care for hard words spoken to him. He almost expected that people's words should be hard and painful. He did not look for the comfort of affectionate soft greetings, and perhaps would not have appreciated them had they come to him. He caught Mrs. Dale walking in the garden, and brought her into his own room, feeling that he had a better chance there than in her own house. She, with an old dislike to being lectured in that room, had endeavoured to avoid the interview, but had failed.

'So I met John Eames at the manor,' he had said to her in the garden.

'Ah, yes; and how did he get on there? I cannot conceive poor Johnny keeping holiday with the earl and his sister. How did he behave to them, and how did they behave to him?'

'I can assure you he was very much at home there.'

'Was he, indeed? Well, I hope it will do him good. He is, I'm sure, a very good young man; only rather awkward.'

'I didn't think him awkward at all. You'll find, Mary, that he'll do very well; – a great deal better than his father did.'

'I'm sure I hope he may.' After that Mrs. Dale made her attempt to escape; but the squire had taken her prisoner, and led her captive into the house. 'Mary,' he said, as soon as he had induced her to sit down, 'it is time that this should be settled between my nephew and niece.'

'I am afraid there will be nothing to settle.'

'What do you mean; – that you disapprove of it?'

'By no means, – personally. I should approve of it very strongly. But that has nothing to do with the question.'

'Yes, it has. I beg your pardon, but it must have, and should have a great deal to do with it. Of course, I am not saying that anybody should now ever be compelled to marry anybody.'

'I hope not.'

'I never said that they ought, and never thought so. But I do think that the wishes of all her family should have very great weight with a girl that has been well brought up.'

'I don't know whether Bell has been well brought up; but in such a matter as this nobody's wishes would weigh a feather with her; and, indeed, I could not take upon myself even to express a wish. To you I can say that I should have been very happy if she could have regarded her cousin as you wish her to do.'

'You mean that you are afraid to tell her so?'

'I am afraid to do what I think is wrong, if you mean that.'

'I don't think it would be wrong, and therefore I shall speak to her myself.'

'You must do as you like about that, Mr. Dale; I can't prevent you. I shall think you wrong to harass her on such a matter, and I fear also that her answer will not be satisfactory to you. If you choose to tell her your opinion, you must do so. Of course I shall think you wrong, that's all.'

Mrs. Dale's voice as she said this was stern enough, and so was her countenance. She could not forbid the uncle to speak his mind to his niece, but she especially disliked the idea of any interference with her daughter. The squire got up and walked about the room, trying to compose himself that he might answer her rationally, but without anger.

'May I go now?' said Mrs. Dale.

'May you go? Of course you may go if you like it. If you think that I am intruding upon you in speaking to you of the welfare of your two girls, whom I endeavoured to regard as my own daughters, – except in this, that I know they have never been taught to love me, – if you think that is an interference on my part to show anxiety for their welfare, of course you may go.'

'I did not mean to say anything to hurt you, Mr. Dale.'

'Hurt me! What does it signify whether I am hurt or not? I have no children of my own, and of course my only business in life is to provide for my nephews and nieces. I am an old fool if I expect that they are to love me in return, and if I venture to express a wish I am interfering and doing wrong! It is hard, – very hard. I know well that they have been brought up to dislike me, and yet I am endeavouring to do my duty by them.'

'Mr. Dale, that accusation has not been deserved. They have not been brought up to dislike you. I believe that they have both loved and respected you as their uncle; but such love and respect will not give you a right to dispose of their hands.'

'Who wants to dispose of their hands?'

'There are some things in which I think no uncle, – no parent, – should interfere, and of all such things this is the chief. If after that you may choose to tell her your wishes, of course you can do so.'

'It will not be much good after you have set her against me.'

'Mr. Dale, you have no right to say such things to me, and you are very unjust in doing so. If you think that I have set my girls against you, it will be much better that we should leave Allington altogether. I have been placed in circumstances which have made it difficult for me to do my duty to my children; but I have endeavoured to do it, not regarding my own personal wishes. I am quite sure, however, that it would be wrong in me to keep them here, if I am to be told by you that I have taught them to regard you unfavourably. Indeed, I cannot suffer such a thing to be said to me.'

All this Mrs. Dale said with an air of decision, and with a voice expressing a sense of injury received, which made the squire feel that she was very much in earnest.

'Is it not true,' he said, defending himself, 'that in all that relates to the girls you have ever regarded me with suspicion?'

'No, it is not true.' And then she corrected herself, feeling that there was something of truth in the squire's last assertion. 'Certainly not with suspicion,' she said. 'But as this matter has gone so far, I will explain what my real feelings have been. In worldly matters you can do much for my girls, and have done much.'

'And wish to do more,' said the squire.

'I am sure you do. But I cannot on that account give up my place as their only living parent. They are my children, and not yours. And even could I bring myself to allow you to act as their guardian and natural protector, they would not consent to such an arrangement. You cannot call that suspicion.'

'I can call it jealousy.'

'And should not a mother be jealous of her children's love?'

During all this time the squire was walking up and down the room with his hands in his trousers pockets. And when Mrs. Dale had last spoken, he continued his walk for some time in silence.

'Perhaps it is well that you should have spoken out,' he said.

'The manner in which you accused me made it necessary.'

'I did not intend to accuse you, and I do not do so now; but I think that you have been, and that you are, very hard to me, – very hard indeed. I have endeavoured to make your children, and yourself also, sharers with me in such prosperity as has been mine. I have striven to add to your comfort and to their happiness. I am most anxious to secure their future welfare. You would have been very wrong had you declined to accept this on their behalf; but I think that in return for it you need not have begrudged me the affection and obedience which generally follows from such good offices.'

'Mr. Dale, I have begrudged you nothing of this.'

'I am hurt; – I am hurt,' he continued. And she was surprised by his look of pain even more than by the unaccustomed warmth of his words. 'What you have said has, I have known, been the case all along. But though I had felt it to be so, I own that I am hurt by your open words.'

'Because I have said that my own children must ever be my own?'

'Ah, you have said more than that. You and the girls have been living here, close to me, for – how many years is it now? – and during all those years there has grown up for me no kindly feeling. Do you think that I cannot hear, and see, and feel? Do you suppose that I am a fool and do not know? As for yourself you would never enter this house if you did not feel yourself constrained to do so for the sake of appearances. I suppose it is all as it should be. Having no children of my own, I owe the duty of a parent to my nieces; but I have no right to expect from them in return either love, regard, or obedience. I know I am keeping you here against your will, Mary. I won't do so any longer.' And he made a sign to her that she was to depart.

As she rose from her seat her heart was softened towards him. In these latter days he had shown much kindness to the girls, – a kindness that was more akin to the gentleness of love than had ever come from him before. Lily's fate had seemed to melt even his sternness, and he had striven to be tender in his words and ways. And now he spoke as though he had loved the girls, and had loved them in vain. Doubtless he had been a disagreeable neighbour to his sister-in-law, making her feel that it was never for her personally that he had opened his hand. Doubtless he had been moved by an unconscious desire to undermine and take upon himself her authority with her own children. Doubtless he had looked askance at her from the first day of her marriage with his brother. She had been keenly alive to all this since she had first known him, and more keenly alive to it than ever since the failure of those efforts she had made to live with him on terms of affection, made during the first year or two of her residence at the Small House. But, nevertheless, in spite of all, her heart bled for him now. She had gained her victory over him, having fully held her own position with her children; but now that he complained that he had been beaten in the struggle, her heart bled for him.

'My brother,' she said, and as she spoke she offered him her hands, 'it may be that we have not thought as kindly of each other as we should have done.'

'I have endeavoured,' said the old man. 'I have endeavoured—' And then he stopped, either hindered by some excess of emotion, or unable to find the words which were necessary for the expression of his meaning.

'Let us endeavour once again, – both of us.'

'What, begin again at near seventy! No, Mary, there is no more beginning again for me. All this shall make no difference to the girls. As long as I am here they shall have the house. If they marry, I will do for them what I can. I believe Bernard is much in earnest in his suit, and if Bell will listen to him, she shall still be welcomed here as mistress of Allington. What you have said shall make no difference; – but as to beginning again, it is simply impossible.'

After that Mrs. Dale walked home through the garden by herself. He had studiously told her that that house in which they lived should be lent, not to her, but to her children, during his lifetime. He had positively declined the offer of her warmer regard. He had made her understand that they were to look on each other almost as enemies; but that she, enemy as she was, should still be allowed the use of his munificence, because he chose to do his duty by his nieces!

'It will be better for us that we shall leave it,' she said to herself as she seated herself in her own armchair over the drawing-room fire.

CHAPTER XXXVIII

DR. CROFTS IS CALLED IN

MRS. DALE had not sat long in her drawing-room before tidings were brought to her which for a while drew her mind away from that question of her removal. 'Mamma,' said Bell, entering the room, 'I really do believe that Jane has got scarlatina.' Jane, the parlour-maid, had been ailing for the last two days, but nothing serious had hitherto been suspected.

Mrs. Dale instantly jumped up. 'Who is with her?' she asked.

It appeared from Bell's answer that both she and Lily had been with the girl, and that Lily was still in the room. Whereupon Mrs. Dale ran upstairs, and there was on the sudden a commotion in the house. In an hour or so the village doctor was there, and he expressed an opinion that the girl's ailment was certainly scarlatina. Mrs. Dale, not satisfied with this, sent off a boy to Guestwick for Dr. Crofts, having herself maintained an opposition of many years' standing against the medical reputation of the apothecary, and gave a positive order to the two girls not to visit poor Jane again. She herself had had scarlatina, and might do as she pleased. Then, too, a nurse was hired.

All this changed for a few hours the current of Mrs. Dale's thoughts: but in the evening she went back to the subject of her morning conversation, and before the three ladies went to bed, they held together an open council of war upon the subject. Dr. Crofts had been found to be away from Guestwick, and word had been sent on his behalf that he would be over at Allington early on the following morning. Mrs. Dale had almost made up her mind that the malady of her favourite maid was not scarlatina, but had not on that account relaxed her order as to the absence of her daughters from the maid's bedside.

'Let us go at once,' said Bell, who was even more opposed to any domination on the part of her uncle than was her mother. In the discussion which had been taking place between them the whole matter of Bernard's courtship had come upon the carpet. Bell had kept her cousin's offer to herself as long as she had been able to do so; but since her uncle had pressed the subject upon Mrs. Dale, it was impossible for Bell to remain silent any longer. 'You do not want me to marry him, mamma; do you?' she had said, when her mother had spoken with some show of kindness towards Bernard. In answer to this, Mrs. Dale had protested vehemently that she had no such wish, and Lily, who still held to her belief in Dr. Crofts, was almost equally animated. To them all, the idea that their uncle should in any way interfere in their own views of life, on the strength of the pecuniary assistance which they had received from him, was peculiarly distasteful. But it was especially distasteful that he should presume to have even an opinion as to their disposition in marriage. They declared to each other that their uncle could have no right to object to any marriage which either of them might contemplate as long as their mother should approve of it. The poor old squire had been right in saying that he was regarded with suspicion. He was so regarded. The fault had certainly been his own, in having endeavoured to win the daughters without thinking it worth his while to win the mother. The girls had unconsciously felt that the attempt was made, and had vigorously rebelled against it. It had not been their fault that they had been brought to live in their uncle's house, and made to ride on his ponies, and to eat partially of his bread. They had so eaten, and so lived, and declared themselves to be grateful. The squire was good in his way, and they recognized his goodness; but not on that account would they transfer to him one jot of the allegiance which as children they owed to their mother. When she told them her tale, explaining to them the words which their uncle had spoken that morning, they expressed their regret that he should be so grieved; but they were strong in assurances to their mother that she had been sinned against, and was not sinning.

'Let us go at once,' said Bell.

'It is much easier said than done, my dear.'

'Of course it is, mamma; else we shouldn't be here now. What I mean is this, – let us take some necessary first step at once. It is clear that my uncle thinks that our remaining here should give him some right over us. I do not say that he is wrong to think so. Perhaps it is natural. Perhaps, in accepting his kindness, we ought to submit ourselves to him. If that be so, it is a conclusive reason for our going.'

'Could we not pay him rent for the house,' said Lily, 'as Mrs. Hearn does? You would like to remain here, mamma, if you could do that?'

'But we could not do that, Lily. We must choose for ourselves a smaller house than this, and one that is not burdened with the expense of a garden. Even if we paid but a moderate rent for this place, we should not have the means of living here.'

'Not if we lived on toast and tea?' said Lily, laughing.

'But I should hardly wish you to live upon toast and tea; and indeed I fancy that I should get tired of such a diet myself.'

'Never, mamma,' said Lily. 'As for me, I confess to a longing after mutton chops; but I don't think you would ever want such vulgar things.'

'At any rate, it would be impossible to remain here,' said Bell. 'Uncle Christopher would not take rent from mamma; and even if he did, we should not know how to go on with our other arrangements after such a change. No; we must give up the dear old Small House.'

'It is a dear old house,' said Lily, thinking, as she spoke, more of those late scenes in the garden, when Crosbie had been with them in the autumn months, than of any of the former joys of her childhood.

'After all, I do not know that I should be right to move,' said Mrs. Dale, doubtingly.

'Yes, yes,' said both the girls at once. 'Of course you will be right, mamma; there cannot be a doubt about it, mamma. If we can get any cottage, or even lodgings, that would be better than remaining here, now that we know what uncle Christopher thinks of it.'

'It will make him very unhappy,' said Mrs. Dale.

But even this argument did not in the least move the girls. They were very sorry that their uncle should be unhappy. They would endeavour to show him by some increased show of affection that their feelings towards him were not unkind. Should he speak to them they would endeavour to explain to him that their thoughts towards him were altogether affectionate. But they could not remain at Allington increasing their load of gratitude, seeing that he expected a certain payment which they did not feel themselves able to render.

'We should be robbing him, if we stayed here,' Bell declared; – 'wilfully robbing him of what he believes to be his just share of the bargain.'

So it was settled among them that notice should be given to their uncle of their intention to quit the Small House of Allington.

And then came the question as to their new home. Mrs. Dale was aware that her income was at any rate better than that possessed by Mrs. Eames, and therefore she had fair ground for presuming that she could afford to keep a house at Guestwick. 'If we do go away, that is what we must do,' she said.

'And we shall have to walk out with Mary Eames, instead of Susan Boyce,' said Lily. 'It won't make so much difference after all.'

'In that respect we shall gain as much as we lose,' said Bell.

'And then it will be so nice to have the shops,' said Lily, ironically.

'Only we shall never have any money to buy anything,' said Bell.

'But we shall see more of the world,' said Lily. 'Lady Julia's carriage comes into town twice a week, and the Miss Gruffens drive about in great style. Upon the whole, we shall gain a great deal; only for the poor old garden. Mamma, I do think I shall break my heart at parting with Hopkins; and as to him, I shall be disappointed in mankind if he ever holds his head up again after I am gone.'

But in truth there was very much of sadness in their resolution, and to Mrs. Dale it seemed as though she were managing matters badly for her daughters, and allowing poverty and misfortune to come upon them through her own fault. She well knew how great a load of sorrow was lying on Lily's heart, hidden beneath those little attempts at pleasantry which she made. When she spoke of being disappointed in mankind, Mrs. Dale could hardly repress an outward shudder that would betray her thoughts. And now she was consenting to take them forth from their comfortable home, from the luxury of their lawns and gardens, and to bring them to some small dingy corner of a provincial town, – because she had failed to make herself happy with her brother-in-law. Could she be right to give up all the advantages which they enjoyed at Allington, – advantages which had come to them from so legitimate a source, – because her own feelings had been wounded? In all their future want of comfort, in the comfortless dowdiness of the new home to which she would remove them, would she not always blame herself for having brought them to that by her own false pride? And yet it seemed to her that she now had no alternative. She could not now teach her daughters to obey their uncle's wishes in all things. She could not make Bell understand that it would be well that she should marry Bernard because the squire had set his heart on such a marriage. She had gone so far that she could not now go back.

'I suppose we must move at Lady-day?'* said Bell, who was in favour of instant action. 'If so, had you not better let uncle Christopher know at once?'

'I don't think that we can find a house by that time.'

'We can get in somewhere,' continued Bell. 'There are plenty of lodgings in Guestwick, you know.' But the sound of the word lodgings was uncomfortable in Mrs. Dale's ears.

'If we are to go, let us go at once,' said Lily. 'We need not stand much upon the order of our going.'

'Your uncle will be very much shocked,' said Mrs. Dale.

'He cannot say that it is your fault,' said Bell.

It was thus agreed between them that the necessary information should be at once given to the squire, and that the old, well-loved house should be left for ever. It would be a great fall in a worldly point of view, – from the Allington Small House to an abode in some little street of Guestwick. At Allington they had been county people, – raised to a level with their own squire and other squires by the circumstance of their residence; but at Guestwick they would be small even among the people of the town. They would be on an equality with the Eameses, and much looked down upon by the Gruffens. They would hardly dare to call any more at Guestwick Manor, seeing that they certainly could not expect Lady Julia to call upon them at Guestwick. Mrs. Boyce no doubt would patronize them, and they could already anticipate the condolence which would be offered to them by Mrs. Hearn. Indeed such a movement on their part would be tantamount to a confession of failure in the full hearing of so much of the world as was known to them.

I must not allow my readers to suppose that these considerations were a matter of indifference to any of the ladies at the Small House. To some women of strong mind, of highly-strung philosophic tendencies, such considerations might have been indifferent. But Mrs. Dale was not of this nature, nor were her daughters. The good things of the world were good in their eyes, and they valued the privilege of a pleasant social footing among their friends. They were by no means capable of a wise contempt of the advantages which chance had hitherto given to them. They could not go forth rejoicing in the comparative poverty of their altered condition. But then, neither could they purchase those luxuries which they were about to abandon at the price which was asked for them.

'Had you not better write to my uncle?' said one of the girls. But to this Mrs. Dale objected that she could not make a letter on such a subject clearly intelligible, and that therefore she would see the squire on the following morning. 'It will be very dreadful,' she said, 'but it will soon be over. It is not what he will say at the moment that I fear so much, as the bitter reproaches of his face when I shall meet him afterwards.' So, on the following morning, she again made her way, and now without invitation, to the squire's study.

'Mr. Dale,' she began, starting upon her work with some confusion in her manner, and hurry in her speech, 'I have been thinking over what we were saying together yesterday, and I have come to a resolution which I know I ought to make known to you without a moment's delay.'

The squire also had thought of what had passed between them, and had suffered much as he had done so; but he had thought of it without acerbity or anger. His thoughts were ever gentler than his words, and his heart softer than any exponent of his heart that he was able to put forth. He wished to love his brother's children, and to be loved by them; but even failing that, he wished to do good to them. It had not occurred to him to be angry with Mrs. Dale after that interview was over. The conversation had not gone pleasantly with him; but then he hardly expected that things would go pleasantly. No idea had occurred to him that evil could come upon any of the Dale ladies from the words which had then been spoken. He regarded the Small House as their abode and home as surely as the Great House was his own. In giving him his due, it must be declared that any allusion to their holding these as a benefit done to them by him had been very far from his thoughts. Mrs. Hearn, who held her cottage at half its real value, grumbled almost daily at him as her landlord; but it never occurred to him that therefore he should raise her rent, or that in not doing so he was acting with special munificence. It had ever been to him a grumbling, cross-grained, unpleasant world; and he did not expect from Mrs. Hearn, or from his sister-in-law, anything better than that to which he had ever been used.

'It will make me very happy,' said he, 'if it has any bearing on Bell's marriage with her cousin.'

'Mr. Dale, that is out of the question. I would not vex you by saying so if I were not certain of it; but I know my child so well!'

'Then we must leave it to time, Mary.'

'Yes, of course; but no time will suffice to make Bell change her mind. We will, however, leave the subject. And now, Mr. Dale, I have to tell you of something else; – we have resolved to leave the Small House.'

'Resolved on what?' said the squire, turning his eyes full upon her.

'We have resolved to leave the Small House.'

'Leave the Small House!' he said, repeating her words; 'and where on earth do you mean to go?'

'We think we shall go into Guestwick.'

'And why?'

'Ah, that is so hard to explain. If you would only accept the fact as I tell it to you, and not ask for the reasons which have guided me!'

'But that is out of the question, Mary. In such a matter as that I must ask your reasons; and I must tell you also that, in my opinion, you will not be doing your duty to your daughters in carrying out such an intention, unless your reasons are very strong indeed.'

'But they are very strong,' said Mrs. Dale; and then she paused.

'I cannot understand it,' said the squire. 'I cannot bring myself to believe that you are really in earnest. Are you not comfortable there?'

'More comfortable than we have any right to be with our means.'

'But I thought you always did very nicely with your money. You never get into debt.'

'No; I never get into debt. It is not that, exactly. The fact is, Mr. Dale, we have no right to live there without paying rent; but we could not afford to live there if we did pay rent.'

'Who has talked about rent?' he said, jumping up from his chair. 'Some one has been speaking falsehoods of me behind my back.' No gleam of the real truth had yet come to him. No idea had reached his mind that his relatives thought it necessary to leave his house in consequence of any word that he himself had spoken. He had never considered himself to have been in any special way generous to them, and would not have thought it reasonable that they should abandon the house in which they had been living, even if his anger against them had been strong and hot. 'Mary,' he said, 'I must insist upon getting to the bottom of this. As for your leaving the house, it is out of the question. Where can you be better off, or so well? As to going into Guestwick, what sort of life would there be for the girls? I put all that aside as out of the question; but I must know what has induced you to make such a proposition. Tell me honestly, – has any one spoken evil of me behind my back?'

Mrs. Dale had been prepared for opposition and for reproach; but there was a decision about the squire's words, and an air of masterdom in his manner, which made her recognize more fully than she had yet done the difficulty of her position. She almost began to fear that she would lack power to carry out her purpose.

'Indeed, it is not so, Mr. Dale.'

'Then what is it?'

'I know that if I attempt to tell you, you will be vexed, and will contradict me.'

'Vexed I shall be, probably.'

'And yet I cannot help it. Indeed, I am endeavouring to do what is right by you and by the children.'

'Never mind me; your duty is to think of them.'

'Of course it is; and in doing this they most cordially agree with me.'

In using such argument as that, Mrs. Dale showed her weakness, and the squire was not slow to take advantage of it. 'Your duty is to them,' he said; 'but I do not mean by that that your duty is to let them act in any way that may best please them for the moment. I can understand that they should be run away with by some romantic nonsense, but I cannot understand it of you.'

'The truth is this, Mr. Dale. You think that my children owe to you that sort of obedience which is due to a parent, and as long as they remain here, accepting from your hands so large a part of their daily support, it is perhaps natural that you should think so. In this unhappy affair about Bell—'

'I have never said anything of the kind,' said the squire, interrupting her.

'No; you have not said so. And I do not wish you to think that I make any complaint. But I feel that it is so, and they feel it. And, therefore, we have made up our minds to go away.'

Mrs. Dale, as she finished, was aware that she had not told her story well, but she had acknowledged to herself that it was quite out of her power to tell it as it should be told. Her main object was to make her brother-in-law understand that she certainly would leave his house, and to make him understand this with as little pain to himself as possible. She did not in the least mind his thinking her foolish, if only she could so carry her point as to be able to tell her daughters on her return that the matter was settled. But the squire, from his words and manners, seemed indisposed to give her this privilege.

'Of all the propositions which I ever heard,' said he, 'it is the most unreasonable. It amounts to this, that you are too proud to live rent-free in a house which belongs to your husband's brother, and therefore you intend to subject yourself and your children to the great discomfort of a very straitened income. If you yourself only were concerned I should have no right to say anything; but I think myself bound to tell you that, as regards the girls, everybody that knows you will think you to have been very wrong. It is in the natural course of things that they should live in that house. The place has never been let. As far as I know, no rent has ever been paid for the house since it was built. It has always been given to some member of the family, who has been considered as having the best right to it. I have considered your footing there as firm as my own here. A quarrel between me and your children would be to me a great calamity, though, perhaps they might be indifferent to it. But if there were such a quarrel it would afford no reason for their leaving that house. Let me beg you to think over the matter again.'

The squire could assume an air of authority on certain occasions, and he had done so now. Mrs. Dale found that she could only answer him by a simple repetition of her own intention; and, indeed, failed in making him any serviceable answer whatsoever.

'I know that you are very good to my girls,' she said.

'I will say nothing about that,' he answered; not thinking at that moment of the Small House, but of the full possession which he had desired to give to the elder of all the privileges which should belong to the mistress of Allington, – thinking also of the means by which he was hoping to repair poor Lily's shattered fortunes. What words were further said had no great significance, and Mrs. Dale got herself away, feeling that she had failed. As soon as she was gone the squire arose, and putting on his greatcoat, went forth with his hat and stick to the front of the house. He went out in order that his thoughts might be more free, and that he might indulge in that solace which an injured man finds in contemplating his injury. He declared to himself that he was very hardly used, – so hardly used, that he almost began to doubt himself and his own motives. Why was it that the people around him disliked him so strongly, – avoided him and thwarted him in the efforts which he made for their welfare? He offered to his nephew all the privileges of a son, – much more indeed than the privileges of a son, – merely asking in return that he would consent to live permanently in the house which was to be his own. But his nephew refused. 'He cannot bear to live with me,' said the old man to himself sorely. He was prepared to treat his nieces with more generosity than the daughters of the House of Allington had usually received from their fathers; and they repelled his kindness, running away from him, and telling him openly that they would not be beholden to him. He walked slowly up and down the terrace, thinking of this very bitterly.

He did not find in the contemplation of his grievance all that solace which a grievance usually gives, because he accused himself in his thoughts rather than others. He declared to himself that he was made to be hated, and protested to himself that it would be well that he should die and be buried out of memory, so that the remaining Dales might have a better chance of living happily; and then as he thus discussed all this within his own bosom, his thoughts were very tender, and though he was aggrieved, he was most affectionate to those who had most injured him. But it was absolutely beyond his power to reproduce outwardly, with words and outward signs, such thoughts and feelings.

It was now very nearly the end of the year, but the weather was still soft and open. The air was damp rather than cold, and the lawns and fields still retained the green tints of new vegetation. As the squire was walking on the terrace Hopkins came up to him, and touching his hat, remarked that they should have frost in a day or two.

'I suppose we shall,' said the squire.

'We must have the mason to the flues of that little grape-house, sir, before I can do any good with a fire there.'

'Which grape-house?' said the squire, crossly.

'Why, the grape-house in the other garden, sir. It ought to have been done last year by rights.' This Hopkins said to punish his master for being cross to him. On that matter of the flues of Mrs. Dale's grape-house he had, with much consideration, spared his master during the last winter, and he felt that this ought to be remembered now. 'I can't put any fire in it, not to do any real good, till something's done. That's sure.'

'Then don't put any fire in it,' said the squire.

Now the grapes in question were supposed to be peculiarly fine, and were the glory of the garden of the Small House. They were always forced, though not forced so early as those at the Great House, and Hopkins was in a state of great confusion.

'They'll never ripen, sir; not the whole year through.'

'Then let them be unripe,' said the squire, walking about.

Hopkins did not at all understand it. The squire in his natural course was very unwilling to neglect any such matter as this, but would be specially unwilling to neglect anything touching the Small House. So Hopkins stood on the terrace, raising his hat and scratching his head. 'There's something wrong amongst them,' said he to himself, sorrowfully.

But when the squire had walked to the end of the terrace and had turned upon the path which led round the side of the house, he stopped and called to Hopkins.

'Have what is needful done to the flue,' he said.

'Yes, sir; very well, sir. It'll only be re-setting the bricks. Nothing more ain't needful, just this winter.'

'Have the place put in perfect order while you're about it,' said the squire, and then he walked away.

CHAPTER XXXIX

DR. CROFTS IS TURNED OUT

'Have you heard the news, my dear, from the Small House?' said Mrs. Boyce to her husband, some two or three days after Mrs. Dale's visit to the squire. It was one o'clock, and the parish pastor had come in from his ministrations to dine with his wife and children.

'What news' said Mr. Boyce, for he had heard none.

'Mrs. Dale and the girls are going to leave the Small House; they're going into Guestwick to live.'

'Mrs. Dale going away; nonsense!' said the vicar. 'What on earth should take her into Guestwick? She doesn't pay a shilling of rent where she is.'

'I can assure you it's true, my dear. I was with Mrs. Hearn just now, and she had it direct from Mrs. Dale's own lips. Mrs. Hearn said she'd never been taken so much aback in her whole life. There's been some quarrel, you may be sure of that.'

Mr. Boyce sat silent, pulling off his dirty shoes preparatory to his dinner. Tidings so important, as touching the social life of his parish, had not come to him for many a day, and he could hardly bring himself to credit them at so short a notice.

'Mrs. Hearn says that Mrs. Dale spoke ever so firmly about it, as though determined that nothing should change her.'

'And did she say why?'

'Well, not exactly. But Mrs. Hearn said she could understand there had been words between her and the squire. It couldn't be anything else, you know. Probably it had something to do with that man Crosbie.'

'They'll be very pushed about money,' said Mr. Boyce, thrusting his feet into his slippers.

'That's just what I said to Mrs. Hearn. And those girls have never been used to anything like real economy. What's to become of them I don't know;' and Mrs. Boyce, as she expressed her sympathy for her dear friends, received considerable comfort from the prospect of their future poverty. It always is so, and Mrs. Boyce was not worse than her neighbours.

'You'll find they'll make it up before the time comes,' said Mr. Boyce, to whom the excitement of such a change in affairs was almost too good to be true.

'I'm afraid not,' said Mrs. Boyce; 'I'm afraid not. They are both so determined. I always thought that riding and giving the girls hats and habits was injurious. It was treating them as though they were the squire's daughters, and they were not the squire's daughters.'

'It was almost the same thing.'

'But now we see the difference,' said the judicious Mrs. Boyce. 'I often said that dear Mrs. Dale was wrong, and it turns out that I was right. It will make no difference to me, as regards calling on them and that sort of thing.'

'Of course it won't.'

'Not but what there must be a difference, and a very great difference too. It will be a terrible come down for poor Lily with the loss of her fine husband and all.'

After dinner, when Mr. Boyce had again gone forth upon his labours, the same subject was discussed between Mrs. Boyce and her daughters, and the mother was very careful to teach her children that Mrs. Dale would be just as good a person as ever she had been, and quite as much a lady, even though she should live in a very dingy house at Guestwick; from which lesson the Boyce girls learned plainly that Mrs Dale, with Bell and Lily, were about to have a fall in the world, and that they were to be treated accordingly.

From all this, it will be discovered that Mrs. Dale had not given way to the squire's arguments, although she had found herself unable to answer them. As she had returned home she had felt herself to be almost vanquished, and had spoken to the girls with the air and tone of a woman who hardly knew in which course lay the line of her duty. But they had not seen the squire's manner on the occasion, nor heard his words, and they could not understand that their own purpose should be abandoned because he did not like it. So they talked their mother into fresh resolves, and on the following morning she wrote a note to her brother-in-law, assuring him that she had thought much of all that he had said, but again declaring that she regarded herself as bound in duty to leave the Small House. To this he had returned no answer, and she had communicated her intention to Mrs. Hearn, thinking it better that there should be no secret in the matter.

'I am sorry to hear that your sister-in-law is going to leave us,' Mr. Boyce said to the squire that same afternoon.

'Who told you that?' asked the squire, showing by his tone that he by no means liked the topic of conversation which the parson had chosen.

'Well, I had it from Mrs. Boyce, and I think Mrs. Hearn told her.'

'I wish Mrs. Hearn would mind her own business, and not spread idle reports.'

The squire said nothing more, and Mr. Boyce felt that he had been very unjustly snubbed.

Dr. Crofts had come over and pronounced as a fact that it was scarlatina. Village apothecaries are generally wronged by the doubts which are thrown upon them, for the town doctors when they come always confirm what the village apothecaries have said.

'There can be no doubt as to its being scarlatina,' the doctor declared; 'but the symptoms are all favourable.'

There was, however, much worse coming than this. Two days afterwards Lily found herself to be rather unwell. She endeavoured to keep it to herself, fearing that she should be brought under the doctor's notice as a patient; but her efforts were unavailing, and on the following morning it was known that she had also taken the disease. Dr. Crofts declared that everything was in her favour. The weather was cold. The presence of the malady in the house had caused them all to be careful, and, moreover, good advice was at hand at once. The doctor begged Mrs. Dale not to be uneasy, but he was very eager in begging that the two sisters might not be allowed to be together. 'Could you not send Bell into Guestwick, – to Mrs. Eames's?' said he. But Bell did not choose to be sent to Mrs. Eames's, and was with great difficulty kept out of her mother's bedroom, to which Lily as an invalid was transferred.

'If you will allow me to say so,' he said to Bell, on the second day after Lily's complaint had declared itself, 'you are wrong to stay here in the house.'

'I certainly shall not leave mamma, when she has got so much upon her hands,' said Bell.

'But if you should be taken ill she would have more on her hands,' pleaded the doctor.

'I could not do it,' Bell replied. 'If I were taken over to Guestwick, I should be so uneasy that I should walk back to Allington the first moment that I could escape from the house.'

'I think your mother would be more comfortable without you.'

'And I think she would be more comfortable with me. I

don't ever like to hear of a woman running away from illness; but when a sister or a daughter does so, it is intolerable.' So Bell remained, without permission indeed to see her sister, but performing various outside administrations which were much needed.

And thus all manner of trouble came upon the inhabitants of the Small House, falling upon them as it were in a heap together. It was as yet barely two months since those terrible tidings had come respecting Crosbie; tidings which, it was felt at the time, would of themselves be sufficient to crush them; and now to that misfortune other misfortunes had been added, – one quick upon the heels of another. In the teeth of the doctor's kind prophecy Lily became very ill, and after a few days was delirious. She would talk to her mother about Crosbie, speaking of him as she used to speak in the autumn that was passed. But even in her madness she remembered that they had resolved to leave their present home; and she asked the doctor twice whether their lodgings in Guestwick were ready for them.

It was thus that Crofts first heard of their intention. Now, in these days of Lily's worst illness, he came daily over to Allington, remaining there, on one occasion, the whole night. For all this he would take no fee; – nor had he ever taken a fee from Mrs. Dale. 'I wish you would not come so often,' Bell said to him one evening, as he stood with her at the drawing-room fire, after he had left the patient's room; 'you are overloading us with obligations.' On that day Lily was over the worst of the fever, and he had been able to tell Mrs. Dale that he did not think that she was now in danger.

'It will not be necessary much longer,' he said; 'the worst of it is over.'

'It is such a luxury to hear you say so. I suppose we shall owe her life to you; but nevertheless—'

'Oh, no; scarlatina is not such a terrible thing now as it used to be.'

'Then why should you have devoted your time to her as you have done? It frightens me when I think of the injury we must have done you.'

'My horse has felt it more than I have,' said the doctor,

laughing. 'My patients at Guestwick are not so very numerous.' Then, instead of going, he sat himself down. 'And it is really true,' he said, 'that you are all going to leave this house?'

'Quite true. We shall do so at the end of March, if Lily is well enough to be moved.'

'Lily will be well long before that, I hope; not, indeed, that she ought to be moved out of her own rooms for many weeks to come yet.'

'Unless we are stopped by her we shall certainly go at the end of March.' Bell now had also sat down, and they both remained for some time looking at the fire in silence.

'And why is it, Bell?' he said, at last. 'But I don't know whether I have a right to ask.'

'You have a right to ask any question about us,' she said. 'My uncle is very kind. He is more than kind; he is generous. But he seems to think that our living here gives him a right to interfere with mamma. We don't like that, and, therefore, we are going.'

The doctor still sat on one side of the fire, and Bell still sat opposite to him; but the conversation did not form itself very freely between them. 'It is bad news,' he said, at last.

'At any rate when we are ill you will not have so far to come and see us.'

'Yes, I understand. That means that I am ungracious not to congratulate myself on having you all so much nearer to me; but I do not in the least. I cannot bear to think of you as living anywhere but here at Allington. Dales will be out of their place in a street at Guestwick.'

'That's hard upon the Dales, too.'

'It is hard upon them. It's a sort of offshoot from that very tyrannical law of noblesse oblige. I don't think you ought to go away from Allington, unless the circumstances are very imperative.'

'But they are very imperative.'

'In that case, indeed!' And then again he fell into silence.

'Have you never seen that mamma is not happy here?' she said, after another pause. 'For myself, I never quite understood it all before as I do now; but now I see it.'

'And I have seen it; – have seen at least what you mean. She has led a life of restraint; but then, how frequently is such restraint the necessity of a life? I hardly think that your mother would move on that account.'

'No. It is on our account. But this restraint, as you call it, makes us unhappy, and she is governed by seeing that. My uncle is generous to her as regards money; but in other things, – in matters of feeling, – I think he has been ungenerous.'

'Bell,' said the doctor; and then he paused.

She looked up at him, but made no answer. He had always called her by her Christian name, and they two had ever regarded each other as close friends. At the present moment she had forgotten all else besides this, and yet she had infinite pleasure in sitting there and talking to him.

'I am going to ask you a question which perhaps I ought not to ask, only that I have known you so long that I almost feel that I am speaking to a sister.'

'You may ask me what you please,' said she.

'It is about your cousin Bernard.'

'About Bernard!' said Bell.

It was now dusk; and as they were sitting without other light than that of the fire, she knew that he could not discern the colour which covered her face as her cousin's name was mentioned. But, had the light of day pervaded the whole room, I doubt whether Crofts would have seen that blush, for he kept his eyes firmly upon the fire.

'Yes, about Bernard? I don't know whether I ought to ask you.'

'I'm sure I can't say,' said Bell, speaking words of the nature of which she was not conscious.

'There has been a rumour in Guestwick that he and you—'

'It is untrue,' said Bell; 'quite untrue. If you hear it repeated, you should contradict it. I wonder why people should say such things.'

'It would have been an excellent marriage; – all your friends must have approved it.'

'What do you mean, Dr. Crofts? How I do hate those

words, "an excellent marriage". In them is contained more of wicked worldliness than any other words that one ever hears spoken. You want me to marry my cousin simply because I should have a great house to live in, and a coach. I know that you are my friend; but I hate such friendship as that.'

'I think you misunderstand me, Bell. I mean that it would have been an excellent marriage, provided you had both loved each other.'

'No, I don't misunderstand you. Of course it would be an excellent marriage, if we loved each other. You might say the same if I loved the butcher or the baker. What you mean is, that it makes a reason for loving him.'

'I don't think I did mean that.'

'Then you mean nothing.'

After that, there were again some minutes of silence during which Dr. Crofts got up to go away. 'You have scolded me very dreadfully,' he said, with a slight smile, 'and I believe I have deserved it for interfering—'

'No; not at all for interfering.'

'But at any rate you must forgive me before I go.'

'I won't forgive you at all, unless you repent of your sins, and alter altogether the wickedness of your mind. You will become very soon as bad as Dr. Gruffen.'

'Shall I?'

'Oh, but I will forgive you; for after all, you are the most generous man in the world.'

'Oh, yes; of course I am. Well, – good-bye.'

'But, Dr. Crofts, you should not suppose others to be so much more worldly than yourself. You do not care for money so very much—'

'But I do care very much.'

'If you did you would not come here for nothing day after day.'

'I do care for money very much. I have sometimes nearly broken my heart because I could not get opportunities of earning it. It is the best friend that a man can have—'

'Oh, Dr. Crofts!'

'—the best friend that a man can have, if it be honestly

come by. A woman can hardly realize the sorrow which may fall upon a man from the want of such a friend.'

'Of course a man likes to earn a decent living by his profession; and you can do that.'

'That depends upon one's ideas of decency.'

'Ah! mine never ran very high. I've always had a sort of aptitude for living in a pigsty; – a clean pigsty, you know, with nice fresh bean straw to lie upon. I think it was a mistake when they made a lady of me. I do, indeed.'

'I do not,' said Dr. Crofts.

'That's because you don't quite know me yet. I've not the slightest pleasure in putting on three different dresses a day. I do it very often because it comes to me to do it, from the way in which we have been taught to live. But when we get to Guestwick I mean to change all that; and if you come in to tea, you'll see me in the same brown frock that I wear in the morning, – unless, indeed, the morning work makes the brown frock dirty. Oh, Dr. Crofts! you'll have it pitch-dark riding home under the Guestwick elms.'

'I don't mind the dark,' he said; and it seemed as though he hardly intended to go even yet.

'But I do,' said Bell, 'and I shall ring for candles.' But he stopped her as she put her hand out to the bell-pull.

'Stop a moment, Bell. You need hardly have the candles before I go, and you need not begrudge my staying either, seeing that I shall be all alone at home.'

'Begrudge your staying!'

'But, however, you shall begrudge it, or else make me very welcome.' He still held her by the wrist, which he had caught as he prevented her from summoning the servant.

'What do you mean?' said she. 'You know you are welcome to us as flowers in May. You always were welcome; but now, when you have come to us in our trouble— At any rate, you shall never say that I turn you out.'

'Shall I never say so?' And still he held her by the wrist. He had kept his chair throughout, but she was standing before him, – between him and the fire. But she, though he held her in this way, thought little of his words, or of his action. They had known each other with great intimacy, and though Lily would still laugh at her, saying that Dr. Crofts was her lover, she had long since taught herself that no such feeling as that would ever exist between them.

'Shall I never say so, Bell? What if so poor a man as I ask for the hand that you will not give to so rich a man as your cousin Bernard?'

She instantly withdrew her arm and moved back very quickly a step or two across the rug. She did it almost with the motion which she might have used had he insulted her; or had a man spoken such words who would not, under any circumstances, have a right to speak them.

'Ah, yes! I thought it would be so,' he said. 'I may go now, and may know that I have been turned out.'

'What is it you mean, Dr. Crofts? What is it you are saying? Why do you talk that nonsense, trying to see if you can provoke me?'

'Yes; it is nonsense. I have no right to address you in that way, and certainly should not have done it now that I am in your house in the way of my profession. I beg your pardon.' Now he also was standing, but he had not moved from his side of the fireplace. 'Are you going to forgive me before I go?'

'Forgive you for what?' said she.

'For daring to love you; for having loved you almost as long as you can remember; for loving you better than all beside. This alone you should forgive; but will you forgive me for having told it?'

He had made her no offer, nor did she expect that he was about to make one. She herself had hardly yet realized the meaning of his words, and she certainly had asked herself no question as to the answer which she should give to them. There are cases in which lovers present themselves in so unmistakeable a guise that the first word of open love uttered by them tells their whole story, and tells it without the possibility of a surprise. And it is generally so when the lover has not been an old friend, when even his acquaintance has been of modern date. It had been so essentially in the case of Crosbie and Lily Dale. When Crosbie came to Lily and made his offer, he did it with perfect ease and thorough self-possession, for he almost knew that it was expected. And Lily, though she had been flurried for a moment, had her answer pat enough. She already loved the man with all her heart, delighted in his presence, basked in the sunshine of his manliness, rejoiced in his wit, and had tuned her ears to the tone of his voice. It had all been done, and the world expected it. Had he not made his offer, Lily would have been ill-treated; – though, alas, alas, there was future ill-treatment, so much heavier, in store for her! But there are other cases in which a lover cannot make himself known as such without great difficulty, and when he does do so, cannot hope for an immediate answer in his favour. It is hard upon old friends that this difficulty should usually fall the heaviest upon them. Crofts had been so intimate with the Dale family that very many persons had thought it probable that he would marry one of the girls. Mrs. Dale herself had thought so, and had almost hoped it. Lily had certainly done both. These thoughts and hopes had somewhat faded away, but yet their former existence should have been in the doctor's favour. But now, when he had in some way spoken out, Bell started back from him and would not believe that he was in earnest. She probably loved him better than any man in the world, and yet, when he spoke to her of love, she could not bring herself to understand him.

'I don't know what you mean, Dr. Crofts; indeed, I do not,' she said.

'I had meant to ask you to be my wife; simply that. But you shall not have the pain of making me a positive refusal. As I rode here to-day I thought of it. During my frequent rides of late I have thought of little else. But I told myself that I had no right to do it. I have not even a house in which it would be fit that you should live.'

'Dr. Crofts, if I loved you, – if I wished to marry you—' and then she stopped herself.

'But you do not?'

'No; I think not. I suppose not. No. But in any way no consideration about money has anything to do with it.'

'But I am not that butcher or that baker whom you could love?'

'No,' said Bell; and then she stopped herself from further speech, not as intending to convey all her answer in that one word, but as not knowing how to fashion any further words.

'I knew it would be so,' said the doctor.

It will, I fear, be thought by those who condescend to criticize this lover's conduct and his mode of carrying on his suit, that he was very unfit for such work. Ladies will say that he wanted courage, and men will say that he wanted wit. I am inclined, however, to believe that he behaved as well as men generally do behave on such occasions, and that he showed himself to be a good average lover. There is your bold lover, who knocks his lady-love over as he does a bird, and who would anathematize himself all over, and swear that his gun was distraught, and look about as though he thought the world was coming to an end, if he missed to knock over his bird. And there is your timid lover, who winks his eyes when he fires, who has felt certain from the moment in which he buttoned on his knickerbockers that he would kill nothing, and who, when he hears the loud congratulations of his friends, cannot believe that he really did bag that beautiful winged thing by his own prowess. The beautiful winged thing which the timid man carries home in his bosom, declining to have it thrown into a miscellaneous cart, so that it may never be lost in a common crowd of game, is better to him than are the slaughtered hecatombs* to those who kill their birds by the hundred.

But Dr. Crofts had so winked his eye, that he was not in the least aware whether he had winged his bird or not. Indeed, having no one at hand to congratulate him, he was quite sure that the bird had flown away uninjured into the next field. 'No' was the only word which Bell had given in answer to his last sidelong question, and No is not a comfortable word to lovers. But there had been that in Bell's No which might have taught him that the bird was not escaping without a wound, if he had still had any of his wits about him.

'Now I will go,' said he. Then he paused for an answer, but none came. 'And you will understand what I meant when I spoke of being turned out.'

'Nobody – turns you out.' And Bell, as she spoke, had almost descended to a sob.

'But it is time that I should go; is it not? And, Bell, don't suppose that this little scene will keep me away from your sister's bedside. I shall be here to-morrow, and you will find that you will hardly know me again for the same person.' Then in the dark he put out his hand to her.

'Good-bye,' she said, giving him her hand. He pressed hers very closely, but she, though she wished to do so, could not bring herself to return the pressure. Her hand remained passive in his, showing no sign of offence; but it was absolutely passive.

'Good-bye, dearest friend,' he said.

'Good-bye,' she answered, – and then he was gone.

She waited quite still till she heard the front-door close after him, and then she crept silently up to her own bedroom, and sat herself down in a low rocking-chair over the fire. It was in accordance with a custom already established that her mother should remain with Lily till the tea was ready downstairs; for in these days of illness such dinners as were provided were eaten early. Bell, therefore, knew that she had still some half-hour of her own, during which she might sit and think undisturbed.

And what naturally should have been her first thoughts? – that she had ruthlessly refused a man who, as she now knew, loved her well, and for whom she had always felt the warmest friendship? Such were not her thoughts, nor were they in any way akin to this. They ran back instantly to years gone by, – over long years, as her few years were counted, – and settled themselves on certain halcyon days, in which she had dreamed that he had loved her, and had fancied that she had loved him. How she had schooled herself for those days since that, and taught herself to know that her thoughts had been overbold! And now it had all come round. The only man that she had ever liked had loved her. Then there came to her a memory of a certain day, in which she had been almost proud to think that Crosbie had admired her, in which she had almost hoped that it might be so; and as she thought of this she blushed, and struck her foot twice upon the floor. 'Dear Lily,' she said to herself – 'poor Lily!' But the feeling which induced her then to think of her sister had had no relation to that which had first brought Crosbie into her mind.

And this man had loved her through it all, – this priceless, peerless man, – this man who was as true to the backbone as that other man had shown himself to be false; who was as sound as the other man had proved himself to be rotten. A smile came across her face as she sat looking at the fire, thinking of this. A man had loved her, whose love was worth possessing. She hardly remembered whether or no she had refused him or accepted him. She hardly asked herself what she would do. As to all that it was necessary that she should have many thoughts, but the necessity did not press upon her quite immediately. For the present she might sit and triumph; – and thus triumphant she sat there till the old nurse came in and told her that her mother was waiting for her below.

CHAPTER XL

PREPARATIONS FOR THE WEDDING

THE fourteenth of February was finally settled as the day on which Mr. Crosbie was to be made the happiest of men. A later day had been at first named, the twenty-seventh or twenty-eighth having been suggested as an improvement over the first week in March; but Lady Amelia had been frightened by Crosbie's behaviour on that Sunday evening, and had made the countess understand that there should be no unnecessary delay. 'He doesn't scruple at that kind of thing,' Lady Amelia had said in one of her letters, showing perhaps less trust in the potency of her own rank than might have been expected from her. The countess, however, had agreed with her, and when Crosbie received from his mother-in-law a very affectionate epistle, setting forth all the reasons which would make the fourteenth so much more convenient a day than the twenty-eighth, he was unable to invent an excuse for not being made happy a fortnight earlier than the time named in the bargain. His first impulse had been against yielding, arising from some feeling which made him think that more than the bargain ought not to be exacted. But what was the use to him of quarrelling? What the use, at least, of quarrelling just then? He believed that he could more easily enfranchise himself from the De Courcy tyranny when he should be once married than he could do now. When Lady Alexandrina should be his own he would let her know that he intended to be her master. If in doing so it would be necessary that he should divide himself altogether from the De Courcys, such division should be made. At the present moment he would yield to them, at any rate in this matter. And so the fourteenth of February was fixed for the marriage.

In the second week in January Alexandrina came up to look after her things; or, in more noble language, to fit herself with becoming bridal appanages.* As she could not properly do all this work alone, or even under the surveillance and with the assistance of a sister, Lady De Courcy was to come up also. But Alexandrina came first, remaining with her sister in St. John's Wood till the countess should arrive. The countess had never yet condescended to accept of her son-in-law's hospitality, but always went to the cold, comfortless house in Portman Square, – the house which had been the De Courcy town family mansion for many years, and which the countess would long since have willingly exchanged for some abode on the other side of Oxford Street; but the earl had been obdurate; his clubs and certain lodgings which he had occasionally been wont to occupy, were on the right side of Oxford Street; why should he change his old family residence? So the countess was coming up to Portman Square, not having been even asked on this occasion to St. John's Wood.

'Don't you think we'd better,' Mr. Gazebee had said to his wife, almost trembling at the renewal of his own proposition.

'I think not, my dear,' Lady Amelia had answered. 'Mamma is not very particular; but there are little things, you know—'

'Oh, yes, of course,' said Mr. Gazebee; and then the conversation had been dropped. He would most willingly have entertained his august mother-in-law during her visit to the metropolis, and yet her presence in his house would have made him miserable as long as she remained there.

But for a week Alexandrina sojourned under Mr. Gazebee's roof, during which time Crosbie was made happy with all the delights of an expectant bridegroom. Of course he was given to understand that he was to dine at the Gazebees' every day, and spend all his evenings there; and, under the circumstances, he had no excuse for not doing so. Indeed, at the present moment, his hours would otherwise have hung heavily enough upon his hands. In spite of his bold resolution with reference to his eye, and his intention not to be debarred from the pleasures of society by the marks of the late combat, he had not, since that occurrence, frequented his club very closely; and though London was now again becoming fairly full, he did not find himself going out so much as had been his wont. The brilliance of his coming marriage did not seem to have added much to his popularity; in fact, the world, – his world, – was beginning to look coldly at him. Therefore that daily attendance at St. John's Wood was not felt to be so irksome as might have been expected.

A residence had been taken for the couple in a very fashionable row of buildings abutting upon the Bayswater Road, called Princess Royal Crescent. The house was quite new, and the street being unfinished had about it a strong smell of mortar, and a general aspect of builders' poles and brickbats; but, nevertheless, it was acknowledged to be a quite correct locality. From one end of the crescent a corner of Hyde Park could be seen, and the other abutted on a very handsome terrace indeed, in which lived an ambassador, – from South America, – a few bankers' senior clerks, and a peer of the realm. We know how vile is the sound of Baker Street, and how absolutely foul to the polite ear is the name of Fitzroy Square. The houses, however, in those purlieus are substantial, warm, and of good size. The house in Princess Royal Crescent was certainly not substantial, for in these days substantially-built houses do not pay. It could hardly have been warm, for, to speak the truth, it was even yet not finished throughout; and as for the size, though the drawing-room was a noble apartment, consisting of a section of the whole house, with a corner cut out for the staircase, it was very much cramped in its other parts, and was made like a cherub, in this respect, that it had no rear belonging to it. 'But if you have no private fortune of your own, you cannot have everything,' as the countess observed when Crosbie objected to the house because a closet under the staircase was to be assigned to him as his own dressing-room.

When the question of the house was first debated, Lady Amelia had been anxious that St. John's Wood should be selected as the site, but to this Crosbie had positively objected.

'I think you don't like St. John's Wood,' Lady Amelia had said to him somewhat sternly, thinking to awe him into a declaration that he entertained no general enmity to the neighbourhood. But Crosbie was not weak enough for this.

'No; I do not,' he said. 'I have always disliked it. It amounts to a prejudice, I daresay. But if I were made to live here I am convinced I should cut my throat in the first six months.'

Lady Amelia had then drawn herself up, declaring her sorrow that her house should be so hateful to him.

'Oh, dear, no,' said he. 'I like it very much for you, and enjoy coming here of all things. I speak only of the effect which living here myself would have upon me.'

Lady Amelia was quite clever enough to understand it all; but she had her sister's interest at heart, and therefore persevered in her affectionate solicitude for her brother-in-law, giving up that point as to St. John's Wood. Crosbie himself had wished to go to one of the new Pimlico squares down near Vauxhall Bridge and the river, actuated chiefly by consideration of the enormous distance lying between that locality and the northern region in which Lady Amelia lived; but to this Lady Alexandrina had objected strongly. If, indeed, they could have achieved Eaton Square, or a street leading out of Eaton Square, – if they could have crept on to the hem of the skirt of Belgravia, – the bride would have been delighted. And at first she was very nearly being taken in with the idea that such was the proposal made to her. Her geographical knowledge of Pimlico had not been perfect, and she had nearly fallen into a fatal error. But a friend had kindly intervened. 'For heaven's sake, my dear, don't let him take you anywhere beyond Eccleston Square!' had been exclaimed to her in dismay by a faithful married friend. Thus warned, Alexandrina had been firm, and now their tent was to be pitched in Princess Royal Crescent, from one end of which Hyde Park may be seen.

The furniture had been ordered chiefly under the inspection, and by the experience, of the Lady Amelia. Crosbie had satisfied himself by declaring that she could get the things cheaper than he could buy them, and that he had not taste for such employment. Nevertheless, he had felt that he was being made subject to tyranny and brought under the thumb of subjection. He could not go cordially into this matter of beds and chairs, and, therefore, at last deputed the whole matter to the De Courcy faction. And for this there was another reason, not hitherto mentioned. Mr. Mortimer Gazebee was finding the money with which all the furniture was being bought. He, with an honest but almost unintelligible zeal for the De Courcy family, had tied up every shilling on which he could lay his hand as belonging to Crosbie, in the interest of Lady Alexandrina. He had gone to work for her, scraping here and arranging there, strapping the new husband down upon the grindstone of his matrimonial settlement, as though the future bread of his, Gazebee's, own children were dependent on the validity of his legal workmanship. And for this he was not to receive a penny, or gain any advantage, immediate or ulterior. It came from his zeal, – his zeal for the coronet which Lord De Courcy wore. According to his mind an earl and an earl's belongings were entitled to such zeal. It was the theory in which he had been educated, and amounted to a worship which, unconsciously, he practised. Personally, he disliked Lord De Courcy, who ill-treated him. He knew that the earl was a heartless, cruel, bad man. But as an earl he was entitled to an amount of service which no commoner could have commanded from Mr. Gazebee. Mr. Gazebee, having thus tied up all the available funds in favour of Lady Alexandrina's seemingly expected widowhood, was himself providing the money with which the new house was to be furnished. 'You can pay me a hundred and fifty a year with four per cent. till it is liquidated,' he had said to Crosbie; and Crosbie had assented with a grunt. Hitherto, though he had lived in London expensively, and as a man of fashion, he had never owed any one anything. He was now to begin that career of owing. But when a clerk in a public office marries an earl's daughter, he cannot expect to have everything his own way.

Lady Amelia had bought the ordinary furniture – the beds, the stair-carpets, the washing-stands, and the kitchen things. Gazebee had got a bargain of the dinner-table and sideboard. But Lady Alexandrina herself was to come up with reference to the appurtenances of the drawing-room. It was with reference to matters of costume that the countess intended to lend her assistance – matters of costume as to which the bill could not be sent in to Gazebee, and be paid for by him with five per cent. duly charged against the bridegroom. The bridal trousseau must be produced by De Courcy's means, and, therefore, it was necessary that the countess herself should come upon the scene. 'I will have no bills, d'ye hear?' snarled the earl, gnashing and snapping upon his words with one specially ugly black tooth. 'I won't have any bills about this affair.' And yet he made no offer of ready money. It was very necessary under such circumstances that the countess herself should come upon the scene. An ambiguous hint had been conveyed to Mr. Gazebee, during a visit of business which he had lately made to Courcy Castle, that the milliner's bills might as well be pinned on to those of the furniture-makers, the crockery-mongers, and the like. The countess, putting it in her own way, had gently suggested that the fashion of the thing had changed lately, and that such an arrangement was considered to be the proper thing among people who lived really in the world. But Gazebee was a clear-headed, honest man; and he knew the countess. He did not think that such an arrangement could be made on the present occasion. Whereupon the countess pushed her suggestion no further, but made up her mind that she must come up to London herself.

It was pleasant to see the Ladies Amelia and Alexandrina, as they sat within a vast emporium of carpets in Bond Street, asking questions of the four men who were waiting upon them, putting their heads together and whispering, calculating accurately as to extra twopences a yard, and occasioning as much trouble as it was possible for them to give. It was pleasant because they managed their large hoops cleverly among the huge rolls of carpets, because they were enjoying themselves thoroughly, and taking to themselves the homage of the men as clearly their due. But it was not so pleasant to look at Crosbie, who was fidgeting to get away to his office, to whom no power of choosing in the matter was really given, and whom the men regarded as being altogether supernumerary. The ladies had promised to be at the shop by half-past ten, so that Crosbie should reach his office at eleven – or a little after. But it was nearly eleven before they left the Gazebee residence, and it was very evident that half-an-hour among the carpets would be by no means sufficient. It seemed as though miles upon miles of gorgeous colouring were unrolled before them; and then when any pattern was regarded as at all practicable, it was unrolled backwards and forwards till a room was nearly covered by it. Crosbie felt for the men who were hauling about the huge heaps of material; but Lady Amelia sat as composed as though it were her duty to inspect every yard of stuff in the warehouse. 'I think we'll look at that one at the bottom again.' Then the men went to work and removed a mountain. 'No, my dear, that green in the scroll-work won't do. It would fly directly, if any hot water were spilt.' The man smiling ineffably, declared that that particular green never flew anywhere. But Lady Amelia paid no attention to him, and the carpet for which the mountain had been removed became part of another mountain.

'That might do,' said Alexandrina, gazing upon a magnificent crimson ground through which rivers of yellow meandered, carrying with them in their streams an infinity of blue flowers. And as she spoke she held her head gracefully on one side, and looked down upon the carpet doubtingly. Lady Amelia poked it with her parasol as though to test its durability, and whispered something about yellows showing the dirt. Crosbie took out his watch and groaned.

'It's a superb carpet, my lady, and about the newest thing we have. We put down four hundred and fifty yards of it for the Duchess of South Wales, at Cwddglwlch Castle, only last month. Nobody has had it since, for it has not been in stock.' Whereupon Lady Amelia again poked it, and then got up and walked upon it. Lady Alexandrina held her head a little more on one side.

'Five and three?' said Lady Amelia.

'Oh, no, my lady; five and seven; and the cheapest carpet we have in the house. There is twopence a yard more in the colour; there is, indeed.'

'And the discount?' asked Lady Amelia.

'Two and a half, my lady.'

'Oh dear, no,' said Lady Amelia. 'I always have five per cent. for immediate payment – quite immediate, you know.' Upon which the man declared the question must be referred to his master. Two and a half was the rule of the house. Crosbie, who had been looking out of the window, said that upon his honour he couldn't wait any longer.

'And what do you think of it, Adolphus?' asked Alexandrina.

'Think of what?'

'Of the carpet – this one, you know?'

'Oh – what do I think of the carpet? I don't think I quite like all these yellow bands; and isn't it too red? I should have thought something brown with a small pattern would have been better. But, upon my word, I don't much care.'

'Of course he doesn't,' said Lady Amelia. Then the two ladies put their heads together for another five minutes, and the carpet was chosen – subject to that question of the discount. 'And now about the rug,' said Lady Amelia. But here Crosbie rebelled, and insisted that he must leave them and go to his office. 'You can't want me about the rug,' he said. 'Well, perhaps not,' said Lady Amelia. But it was manifest that Alexandrina did not approve of being thus left by her male attendant.

The same thing happened in Oxford Street with reference to the chairs and sofas, and Crosbie began to wish that he were settled, even though he should have to dress himself in the closet under the staircase. He was learning to hate the whole household in St. John's Wood, and almost all that belonged to it. He was introduced there to little family economies of which hitherto he had known nothing, and which were disgusting to him, and the necessity for which was especially explained to him. It was to men placed as he was about to place himself that these economies were so vitally essential – to men who with limited means had to maintain a decorous outward face towards the fashionable world. Ample supplies of butchers' meat and unlimited washing-bills might be very well upon fifteen hundred a year to those who went out but seldom, and who could use the first cab that came to hand when they did go out. But there were certain things that Lady Alexandrina must do, and therefore the strictest household economy became necessary. Would Lily Dale have required the use of a carriage, got up to look as though it were private, at the expense of her husband's beefsteaks and clean shirts? That question and others of that nature were asked by Crosbie within his own mind, not unfrequently.

But, nevertheless, he tried to love Alexandrina, or rather to persuade himself that he loved her. If he could only get her away from the De Courcy faction, and especially from the Gazebee branch of it, he would break her of all that. He would teach her to sit triumphantly in a street cab, and to cater for her table with a plentiful hand. Teach her! – at some age over thirty; and with such careful training as she had already received! Did he intend to forbid her ever again to see her relations, ever to go to St. John's Wood, or to correspond with the countess and Lady Margaretta? Teach her, indeed! Had he yet to learn that he could not wash a blackamoor white? – that he could not have done so even had he himself been well adapted for the attempt, whereas he was in truth nearly as ill adapted as a man might be? But who could pity him? Lily, whom he might have had in his bosom, would have been no blackamoor.

Then came the time of Lady De Courcy's visit to town, and Alexandrina moved herself off to Portman Square. There was some apparent comfort in this to Crosbie, for he would thereby be saved from those daily dreary journeys up to the north-west. I may say that he positively hated that windy corner near the church, round which he had to walk in getting to the Gazebee residence, and that he hated the lamp which guided him to the door, and the very door itself. This door stood buried as it were in a wall, and opened on to a narrow passage which ran across a so-called garden, or front yard, containing on each side two iron receptacles for geraniums, painted to look like Palissy ware,* and a naked female on a pedestal. No spot in London was, as he thought, so cold as the bit of pavement immediately in front of that door. And there he would be kept five, ten, fifteen minutes as he declared – though I believe in my heart that the time never exceeded three, – while Richard was putting off the trappings of his work and putting on the trappings of his grandeur.

If people would only have their doors opened to you by such assistance as may come most easily and naturally to the work! I stood lately for some minutes on a Tuesday afternoon at a gallant portal, and as I waxed impatient a pretty maiden came and opened it. She was a pretty maiden, though her hands and face and apron told tales of the fire-grates. 'Laws, sir,' she said, 'the visitors' day is Wednesday; and if you would come then, there would be the man in livery!' She took my card with the corner of her apron, and did just as well as the man in livery; but what would have happened to her had her little speech been overheard by her mistress?

Crosbie hated the house in St. John's Wood, and therefore the coming of the countess was a relief to him. Portman Square was easily to be reached, and the hospitalities of the countess would not be pressed upon him so strongly as those of the Gazebees. When he first called he was shown into the great family dining-room, which looked out towards the back of the house. The front windows were, of course, closed, as the family was not supposed to be in London. Here he remained in the room for some quarter of an hour, and then the countess descended upon him in all her grandeur. Perhaps he had never before seen her so grand. Her dress was very large, and rustled through the broad doorway, as if demanding even a broader passage. She had on a wonder of a bonnet, and a velvet mantle that was nearly as expansive as her petticoats. She threw her head a little back as she accosted him, and he instantly perceived that he was enveloped in the fumes of an affectionate, but somewhat contemptuous patronage. In old days he had liked the countess, because her manner to him had always been flattering. In his intercourse with her he had been able to feel that he gave quite as much as he got, and that the countess was aware of the fact. In all the circumstances of their acquaintance the ascendancy had been with him, and therefore the acquaintance had been a pleasant one. The countess had been a good-natured, agreeable woman, whose rank and position had made her house pleasant to him; and therefore he had consented to shine upon her with such a light as he had to give. Why was it that the matter was reversed, now that there was so much stronger a cause for good feeling between them? He knew that there was such change, and with bitter internal upbraidings he acknowledged to himself that this woman was getting the mastery over him. As the friend of the countess he had been a great man in her eyes; – in all her little words and looks she had acknowledged his power; but now, as her son-in-law, he was to become a very little man, – such as was Mortimer Gazebee!

'My dear Adolphus,' she said, taking both his hands, 'the day is coming very near now; is it not?'

'Very near, indeed,' he said.

'Yes, it is very near. I hope you feel yourself a happy man.'

'Oh, yes, that's of course.'

'It ought to be. Speaking very seriously, I mean that it ought to be a matter of course. She is everything that a man should desire in a wife. I am not alluding now to her rank, though of course you feel what a great advantage she gives you in this respect.'

Crosbie muttered something as to his consciousness of

having drawn a prize in the lottery; but he so muttered it as not to convey to the lady's ears a proper sense of his dependent gratitude. 'I know of no man more fortunate than you have been,' she continued; 'and I hope that my dear girl will find that you are fully aware that it is so. I think that she is looking rather fagged. You have allowed her to do more than was good for her in the way of shopping.'

'She has done a good deal, certainly,' said Crosbie.

'She is so little used to anything of that kind! But of course, as things have turned out, it was necessary that she should see to these things herself.'

'I rather think she liked it,' said Crosbie.

'I believe she will always like doing her duty. We are just going now to Madame Millefranc's, to see some silks; – perhaps you would wish to go with us?'

Just at this moment Alexandrina came into the room, and looked as though she were in all respects a smaller edition of her mother. They were both well-grown women, with handsome, large figures, and a certain air about them which answered almost for beauty. As to the countess, her face, on close inspection, bore, as it was entitled to do, deep signs of age; but she so managed her face that any such close inspection was never made; and her general appearance for her time of life was certainly good. Very little more than this could be said in favour of her daughter.

'Oh, dear, no, mamma,' she said, having heard her mother's last words. 'He's the worst person in a shop in the world. He likes nothing, and dislikes nothing. Do you, Adolphus?'

'Indeed I do. I like all the cheap things, and dislike all the dear things.'

'Then you certainly shall not go with us to Madame Millefranc's,' said Alexandrina.

'It would not matter to him there, you know, my dear,' said the countess, thinking perhaps of the suggestion she had lately made to Mr. Gazebee.

On this occasion Crosbie managed to escape, simply promising to return to Portman Square in the evening after dinner. 'By-the-by, Adolphus,' said the countess, as he handed her into the hired carriage which stood at the door, 'I wish you would go to Lambert's, on Ludgate Hill, for me. He has had a bracelet of mine for nearly three months. Do, there's a good creature. Get it if you can, and bring it up this evening.'

Crosbie, as he made his way back to his office, swore that he would not do the bidding of the countess. He would not trudge off into the city after her trinkets. But at five o'clock, when he left his office, he did go there. He apologized to himself by saying that he had nothing else to do, and bethought himself that at the present moment his lady mother-in-law's smiles might be more convenient than her frowns. So he went to Lambert's, on Ludgate Hill, and there learned that the bracelet had been sent down to Courcy Castle full two months since.

After that he dined at his club, at Sebright's. He dined alone, sitting by no means in bliss with his half-pint of sherry on the table before him. A man now and then came up and spoke to him, one a few words, and another a few, and two or three congratulated him as to his marriage; but the club was not the same thing to him as it had formerly been. He did not stand in the centre of the rug, speaking indifferently to all or any around him, ready with his joke, and loudly on the alert with the last news of the day. How easy it is to be seen when any man has fallen from his pride of place, though the altitude was ever so small, and the fall ever so slight. Where is the man who can endure such a fall without showing it in his face, in his voice, in his step, and in every motion of every limb? Crosbie knew that he had fallen, and showed that he knew it by the manner in which he ate his mutton-chop.

At half-past eight he was again in Portman Square, and found the two ladies crowding over a small fire in a small back drawing-room. The furniture was all covered with brown holland, and the place had about it that cold comfortless feeling which uninhabited rooms always produce. Crosbie, as he had walked from the club up to Portman Square, had indulged in some serious thoughts. The kind of life which he had hitherto led had certainly passed away from him. He could never again be the pet of a club, or indulged as one to whom all good things were to be given without any labour at earning them on his own part. Such for some years had been his good fortune, but such could be his good fortune no longer. Was there anything within his reach which he might take in lieu of that which he had lost? He might still be victorious at his office, having more capacity for such victory than others around him. But such success alone would hardly suffice for him. Then he considered whether he might not even yet be happy in his own home, – whether Alexandrina, when separated from her mother, might not become such a wife as he could love. Nothing softens a man's feelings so much as failure, or makes him turn so anxiously to an idea of home as buffetings from those he meets abroad. He had abandoned Lily because his outer world had seemed to him too bright to be deserted. He would endeavour to supply her place with Alexandrina, because his outer world had seemed to him too harsh to be supported. Alas! alas! a man cannot so easily repent of his sins, and wash himself white from their stains!

When he entered the room the two ladies were sitting over the fire, as I have stated, and Crosbie could immediately perceive that the spirit of the countess was not serene. In fact there had been a few words between the mother and child on that matter of the trousseau, and Alexandrina had plainly told her mother that if she were to be married at all she would be married with such garments belonging to her as were fitting for an earl's daughter. It was in vain that her mother had explained with many circumlocutional phrases, that the fitness in this respect should be accommodated rather to the plebeian husband than to the noble parent. Alexandrina had been very firm, and had insisted on her rights, giving the countess to understand that if her orders for finery were not complied with, she would return as a spinster to Courcy, and prepare herself for partnership with Rosina.

'My dear,' said the countess, piteously, 'you can have no idea of what I shall have to go through with your father. And, of course, you could get all these things afterwards.'

'Papa has no right to treat me in such a way. And if he would not give me any money himself, he should have let me have some of my own.'

'Ah, my dear, that was Mr. Gazebee's fault.'

'I don't care whose fault it was. It certainly was not mine. I won't have him to tell me' – him was intended to signify Adolphus Crosbie – 'that he had to pay for my wedding-clothes.'

'Of course not that, my dear.'

'No; nor yet the things which I wanted immediately. I'd much rather go and tell him at once that the marriage must be put off.'

Alexandrina of course carried her point, the countess reflecting with a maternal devotion equal almost to that of the pelican,* that the earl could not do more than kill her. So the things were ordered as Alexandrina chose to order them, and the countess desired that the bills might be sent in to Mr. Gazebee. Much self-devotion had been displayed by the mother, but the mother thought that none had been displayed by the daughter, and therefore she had been very cross with Alexandrina.

Crosbie, taking a chair, sat himself between them, and in a very good-humoured tone explained the little affair of the bracelet. 'Your ladyship's memory must have played you false,' said he, with a smile.

'My memory is very good,' said the countess; 'very good indeed. If Twitch got it, and didn't tell me, that was not my fault.' Twitch was her ladyship's lady's-maid. Crosbie, seeing how the land lay, said nothing more about the bracelet.

After a minute or two he put out his hand to take that of Alexandrina. They were to be married now in a week or two, and such a sign of love might have been allowed to him, even in the presence of the bride's mother. He did succeed in getting hold of her fingers, but found in them none of the softness of a response. 'Don't,' said Lady Alexandrina, withdrawing her hand; and the tone of her voice as she spoke the word was not sweet to his ears. He remembered at the moment a certain scene which took place one evening at the little bridge at Allington, and Lily's voice, and Lily's words, and Lily's passion, as he caressed her: 'Oh, my love, my love, my love!'

'My dear,' said the countess, 'they know how tired I am. I wonder whether they are going to give us any tea.' Whereupon Crosbie rang the bell, and, on resuming his chair, moved it a little farther away from his lady-love.

Presently the tea was brought to them by the housekeeper's assistant, who did not appear to have made herself very smart for the occasion, and Crosbie thought that he was *de trop.* This, however, was a mistake on his part. As he had been admitted into the family, such little matters were no longer subject of care. Two or three months since, the countess would have fainted at the idea of such a domestic appearing with a tea-tray before Mr. Crosbie. Now, however, she was utterly indifferent to any such consideration. Crosbie was to be admitted into the family, thereby becoming entitled to certain privileges, – and thereby also becoming subject to certain domestic drawbacks. In Mrs. Dale's little household there had been no rising to grandeur; but then, also, there had never been any bathos of dirt. Of this also Crosbie thought as he sat with his tea in his hand.

He soon, however, got himself away. When he rose to go Alexandrina also rose, and he was permitted to press his nose against her cheekbone by way of a salute.

'Good-night, Adolphus,' said the countess, putting out her hand to him. 'But stop a minute; I know there is something I want you to do for me. But you will look in as you go to your office to-morrow morning.'

CHAPTER XLI

DOMESTIC TROUBLES

WHEN Crosbie was making his ineffectual inquiry after Lady De Courcy's bracelet at Lambert's, John Eames was in the act of entering Mrs. Roper's front door in Burton Crescent.

'Oh, John, where's Mr. Cradell?' were the first words which greeted him, and they were spoken by the divine Amelia.

Now, in her usual practice of life, Amelia did not interest herself much as to the whereabouts of Mr. Cradell.

'Where's Cradell?' said Eames, repeating the question. 'Upon my word, I don't know. I walked to the office with him, but I haven't seen him since. We don't sit in the same room, you know.'

'John!' and then she stopped.

'What's up now?' said John.

'John! That woman's off and left her husband. As sure as your name's John Eames, that foolish fellow has gone off with her.'

'What, Cradell? I don't believe it.'

'She went out of this house at two o'clock in the afternoon, and has never been back since.' That, certainly, was only four hours from the present time, and such an absence from home in the middle of the day was but weak evidence on which to charge a married woman with the great sin of running off with a lover. This Amelia felt, and therefore she went on to explain. 'He's there upstairs in the drawing-room, the very picture of disconsolateness.'

'Who, – Cradell?'

'Lupex is. He's been drinking a little, I'm afraid; but he's very unhappy, indeed. He had an appointment to meet his wife here at four o'clock, and when he came he found her gone. He rushed up into their room, and now he says she has broken open a box he had and taken off all his money.'

'But he never had any money.'

'He paid mother some the day before yesterday.'

'That's just the reason he shouldn't have any to-day.'

'She certainly has taken things she wouldn't have taken if she'd merely gone out shopping or anything like that, for I've been up in the room and looked about it. She'd three necklaces. They weren't much account; but she must have them all on, or else have got them in her pocket.'

'Cradell has never gone off with her in that way. He may be a fool—'

'Oh, he is, you know. I've never seen such a fool about a woman as he has been.'

'But he wouldn't be a party to stealing a lot of trumpery

trinkets, or taking her husband's money. Indeed, I don't think he has anything to do with it.' Then Eames thought over the circumstances of the day, and remembered that he had certainly not seen Cradell since the morning. It was that public servant's practice to saunter into Eames's room in the middle of the day, and there consume bread and cheese and beer, – in spite of an assertion which Johnny had once made as to crumbs of biscuit bathed in ink. But on this special day he had not done so. 'I can't think he has been such a fool as that,' said Johnny.

'But he has,' said Amelia. 'It's dinner-time now, and where is he? Had he any money left, Johnny?'

So interrogated, Eames disclosed a secret confided to him by his friend which no other circumstances would have succeeded in dragging from his breast.

'She borrowed twelve pounds from him about a fortnight since immediately after quarter-day. And she owed him money, too, before that.'

'Oh, what a soft!' exclaimed Amelia; 'and he hasn't paid mother a shilling for the last two months!'

'It was his money, perhaps, that Mrs. Roper got from Lupex the day before yesterday. If so, it comes to the same thing as far as she is concerned, you know.'

'And what are we to do now?' said Amelia, as she went before her lover upstairs. 'Oh, John, what will become of me if ever you serve me in that way? What would I do if you were to go off with another lady?'

'Lupex hasn't gone off,' said Eames, who hardly knew what to say when the matter was brought before him with so closely personal a reference.

'But it's the same thing,' said Amelia. 'Hearts is divided. Hearts that have been joined together ought never to be divided; ought they?' And then she hung upon his arm just as they got to the drawing-room door.

'Hearts and darts are all my eye,' said Johnny. 'My belief is that a man had better not marry at all. How d'you do, Mr. Lupex? Is anything the matter?'

Mr. Lupex was seated on a chair in the middle of the room, and was leaning with his head over the back of it. So despondent was he in his attitude that his head would have fallen off and rolled on to the floor, had it followed the course which its owner seemed to intend that it should take. His hands hung down also along the back legs of the chair, till his fingers almost touched the ground, and altogether his appearance was pendent, drooping, and woe-begone. Miss Spruce was seated in one corner of the room, with her hands folded in her lap before her, and Mrs. Roper was standing on the rug with a look of severe virtue on her brow, – of virtue which, to judge by its appearance, was very severe. Nor was its severity intended to be exercised solely against Mrs. Lupex. Mrs. Roper was becoming very tired of Mr. Lupex also, and would not have been unhappy if he also had run away, – leaving behind him so much of his property as would have paid his bill.

Mr. Lupex did not stir when first addressed by John Eames, but a certain convulsive movement was to be seen on the back of his head, indicating that this new arrival in the drawing-room had produced a fresh occasion of agony. The chair, too, quivered under him, and his fingers stretched themselves nearer to the ground and shook themselves.

'Mr. Lupex, we're going to dinner immediately,' said Mrs. Roper. 'Mr. Eames, where is your friend, Mr. Cradell?'

'Upon my word I don't know,' said Eames.

'But I know,' said Lupex, jumping up and standing at his full height, while he knocked down the chair which had lately supported him. 'The traitor to domestic bliss! I know. And wherever he is, he has that false woman in his arms. Would he were here!' And as he expressed the last wish he went through a motion with his hands and arms which seemed intended to signify that if that unfortunate young man were in the company he would pull him in pieces and double him up, and pack him close, and then despatch his remains off, through infinite space, to the Prince of Darkness. 'Traitor,' he exclaimed, as he finished the process, 'False traitor! Foul traitor! And she too!' Then, as he thought of this softer side of the subject, he prepared himself to relapse again on to the chair. Finding it on the ground he had to pick it up. He did pick it up, and once more flung away his head over the back of it, and stretched his finger-nails almost down to the carpet.

'James,' said Mrs. Roper to her son, who was now in the room, 'I think you'd better stay with Mr. Lupex while we are at dinner. Come, Miss Spruce, I'm very sorry that you should be annoyed by this kind of thing.'

'It don't hurt me,' said Miss Spruce, preparing to leave the room. 'I'm only an old woman.'

'Annoyed!' said Lupex, raising himself again from his chair, not perhaps altogether disposed to remain upstairs while the dinner, for which it was intended that he should some day pay, was being eaten below. 'Annoyed! It is a profound sorrow to me that any lady should be annoyed by my misfortunes. As regards Miss Spruce, I look upon her character with profound veneration.'

'You needn't mind me; I'm only an old woman,' said Miss Spruce.

'But, by heavens, I do mind!' exclaimed Lupex; and hurrying forward he seized Miss Spruce by the hand. 'I shall always regard age as entitled—' But the special privileges which Mr. Lupex would have accorded to age were never made known to the inhabitants of Mrs. Roper's boarding-house, for the door of the room was again opened at this moment, and Mr. Cradell entered.

'Here you are, old fellow, to answer for yourself,' said Eames.

Cradell, who had heard something as he came in at the front door, but had not heard that Lupex was in the drawing-room, made a slight start backwards when he saw that gentleman's face. 'Upon my word and honour,' he began; – but he was able to carry his speech no further. Lupex, dropping the hand of the elderly lady whom he reverenced, was upon him in an instant, and Cradell was shaking beneath his grasp like an aspen leaf, – or rather not like an aspen leaf, unless an aspen leaf when shaken is to be seen with its eyes shut, its mouth open, and its tongue hanging out.

'Come, I say,' said Eames, stepping forward to his friend's assistance; 'this won't do at all, Mr. Lupex. You've been drinking. You'd better wait till to-morrow morning, and speak to Cradell then.'

'To-morrow morning, viper,' shouted Lupex, still holding his prey, but looking back at Eames over his shoulder. Who the viper was had not been clearly indicated. 'When will he restore to me my wife? When will he restore to me my honour?'

'Upon-on-on-on my—' It was for the moment in vain that poor Mr. Cradell endeavoured to asseverate his innocence, and to stake his honour upon his own purity as regarded Mrs. Lupex. Lupex still held to his enemy's cravat, though Eames had now got him by the arm, and so far impeded his movements as to hinder him from proceeding to any graver attack.

'Jemima, Jemima, Jemima!' shouted Mrs. Roper. 'Run for the police; run for the police!' But Amelia, who had far more presence of mind than her mother, stopped Jemima as she was making to one of the front windows. 'Keep where you are,' said Amelia. 'They'll come quiet in a minute or two.' And Amelia no doubt was right. Calling for the police when there is a row in the house is like summoning the water-engines when the soot is on fire in the kitchen chimney. In such cases good management will allow the soot to burn itself out, without aid from the water-engines. In the present instance the police were not called in, and I am inclined to think that their presence would not have been advantageous to any of the party.

'Upon-my-honour – I know nothing about her,' were the first words which Cradell was able to articulate, when Lupex, under Eames's persuasion, at last relaxed his hold.

Lupex turned round to Miss Spruce with a sardonic grin. 'You hear his words, – this enemy to domestic bliss, – Ha, ha! man, tell me whither you have conveyed my wife!'

'If you were to give me the Bank of England I don't know,' said Cradell.

'And I'm sure he does not know,' said Mrs. Roper, whose suspicions against Cradell were beginning to subside. But as her suspicions subsided, her respect for him decreased. Such was the case also with Miss Spruce, and with Amelia, and with Jemima. They had all thought him to be a great fool for running away with Mrs. Lupex, but now they were beginning to think him a poor creature because he had not done so. Had he committed that active folly he would have been an interesting fool. But now, if as they all suspected, he knew no more about Mrs. Lupex than they did, he would be a fool without any special interest whatever.

'Of course he doesn't,' said Eames.

'No more than I do,' said Amelia.

'His very looks show him innocent,' said Mrs. Roper.

'Indeed they do,' said Miss Spruce.

Lupex turned from one to the other as they thus defended the man whom he suspected, and shook his head at each assertion that was made. 'And if he doesn't know, who does?' he asked. 'Haven't I seen it all for the last three months? Is it reasonable to suppose that a creature such as she, used to domestic comforts all her life, should have gone off in this way, at dinner-time, taking with her my property and all her jewels, and that nobody should have instigated her; nobody assisted her? Is that a story to tell to such a man as me? You may tell it to the marines!' Mr. Lupex, as he made this speech, was walking about the room, and as he finished it he threw his pocket-handkerchief with violence on to the floor. 'I know what to do, Mrs. Roper,' he said. 'I know what steps to take. I shall put the affair into the hands of my lawyer to-morrow morning.' Then he picked up his handkerchief and walked down into the dining-room.

'Of course you know nothing about it?' said Eames to his friend, having run upstairs for the purpose of saying a word to him while he washed his hands.

'What, – about Maria? I don't know where she is, if you mean that.'

'Of course I mean that. What else should I mean? And what makes you call her Maria?'

'It is wrong. I admit it's wrong. The word will come out, you know.'

'Will come out! I'll tell you what it is, old fellow, you'll get yourself into a mess, and all for nothing. That fellow will have you up before the police for stealing his things—'

'But, Johnny—'

'I know all about it. Of course you have not stolen them, and of course there was nothing to steal. But if you go on calling her Maria you'll find that he'll have a pull on you. Men don't call other men's wives names for nothing.'

'Of course we've been friends,' said Cradell, who rather liked this view of the matter.

'Yes, – you have been friends! She's diddled you out of your money, and that's the beginning and the end of it. And now, if you go on showing off your friendship, you'll be out of more money. You're making an ass of yourself. That's the long and the short of it.'

'And what have you made of yourself with that girl? There are worse asses than I am yet, Master Johnny.' Eames, as he had no answer ready to this counter attack, left the room and went downstairs. Cradell soon followed him, and in a few minutes they were all eating their dinner together at Mrs. Roper's hospitable table.

Immediately after dinner Lupex took himself away, and the conversation upstairs became general on the subject of the lady's departure.

'If I was him I'd never ask a question about her, but let her go,' said Amelia.

'Yes; and then have all her bills following you, wherever you went,' said Amelia's brother.

'I'd sooner have her bills than herself,' said Eames.

'My belief is, that she's been an ill-used woman,' said Cradell. 'If she had a husband that she could respect and have loved, and all that sort of thing, she would have been a charming woman.'

'She's every bit as bad as he is,' said Mrs. Roper.

'I can't agree with you, Mrs. Roper,' continued the lady's champion. 'Perhaps I ought to understand her position better than any one here, and—'

'Then that's just what you ought not to do, Mr. Cradell,' said Mrs. Roper. And now the lady of the house spoke out her mind with much maternal dignity and with some feminine severity. 'That's just what a young man like you has no business to know. What's a married woman like that to you, or you to her; or what have you to do with understanding her position? When you've a wife of your own, if ever you do have one, you'll find you'll have trouble enough then without anybody else interfering with you. Not but what I believe you're innocent as a lamb about Mrs. Lupex; that is, as far as any harm goes. But you've got yourself into all this trouble by meddling, and was like enough to get yourself choked upstairs by that man. And who's to wonder when you go on pretending to be in love with a woman in that way, and she old enough to be your mother? What would your mamma say if she saw you at it?'

'Ha, ha, ha!' laughed Cradell.

'It's all very well your laughing, but I hate such folly. If I see a young man in love with a young woman, I respect him for it;' and then she looked at Johnny Eames. 'I respect him for it, – even though he may now and then do things as he shouldn't. They most of 'em does that. But to see a young man like you, Mr. Cradell, dangling after an old married woman, who doesn't know how to behave herself; and all just because she lets him do it; – ugh! – an old broomstick with a petticoat on would do just as well. It makes me sick to see it, and that's the truth of it. I don't call it manly; and it ain't manly, is it, Miss Spruce?'

'Of course I know nothing about it,' said the lady to whom the appeal was thus made. 'But a young gentleman should keep himself to himself till the time comes to speak out, – begging your pardon all the same, Mr. Cradell.'

'I don't see what a married woman should want with any one after her but her own husband,' said Amelia.

'And perhaps not always that,' said John Eames.

It was about an hour after this when the front-door bell was rung, and a scream from Jemima announced to them all that some critical moment had arrived. Amelia, jumping up, opened the door, and then the rustle of a woman's dress was heard on the lower stairs. 'Oh, laws, ma'am, you have given us sich a turn,' said Jemima. 'We all thought you was run away.'

'It's Mrs. Lupex,' said Amelia. And in two minutes more that ill-used lady was in the room.

'Well, my dears,' said she, gaily, 'I hope nobody has waited dinner.'

'No; we didn't wait dinner,' said Mrs. Roper, very gravely.

'And where's my Orson? Didn't he dine at home? Mr. Cradell, will you oblige me by taking my shawl? But perhaps you had better not. People are so censorious; ain't they, Miss Spruce? Mr. Eames shall do it; and everybody knows that that will be quite safe. Won't it, Miss Amelia?'

'Quite, I should think,' said Amelia. And Mrs. Lupex knew that she was not to look for an ally in that quarter on the present occasion. Eames got up to take the shawl, and Mrs. Lupex went on.

'And didn't Orson dine at home? Perhaps they kept him down at the theatre. But I've been thinking all day what fun it would be when he thought his bird was flown.'

'He did dine at home,' said Mrs. Roper; 'and he didn't seem to like it. There wasn't much fun, I can assure you.'

'Ah, wasn't there, though? I believe that man would like to have me tied to his button-hole. I came across a few friends, – lady friends, Mr. Cradell, though two of them had their husbands; so we made a party, and just went down to Hampton Court. So my gentleman has gone again, has he? That's what I get for gadding about myself, isn't it, Miss Spruce?'

Mrs. Roper, as she went to bed that night, made up her mind that, whatever might be the cost and trouble of doing so, she would lose no further time in getting rid of her married guests.

CHAPTER XLII

LILY'S BEDSIDE

Lily Dale's constitution was good, and her recovery was retarded by no relapse or lingering debility; but, nevertheless, she was forced to keep her bed for many days after the fever had left her. During all this period Dr. Crofts came every day. It was in vain that Mrs. Dale begged him not to do so; telling him in simple words that she felt herself bound not to accept from him all this continuation of his unremunerated labours now that the absolute necessity for them was over. He answered her only by little jokes, or did not answer her at all; but still he came daily, almost always at the same hour, just as the day was waning, so that he could sit for a quarter of an hour in the dusk, and then ride home to Guestwick in the dark. At this time Bell had been admitted into her sister's room, and she would always meet Dr. Crofts at Lily's bedside; but she never sat with him alone, since the day on which he had offered her his love with half-articulated words, and she had declined it with words also half-articulated. She had seen him alone since that, on the stairs, or standing in the hall, but she had not remained with him, talking to him after her old fashion, and no further word of his love had been spoken in speech either half or wholly articulate.

Nor had Bell spoken of what had passed to any one else. Lily would probably have told both her mother and sister instantly; but then no such scene as that which had taken place with Bell would have been possible with Lily. In whatever way the matter might have gone with her, there would certainly have been some clear tale to tell when the interview was over. She would have known whether or no she loved the man, or could love him, and would have given him some true and intelligible answer. Bell had not done so, but had given him an answer which, if true, was not intelligible, and if intelligible was not true. And yet, when she had gone away to think over what had passed, she had been happy and satisfied, and almost triumphant. She had never yet asked herself whether she expected anything further from Dr. Crofts, nor what that something further might be, – and yet she was happy!

Lily had now become pert and saucy in her bed, taking upon herself the little airs which are allowed to a convalescent invalid as compensation for previous suffering and restraint. She pretended to much anxiety on the subject of her dinner, and declared that she would go out on such or such a day, let Dr. Crofts be as imperious as he might. 'He's an old savage, after all,' she said to her sister, one evening, after he was gone, 'and just as bad as the rest of them.'

'I do not know who the rest of them are,' said Bell, 'but at any rate he's not very old.'

'You know what I mean. He's just as grumpy as Dr. Gruffen, and thinks everybody is to do what he tells them. Of course, you take his part.'

'And of course you ought, seeing how good he has been.'

'And of course I should, to anybody but you. I do like to abuse him to you.'

'Lily. Lily!'

'So I do. It's so hard to knock any fire out of you, that when one does find the place where the flint lies, one can't help hammering at it. What did he mean by saying that I shouldn't get up on Sunday? Of course I shall get up if I like it.'

'Not if mamma asks you not?'

'Oh, but she won't, unless he interferes and dictates to her. Oh, Bell, what a tyrant he would be if he were married!'

'Would he?'

'And how submissive you would be, if you were his wife! It's a thousand pities that you are not in love with each other; – that is, if you are not.'

'Lily, I thought that there was a promise between us about that.'

'Ah! but that was in other days. Things are all altered since that promise was given, – all the world has been altered.' And as she said this the tone of her voice changed, and it had become almost sad. 'I feel as though I ought to be allowed now to speak about anything I please.'

'You shall, if it pleases you, my pet.'

'You see how it is, Bell; I can never again have anything of my own to talk about.'

'Oh, my darling, do not say that.'

'But it is so, Bell; and why not say it? Do you think I never say it to myself in the hours when I am all alone, thinking over it – thinking, thinking, thinking. You must not, – you must not grudge to let me talk of it sometimes.'

'I will not grudge you anything; – only I cannot believe that it must be so always.'

'Ask yourself, Bell, how it would be with you. But I sometimes fancy that you measure me differently from yourself.'

'Indeed I do, for I know how much better you are.'

'I am not so much better as to be ever able to forget all that. I know I never shall do so. I have made up my mind about it clearly and with an absolute certainty.'

'Lily, Lily, Lily! pray do not say so.'

'But I do say it. And yet I have not been very mopish and melancholy; have I, Bell? I do think I deserve some little credit, and yet, I declare, you won't allow me the least privilege in the world.'

'What privilege would you wish me to give you?'

'To talk about Dr. Crofts.'

'Lily, you are a wicked, wicked tyrant.' And Bell leaned over her, and fell upon her, and kissed her, hiding her own face in the gloom of the evening. After that it came to be an accepted understanding between them that Bell was not altogether indifferent to Dr. Crofts.

'You heard what he said, my darling,' Mrs. Dale said the next day, as the three were in the room together after Dr. Crofts was gone. Mrs. Dale was standing on one side of the bed, and Bell on the other, while Lily was scolding them both. 'You can get up for an hour or two to-morrow, but he thinks you had better not go out of the room.'

'What would be the good of that, mamma? I am so tired of looking always at the same paper. It is such a tiresome paper. It makes one count the pattern over and over again. I wonder how you ever can live here.'

'I've got used to it, you see.'

'I never can get used to that sort of thing; but go on counting, and counting, and counting. I'll tell you what I should like; and I'm sure it would be the best thing, too.'

'And what would you like?' said Bell.

'Just to get up at nine o'clock to-morrow, and go to church as though nothing had happened. Then, when Dr. Crofts came in the evening, you would tell him I was down at the school.'

'I wouldn't quite advise that,' said Mrs. Dale.

'It would give him such a delightful start. And when he

found I didn't die immediately, as of course I ought to do according to rule, he would be so disgusted.'

'It would be very ungrateful, to say the least of it,' said Bell.

'No, it wouldn't, a bit. He needn't come, unless he likes it. And I don't believe he comes to see me at all. It's all very well, mamma, your looking in that way; but I'm sure it's true. And I'll tell you what I'll do, I'll pretend to be bad again, otherwise the poor man will be robbed of his only happiness.'

'I suppose we must allow her to say what she likes till she gets well,' said Mrs. Dale, laughing. It was now nearly dark, and Mrs. Dale did not see that Bell's hand had crept under the bed-clothes, and taken hold of that of her sister. 'It's true, mamma,' continued Lily, 'and I defy her to deny it. I would forgive him for keeping me in bed if he would only make her fall in love with him.'

'She has made a bargain, mamma,' said Bell, 'that she is to say whatever she likes till she gets well.'

'I am to say whatever I like always; that was the bargain; and I mean to stand to it.'

On the following Sunday Lily did get up, but did not leave her mother's bedroom. There she was, seated in that half-dignified and half-luxurious state which belongs to the first getting up of an invalid, when Dr. Crofts called. There she had eaten her tiny bit of roast mutton, and had called her mother a stingy old creature, because she would not permit another morsel; and there she had drunk her half glass of port-wine, pretending that it was very bad, and twice worse than the doctor's physic; and there, Sunday though it was, she had fully enjoyed the last hour of daylight, reading that exquisite new novel which had just completed itself, amidst the jarring criticisms of the youth and age of the reading public.

'I am quite sure she was right in accepting him, Bell,' she said, putting down the book as the light was fading, and beginning to praise the story.

'It was a matter of course,' said Bell. 'It always is right in the novels. That's why I don't like them. They are too sweet.'

'That's why I do like them, because they are so sweet. A sermon is not to tell you what you are, but what you ought to be, and a novel should tell you not what you are to get, but what you'd like to get.'

'If so, then, I'd go back to the old school, and have the heroine really a heroine, walking all the way up from Edinburgh to London, and falling among thieves; or else nursing a wounded hero, and describing the battle from the window. We've got tired of that; or else the people who write can't do it now-a-days. But if we are to have real life, let it be real.'

'No, Bell, no!' said Lily. 'Real life sometimes is so painful.' Then her sister, in a moment, was down on the floor at her feet, kissing her hand and caressing her knees, and praying that the wound might be healed.

On that morning Lily had succeeded in inducing her sister to tell her all that had been said by Dr. Crofts. All that had been said by herself also, Bell had intended to tell; but when she came to this part of the story, her account was very lame. 'I don't think I said anything,' she said. 'But silence always gives consent. He'll know that,' Lily had rejoined. 'No, he will not; my silence didn't give any consent; I'm sure of that. And he didn't think that it did.' 'But you didn't mean to refuse him?' 'I think I did. I don't think I knew what I meant; and it was safer, therefore, to look no, than to look yes. If I didn't say it, I'm sure I looked it.' 'But you wouldn't refuse him now?' asked Lily. 'I don't know,' said Bell. 'It seems as though I should want years to make up my mind; and he won't ask me again.'

Bell was still at her sister's feet, caressing them, and praying with all her heart that the wound might be healed in due time, when Mrs. Dale came in and announced the doctor's daily visit. 'Then I'll go,' said Bell.

'Indeed you won't,' said Lily. 'He's coming simply to make a morning call, and nobody need run away. Now, Dr. Crofts, you need not come and stand over me with your watch, for I won't let you touch my hand except to shake hands with me;' and then she held her hand out to him. 'And all you'll know of my tongue you'll learn from the sound.'

'I don't care in the least for your tongue.'

'I dare say not, and yet you may some of these days. I can speak out, if I like it; can't I, mamma?'

'I should think Dr. Crofts knows that by this time, my dear.'

'I don't know. There are some things gentlemen are very slow to learn. But you must sit down, Dr. Crofts, and make yourself comfortable and polite; for you must understand that you are not master here any longer. I am out of bed now, and your reign is over.'

'That's the gratitude of the world, all through,' said Mrs. Dale.

'Who is ever grateful to a doctor? He only cures you that he may triumph over some other doctor, and declare, as he goes by Dr. Gruffen's door, "There, had she called you in, she'd have been dead before now; or else would have been ill for twelve months." Don't you jump for joy when Dr. Gruffen's patients die?'

'Of course I do – out in the market-place, so that everybody shall see me,' said the doctor.

'Lily, how can you say such shocking things?' said her sister.

Then the doctor did sit down, and they were all very cosy together over the fire, talking about things which were not medical, or only half medical in their appliance. By degrees the conversation came round to Mrs. Eames and to John Eames. Two or three days since, Crofts had told Mrs. Dale of that affair at the railway station, of which up to that time she had heard nothing. Mrs. Dale, when she was assured that young Eames had given Crosbie a tremendous thrashing – the tidings of the affair which had got themselves substantiated at Guestwick so described the nature of the encounter – could not withhold some meed of applause.

'Dear boy!' she said, almost involuntarily. 'Dear boy! it came from the honestness of his heart!' And then she gave special injunctions to the doctor – injunctions which were surely unnecessary – that no word of the matter should be whispered before Lily.

'I was at the manor, yesterday,' said the doctor, 'and the earl would talk about nothing but Master Johnny. He says he's the finest fellow going.' Whereupon Mrs. Dale touched him with her foot, fearing that the conversation might be led away in the direction of Johnny's prowess.

'I am so glad,' said Lily. 'I always knew that they'd find John out at last.'

'And Lady Julia is just as fond of him,' said the doctor.

'Dear me!' said Lily. 'Suppose they were to make up a match!'

'Lily, how can you be so absurd?'

'Let me see; what relation would he be to us? He would certainly be Bernard's uncle, and uncle Christopher's half brother-in-law. Wouldn't it be odd?'

'It would rather,' said Mrs. Dale.

'I hope he'll be civil to Bernard. Don't you, Bell? Is he to give up the Income-tax Office, Dr. Crofts?'

'I didn't hear that that was settled yet.' And so they went on talking about John Eames.

'Joking apart,' said Lily, 'I am very glad that Lord De Guest has taken him by the hand. Not that I think an earl is better than anybody else, but because it shows that people are beginning to understand that he has got something in him. I always said that they who laughed at John would see him hold up his head yet.' All which words sank deep into Mrs. Dale's mind. If only, in some coming time, her pet might be taught to love this new young hero! But then would not that last heroic deed of his militate most strongly against any possibility of such love?

'And now I may as well be going,' said the doctor, rising from his chair. At this time Bell had left the room, but Mrs. Dale was still there.

'You need not be in such a hurry, especially this evening,' said Lily.

'Why especially this evening?'

'Because it will be the last. Sit down again, Dr. Crofts. I've got a little speech to make to you. I've been preparing it all the morning, and you must give me an opportunity of speaking it.'

'I'll come the day after to-morrow, and I'll hear it then.'

'But I choose, sir, that you should hear it now. Am I not

to be obeyed when I first get up on to my own throne? Dear, dear Dr. Crofts, how am I to thank you for all that you have done?'

'How are any of us to thank him?' said Mrs. Dale.

'I hate thanks,' said the doctor. 'One kind glance of the eye is worth them all, and I've had many such in this house.'

'You have our hearts' love, at any rate,' said Mrs. Dale.

'God bless you all!' said he, as he prepared to go.

'But I haven't made my speech yet,' said Lily. 'And to tell the truth, mamma, you must go away, or I shall never be able to make it. It's very improper, is it not, turning you out, but it shall only take three minutes.' Then Mrs. Dale, with some little joking word, left the room; but, as she left it, her mind was hardly at ease. Ought she to have gone, leaving it to Lily's discretion to say what words she might think fit to Dr. Crofts? Hitherto she had never doubted her daughters – not even their discretion; and therefore it had been natural to her to go when she was bidden. But as she went downstairs she had her doubts whether she was right or no.

'Dr. Crofts,' said Lily, as soon as they were alone. 'Sit down there, close to me. I want to ask you a question. What was it you said to Bell when you were alone with her the other evening in the parlour?'

The doctor sat for a moment without answering, and Lily, who was watching him closely, could see by the light of the fire that he had been startled – had almost shuddered as the question was asked him.

'What did I say to her?' and he repeated her words in a very low voice. 'I asked her if she could love me, and be my wife.'

'And what answer did she make to you?'

'What answer did she make? She simply refused me.'

'No, no, no; don't believe her, Dr. Crofts. It was not so; – I think it was not so. Mind you, I can say nothing as coming from her. She has not told me her own mind. But if you really love her, she will be mad to refuse you.'

'I do love her, Lily; that at any rate is true.'

'Then go to her again. I am speaking for myself now. I cannot afford to lose such a brother as you would be. I love you so dearly that I cannot spare you. And she, – I think she'll learn to love you as you would wish to be loved. You know her nature, how silent she is, and averse to talk about herself. She has confessed nothing to me but this, – that you spoke to her and took her by surprise. Are we to have another chance? I know how wrong I am to ask such a question. But, after all, is not the truth the best?'

'Another chance!'

'I know what you mean, and I think she is worthy to be your wife. I do, indeed; and if so, she must be very worthy. You won't tell of me, will you now, doctor?'

'No; I won't tell of you.'

'And you'll try again?'

'Yes; I'll try again.'

'God bless you, my brother! I hope, – I hope you'll be my brother.' Then, as he put out his hand to her once more, she raised her head towards him, and he, stooping down, kissed her forehead. 'Make mamma come to me,' were the last words she spoke as he went out at the door.

'So you've made your speech,' said Mrs. Dale.

'Yes, mamma.'

'I hope it was a discreet speech.'

'I hope it was, mamma. But it has made me so tired, and I believe I'll go to bed. Do you know I don't think I should have done much good down at the school to-day?'

Then Mrs. Dale, in her anxiety to repair what injury might have been done to her daughter by over-exertion, omitted any further mention of the farewell speech.

Dr. Crofts as he rode home enjoyed but little of the triumph of a successful lover. 'It may be that she's right,' he said to himself; 'and, at any rate, I'll ask again.' Nevertheless, that 'No' which Bell had spoken, and had repeated, still sounded in his ears harsh and conclusive. There are men to whom a peal of noes rattling about their ears never takes the sound of a true denial, and others to whom the word once pronounced, be it whispered ever so softly, comes as though it were an unchangeable verdict from the supreme judgment-seat.

CHAPTER XLIII

FIE, FIE!

WILL any reader remember the loves, – no, not the loves; that word is so decidedly ill-applied as to be incapable of awakening the remembrance of any reader; but the flirtations – of Lady Dumbello and Mr. Plantagenet Palliser? Those flirtations, as they had been carried on at Courcy Castle, were laid bare in all their enormities to the eye of the public, and it must be confessed that if the eye of the public was shocked, that eye must be shocked very easily.

But the eye of the public was shocked, and people who were particular as to their morals said very strange things. Lady De Courcy herself said very strange things indeed, shaking her head, and dropping mysterious words; whereas Lady Clandidlem spoke much more openly, declaring her opinion that Lady Dumbello would be off before May. They both agreed that it would not be altogether bad for Lord Dumbello that he should lose his wife, but shook their heads very sadly when they spoke of poor Plantagenet Palliser. As to the lady's fate, that lady whom they had both almost worshipped during the days at Courcy Castle, – they did not seem to trouble themselves about that.

And it must be admitted that Mr. Palliser had been a little imprudent, – imprudent, that is, if he knew anything about the rumours afloat, – seeing that soon after his visit at Courcy Castle he had gone down to Lady Hartletop's place in Shropshire, at which the Dumbellos intended to spend the winter, and on leaving it had expressed his intention of returning in February. The Hartletop people had pressed him very much, – the pressure having come with peculiar force from Lord Dumbello. Therefore it is reasonable to suppose that the Hartletop people had not heard of the rumour.

Mr. Plantagenet Palliser spent his Christmas with his uncle, the Duke of Omnium, at Gatherum Castle. That is to say, he reached the castle in time for dinner on Christmas eve, and left it on the morning after Christmas day. This was in accordance with the usual practice of his life, and the tenants, dependants, and followers of the Omnium interest were always delighted to see this manifestation of a healthy English domestic family feeling between the duke and his nephew. But the amount of intercourse on such occasions between them was generally trifling. The duke would smile as he put out his right hand to his nephew, and say, –

'Well, Plantagenet, – very busy, I suppose?'

The duke was the only living being who called him Plantagenet to his face, though there were some scores of men who talked of Planty Pal behind his back. The duke had been the only living being so to call him. Let us hope that it still was so, and that there had arisen no feminine exception, dangerous in its nature and improper in its circumstances.

'Well, Plantagenet,' said the duke, on the present occasion, 'very busy, I suppose?'

'Yes, indeed, duke,' said Mr. Palliser. 'When a man gets the harness on him he does not easily get quit of it.'

The duke remembered that his nephew had made almost the same remark at his last Christmas visit.

'By-the-by,' said the duke, 'I want to say a word or two to you before you go.'

Such a proposition on the duke's part was a great departure from his usual practice, but the nephew of course undertook to obey his uncle's behests.

'I'll see you before dinner to-morrow,' said Plantagenet.

'Ah, do,' said the duke. 'I'll not keep you five minutes.' And at six o'clock on the following afternoon the two were closeted together in the duke's private room.

'I don't suppose there is much in it,' began the duke, 'but people are talking about you and Lady Dumbello.'

'Upon my word, people are very kind.' And Mr. Palliser bethought himself of the fact, – for it certainly was a fact, – that people for a great many years had talked about his uncle and Lady Dumbello's mother-in-law.

'Yes; kind enough; are they not? You've just come from Hartlebury, I believe.' Hartlebury was the Marquis of Hartletop's seat in Shropshire.

'Yes, I have. And I'm going there again in February.'

'Ah, I'm sorry for that. Not that I mean, of course, to interfere with your arrangements. You willl acknowledge that I have not often done so, in any matter whatever.'

'No; you have not,' said the nephew, comforting himself with an inward assurance that no such interference on his uncle's part could have been possible.

'But in this instance it would suit me, and I really think it would suit you too, that you should be as little at Hartlebury as possible. You have said you would go there, and of course you will go. But if I were you, I would not stay above a day or two.'

Mr. Plantagenet Palliser received everything he had in the world from his uncle. He sat in Parliament through his uncle's interest, and received an allowance of ever so many thousand a year which his uncle could stop to-morrow by his mere word. He was his uncle's heir, and the dukedom, with certain entailed properties, must ultimately fall to him, unless his uncle should marry and have a son. But by far the greater portion of the duke's property was unentailed; the duke might probably live for the next twenty years or more; and it was quite possible that, if offended, he might marry and become a father. It may be said that no man could well be more dependent on another than Plantagenet Palliser was upon his uncle; and it may be said also that no father or uncle ever troubled his heir with less interference. Nevertheless, the nephew felt himself aggrieved by this allusion to his private life, and resolved at once that he would not submit to such surveillance.

'I don't know how long I shall stay,' said he; 'but I cannot say that my visit will be influenced one way or the other by such a rumour as that.'

'No; probably not. But it may perhaps be influenced by my request.' And the duke, as he spoke, looked a little savage.

'You wouldn't ask me to regard a report that has no foundation.'

'I am not asking about its foundation. Nor do I in the least wish to interfere with your manner in life.' By which last observation the duke intended his nephew to understand that he was quite at liberty to take away any other gentleman's wife, but that he was not at liberty to give occasion even for a surmise that he wanted to take Lord Dumbello's wife. 'The fact is this, Plantagenet. I have for many years been intimate with that family. I have not many intimacies, and shall probably never increase them. Such friends as I have, I wish to keep, and you will easily perceive that any such report as that which I have mentioned, might make it unpleasant for me to go to Hartlebury, or for the Hartlebury people to come here.' The duke certainly could not have spoken plainer, and Mr. Palliser understood him thoroughly. Two such alliances between the two families could not be expected to run pleasantly together, and even the rumour of any such second alliance might interfere with the pleasantness of the former one.

'That's all,' said the duke.

'It's a most absurd slander,' said Mr. Palliser.

'I dare say. Those slanders always are absurd; but what can we do? We can't tie up people's tongues.' And the duke looked as though he wished to have the subject considered as finished, and to be left alone.

'But we can disregard them,' said the nephew, indiscreetly.

'You may. I have never been able to do so. And yet, I believe, I have not earned for myself the reputation of being subject to the voices of men. You think that I am asking much of you; but you should remember that hitherto I have given much and have asked nothing. I expect you to oblige me in this matter.'

Then Mr. Plantagenet Palliser left the room, knowing that he had been threatened. What the duke had said amounted to this. – If you go on dangling after Lady Dumbello, I'll stop the seven thousand a year which I give you. I'll oppose your next return at Silverbridge, and I'll make a will and leave away from you Matching and The Horns, – a beautiful little place in Surrey, the use of which had been already offered to Mr. Palliser in the event of his marriage; all the Littlebury estate in Yorkshire, and the enormous Scotch property. Of my personal goods, and money invested in loans, shares, and funds, you shall never touch a shilling, or the value of a shilling. And, if I find that I can suit myself, it may be that I'll leave you plain Mr. Plantagenet Palliser, with a little first cousin for the head of your family.

The full amount of this threat Mr. Palliser understood, and, as he thought of it, he acknowledged to himself that he had never felt for Lady Dumbello anything like love. No conversation between them had ever been warmer than that of which the reader has seen a sample. Lady Dumbello had been nothing to him. But now, – now that the matter had been put before him in this way, might it not become him, as a gentleman, to fall in love with so very beautiful a woman, whose name had already been linked with his own? We all know that story of the priest, who, by his question in the confessional, taught the ostler to grease the horses' teeth. 'I never did yet,' said the ostler, 'but I'll have a try at it.' In this case, the duke had acted the part of the priest, and Mr. Palliser, before the night was over, had almost become as ready a pupil as the ostler. As to the threat, it would ill become him, as a Palliser and a Plantagenet, to regard it. The duke would not marry. Of all men in the world he was the least likely to spite his own face by cutting off his own nose; and, for the rest of it, Mr. Palliser would take his chance. Therefore he went down to Hartlebury early in February, having fully determined to be very particular in his attentions to Lady Dumbello.

Among a houseful of people at Hartlebury, he found Lord Porlock, a slight, sickly, worn-out looking man, who had something about his eye of his father's hardness, but nothing in his mouth of his father's ferocity.

'So your sister is going to be married?' said Mr. Palliser.

'Yes. One has no right to be surprised at anything they do, when one remembers the life their father leads them.'

'I was going to congratulate you.'

'Don't do that.'

'I met him at Courcy, and rather liked him.'

Mr. Palliser had barely spoken to Mr. Crosbie at Courcy, but then in the usual course of his social life he seldom did more than barely speak to anybody.

'Did you?' said Lord Porlock. 'For the poor girl's sake I hope he's not a ruffian. How any man should propose to my father to marry a daughter out of his house, is more than I can understand. How was my mother looking?'

'I didn't see anything amiss about her.'

'I expect that he'll murder her some day.' Then that conversation came to an end.

Mr. Palliser himself perceived – as he looked at her he could not but perceive – that a certain amount of social energy seemed to enliven Lady Dumbello when he approached her. She was given to smile when addressed, but her usual smile was meaningless, almost leaden, and never in any degree flattering to the person to whom it was accorded. Very many women smile as they answer the words which are spoken to them, and most who do so flatter by their smile. The thing is so common that no one thinks of it. The flattering pleases, but means nothing. The impression unconsciously taken simply conveys a feeling that the woman has made herself agreeable, as it was her duty to do, – agreeable, as far as that smile went, in some very infinitesimal degree. But she has thereby made her little contribution to society. She will make the same contribution a hundred times in the same evening. No one knows that she has flattered anybody; she does not know it herself; and the world calls her an agreeable woman. But Lady Dumbello put no flattery into her customary smiles. They were cold, unmeaning, accompanied by no special glance of the eye, and seldom addressed to the individual. They were given to the room at large; and the room at large, acknowledging her great pretensions, accepted them as sufficient. But when Mr. Palliser came near to her she would turn herself slightly, ever so slightly, on her seat, and would allow her eyes to rest for a moment upon his face. Then when he remarked that it had been rather cold, she would smile actually upon him as she acknowledged the truth of his observation. All this Mr. Palliser taught himself to observe, having been instructed by his foolish uncle in that lesson as to the greasing of the horses' teeth.

But, nevertheless, during the first week of his stay at

Hartlebury, he did not say a word to her more tender than his observation about the weather. It is true that he was very busy. He had undertaken to speak upon the address, and as Parliament was now about to be opened, and as his speech was to be based upon statistics, he was full of figures and papers. His correspondence was pressing, and the day was seldom long enough for his purposes. He felt that the intimacy to which he aspired was hindered by the laborious routine of his life; but nevertheless he would do something before he left Hartlebury, to show the special nature of his regard. He would say something to her, that should open to her view the secret of – shall we say his heart? Such was his resolve, day after day. And yet day after day went by, and nothing was said. He fancied that Lord Dumbello was somewhat less friendly in his manner than he had been, that he put himself in the way and looked cross; but, as he declared to himself, he cared very little for Lord Dumbello's looks.

'When do you go to town?' he said to her one evening.

'Probably in April. We certainly shall not leave Hartlebury before that.'

'Ah, yes. You stay for the hunting.'

'Yes; Lord Dumbello always remains here through March. He may run up to town for a day or two.'

'How comfortable! I must be in London on Thursday, you know.'

'When Parliament meets, I suppose?'

'Exactly. It is such a bore; but one has to do it.'

'When a man makes a business of it, I suppose he must.'

'Oh, dear, yes; it's quite imperative.' Then Mr. Palliser looked round the room, and thought he saw Lord Dumbello's eye fixed upon him. It was really very hard work. If the truth must be told, he did not know how to begin. What was he to say to her? How was he to commence a conversation that should end by being tender? She was very handsome certainly, and for him she could look interesting; but for his very life he did not know how to begin to say anything special to her. A liaison with such a woman as Lady Dumbello, – platonic, innocent, but nevertheless very intimate

– would certainly lend a grace to his life, which, under its present circumstances, was rather dry. He was told, – told by public rumour which had reached him through his uncle, – that the lady was willing. She certainly looked as though she liked him; but how was he to begin? The art of startling the House of Commons and frightening the British public by the voluminous accuracy of his statistics he had already learned; but what was he to say to a pretty woman?

'You'll be sure to be in London in April?'

This was on another occasion.

'Oh, yes; I think so.'

'In Carlton Gardens, I suppose.'

'Yes; Lord Dumbello has got a lease of the house now.'

'Has he, indeed? Ah, it's an excellent house. I hope I shall be allowed to call there sometimes.'

'Certainly, – only I know you must be so busy.'

'Not on Saturdays and Sundays.'

'I always receive on Sundays,' said Lady Dumbello. Mr. Palliser felt that there was nothing peculiarly gracious in this. A permission to call when all her other acquaintances would be there, was not much; but still, perhaps, it was as much as he could expect to obtain on that occasion. He looked up and saw that Lord Dumbello's eyes were again upon him, and that Lord Dumbello's brow was black. He began to doubt whether a country house, where all the people were thrown together, was the best place in the world for such manœuvring. Lady Dumbello was very handsome, and he liked to look at her, but he could not find any subject on which to interest her in that drawing-room at Hartlebury. Later in the evening he found himself saying something to her about the sugar duties, and then he knew that he had better give it up. He had only one day more, and that was required imperatively for his speech. The matter would go much easier in London, and he would postpone it till then. In the crowded rooms of London private conversation would be much easier, and Lord Dumbello wouldn't stand over and look at him. Lady Dumbello had taken his remarks about the sugar very kindly, and had asked for a definition of an ad valorem* duty. It was a nearer approach to a real conversation than he had ever before made; but the subject had been unlucky, and could not, in his hands, be brought round to anything tender; so he resolved to postpone his gallantry till the London spring should make it easy, and felt as he did so, that he was relieved for the time from a heavy weight.

'Good-bye, Lady Dumbello,' he said, on the next evening. 'I start early to-morrow morning.'

'Good-bye, Mr. Palliser.'

As she spoke she smiled ever so sweetly, but she certainly had not learned to call him Plantagenet as yet. He went up to London and immediately got himself to work. The accurate and voluminous speech came off with considerable credit to himself, – credit of that quiet, enduring kind which is accorded to such men. The speech was respectable, dull, and correct. Men listened to it, or sat with their hats over their eyes, asleep, pretending to do so; and the daily Jupiter in the morning had a leading article about it, which, however, left the reader at its close altogether in doubt whether Mr. Palliser might be supposed to be a great financial pundit or no. Mr. Palliser might become a shining light to the moneyed world, and a glory to the banking interests; he might be a future Chancellor of the Exchequer. But then again, it might turn out that, in these affairs, he was a mere ignis fatuus,* a blind guide, – a man to be laid aside as very respectable, but of no depth. Who, then, at the present time, could judiciously risk his credit by declaring whether Mr. Palliser understood his subject or did not understand it? We are not content in looking to our newspapers for all the information that earth and human intellect can afford; but we demand from them what we might demand if a daily sheet could come to us from the world of spirits. The result, of course, is this, – that the papers do pretend that they have come daily from the world of spirits; but the oracles are very doubtful, as were those of old.

Plantagenet Palliser, though he was contented with this article, felt, as he sat in his chambers in the Albany, that something else was wanting to his happiness. This sort of life was all very well. Ambition was a grand thing, and it became him, as a Palliser and a future peer, to make politics his profession. But might he not spare an hour or two for Amaryllis in the shade? * Was it not hard, this life of his? Since he had been told that Lady Dumbello smiled upon him, he had certainly thought more about her smiles than had been good for his statistics. It seemed as though a new vein in his body had been brought into use, and that blood was running where blood had never run before. If he had seen Lady Dumbello before Dumbello had seen her, might he not have married her? Ah! in such case as that, had she been simply Miss Grantly, or Lady Griselda Grantly, as the case might have been, he thought he might have been able to speak to her with more ease. As it was, he certainly had found the task difficult, down in the country, though he had heard of men of his class doing the same sort of thing all his life. For my own part, I believe, that the reputed sinners are much more numerous than the sinners.

As he sat there, a certain Mr. Fothergill came in upon him. Mr. Fothergill was a gentleman who managed most of his uncle's ordinary affairs, – a clever fellow, who knew on which side his bread was buttered. Mr. Fothergill was naturally anxious to stand well with the heir; but to stand well with the owner was his business of life, and with that business he never allowed anything to interfere. On this occasion Mr. Fothergill was very civil, complimenting his future possible patron on his very powerful speech, and predicting for him political power with much more certainty than the newspapers which had, or had not, come from the world of spirits. Mr. Fothergill had come in to say a word or two about some matter of business. As all Mr. Palliser's money passed through Mr. Fothergill's hands, and as his electioneering interests were managed by Mr. Fothergill, Mr. Fothergill not unfrequently called to say a necessary word or two. When this was done he said another word or two, which might be necessary or not, as the case might be.

'Mr. Palliser,' said he, 'I wonder you don't think of marrying. I hope you'll excuse me.'

Mr. Palliser was by no means sure that he would excuse him, and sat himself suddenly upright in his chair in a manner that was intended to exhibit a first symptom of outraged dignity. But, singularly enough, he had himself been thinking of marriage at that moment. How would it have been with him had he known the beautiful Griselda before the Dumbello alliance had been arranged? Would he have married her? Would he have been comfortable if he had married her? Of course he could not marry now, seeing that he was in love with Lady Dumbello, and that the lady in question, unfortunately, had a husband of her own; but though he had been thinking of marrying, he did not like to have the subject thus roughly thrust before his eyes, and, as it were, into his very lap by his uncle's agent. Mr. Fothergill, no doubt, saw the first symptom of outraged dignity, for he was a clever, sharp man. But, perhaps, he did not in truth much regard it. Perhaps he had received instructions which he was bound to regard above all other matters.

'I hope you'll excuse me, Mr. Palliser, I do, indeed; but I say it because I am half afraid of some, – some, – some diminution of good feeling, perhaps, I had better call it, between you and your uncle. Anything of that kind would be such a monstrous pity.'

'I am not aware of any such probability.'

This Mr. Palliser said with considerable dignity; but when the words were spoken he bethought himself whether he had not told a fib.

'No; perhaps not. I trust there is not such probability. But the duke is a very determined man if he takes anything into his head; – and then he has so much in his power.'

'He has not me in his power, Mr. Fothergill.'

'No, no, no. One man does not have another in his power in this country, – not in that way; but then you know, Mr. Palliser, it would hardly do to offend him; would it?'

'I would rather not offend him, as is natural. Indeed, I do not wish to offend any one.'

'Exactly so; and least of all the duke, who has the whole property in his own hands. We may say the whole, for he can marry to-morrow if he pleases. And then his life is so good. I don't know a stouter man of his age, anywhere.'

'I'm very glad to hear it.'

'I'm sure you are, Mr. Palliser. But if he were to take offence, you know?'

'I should put up with it.'

'Yes, exactly; that's what you would do. But it would be worth while to avoid it, seeing how much he has in his power.'

'Has the duke sent you to me now, Mr. Fothergill?'

'No, no, no, – nothing of the sort. But he dropped words the other day which made me fancy that he was not quite, – quite, – quite at ease about you. I have long known that he would be very glad indeed to see an heir born to the property. The other morning, – I don't know whether there was anything in it, – but I fancied he was going to make some change in the present arrangements. He did not do it, and it might have been fancy. Only think, Mr. Palliser, what one word of his might do! If he says a word, he never goes back from it.' Then, having said so much, Mr. Fothergill went his way.

Mr. Palliser understood the meaning of all this very well. It was not the first occasion on which Mr. Fothergill had given him advice, – advice such as Mr. Fothergill himself had no right to give him. He always received such counsel with an air of half-injured dignity, intending thereby to explain to Mr. Fothergill that he was intruding. But he knew well whence the advice came; and though, in all such cases, he had made up his mind not to follow such counsel, it had generally come to pass that Mr. Palliser's conduct had more or less accurately conformed itself to Mr. Fothergill's advice. A word from the duke might certainly do a great deal! Mr. Palliser resolved that in that affair of Lady Dumbello he would follow his own devices. But, nevertheless, it was undoubtedly true that a word from the duke might do a great deal!

We, who are in the secret, know how far Mr. Palliser had already progressed in his iniquitous passion before he left Hartlebury. Others, who were perhaps not so well informed, gave him credit for a much more advanced success. Lady Clandidlem, in her letter to Lady De Courcy, written immediately after the departure of Mr. Palliser, declared that.

having heard of that gentleman's intended matutinal departure, she had confidently expected to learn at the breakfast-table that Lady Dumbello had flown with him. From the tone of her ladyship's language, it seemed as though she had been robbed of an anticipated pleasure by Lady Dumbello's prolonged sojourn in the halls of her husband's ancestors. 'I feel, however, quite convinced,' said Lady Clandidlem, 'that it cannot go on longer than the spring. I never yet saw a man so infatuated as Mr. Palliser. He did not leave her for one moment all the time he was here. No one but Lady Hartletop would have permitted it. But, you know, there is nothing so pleasant as good old family friendships.'

CHAPTER XLIV

VALENTINE'S DAY AT ALLINGTON

LILY had exacted a promise from her mother before her illness, and during the period of her convalescence often referred to it, reminding her mother that that promise had been made, and must be kept. Lily was to be told the day on which Crosbie was to be married. It had come to the knowledge of them all that the marriage was to take place in February. But this was not sufficient for Lily. She must know the day.

And as the time drew nearer, – Lily becoming stronger the while, and less subject to medical authority, – the marriage of Crosbie and Alexandrina was spoken of much more frequently at the Small House. It was not a subject which Mrs. Dale or Bell would have chosen for conversation; but Lily would refer to it. She would begin by doing so almost in a drolling strain, alluding to herself as a forlorn damsel in a play-book; and then she would go on to speak of his interests as a matter which was still of great moment to her. But in the course of such talking she would too often break down, showing by some sad word or melancholy tone how great was the burden on her heart. Mrs. Dale and Bell would willingly have avoided the subject, but Lily would not have it avoided. For them it was a very difficult matter on which to speak in her hearing. It was not permitted to them to say a word of abuse against Crosbie, as to whom they thought that no word of condemnation could be sufficiently severe; and they were forced to listen to such excuses for his conduct as Lily chose to manufacture, never daring to point out how vain those excuses were.

Indeed, in those days Lily reigned as a queen at the Small House. Ill-usage and illness together falling into her hands had given her such power, that none of the other women were able to withstand it. Nothing was said about it; but it was understood by them all, Jane and the cook included, that Lily was for the time paramount. She was a dear, gracious, loving, brave queen, and no one was anxious to rebel; – only that those praises of Crosbie were so very bitter in the ears of her subjects. The day was named soon enough, and the tidings came down to Allington. On the fourteenth of February, Crosbie was to be made a happy man. This was not known to the Dales till the twelfth, and they would willingly have spared the knowledge then, had it been possible to spare it. But it was not so, and on that evening Lily was told.

During these days, Bell used to see her uncle daily. Her visits were made with the pretence of taking to him information as to Lily's health; but there was perhaps at the bottom of them a feeling that, as the family intended to leave the Small House at the end of March, it would be well to let the squire know that there was no enmity in their hearts against him. Nothing more had been said about their moving, – nothing, that is, from them to him. But the matter was going on, and he knew it. Dr. Crofts was already in treaty on their behalf for a small furnished house at Guestwick. The squire was very sad about it, – very sad indeed. When Hopkins spoke to him on the subject, he sharply desired that faithful gardener to hold his tongue, giving it to be understood that such things were not to be made matter of talk by the Allington dependants till they had been officially announced. With Bell during these visits he never alluded to the matter. She was the chief sinner, in that she had refused to marry her cousin, and had declined even to listen to rational counsel upon the matter. But the squire felt that he could not discuss the subject with her, seeing that he had been specially informed by Mrs. Dale that his interference would not be permitted; and then he was perhaps aware that if he did discuss the subject with Bell, he would not gain much by such discussion. Their conversation, therefore, generally fell upon Crosbie, and the tone in which he was mentioned in the Great House was very different from that assumed in Lily's presence.

'He'll be a wretched man,' said the squire, when he told Bell of the day that had been fixed.

'I don't want him to be wretched,' said Bell. 'But I can hardly think that he can act as he has done without being punished.'

'He will be a wretched man. He gets no fortune with her, and she will expect everything that fortune can give. I believe, too, that she is older than he is. I cannot understand it. Upon my word, I cannot understand how a man can be such a knave and such a fool. Give my love to Lily. I'll see her to-morrow or the next day. She's well rid of him; I'm sure of that; – though I suppose it would not do to tell her so.'

The morning of the fourteenth came upon them at the Small House, as comes the morning of those special days which have been long considered, and which are to be long remembered. It brought with it a hard, bitter frost, – a black, biting frost, – such a frost as breaks the water-pipes, and binds the ground to the hardness of granite. Lily, queen as she was, had not yet been allowed to go back to her own chamber, but occupied the larger bed in her mother's room, her mother sleeping on a smaller one.

'Mamma,' she said, 'how cold they'll be!' Her mother had announced to her the fact of the black frost, and these were the first words she spoke.

'I fear their hearts will be cold also,' said Mrs. Dale. She ought not to have said so. She was transgressing the acknowledged rule of the house in saying any word that could be construed as being inimical to Crosbie or his bride. But her feeling on the matter was too strong, and she could not restrain herself.

'Why should their hearts be cold? Oh, mamma, that is a terrible thing to say. Why should their hearts be cold?'

'I hope it may not be so.'

'Of course you do; of course we all hope it. He was not cold-hearted, at any rate. A man is not cold-hearted, because he does not know himself. Mamma, I want you to wish for their happiness.'

Mrs. Dale was silent for a minute or two before she answered this, but then she did answer it. 'I think I do,' said she. 'I think I do wish for it.'

'I am very sure that I do,' said Lily.

At this time Lily had her breakfast upstairs, but went down into the drawing-room in the course of the morning.

'You must be very careful in wrapping yourself as you go downstairs,' said Bell, who stood by the tray on which she had brought up the toast and tea. 'The cold is what you would call awful.'

'I should call it jolly,' said Lily, 'if I could get up and go out. Do you remember lecturing me about talking slang the day that he first came?'

'Did I, my pet?'

'Don't you remember, when I called him a swell? Ah, dear! so he was. That was the mistake, and it was all my own fault, as I had seen it from the first.'

Bell for a moment turned her face away, and beat with her foot against the ground. Her anger was more difficult of restraint than was even her mother's, – and now, not restraining it, but wishing to hide it, she gave it vent in this way.

'I understand, Bell. I know what your foot means when it goes in that way; and you shan't do it. Come here, Bell, and let me teach you Christianity. I'm a fine sort of teacher, am I not? And I did not quite mean that.'

'I wish I could learn it from some one,' said Bell. 'There are circumstances in which what we call Christianity seems to me to be hardly possible.'

'When your foot goes in that way it is a very unchristian foot, and you ought to keep it still. It means anger against him, because he discovered before it was too late that he would not be happy, – that is, that he and I would not be happy together if we were married.'

'Don't scrutinize my foot too closely, Lily.'

'But your foot must bear scrutiny, and your eyes, and your voice. He was very foolish to fall in love with me. And so was I very foolish to let him love me, at a moment's notice, – without a thought as it were. I was so proud of having him, that I gave myself up to him all at once, without giving him a chance of thinking of it. In a week or two it was done. Who could expect that such an engagement should be lasting?'

'And why not? That is nonsense, Lily. But we will not talk about it.'

'Ah, but I want to talk about it. It was as I have said, and if so, you shouldn't hate him because he did the only thing which he honestly could do when he found out his mistake.'

'What; become engaged again within a week!'

'There had been a very old friendship, Bell; you must remember that. But I was speaking of his conduct to me, and not of his conduct to—' And then she remembered that that other lady might at this very moment possess the name which she had once been so proud to think that she would bear herself. Bell,' she said, stopping her other speech suddenly, 'at what o'clock do people get married in London?'

'Oh, at all manner of hours, – any time before twelve. They will be fashionable, and will be married late.'

'You don't think she's Mrs. Crosbie yet, then?'

'Lady Alexandrina Crosbie,' said Bell, shuddering.

'Yes, of course; I forgot. I should so like to see her. I feel such an interest about her. I wonder what coloured hair she has. I suppose she is a sort of Juno of a woman, – very tall and handsome. I'm sure she has not got a pug-nose like me. Do you know what I should really like, only of course it's not possible; – to be godmother to his first child.'

'Oh, Lily!'

'I should. Don't you hear me say that I know it's not possible? I'm not going up to London to ask her. She'll have all manner of grandees for her godfathers and godmothers. I wonder what those grand people are really like.'

'I don't think there's any difference. Look at Lady Julia.'

'Oh, she's not a grand person. It isn't merely having a title. Don't you remember that he told us that Mr. Palliser is about the grandest grandee of them all. I suppose people do learn to like them. He always used to say that he had been so long among people of that sort, that it would be very difficult for him to divide himself off from them. I should never have done for that kind of thing; should I?'

'There is nothing I despise so much as what you call that kind of thing.'

'Do you? I don't. After all, think how much work they do. He used to tell me of that. They have all the governing in their hands, and get very little money for doing it.'

'Worse luck for the country.'

'The country seems to do pretty well. But you're a Radical, Bell. My belief is, you wouldn't be a lady if you could help it.'

'I'd sooner be an honest woman.'

'And so you are, – my own dear, dearest, honest Bell, – and the fairest lady that I know. If I were a man, Bell, you are just the girl that I should worship.'

'But you are not a man; so it's no good.'

'But you mustn't let your foot go astray in that way; you mustn't, indeed. Somebody said, that whatever is, is right,* and I declare I believe it.'

'I'm sometimes inclined to think, that whatever is, is wrong.'

'That's because you're a Radical. I think I'll get up now, Bell; only it's so frightfully cold that I'm afraid.'

'There's a beautiful fire,' said Bell.

'Yes; I see. But the fire won't go all around me, like the bed does. I wish I could know the very moment when they're at the altar. It's only half-past ten yet.'

'I shouldn't be at all surprised if it's over.'

'Over! What a word that is! A thing like that is over, and

then all the world cannot put it back again. What if he should be unhappy after all?'

'He must take his chance,' said Bell, thinking within her own mind that that chance would be a very bad one.

'Of course he must take his chance. Well, – I'll get up now.' And then she took her first step out into the cold world beyond her bed. 'We must all take our chance. I have made up my mind that it will be at half-past eleven.'

When half-past eleven came, she was seated in a large easy chair over the drawing-room fire, with a little table by her side, on which a novel was lying. She had not opened her book that morning, and had been sitting for some time perfectly silent, with her eyes closed, and her watch in her hand.

'Mamma,' she said at last, 'it is over now, I'm sure.'

'What is over, my dear?'

'He has made that lady his wife. I hope God will bless them, and I pray that they may be happy.' As she spoke these words, there was an unwonted solemnity in her tone which startled Mrs. Dale and Bell.

'I also will hope so,' said Mrs. Dale. 'And now, Lily, will it not be well that you should turn your mind away from the subject, and endeavour to think of other things?'

'But I can't, mamma. It is so easy to say that; but people can't choose their own thoughts.'

'They can usually direct them as they will, if they make the effort.'

'But I can't make the effort. Indeed, I don't know why I should. It seems natural to me to think about him, and I don't suppose it can be very wrong. When you have had so deep an interest in a person, you can't drop him all of a sudden.' Then there was again silence, and after a while Lily took up her novel. She made that effort of which her mother had spoken, but she made it altogether in vain. 'I declare, Bell,' she said, 'it's the greatest rubbish I ever attempted to read.' This was specially ungrateful, because Bell had recommended the book. 'All the books have got to be so stupid! I think I'll read Pilgrim's Progress again.'

'What do you say to Robinson Crusoe?' said Bell.

'Or Paul and Virginia?'* said Lily. 'But I believe I'll have Pilgrim's Progress. I never can understand it, but I rather think that makes it nicer.'

'I hate books I can't understand,' said Bell. 'I like a book to be clear as running water, so that the whole meaning may be seen at once.'

'The quick seeing of the meaning must depend a little on the reader, must it not?' said Mrs. Dale.

'The reader mustn't be a fool, of course,' said Bell.

'But then so many readers are fools,' said Lily. 'And yet they get something out of their reading. Mrs. Crump is always poring over the Revelations, and nearly knows them by heart. I don't think she could interpret a single image, but she has a hazy, misty idea of the truth. That's why she likes it, – because it's too beautiful to be understood; and that's why I like Pilgrim's Progress.' After which Bell offered to get the book in question.

'No, not now,' said Lily. 'I'll go on with this, as you say it's so grand. The personages are always in their tantrums, and go on as though they were mad. Mamma, do you know where they're going for the honeymoon?'

'No, my dear.'

'He used to talk to me about going to the lakes.' And then there was another pause, during which Bell observed that her mother's face became clouded with anxiety. 'But I won't think of it any more,' continued Lily; 'I will fix my mind to something.' And then she got up from her chair. 'I don't think it would have been so difficult if I had not been ill.'

'Of course it would not, my darling.'

'And I'm going to be well again now, immediately. Let me see: I was told to read Carlyle's History of the French Revolution, and I think I'll begin now.' It was Crosbie who had told her to read the book, as both Bell and Mrs. Dale were well aware. 'But I must put it off till I can get it down from the other house.'

'Jane shall fetch it, if you really want it,' said Mrs. Dale.

'Bell shall get it, when she goes up in the afternoon; will you, Bell? And I'll try to get on with this stuff in the meantime.' Then again she sat with her eyes fixed upon the pages of the book. 'I'll tell you what, mamma, – you may have some comfort in this: that when to-day's gone by, I shan't make a fuss about any other day.'

'Nobody thinks that you are making a fuss, Lily.'

'Yes, but I am. Isn't it odd, Bell, that it should take place on Valentine's day? I wonder whether it was so settled on purpose, because of the day. Oh, dear, I used to think so often of the letter that I should get from him on this day, when he would tell me that I was his valentine. Well; he's got another – valen – tine – now.' So much she said with articulate voice, and then she broke down, bursting out into convulsive sobs, and crying in her mother's arms as though she would break her heart. And yet her heart was not broken, and she was still strong in that resolve which she had made, that her grief should not overpower her. As she had herself said, the thing would not have been so difficult, had she not been weakened by illness.

'Lily, my darling; my poor, ill-used darling.'

'No, mamma, I won't be that.' And she struggled grievously to get the better of the hysterical attack which had overpowered her. 'I won't be regarded as ill-used; not as specially ill-used. But I am your darling, your own darling. Only I wish you'd beat me and thump me when I'm such a fool, instead of pitying me. It's a great mistake being soft to people when they make fools of themselves. There, Bell; there's your stupid book, and I won't have any more of it. I believe it was that that did it.' And she pushed the book away from her.

After this little scene she said no further word about Crosbie and his bride on that day, but turned the conversation towards the prospect of their new house at Guestwick.

'It will be a great comfort to be nearer Dr. Crofts; won't it, Bell?'

'I don't know,' said Bell.

'Because if we are ill, he won't have such a terrible distance to come?'

'That will be a comfort for him, I should think,' said Bell, very demurely.

In the evening the first volume of the French Revolution

had been procured, and Lily stuck to her reading with laudable perseverance; till at eight her mother insisted on her going to bed, queen as she was.

'I don't believe a bit, you know, that the king was such a bad man as that,' she said.

'I do,' said Bell.

'Ah, that's because you're a Radical. I never will believe that kings are so much worse than other people. As for Charles the First, he was about the best man in history.'

This was an old subject of dispute; but Lily on the present occasion was allowed her own way, – as being an invalid.

CHAPTER XLV

VALENTINE'S DAY IN LONDON

THE fourteenth of February in London was quite as black, and cold, and as wintersome as it was at Allington, and was, perhaps, somewhat more melancholy in its coldness. Nevertheless Lady Alexandrina De Courcy looked as bright as bridal finery could make her, when she got out of her carriage and walked into St. James's church at eleven o'clock on that morning.

It had been finally arranged that the marriage should take place in London. There were certainly many reasons which would have made a marriage from Courcy Castle more convenient. The De Courcy family were all assembled at their country family residence, and could therefore have been present at the ceremony without cost or trouble. The castle too was warm with the warmth of life; and the pleasantness of home would have lent a grace to the departure of one of the daughters of the house. The retainers and servants were there, and something of the rich mellowness of a noble alliance might have been felt, at any rate by Crosbie, at a marriage so celebrated. And it must have been acknowledged, even by Lady De Courcy, that the house in Portman Square was very cold – that a marriage from thence would be cold, – that there could be no hope of attaching to it any honour and glory, or of making it resound with fashionable éclat in the columns of the *Morning Post*. But then, had they been married in the country, the earl would have been there; whereas there was no probability of his travelling up to London for the purpose of being present on such an occasion.

The earl was very terrible in these days, and Alexandrina, as she became confidential in her communications with her future husband, spoke of him as of an ogre, who could not by any means be avoided in all the concerns of life, but whom one might shun now and again by some subtle device and careful arrangement of favourable circumstances. Crosbie had more than once taken upon himself to hint that he did not specially regard the ogre, seeing that for the future he could keep himself altogether apart from the malicious monster's dominions.

'He will not come to me in our new home,' he had said to his love, with some little touch of affection. But to this view of the case Lady Alexandrina had demurred. The ogre in question was not only her parent, but was also a noble peer, and she could not agree to any arrangement by which their future connection with the earl, and with nobility in general, might be endangered. Her parent, doubtless, was an ogre, and in his ogreship could make himself very terrible to those near him; but then might it not be better for them to be near to an earl who was an ogre, than not to be near to any earl at all? She had therefore signified to Crosbie that the ogre must be endured.

But, nevertheless, it was a great thing to be rid of him on that happy occasion. He would have said very dreadful things, – things so dreadful that there might have been a question whether the bridegroom could have borne them. Since he had heard of Crosbie's accident at the railway station, he had constantly talked with fiendish glee of the beating which had been administered to his son-in-law. Lady De Courcy in taking Crosbie's part, and maintaining that the match was fitting for her daughter, had ventured to declare before her husband that Crosbie was a man of fashion, and the earl would now ask, with a loathsome grin, whether the bridegroom's fashion had been improved by his little adventure at Paddington. Crosbie, to whom all this was not repeated, would have preferred a wedding in the country. But the countess and Lady Alexandrina knew better.

The earl had strictly interdicted any expenditure, and the countess had of necessity construed this as forbidding any unnecessary expense. 'To marry a girl without any immediate cost was a thing which nobody could understand,' as the countess remarked to her eldest daughter.

'I would really spend as little as possible,' Lady Amelia had answered. 'You see, mamma, there are circumstances about it which one doesn't wish to have talked about just at present. There's the story of that girl, – and then that fracas at the station. I really think it ought to be as quiet as possible.' The good sense of Lady Amelia was not to be disputed, as her mother acknowledged. But then if the marriage were managed in any notoriously quiet way, the very notoriety of that quiet would be as dangerous as an attempt at loud glory. 'But it won't cost as much,' said Amelia. And thus it had been resolved that the wedding should be very quiet.

To this Crosbie had assented very willingly, though he had not relished the manner in which the countess had explained to him her views.

'I need not tell you, Adolphus,' she had said, 'how thoroughly satisfied I am with this marriage. My dear girl feels that she can be happy as your wife, and what more can I want? I declared to her and to Amelia that I was not ambitious, for their sakes, and have allowed them both to please themselves.'

'I hope they have pleased themselves,' said Crosbie.

'I trust so; but nevertheless, – I don't know whether I make myself understood?'

'Quite so, Lady De Courcy. If Alexandrina were going to marry the eldest son of a marquis, you would have a longer procession to church than will be necessary when she marries me.'

'You put it in such an odd way, Adolphus.'

'It's all right so long as we understand each other. I can assure you I don't want any procession at all. I should be quite contented to go down with Alexandrina, arm in arm, like Darby and Joan, and let the clerk give her away.'

We may say that he would have been much better contented could he have been allowed to go down the street without any encumbrance on his arm. But there was no possibility now for such deliverance as that.

Both Lady Amelia and Mr. Gazebee had long since discovered the bitterness of his heart and the fact of his repentance, and Gazebee had ventured to suggest to his wife that his noble sister-in-law was preparing for herself a life of misery.

'He'll become quiet and happy when he's used to it,' Lady Amelia had replied, thinking, perhaps, of her own experiences.

'I don't know, my dear; he's not a quiet man. There's something in his eye which tells me that he could be very hard to a woman.'

'It has gone too far now for any change,' Lady Amelia had answered.

'Well; perhaps it has.'

'And I know my sister so well; she would not hear of it. I really think they will do very well when they become used to each other.'

Mr. Gazebee, who also had had his own experiences, hardly dared to hope so much. His home had been satisfactory to him, because he had been a calculating man, and having made his calculation correctly was willing to take the net result. He had done so all his life with success. In his house his wife was paramount, – as he very well knew. But no effort on his wife's part, had she wished to make such effort, could have forced him to spend more than two-thirds of his income. Of this she also was aware, and had trimmed her sails accordingly, likening herself to him in this respect. But of such wisdom, and such trimmings, and such adaptability, what likelihood was there with Mr. Crosbie and Lady Alexandrina?

'At any rate, it is too late now,' said Lady Amelia, thus concluding the conversation.

But nevertheless, when the last moment came, there was

some little attempt at glory. Who does not know the way in which a lately married couple's little dinner-party stretches itself out from the pure simplicity of a fried sole and a leg of mutton to the attempt at clear soup, the unfortunately cold dish of round balls which is handed about after the sole, and the brightly red jelly, and beautifully pink cream, which are ordered, in the last agony of ambition, from the next pastrycook's shop?

'We cannot give a dinner, my dear, with only cook and Sarah.'

It has thus begun, and the husband has declared that he has no such idea. 'If Phipps and Dowdney can come here and eat a bit of mutton, they are very welcome; if not, let them stay away. And you might as well ask Phipps's sister; just to have some one to go with you into the drawing-room.'

'I'd much rather go alone, because then I can read,' – or sleep, we may say.

But her husband has explained that she would look friendless in this solitary state, and therefore Phipps's sister has been asked. Then the dinner has progressed down to those costly jellies which have been ordered in a last agony. There has been a conviction on the minds of both of them that the simple leg of mutton would have been more jolly for them all. Had those round balls not been carried about by a hired man; had simple mutton with hot potatoes been handed to Miss Phipps by Sarah, Miss Phipps would not have simpered with such unmeaning stiffness when young Dowdney spoke to her. They would have been much more jolly. 'Have a bit more mutton, Phipps; and where do you like it?' How pleasant it sounds! But we all know that it is impossible. My young friend had intended this, but his dinner had run itself away to cold round balls and coloured forms from the pastrycook. And so it was with the Crosbie marriage.

The bride must leave the church in a properly appointed carriage, and the postboys must have wedding favours. So the thing grew; not into noble proportions, not into proportions of true glory, justifying the attempt and making good the gala. A well-cooked rissole,* brought pleasantly to you, is good eating. A gala marriage, when everything is in keeping, is excellent sport. Heaven forbid that we should have no gala marriages! But the small spasmodic attempt, made in opposition to manifest propriety, made with an inner conviction of failure, – that surely should be avoided in marriages, in dinners, and in all affairs of life.

There were bridesmaids and there was a breakfast. Both Margaretta and Rosina came up to London for the occasion, as did also a first cousin of theirs, one Miss Gresham, a lady whose father lived in the same county. Mr. Gresham had married a sister of Lord De Courcy's, and his services were also called into requisition. He was brought up to give away the bride, because the earl, – as the paragraph in the newspaper declared, – was confined at Courcy Castle by his old hereditary enemy, the gout. A fourth bridesmaid also was procured, and thus there was a bevy, though not so large a bevy as is now generally thought to be desirable. There were only three or four carriages at the church, but even three or four were something. The weather was so frightfully cold that the light-coloured silks of the ladies carried with them a show of discomfort. Girls should be very young to look nice in light dresses on a frosty morning, and the bridesmaids at Lady Alexandrina's wedding were not very young. Lady Rosina's nose was decidedly red. Lady Margaretta was very wintry, and apparently very cross. Miss Gresham was dull, tame, and insipid; and the Honourable Miss O'Flaherty, who filled the fourth place, was sulky at finding that she had been invited to take a share in so very lame a performance.

But the marriage was made good, and Crosbie bore up against his misfortunes like a man. Montgomerie Dobbs and Fowler Pratt both stood by him, giving him, let us hope, some assurance that he was not absolutely deserted by all the world, – that he had not given himself up, bound hand and foot, to the De Courcys, to be dealt with in all matters as they might please. It was that feeling which had been so grievous to him, – and that other feeling, cognate to it, that if he should ultimately succeed in rebelling against the De Courcys, he would find himself a solitary man.

'Yes; I shall go,' Fowler Pratt had said to Montgomerie Dobbs. 'I always stick to a fellow if I can. Crosbie has behaved like a blackguard, and like a fool also; and he knows that I think so. But I don't see why I should drop him on that account. I shall go as he has asked me.'

'So shall I,' said Montgomerie Dobbs, who considered that he would be safe in doing whatever Fowler Pratt did, and who remarked to himself that after all Crosbie was marrying the daughter of an earl.

Then, after the marriage, came the breakfast, at which the countess presided with much noble magnificence. She had not gone to church, thinking, no doubt, that she would be better able to maintain her good humour at the feast, if she did not subject herself to the chance of lumbago in the church. At the foot of the table sat Mr. Gresham, her brother-in-law, who had undertaken to give the necessary toast and make the necessary speech. The Honourable John was there, saying all manner of ill-natured things about his sister and new brother-in-law, because he had been excluded from his proper position at the foot of the table. But Alexandrina had declared that she would not have the matter entrusted to her brother. The Honourable George would not come, because the countess had not asked his wife.

'Maria may be slow, and all that sort of thing,' George had said; 'but she is my wife. And she had got what they haven't. Love me, love my dog, you know.' So he had stayed down at Courcy, – very properly as I think.

Alexandrina had wished to go away before breakfast, and Crosbie would not have cared how early an escape had been provided for him; but the countess had told her daughter that if she would not wait for the breakfast, there should be no breakfast at all, and in fact no wedding; nothing but a simple marriage. Had there been a grand party, that going away of the bride and bridegroom might be very well; but the countess felt that on such an occasion as this nothing but the presence of the body of the sacrifice could give any reality to the festivity. So Crosbie and Lady Alexandrina Crosbie heard Mr. Gresham's speech, in which he prophesied for the young couple an amount of happiness and prosperity almost greater than is compatible with the circumstances of humanity. His young friend Crosbie, whose acquaintance he had been delighted to make, was well known as one of the rising pillars of the State. Whether his future career might be parliamentary, or devoted to the permanent Civil Service of the country, it would be alike great, noble, and prosperous. As to his dear niece, who was now filling that position in life which was most beautiful and glorious for a young woman, – she could not have done better. She had preferred genius to wealth, – so said Mr. Gresham, – and she would find her fitting reward. As to her finding her fitting reward, whatever her preferences may have been, there Mr. Gresham was no doubt quite right. On that head I myself have no doubt whatever. After that, Crosbie returned thanks, making a much better speech than nine men do out of ten on such occasions, and then the thing was over. No other speaking was allowed, and within half an hour from that time, he and his bride were in the post-chaise, being carried away to the Folkestone railway station; for that place had been chosen as the scene of their honeymoon. It had been at one time intended that the journey to Folkestone should be made simply as the first stage to Paris, but Paris and all foreign travelling had been given up by degrees.

'I don't care a bit about France, – we have been there so often,' Alexandrina said.

She had wished to be taken to Naples, but Crosbie had made her understand at the first whispering of the word, that Naples was quite out of the question. He must look now in all things to money. From the very first outset of his career he must save a shilling wherever a shilling could be saved. To this view of life no opposition was made by the De Courcy interest. Lady Amelia had explained to her sister that they ought so to do their honeymooning that it should not cost more than if they began keeping house at once. Certain things must be done which, no doubt, were costly in their nature. The bride must take with her a well-dressed lady's-maid. The rooms at the Folkestone hotel must be large, and on the first floor. A carriage must be hired for her use while she remained; but every shilling must be saved the spending of which would not make itself apparent to the outer world. Oh, deliver us from the poverty of those who, with small means, affect a show of wealth! There is no whitening equal to that of sepulchres whited as they are whited!

By the proper administration of a slight bribe Crosbie secured for himself and his wife a compartment in the railway carriage to themselves. And as he seated himself opposite to Alexandrina, having properly tucked her up with all her bright-coloured trappings, he remembered that he had never in truth been alone with her before. He had danced with her frequently, and been left with her for a few minutes between the figures. He had flirted with her in crowded drawing-rooms, and had once found a moment at Courcy Castle to tell her that he was willing to marry her, in spite of his engagement with Lilian Dale. But he had never walked with her for hours together as he had walked with Lily. He had never talked to her about government, and politics, and books, nor had she talked to him of poetry, of religion, and of the little duties and comforts of life. He had known the Lady Alexandrina for the last six or seven years; but he had never known her, – perhaps never would know her, – as he had learned to know Lily Dale within the space of two months.

And now that she was his wife, what was he to say to her? They two had commenced a partnership which was to make of them for the remaining term of their lives one body and one flesh. They were to be all-in-all to each other. But how was he to begin this all-in-all partnership? Had the priest, with his blessing, done it so sufficiently that no other doing on Crosbie's own part was necessary? There she was, opposite to him, his very actual wife, – bone of his bone; and what was he to say to her? As he settled himself on his seat, taking over his own knees a part of a fine fur rug trimmed with scarlet, with which he had covered her other mufflings, he bethought himself how much easier it would have been to talk to Lily. And Lily would have been ready with all her ears, and all her mind, and all her wit, to enter quickly upon whatever thoughts had occurred to him. In that respect Lily would have been a wife indeed, – a wife that would have transferred herself with quick mental activity into her husband's mental sphere. Had he begun about his office Lily would have been ready for him, but Alexandrina had never yet asked him a single question about his official life. Had he been prepared with a plan for to-morrow's happiness Lily would have taken it up eagerly, but Alexandrina never cared for such trifles.

'Are you quite comfortable?' he said, at last.

'Oh, yes, quite, thank you. By-the-by, what did you do with my dressing-case?'

And that question she did ask with some energy.

'It is under you. You can have it as foot-stool if you like it.'

'Oh, no; I should scratch it. I was afraid that if Hannah had it, it might be lost.' Then again there was silence, and Crosbie again considered as to what he would next say to his wife.

We all know the advice given us of old as to what we should do under such circumstances; and who can be so thoroughly justified in following that advice as a newly-married husband? So he put out his hand for hers and drew her closer to him.

'Take care of my bonnet,' she said, as she felt the motion of the railway carriage when he kissed her. I don't think he kissed her again till he had landed her and her bonnet safely at Folkestone. How often would he have kissed Lily, and how pretty would her bonnet have been when she reached the end of her journey, and how delightfully happy would she have looked when she scolded him for bending it! But Alexandrina was quite in earnest about her bonnet; by far too much in earnest for any appearance of happiness.

So he sat without speaking, till the train came to the tunnel.

'I do so hate tunnels,' said Alexandrina.

He had half intended to put out his hand again, under some mistaken idea that the tunnel afforded him an opportunity. The whole journey was one long opportunity, had he desired it; but his wife hated tunnels, so he drew his hand back again. Lily's little fingers would have been ready for his touch. He thought of this, and could not help thinking of it.

He had *The Times* newspaper in his dressing-bag. She also had a novel with her. Would she be offended if he took out the paper and read it? The miles seemed to pass by very slowly, and there was still another hour down to Folkestone. He longed for his *Times*, but resolved at last that he would not read unless she read first. She also had remembered her novel; but by nature she was more patient than he, and she thought that on such a journey any reading might perhaps be almost improper. So she sat tranquilly, with her eyes fixed on the netting over her husband's head.

At last he could stand it no longer, and he dashed off into a conversation, intended to be most affectionate and serious.

'Alexandrina,' he said, and his voice was well-tuned for the tender serious manner, had her ears been alive to such tuning. 'Alexandrina, this is a very important step that you and I have taken today.'

'Yes; it is, indeed,' said she.

'I trust we shall succeed in making each other happy.'

'Yes; I hope we shall.'

'If we both think seriously of it, and remember that that is our chief duty, we shall do so.'

'Yes, I suppose we shall. I only hope we shan't find the house very cold. It is so new, and I am so subject to colds in my head. Amelia says we shall find it very cold; but then she was always against our going there.'

'The house will do very well,' said Crosbie. And Alexandrina could perceive that there was something of the master in his tone as he spoke.

'I am only telling you what Amelia said,' she replied.

Had Lily been his bride, and had he spoken to her of their future life and mutual duties, how she would have kindled to the theme! She would have knelt at his feet on the floor of the carriage, and, looking up into his face, would have promised him to do her best, – her best, – her very best. And with what an eagerness of inward resolution would she have determined to keep her promise. He thought of all this now, but he knew that he ought not to think of it. Then, for some quarter of an hour, he did take out his newspaper, and she, when she saw him do so, did take out her novel.

He took out his newspaper, but he could not fix his mind upon the politics of the day. Had he not made a terrible mistake? Of what use to him in life would be that thing of a woman that sat opposite to him? Had not a great punishment come upon him, and had he not deserved the punishment? In truth, a great punishment had come upon him. It was not only that he had married a woman incapable of understanding the higher duties of married life, but that he himself would have been capable of appreciating the value of a woman who did understand them. He would have been happy with Lily Dale; and therefore we may surmise that his unhappiness with Lady Alexandrina would be the greater. There are men who, in marrying such as Lady Alexandrina De Courcy, would get the article best suited to them, as Mortimer Gazebee had done in marrying her sister. Miss Griselda Grantly, who had become Lady Dumbello, though somewhat colder and somewhat cleverer than Lady Alexandrina, had been of the same sort. But in marrying her, Lord Dumbello had got the article best suited to him; – if only the ill-natured world would allow him to keep the article. It was in this that Crosbie's failure had been so grievous, – that he had seen and approved the better course, but had chosen for himself to walk in that which was worse. During that week at Courcy Castle, – the week which he passed there immediately after his visit to Allington, – he had deliberately made up his mind that he was more fit for the bad course than for the good one. The course was now before him, and he had no choice but to walk in it.

It was very cold when they got to Folkestone, and Lady Alexandrina shivered as she stepped into the private-looking carriage which had been sent to the station for her use.

'We shall find a good fire in the parlour at the hotel,' said Crosbie.

'Oh, I hope so,' said Alexandrina, 'and in the bedroom too.'

The young husband felt himself to be offended, but he hardly knew why. He felt himself to be offended, and with difficulty induced himself to go through all those little ceremonies the absence of which would have been remarked by everybody. He did his work, however, seeing to all her shawls and wrappings, speaking with good-nature to Hannah, and paying special attention to the dressing-case.

'What time would you like to dine?' he asked, as he prepared to leave her alone with Hannah in the bedroom.

'Whenever you please; only I should like some tea and bread-and-butter presently.'

Crosbie went into the sitting-room, ordered the tea and bread-and-butter, ordered also the dinner, and then stood himself up with his back to the fire, in order that he might think a little of his future career.

He was a man who had long since resolved that his life should be a success. It would seem that all men would so resolve, if the matter were simply one of resolution. But the majority of men, as I take it, make no such resolution, and very many men resolve that they will be unsuccessful. Crosbie, however, had resolved on success, and had done much towards carrying out his purpose. He had made a name for himself, and had acquired a certain fame. That, however, was, as he acknowledged to himself, departing from him. He looked the matter straight in the face, and told himself that his fashion must be abandoned; but the office remained to him. He might still rule over Mr. Optimist, and make a subservient slave of Butterwell. That must be his line in life now, and to that line he would endeavour to be true. As to his wife, and his home, – he would look to them for his breakfast, and perhaps his dinner. He would have a comfortable arm-chair, and if Alexandrina should become a mother, he would endeavour to love his children; but above all things he would never think of Lily. After that he stood and thought of her for half an hour.

'If you please, sir, my lady wants to know at what time you have ordered dinner.'

'At seven, Hannah.'

'My lady says she is very tired, and will lie down till dinner time.'

'Very well, Hannah. I will go into her room when it is time to dress. I hope they are making you comfortable downstairs?'

Then Crosbie strolled out on the pier in the dusk of the cold winter evening.

CHAPTER XLVI

JOHN EAMES AT HIS OFFICE

Mr. Crosbie and his wife went upon their honeymoon tour to Folkestone in the middle of February, and returned to London about the end of March. Nothing of special moment to the interests of our story occurred during those six weeks, unless the proceedings of the young married couple by the sea-side may be thought to have any special interest. With regard to those proceedings I can only say that Crosbie was very glad when they were brought to a close. All holiday-making is hard work, but holiday-making with nothing to do is the hardest work of all. At the end of March they went into their new house, and we will hope that Lady Alexandrina did not find it very cold.

During this time Lily's recovery from her illness was being completed. She had no relapse, nor did anything occur to create a new fear on her account. But, nevertheless, Dr. Crofts gave it as his opinion that it would be expedient to move her into a fresh house at Lady-day. March is not a kindly month for invalids; and therefore with some regret on the part of Mrs. Dale, with much impatience on that of Bell, and with considerable outspoken remonstrance from Lily herself, the squire was requested to let them remain through the month of April. How the squire received this request, and in what way he assented to the doctor's reasoning, will be told in the course of a chapter or two.

In the meantime John Eames had continued his career in London without much immediate satisfaction to himself, or to the lady who boasted to be his heart's chosen queen. Miss Amelia Roper, indeed, was becoming very cross, and in her ill-temper was playing a game that was tending to create a frightful amount of hot water in Burton Crescent. She was devoting herself to a flirtation with Mr. Cradell, not only under the immediate eyes of Johnny Eames, but also under those of Mrs. Lupex. John Eames, the blockhead, did not like it. He was above all things anxious to get rid of Amelia and her claims; so anxious, that on certain moody occasions he would threaten himself with divers tragical terminations to his career in London. He would enlist. He would go to Australia. He would blow out his brains. He would have 'an explanation' with Amelia, tell her that she was a vixen, and proclaim his hatred. He would rush down to Allington and throw himself in despair at Lily's feet. Amelia was the bugbear of his life. Nevertheless, when she flirted with Cradell, he did not like it, and was ass enough to speak to Cradell about it.

'Of course I don't care,' he said, 'only it seems to me that you are making a fool of yourself.'

'I thought you wanted to get rid of her.'

'She's nothing on earth to me; only it does, you know—'

'Does do what?' asked Cradell.

'Why, if I was to be fal-lalling with that married woman, you wouldn't like it. That's all about it. Do you mean to marry her?'

'What! – Amelia?'

'Yes; Amelia.'

'Not if I know it.'

'Then if I were you I would leave her alone. She's only making a fool of you.'

Eames' advice may have been good, and the view taken by him of Amelia's proceedings may have been correct; but as regarded his own part in the affair, he was not wise. Miss Roper, no doubt, wished to make him jealous; and she succeeded in the teeth of his aversion to her and of his love elsewhere. He had not desire to say soft things to Miss Roper. Miss Roper, with all her skill, could not extract a word pleasantly soft from him once a week. But, nevertheless, soft words to her and from her in another quarter made him uneasy. Such being the case, must we not acknowledge that John Eames was still floundering in the ignorance of his hobbledehoyhood?

The Lupexes at this time still held their ground in the Crescent, although repeated warnings to go had been given them. Mrs. Roper, though she constantly spoke of sacrificing all that they owed her, still hankered, with a natural hankering, after her money. And as each warning was accompanied by a demand for payment, and usually produced some slight subsidy on account, the thing went on from week to week; and at the beginning of April Mr. and Mrs. Lupex were still boarders at Mrs. Roper's house.

Eames had heard nothing from Allington since the time of his Christmas visit, and his subsequent correspondence with Lord De Guest. In his letters from his mother he was told that game came frequently from Guestwick Manor, and in this way he knew that he was not forgotten by the earl. But of Lily he had heard not a word, – except, indeed, the rumour, which had now become general, that the Dales from the Small House were about to move themselves into Guestwick. When first he learned this he construed the tidings as favourable to himself, thinking that Lily, removed from the grandeur of Allington, might possibly be more easily within his reach; but, latterly, he had given up any such hope as that, and was telling himself that his friend at the Manor had abandoned all idea of making up the marriage. Three months had already elapsed since his visit. Five months had passed since Crosbie had surrendered his claim. Surely such a knave as Crosbie might be forgotten in five months! If any steps could have been taken through the squire, surely three months would have sufficed for them! It was very manifest to him that there was no ground of hope for him at Allington, and it would certainly be well for him to go off to Australia. He would go to Australia, but he would thrash Cradell first for having dared to interfere with Amelia Roper. That, generally, was the state of his mind during the first week in April.

Then there came to him a letter from the earl which

instantly effected a great change in all his feelings; which taught him to regard Australia as a dream, and almost put him into a good humour with Cradell. The earl had by no means lost sight of his friend's interests at Allington; and, moreover, those interests were now backed by an ally who in this matter must be regarded as much more powerful than the earl. The squire had given in his consent to the Eames alliance.

The earl's letter was as follows:—

'GUESTWICK MANOR, April 7, 18—.

'My dear John,

'I told you to write to me again, and you haven't done it. I saw your mother the other day, or else you might have been dead for anything I knew. A young man always ought to write letters when he is told to do so. [Eames, when he had got so far, felt himself rather aggrieved by this rebuke, knowing that he had abstained from writing to his patron simply from an unwillingness to intrude upon him with his letters. 'By Jove, I'll write to him every week of his life, till he's sick of me,' Johnny said to himself when he found himself thus instructed as to a young man's duties.]

'And now I have got to tell you a long story, and I should like it much better if you were down here, so that I might save myself the trouble; but you would think me ill-natured if I were to keep you waiting. I happened to meet Mr. Dale the other day, and he said that he should be very glad if a certain young lady would make up her mind to listen to a certain young friend of mine. So I asked him what he meant to do about the young lady's fortune, and he declared himself willing to give her a hundred a year during his life, and to settle four thousand pounds upon her after his death. I said that I would do as much on my part by the young man; but as two hundred a year, with your salary, would hardly give you enough to begin with, I'll make mine a hundred and fifty. You'll be getting up in your office soon, and with five hundred a year you ought to be able to get along; especially as you need not insure your life. I should live somewhere near Bloomsbury Square at first, because I'm told you can get a house for nothing. After all, what's fashion worth? You can bring your wife down here in the autumn, and have some shooting. She won't let you go to sleep under the trees, I'll be bound.

'But you must look after the young lady. You will understand that no one has said a word to her about it; or, if they have, I don't know it. You'll find the squire on your side, that's all. Couldn't you manage to come down this Easter? Tell old Buffle, with my compliments, that I want you. I'll write to him if you like it. I did know him at one time, though I can't say I was ever very fond of him. It stands to reason that you can't get on with Miss Lily without seeing her; unless, indeed, you like better to write to her, which always seems to me to be very poor sort of fun. You'd much better come down, and go a-wooing in the regular old-fashioned way. I need not tell you that Lady Julia will be delighted to see you. You are a prime favourite with her since that affair at the railway station. She thinks a great deal more about that than she does about the bull.

'Now, my dear fellow, you know all about it, and I shall take it very much amiss of you if you don't answer my letter soon.

'Your very sincere friend,
De Guest.'

When Eames had finished this letter, sitting at his office-desk, his surprise and elation were so great that he hardly knew where he was or what he ought to do. Could it be the truth that Lily's uncle had not only consented that the match should be made, but that he had also promised to give his niece a considerable fortune? For a few minutes it seemed to Johnny as though all obstacles to his happiness were removed, and that there was no impediment between him and an amount of bliss of which he had hitherto hardly dared to dream. Then, when he considered the earl's munificence, he almost cried. He found that he could not compose his mind to think, or even his hand to write. He did not know whether it would be right in him to accept such pecuniary liberality from any living man, and almost thought that he should feel himself bound to reject the earl's offer. As to the squire's money, that he knew he might accept. All that comes in the shape of a young woman's fortune may be taken by any man.

He would certainly answer the earl's letter, and that at once. He would not leave the office till he had done so. His friend should have cause to bring no further charge against him of that kind. And then again he reverted to the injustice which had been done to him in the matter of letter-writing – as if that consideration were of moment in such a state of circumstances as was now existing. But at last his thoughts brought themselves to the real question at issue. Would Lily Dale accept him? After all, the realization of his good fortune depended altogether upon her feelings; and, as he remembered this, his mind misgave him sorely. It was filled not only with a young lover's ordinary doubts, – with the fear and trembling, incidental to the bashfulness of hobbledehoyhood – but with an idea that that affair with Crosbie would still stand in his way. He did not, perhaps, rightly understand all that Lily had suffered, but he conceived it to be probable that there had been wounds which even the last five months might not yet have cured. Could it be that she would allow him to cure these wounds? As he thought of this he felt almost crushed to the earth by an indomitable bashfulness and conviction of his own unworthiness. What had he to offer worthy of the acceptance of such a girl as Lilian Dale?

I fear that the Crown did not get out of John Eames an adequate return for his salary on that day. So adequate, however, had been the return given by him for some time past, that promotion was supposed throughout the Income-tax Office to be coming in his way, much to the jealousy of Cradell, Fisher, and others, his immediate compeers and cronies. And the place assigned to him by rumour was one which was generally regarded as a perfect Elysium upon earth in the Civil Service world. He was, so rumour said, to become private secretary to the First Commissioner. He would be removed by such a change as this from the large uncarpeted room in which he at present sat; occupying the same desk with another man to whom he had felt himself to be ignominiously bound, as dogs must feel when they are coupled. This room had been the bear-garden of the office. Twelve or fourteen men sat in it. Large pewter pots were brought into it daily at one o'clock, giving it an air that was not aristocratic. The senior of the room, one Mr. Love, who was presumed to have it under his immediate dominion, was a clerk of the ancient stamp, dull, heavy, unambitious, living out on the farther side of Islington, and unknown beyond the limits of his office to any of his younger brethren. He was generally regarded as having given a bad tone to the room. And then the clerks in this room would not unfrequently be blown up, – with very palpable blowings up, – by an official swell, a certain chief clerk, named Kissing, much higher in standing though younger in age than the gentleman of whom we have before spoken. He would hurry in, out of his own neighbouring chamber, with quick step and nose in the air, shuffling in his office slippers, looking on each occasion as though there was some cause to fear that the whole Civil Service were coming to an abrupt termination, and would lay about him with hard words, which some of those in the big room did not find it very easy to bear. His hair was always brushed straight up, his eyes were always very wide open, – and he usually carried a big letter-book with him, keeping in it a certain place with his finger. This book was almost too much for his strength, and he would flop it down, now on this man's desk and now on that man's, and in a long career of such floppings had made himself to be very much hated. On the score of some old grudge he and Mr. Love did not speak to each other; and for this reason, on all occasions of fault-finding, the blown-up young man would refer Mr. Kissing to his enemy.

'I know nothing about it,' Mr. Love would say, not lifting his face from his desk for a moment.

'I shall certainly lay the matter before the Board,' Mr. Kissing would reply, and would then shuffle out of the room with the big book.

Sometimes Mr. Kissing would lay the matter before the Board, and then he, and Mr. Love, and two or three delinquent clerks would be summoned thither. It seldom led to much. The delinquent clerks would be cautioned. One Commissioner would say a word in private to Mr. Love, and another a word in private to Mr. Kissing. Then, when left alone, the Commissioners would have their little jokes, saying that Kissing, they feared, went by favour; and that Love should still be lord of all. But these things were done in the mild days, before Sir Raffle Buffle came to the Board.

There had been some fun in this at first; but of late John Eames had become tired of it. He disliked Mr. Kissing, and the big book out of which Mr. Kissing was always endeavouring to convict him of some official sin, and had got tired of that joke of setting Kissing and Love by the ears together. When the Assistant Secretary first suggested to him that Sir Raffle had an idea of selecting him as private secretary, and when he remembered the cosy little room, all carpeted, with a leathern arm-chair and a separate washing-stand, which in such case would be devoted to his use, and remembered also that he would be put into receipt of an additional hundred a year, and would stand in the way of still better promotion, he was overjoyed. But there were certain drawbacks. The present private secretary, – who had been private secretary also to the late First Commissioner, – was giving up his Elysium because he could not endure the tones of Sir Raffle's voice. It was understood that Sir Raffle required rather more of a private secretary, in the way of obsequious attendance, than was desirable, and Eames almost doubted his own fitness for the place.

'And why should he choose me?' he had asked the Assistant Secretary.

'Well, we have talked it over together, and I think that he prefers you to any other that has been named.'

'But he was so very hard upon me about the affair at the railway station.'

'I think he has heard more about that since; I think that some message has reached him from your friend, Earl De Guest.'

'Oh, indeed!' said Johnny, beginning to comprehend what it was to have an earl for his friend. Since his acquaintance with the nobleman had commenced he had studiously avoided all mention of the earl's name at his office; and yet he received almost daily intimation that the fact was well known there, and not a little considered.

'But he is so very rough,' said Johnny.

'You can put up with that,' said his friend the Assistant Secretary. 'His bark is worse than his bite, as you know; and then a hundred a year is worth having.' Eames was at that moment inclined to take a gloomy view of life in general, and was disposed to refuse the place, should it be offered to him. He had not then received the earl's letter; but now, as he sat with that letter open before him, lying in the drawer beneath his desk so that he could still read it as he leaned back in his chair, he was enabled to look at things in general through a different atmosphere. In the first place, Lilian Dale's husband ought to have a room to himself, with a carpet and an arm-chair; and then that additional hundred a year would raise his income at once to the sum as to which the earl had made some sort of stipulation. But could he get that leave of absence at Easter? If he consented to be Sir Raffle's private secretary, he would make that a part of the bargain.

At this moment the door of the big room was opened, and Mr. Kissing shuffled in with very quick little steps. He shuffled in, and coming direct up to John's desk, flopped his ledger down upon it before its owner had had time to close the drawer which contained the precious letter.

'What have you got in that drawer, Mr. Eames?'

'A private letter, Mr. Kissing.'

'Oh; – a private letter!' said Mr. Kissing, feeling strongly convinced there was a novel hidden there, but not daring to express his belief. 'I have been half the morning, Mr. Eames, looking for this letter to the Admiralty, and you've put it under S!' A bystander listening to Mr. Kissing's tone would have been led to believe that the whole Income-tax Office was jeopardized by the terrible iniquity thus disclosed.

'Somerset House,' pleaded Johnny.

'Psha; – Somerset House! Half the offices in London—'

'You'd better ask Mr. Love,' said Eames. 'It's all done

under his special instructions.' Mr. Kissing looked at Mr. Love, and Mr. Love looked steadfastly at his desk. 'Mr. Love knows all about the indexing,' continued Johnny. 'He's index master general to the department.'

'No, I'm not, Mr. Eames,' said Mr. Love, who rather liked John Eames, and hated Mr. Kissing with his whole heart. 'But I believe the indexes, on the whole, are very well done in this room. Some people don't know how to find letters.'

'Mr. Eames,' began Mr. Kissing, still pointing with a finger of bitter reproach to the misused S, and beginning an oration which was intended for the benefit of the whole room, and for the annihilation of old Mr. Love, 'if you have yet to learn that the word Admiralty begins with A and not with S, you have much to learn which should have been acquired before you first came into this office. Somerset House is not a department.' Then he turned round to the room at large, and repeated the last words, as though they might become very useful if taken well to heart— 'Is not a department. The Treasury is a department; the Home Office is a department; the India Board is a department—'

'No, Mr. Kissing, it isn't,' said a young clerk from the other end of the room.

'You know very well what I mean, sir. The India Office is a department.'

'There's no Board, sir.'

'Never mind; but how any gentleman who has been in the service three months, – not to say three years, – can suppose Somerset House to be a department, is beyond my comprehension. If you have been improperly instructed—'

'We shall know all about it another time,' said Eames. 'Mr. Love will make a memorandum of it.'

'I shan't do anything of the kind,' said Mr. Love.

'If you have been wrongly instructed,—' Mr. Kissing began again, stealing a glance at Mr. Love as he did so; but at this moment the door was again opened, and a messenger summoned Johnny to the presence of the really great man. 'Mr. Eames, to wait upon Sir Raffle.' Upon hearing this Johnny immediately started, and left Mr. Kissing and the big book in possession of his desk. How the battle was waged, and how it raged in the large room, we cannot stop to hear, as it is necessary that we should follow our hero into the presence of Sir Raffle Buffle.

'Ah, Eames, – yes,' said Sir Raffle, looking up from his desk when the young man entered; 'just wait half a minute, will you?' And the knight went to work at his papers, as though fearing that any delay in what he was doing might be very prejudicial to the nation at large. 'Ah, Eames, – well, – yes,' he said again, as he pushed away from him, almost with a jerk, the papers on which he had been writing. 'They tell me that you know the business of this office pretty well.'

'Some of it, sir,' said Eames.

'Well, yes; some of it. But you'll have to understand the whole of it if you come to me. And you must be very sharp about it too. You know that FitzHoward is leaving me?'

'I have heard of it, sir.'

'A very excellent young man, though perhaps not— But we won't mind that. The work is a little too much for him, and he's going back into the office. I believe Lord De Guest is a friend of yours; isn't he?'

'Yes; he is a friend of mine, certainly. He's been very kind to me.'

'Ah, well. I've known the earl for many years, – for very many years; and intimately at one time. Perhaps you may have heard him mention my name?'

'Yes, I have, Sir Raffle.'

'We were intimate once, but those things go off, you know. He's been the country mouse and I've been the town mouse. Ha, ha, ha! You may tell him that I say so. He won't mind that coming from me.'

'Oh, no; not at all,' said Eames.

'Mind you tell him when you see him. The earl is a man for whom I've always had a great respect, – a very great respect, – I may say regard. And now, Eames, what do you say to taking FitzHoward's place? The work is hard. It is fair that I should tell you that. The work will, no doubt, be very hard. I take a greater share of what's going than my predecessors have done; and I don't mind telling you that I have been sent here, because a man was wanted who would do that.' The voice of Sir Raffle, as he continued, became more and more harsh, and Eames began to think how wise FitzHoward had been. 'I mean to do my duty, and I shall expect that my private secretary will do his. But, Mr. Eames, I never forget a man. Whether he be good or bad, I never forget a man. You don't dislike late hours, I suppose.'

'Coming late to the office, you mean? Oh, no, not in the least.'

'Staying late, – staying late. Six or seven o'clock if necessary, putting your shoulder to the wheel when the coach gets into the mud. That's what I've been doing all my life. They've known what I am very well. They've always kept me for the heavy roads. If they paid, in the Civil Service, by the hour, I believe I should have drawn a larger income than any man in it. If you take the vacant chair in the next room you'll find it's no joke. It's only fair that I should tell you that.'

'I can work as hard as any man,' said Eames.

'That's right. That's right. Stick to that and I'll stick to you. It will be a great gratification to me to have by me a friend of my old friend De Guest. Tell him I say so. And now you may as well get into harness at once. FitzHoward is there. You can go in to him, and at half-past four exactly I'll see you both. I'm very exact, mind, – very; – and therefore you must be exact.' Then Sir Raffle looked as though he desired to be left alone.

'Sir Raffle, there's one favour I want to ask of you,' said Johnny.

'And what's that?'

'I am most anxious to be absent for a fortnight or three weeks, just at Easter. I shall want to go in about ten days.'

'Absent for three weeks at Easter, when the parliamentary work is beginning! That won't do for a private secretary.'

'But it's very important, Sir Raffle.'

'Out of the question, Eames; quite out of the question.'

'It's almost life and death to me.'

'Almost life and death. Why, what are you going to do?'

With all his grandeur and national importance, Sir Raffle would be very curious as to little people.

'Well, I can't exactly tell you, and I'm not quite sure myself.'

'Then don't talk nonsense. It's impossible that I should spare my private secretary just at that time of the year. I couldn't do it. The service won't admit of it. You're not entitled to leave at that season. Private secretaries always take their leave in the autumn.'

'I should like to be absent in the autumn, too, but—'

'It's out of the question, Mr. Eames.'

Then John Eames reflected that it behoved him in such an emergency to fire off his big gun. He had a great dislike to firing this big gun, but, as he said to himself, there are occasions which make a big gun very necessary. 'I got a letter from Lord De Guest this morning, pressing me very much to go to him at Easter. It's about business,' added Johnny. 'If there was any difficulty, he said, he should write to you.'

'Write to me,' said Sir Raffle, who did not like to be approached too familiarly in his office, even by an earl.

'Of course I shouldn't tell him to do that. But, Sir Raffle, if I remained out there, in the office,' and Johnny pointed towards the big room with his head, 'I could choose April for my month. And as the matter is so important to me, and to the earl—'

'What can it be?' said Sir Raffle.

'It's quite private,' said John Eames.

Hereupon Sir Raffle became very petulant, feeling that a bargain was being made with him. This young man would only consent to become his private secretary upon certain terms! 'Well, go in to FitzHoward now. I can't lose all my day in this way.'

'But I shall be able to get away at Easter?'

'I don't know. We shall see about it. But don't stand talking there now.' Then John Eames went into FitzHoward's room, and received that gentleman's congratulations on his appointment. 'I hope you like being rung for, like a servant, every minute, for he's always ringing that bell. And he'll roar at you till you're deaf. You must give up all dinner engagements, for, though there is not much to do, he'll never let you go. I don't think anybody ever asks him out to dinner, for he likes being here till seven. And you'll have to write all manner of lies about big people. And, sometimes, when he has sent Rafferty out about his private business, he'll ask you to bring him his shoes.' Now Rafferty was the First Commissioner's messenger.

It must be remembered, however, that this little account was given by an outgoing and discomfited private secretary. 'A man is not asked to bring another man his shoes,' said Eames to himself, 'until he shows himself fit for that sort of business.' Then he made within his own breast a little resolution about Sir Raffle's shoes.

CHAPTER XLVII

THE NEW PRIVATE SECRETARY

'INCOME-TAX OFFICE, April 8, 18—.

'My Dear Lord De Guest,

'I hardly know how to answer your letter, it is so very kind – more than kind. And about not writing before, – I must explain that I have not liked to trouble you with letters. I should have seemed to be encroaching if I had written much. Indeed it didn't come from not thinking about you. And first of all, about the money, – as to your offer, I mean. I really feel that I do not know what I ought to say to you about it, without appearing to be a simpleton. The truth is, I don't know what I ought to do, and can only trust to you not to put me wrong. I have an idea that a man ought not to accept a present of money unless from his father, or somebody like that. And the sum you mention is so very large that it makes me wish that you had not named it. If you choose to be so generous, would it not be better that you should leave it to me in your will?'

'So that he might always want me to be dying,' said Lord De Guest, as he read the letter out loud to his sister.

'I'm sure he wouldn't want that,' said Lady Julia. 'But you may live for twenty-five years, you know.'

'Say fifty,' said the earl. And then he continued the reading of his letter.

'But all that depends so much upon another person, that it is hardly worth while talking about it. Of course I am very much obliged to Mr. Dale, – very much indeed, – and I think that he is behaving very handsomely to his niece. But whether it will do me any good, that is quite another thing. However, I shall certainly accept your kind invitation for Easter, and find out whether I have a chance or not. I must tell you that Sir Raffle Buffle has made me his private secretary, by which I get a hundred a year. He says he was a great crony of yours many years ago, and seems to like talking about you very much. You will understand what that all means. He has sent you ever so many messages, but I don't suppose you will care to get them. I am to go to him to-morrow, and from all I hear I shall have a hard time of it.'

'By George, he will,' said the earl. 'Poor fellow!'

'But I thought a private secretary never had anything to do,' said Lady Julia.

'I shouldn't like to be private secretary to Sir Raffle, myself. But he's young, and a hundred a year is a great thing. How we all of us used to hate that man. His voice sounded like a bell with a crack in it. We always used to be asking for some one to muffle the Buffle. They call him Huffle Scuffle at his office. Poor Johnny!' Then he finished the letter: –

'I told him that I must have leave of absence at Easter, and he at first declared that it was impossible. But I shall carry my point about that. I would not stay away to be made private secretary to the Prime Minister; and yet I almost feel that I might as well stay away for any good that I shall do.

'Give my kind regards to Lady Julia, and tell her how very much obliged to her I am. I cannot express the gratitude which I owe to you. But pray believe me, my dear Lord De Guest, always very faithfully yours,

'John Eames.'

It was late before Eames had finished his letter. He had been making himself ready for his exodus from the big room, and preparing his desk and papers for his successor. About half-past five Cradell came up to him, and suggested that they should walk home together.

'What! You still here?' said Eames. 'I thought you always went at four.' Cradell had remained, hanging about the office, in order that he might walk home with the new private secretary. But Eames did not desire this. He had much of which he desired to think alone, and would fain have been allowed to walk by himself.

'Yes; I had things to do. I say, Johnny, I congratulate you most heartily; I do, indeed.'

'Thank you, old fellow!'

'It is such a grand thing, you know. A hundred a year and all at once! And then such a snug room to yourself, – and that fellow, Kissing, never can come near you. He has been making himself such a beast all day. But, Johnny, I always knew you'd come to something more than common. I always said so.'

'There's nothing uncommon about this; except that Fitz says that old Huffle Scuffle makes himself uncommon nasty.'

'Never mind what Fitz says. It's all jealousy. You'll have it all your own way, if you look sharp. I think you always do have it all your own way. Are you nearly ready?'

'Well, – not quite. Don't wait for me, Caudle.'

'Oh, I'll wait. I don't mind waiting. They'll keep dinner for us if we both stay. Besides, what matters? I'd do more than that for you.'

'I have some idea of working on till eight, and having a chop sent in,' said Johnny. 'Besides – I've got somewhere to call by myself.'

Then Cradell almost cried. He remained silent for two or three minutes, striving to master his emotion; and at last, when he did speak, had hardly succeeded in doing so. 'Oh, Johnny,' he said, 'I know what that means. You are going to throw me over because you are getting up in the world. I have always stuck to you, through everything; haven't I?'

'Don't make yourself a fool, Caudle.'

'Well; so I have. And if they made me private secretary, I

should have been just the same to you as ever. You'd have found no change in me.'

'What a goose you are. Do you say I'm changed, because I want to dine in the city?'

'It's all because you don't want to walk home with me, as we used to do. I'm not such a goose but what I can see. But, Johnny – I suppose I mustn't call you Johnny, now.'

'Don't be such a – con-founded—' Then Eames got up, and walked about the room. 'Come along,' said he, 'I don't care about staying, and don't mind where I dine.' And he bustled away with his hat and gloves, hardly giving Cradell time to catch him before he got out into the streets. 'I tell you what it is, Caudle,' said he, 'all that kind of thing is disgusting.'

'But how would you feel,' whimpered Cradell, who had never succeeded in putting himself quite on a par with his friend, even in his own estimation, since that glorious victory at the railway station. If he could only have thrashed Lupex as Johnny had thrashed Crosbie; then indeed they might have been equal, – a pair of heroes. But he had not done so. He had never told himself that he was a coward, but he considered that circumstances had been specially unkind to him. 'But how would you feel,' he whimpered, 'if the friend whom you liked better than anybody else in the world, turned his back upon you?'

'I haven't turned my back upon you; except that I can't get you to walk fast enough. Come along, old fellow, and don't talk confounded nonsense. I hate all that kind of thing. You never ought to suppose that a man will give himself airs, but wait till he does. I don't believe I shall remain with old Scuffles above a month or two. From all that I can hear that's as much as any one can bear.'

Then Cradell by degrees became happy and cordial, and during the whole walk flattered Eames with all the flattery of which he was master. And Johnny, though he did profess himself to be averse 'to all that kind of thing,' was nevertheless open to flattery. When Cradell told him that though FitzHoward could not manage the Tartar knight, he might probably do so; he was inclined to believe what Cradell said.

'And as to getting him his shoes,' said Cradell, 'I don't suppose he'd ever think of asking you to do such a thing, unless he was in a very great hurry, or something of that kind.'

'Look here, Johnny,' said Cradell, as they got into one of the streets bordering on Burton Crescent, 'you know the last thing in the world I should like to do would be to offend you.'

'All right, Caudle,' said Eames, going on, whereas his companion had shown a tendency towards stopping.

'Look here, now; if I have vexed you about Amelia Roper, I'll make you a promise never to speak to her again.'

'D— Amelia Roper,' said Eames, suddenly stopping himself and stopping Cradell as well. The exclamation was made in a deep angry voice which attracted the notice of one or two who were passing. Johnny was very wrong, – wrong to utter any curse; – very wrong to ejaculate that curse against a human being; and especially wrong to fulminate it against a woman – a woman whom he had professed to love! But he did do so, and I cannot tell my story thoroughly without repeating the wicked word.

Cradell looked up at him and stared. 'I only meant to say,' said Cradell, 'I'll do anything you like in the matter.'

'Then never mention her name to me again. And as to talking to her, you may talk to her till you're both blue in the face, if you please.'

'Oh; – I didn't know. You didn't seem to like it the other day.'

'I was a fool the other day, – a confounded fool. And so I have been all my life. Amelia Roper! Look here, Caudle; if she makes up to you this evening, as I've no doubt she will, for she seems to be playing that game constantly now, just let her have her fling. Never mind me; I'll amuse myself with Mrs. Lupex, or Miss Spruce.'

'But there'll be the deuce to pay with Mrs. Lupex. She's as cross as possible already whenever Amelia speaks to me. You don't know what a jealous woman is, Johnny.' Cradell had got upon what he considered to be his high ground. And on that he felt himself equal to any man. It was no doubt true that Eames had thrashed a man, and that he had not; it was true also that Eames had risen to a very high place in the social world, having become a private secretary; but for a dangerous, mysterious, overwhelming, life-enveloping intrigue – was not he the acknowledged hero of such an affair? He had paid very dearly, both in pocket, and in comfort, for the blessing of Mrs. Lupex's society; but he hardly considered that he had paid too dearly. There are certain luxuries which a man will find to be expensive; but, for all that, they may be worth their price. Nevertheless as he went up the steps of Mrs. Roper's house he made up his mind that he would oblige his friend. The intrigue might in that way become more mysterious, and more life-enveloping; whereas it would not become more dangerous, seeing that Mr. Lupex could hardly find himself to be aggrieved by such a proceeding.

The whole number of Mrs. Roper's boarders were assembled at dinner that day. Mr. Lupex seldom joined that festive board, but on this occasion he was present, appearing from his voice and manner to be in high good-humour. Cradell had communicated to the company in the drawing-room the great good fortune which had fallen upon his friend, and Johnny had thereby become the mark of a certain amount of hero-worship.

'Oh, indeed!' said Mrs. Roper. 'An 'appy woman your mother will be when she hears it. But I always said you'd come down right side uppermost.'

'Handsome is as handsome does,' said Miss Spruce.

'Oh, Mr. Eames!' exclaimed Mrs. Lupex, with graceful enthusiasm, 'I wish you joy from the very depth of my heart. It is such an elegant appointment.'

'Accept the hand of a true and disinterested friend,' said Lupex. And Johnny did accept the hand, though it was very dirty and stained all over with paint.

Amelia stood apart and conveyed her congratulations by a glance, – or, I might better say, by a series of glances. 'And now, – now will you not be mine,' the glances said; 'now that you are rolling in wealth and prosperity?' And then before they went downstairs she did whisper one word to him. 'Oh, I am so happy, John; – so very happy.'

'Bother!' said Johnny, in a tone quite loud enough to reach the lady's ear. Then making his way round the room, he gave his arm to Miss Spruce. Amelia, as she walked down-stairs alone, declared to herself that she would wring his heart. She had been employed in wringing it for some days past, and had been astonished at her own success. It had been clear enough to her that Eames had been piqued by her overtures to Cradell, and she had therefore to play out that game.

'Oh, Mr. Cradell,' she said, as she took her seat next to him. 'The friends I like are the friends that remain always the same. I hate your sudden rises. They do so often make a man upsetting.'

'I should like to try, myself, all the same,' said Cradell.

'Well, I don't think it would make any difference in you; I don't indeed. And of course your time will come too. It's that earl has done it, – he that was worried by the bull. Since we have known an earl we have been so mighty fine.' And Amelia gave her head a little toss, and then smiled archly, in a manner which, to Cradell's eyes, was really very becoming. But he saw that Mrs. Lupex was looking at him from the other side of the table, and he could not quite enjoy the goods which the gods had provided for him.

When the ladies left the dining-room Lupex and the two young men drew their chairs near the fire, and each prepared for himself a moderate potation. Eames made a little attempt at leaving the room, but he was implored by Lupex with such earnest protestations of friendship to remain, and was so weakly fearful of being charged with giving himself airs, that he did as he was desired.

'And here, Mr. Eames, is to your very good health,' said Lupex, raising to his mouth a steaming goblet of gin-and-water, 'and wishing you many years to enjoy your official prosperity.'

'Thank ye,' said Eames. 'I don't know much about the prosperity, but I'm just as much obliged.'

'Yes, sir; when I see a young man of your age beginning to rise in the world, I know he'll go on. Now look at me, Mr. Eames. Mr. Cradell, here's your very good health, and may all unkindness be drowned in the flowing bowl— Look at me, Mr. Eames. I've never risen in the world; I've never done any good in the world, and never shall.'

'Oh, Mr. Lupex, don't say that.'

'Ah, but I do say it. I've always been pulling the devil by the tail, and never yet got as much as a good hold on to that. And I'll tell you why; I never got a chance when I was young. If I could have got any big fellow, a star, you know, to let me paint his portrait when I was your age, – such a one, let us say, as your friend Sir Raffle—'

'What a star!' said Cradell.

'Well, I suppose he's pretty much known in the world, isn't he? Or Lord Derby, or Mr. Spurgeon. You know what I mean. If I'd got such a chance as that when I was young, I should never have been doing jobs of scene-painting at the minor theatres at so much a square yard. You've got the chance now, but I never had it.'

Whereupon Mr. Lupex finished his first measure of gin-and-water.

'It's a very queer thing, – life is,' continued Lupex; and, though he did not at once go to work boldly at the mixing of another glass of toddy, he began gradually, and as if by instinct, to finger the things which would be necessary for that operation. 'A very queer thing. Now, remember, young gentlemen, I'm not denying that success in life will depend upon good conduct; – of course it does; but, then, how often good conduct comes from success! Should I have been what I am now, do you suppose, if some big fellow had taken me by the hand when I was struggling to make an artist of myself? I could have drunk claret and champagne just as well as gin-and-water, and worn ruffles to my shirt as gracefully as many a fellow who used to be very fond of me, and now won't speak to me if he meets me in the streets. I never got a chance, – never.'

'But it's not too late yet, Mr. Lupex,' said Eames.

'Yes, it is, Eames, – yes, it is.' And now Mr. Lupex had grasped the gin-bottle. 'It's too late now. The game's over, and the match is lost. The talent is here. I'm as sure of that now as ever I was. I've never doubted my own ability, – never for a moment. There are men this very day making a thousand a year off their easels who haven't so good and true an eye in drawing as I have, or so good a feeling in colours. I could name them; only I won't.'

'And why shouldn't you try again?' said Eames.

'If I were to paint the finest piece that ever delighted the eye of man, who would come and look at it? Who would have enough belief in me to come as far as this place and see if it were true? No, Eames; I know my own position and my own ways, and I know my own weakness. I couldn't do a day's work now, unless I were certain of getting a certain number of shillings at the end of it. That's what a man comes to when things have gone against him.'

'But I thought men got lots of money by scene-painting?'

'I don't know what you may call lots, Mr. Cradell; I don't call it lots. But I'm not complaining. I know who I have to thank; and if ever I blow my own brains out I shan't be putting the blame on the wrong shoulders. If you'll take my advice,' – and now he turned round to Eames, – 'you'll beware of marrying too soon in life.'

'I think a man should marry early, if he marries well,' said Eames.

'Don't misunderstand me,' continued Lupex. 'It isn't about Mrs. L. I'm speaking. I've always regarded my wife as a very fascinating woman.'

'Hear, hear, hear!' said Cradell, thumping the table.

'Indeed she is,' said Eames.

'And when I caution you against marrying, don't you misunderstand me. I've never said a word against her to any man, and never will. If a man don't stand by his wife, whom will he stand by? I blame no one but myself. But I do say this; I never had a chance; – I never had a chance; – never had a chance.' And as he repeated the words for the third time, his lips were already fixed to the rim of his tumbler.

At this moment the door of the dining-room was opened, and Mrs. Lupex put in her head.

'Lupex,' she said, 'what are you doing?'

'Yes, my dear. I can't say I'm doing anything at the present moment. I was giving a little advice to these young gentlemen.'

'Mr. Cradell, I wonder at you. And, Mr. Eames, I wonder at you, too, – in your position! Lupex, come upstairs at once.' She then stepped into the room and secured the gin-bottle.

'Oh, Mr. Cradell, do come here,' said Amelia, in her liveliest tone, as soon as the men made their appearance above. 'I've been waiting for you this half-hour. I've got such a puzzle for you.' And she made way for him to a chair which was between herself and the wall. Cradell looked half afraid of his fortunes as he took the proffered seat; but he did take it, and was soon secured from any positive physical attack by the strength and breadth of Miss Roper's crinoline.

'Dear me! Here's a change,' said Mrs. Lupex, out loud.

Johnny Eames was standing close, and whispered into her ear, 'Changes are so pleasant sometimes! Don't you think so? I do.'

CHAPTER XLVIII

NEMESIS

CROSBIE had now settled down to the calm realities of married life, and was beginning to think that the odium was dying away which for a week or two had attached itself to him, partly on account of his usage of Miss Dale, but more strongly in consequence of the thrashing which he had received from John Eames. Not that he had in any way recovered his former tone of life, or that he ever hoped to do so. But he was able to go in and out of his club without embarrassment. He could talk with his wonted voice, and act with his wonted authority at his office. He could tell his friends, with some little degree of pleasure in the sound, that Lady Alexandrina would be very happy to see them. And he could make himself comfortable in his own chair after dinner, with his slippers and his newspaper. He could make himself comfortable, or at any rate could tell his wife that he did so.

It was very dull. He was obliged to acknowledge to himself, when he thought over the subject, that the life which he was leading was dull. Though he could go into his club without annoyance, nobody there ever thought of asking him to join them at dinner. It was taken for granted that he was going to dine at home; and in the absence of any provocation to the contrary, he always did dine at home. He had now been in his house for three weeks, and had been asked with his wife to a few bridal dinner-parties, given chiefly by friends of the De Courcy family. Except on such occasions he never passed an evening out of his own house, and had not yet, since his marriage, dined once away from his wife. He told himself that his good conduct in this respect was the result of his own resolution; but, nevertheless, he felt that there was nothing else left for him to do. Nobody asked him to go to the theatre. Nobody begged him to drop in of an evening. Men never asked him why he did not play a rubber. He would generally saunter into Sebright's after he left his office, and lounge about the room for half an hour, talking to a few men. Nobody was uncivil to him. But he knew that the whole thing was changed, and he resolved, with some wisdom, to accommodate himself to his altered circumstances.

Lady Alexandrina also found her new life rather dull, and was sometimes inclined to be a little querulous. She would tell her husband that she never got out, and would declare, when he offered to walk with her, that she did not care for walking in the streets. 'I don't exactly see, then, where you are to walk,' he once replied. She did not tell him that she was fond of riding, and that the Park was a very fitting place for such exercise; but she looked it, and he understood her. 'I'll do all I can for her,' he said to himself; 'but I'll not ruin myself.' 'Amelia is coming to take me for a drive,' she said another time. 'Ah, that'll be very nice,' he answered. 'No; it won't be very nice,' said Alexandrina. 'Amelia is always shopping and bargaining with the tradespeople. But it will be better than being kept in the house without ever stirring out.'

They breakfasted nominally at half-past nine; in truth, it

was always nearly ten, as Lady Alexandrina found it difficult to get herself out of her room. At half-past ten punctually he left his house for his office. He usually got home by six, and then spent the greatest part of the hour before dinner in the ceremony of dressing. He went, at least, into his dressing-room, after speaking a few words to his wife, and there remained, pulling things about, clipping his nails, looking over any paper that came in his way and killing the time. He expected his dinner punctually at seven, and began to feel a little cross if he were kept waiting. After dinner, he drank one glass of wine in company with his wife, and one other by himself, during which latter ceremony he would stare at the hot coals, and think of the thing he had done. Then he would go upstairs, and have, first a cup of coffee, and then a cup of tea. He would read his newspaper, open a book or two, hide his face when he yawned, and try to make believe that he liked it. She had no signs or words of love for him. She never sat on his knee, or caressed him. She never showed him that any happiness had come to her in being allowed to live close to him. They thought that they loved each other: – each thought so; but there was no love, no sympathy, no warmth. The very atmosphere was cold; – so cold that no fire could remove the chill.

In what way would it have been different had Lily Dale sat opposite to him there as his wife, instead of Lady Alexandrina? He told himself frequently that either with one or with the other life would have been the same; that he had made himself for a while unfit for domestic life, and that he must cure himself of that unfitness. But though he declared this to himself in one set of half-spoken thoughts, he would also declare to himself in another set, that Lily would have made the whole house bright with her brightness; that had he brought her home to his hearth, there would have been a sun shining on him every morning and every evening. But, nevertheless, he strove to do his duty, and remembered that the excitement of official life was still open to him. From eleven in the morning till five in the afternoon he could still hold a position which made it necessary that men should regard him with respect, and speak to him with deference.

In this respect he was better off than his wife, for she had no office to which she could betake herself.

'Yes,' she said to Amelia, 'it is all very nice, and I don't mind the house being damp; but I get so tired of being alone.'

'That must be the case with women who are married to men of business.'

'Oh, I don't complain. Of course I knew what I was about. I suppose it won't be so very dull when everybody is up in London.'

'I don't find the season makes much difference to us after Christmas,' said Amelia; 'but no doubt London is gayer in May. You'll find you'll like it better next year; and perhaps you'll have a baby, you know.'

'Psha!' ejaculated Lady Alexandrina; 'I don't want a baby, and don't suppose I shall have one.'

'It's always something to do, you know.'

Lady Alexandrina, though she was not of an energetic temperament, could not but confess to herself that she had made a mistake. She had been tempted to marry Crosbie because Crosbie was a man of fashion, and now she was told that the London season would make no difference to her; – the London season which had hitherto always brought to her the excitement of parties, if it had not given her the satisfaction of amusement. She had been tempted to marry because it appeared to her that a married woman could enjoy society with less restraint than a girl who was subject to her mother or her chaperon; that she would have more freedom of action as a married woman; and now she was told that she must wait for a baby before she could have anything to do. Courcy Castle was sometimes dull, but Courcy Castle would have been better than this.

When Crosbie returned home after this little conversation about the baby, he was told by his wife that they were to dine with the Gazebees on the next Sunday. On hearing this he shook his head with vexation. He knew, however, that he had no right to make complaint, as he had been only taken to St. John's Wood once since they had come home from their marriage trip. There was, however, one point as to which he could grumble. 'Why, on earth, on Sunday?'

'Because Amelia asked me for Sunday. If you are asked for Sunday, you cannot say you'll go on Monday.'

'It is so terrible on a Sunday afternoon. At what hour?'

'She said half-past five.'

'Heavens and earth! What are we to do all the evening?'

'It is not kind of you, Adolphus, to speak in that way of my relations.'

'Come, my love, that's a joke; as if I hadn't heard you say the same thing twenty times. You've complained of having to go up there much more bitterly than I ever did. You know I like your sister, and, in his way, Gazebee is a very good fellow; but after three or four hours, one begins to have had enough of him.'

'It can't be much duller than it is—;' but Lady Alexandrina stopped herself before she finished her speech.

'One can always read at home, at any rate,' said Crosbie.

'One can't always be reading. However, I have said you would go. If you choose to refuse, you must write and explain.'

When the Sunday came the Crosbies of course did go to St. John's Wood, arriving punctually at that door which he so hated at half-past five. One of the earliest resolutions which he made when he first contemplated the De Courcy match, was altogether hostile to the Gazebees. He would see but very little of them. He would shake himself free of that connexion. It was not with that branch of the family that he desired an alliance. But now, as things had gone, that was the only branch of the family with which he seemed to be allied. He was always hearing of the Gazebees. Amelia and Alexandrina were constantly together. He was now dragged there to a Sunday dinner; and he knew that he should often be dragged there, – that he could not avoid such draggings. He already owed money to Mortimer Gazebee, and was aware that his affairs had been allowed to fall into that lawyer's hands in such a way that he could not take them out again. His house was very thoroughly furnished, and he knew that the bills had been paid; but he had not paid them; every shilling had been paid through Mortimer Gazebee.

'Go with your mother and aunt, De Courcy,' the attorney said to the lingering child after dinner; and then Crosbie was left alone with his wife's brother-in-law. This was the period of the St. John's Wood purgatory which was so dreadful to him. With his sister-in-law he could talk, remembering perhaps always that she was an earl's daughter. But with Gazebee he had nothing in common. And he felt that Gazebee, who had once treated him with great deference, had now lost all such feeling. Crosbie had once been a man of fashion in the estimation of the attorney, but that was all over. Crosbie, in the attorney's estimation, was now simply the secretary of a public office, – a man who owed him money. The two had married sisters, and there was no reason why the light of the prosperous attorney should pale before that of the civil servant, who was not very prosperous. All this was understood thoroughly by both the men.

'There's terrible bad news from Courcy,' said the attorney, as soon as the boy was gone.

'Why; what's the matter?'

'Porlock has married; – that woman, you know.'

'Nonsense.'

'He has. The old lady has been obliged to tell me, and she's nearly broken-hearted about it. But that's not the worst of it to my mind. All the world knows that Porlock had gone to the mischief. But he is going to bring an action against his father for some arrears of his allowance, and he threatens to have everything out in court, if he doesn't get his money.'

'But is there money due to him?'

'Yes, there is. A couple of thousand pounds or so. I suppose I shall have to find it. But, upon my honour, I don't know where it's to come from; I don't, indeed. In one way or another, I've paid over fourteen hundred pounds for you.'

'Fourteen hundred pounds!'

'Yes, indeed; – what with the insurance and the furniture, and the bill from our house for the settlements. That's not paid yet, but it's the same thing. A man doesn't get married for nothing, I can tell you.'

'But you've got security.'

'Oh, yes; I've got security. But the thing is the ready

money. Our house has advanced so much on the Courcy property, that they don't like going any further; and therefore it is that I have to do this myself. They'll all have to go abroad, – that'll be the end of it. There's been such a scene between the earl and George. George lost his temper and told the earl that Porlock's marriage was his fault. It has ended in George with his wife being turned out.'

'He has money of his own.'

'Yes, but he won't spend it. He's coming up here, and we shall find him hanging about us. I don't mean to give him a bed here, and I advise you not to do so either. You'll not get rid of him if you do.'

'I have the greatest possible dislike to him.'

'Yes; he's a bad fellow. So is John. Porlock was the best, but he's gone altogether to ruin. They've made a nice mess of it between them; haven't they?'

This was the family for whose sake Crosbie had jilted Lily Dale! His single and simple ambition had been that of being an earl's son-in-law. To achieve that it had been necessary that he should make himself a villain. In achieving it he had gone through all manner of dirt and disgrace. He had married a woman whom he knew he did not love. He was thinking almost hourly of a girl whom he had loved, whom he did love, but whom he had so injured, that, under no circumstances, could he be allowed to speak to her again. The attorney there – who sat opposite to him, talking about his thousands of pounds with that disgusting assumed solicitude which such men put on, when they know very well what they are doing – had made a similar marriage. But he had known what he was about. He had got from his marriage all that he had expected. But what had Crosbie got?

'They're a bad set, – a bad set,' said he in his bitterness.

'The men are,' said Gazebee, very comfortably.

'H – m,' said Crosbie. It was manifest to Gazebee that his friend was expressing a feeling that the women, also, were not all that they should be, but he took no offence, though some portion of the censure might thereby be supposed to attach to his own wife.

'The countess means well,' said Gazebee. 'But she's had a

hard life of it, – a very hard life. I've heard him call her names that would frighten a coal-heaver. I have, indeed. But he'll die soon, and then she'll be comfortable. She has three thousand a year jointure.'

He'll die soon, and then she'll be comfortable! That was one phase of married life. As Crosbie's mind dwelt upon the words, he remembered Lily's promise made in the fields, that she would do everything for him. He remembered her kisses; the touch of her fingers; the low silvery laughing voice; the feel of her dress as she would press close to him. After that he reflected whether it would not be well that he too should die, so that Alexandrina might be comfortable. She and her mother might be very comfortable together, with plenty of money, at Baden Baden!

The squire at Allington, and Mrs. Dale, and Lady Julia De Guest had been, and still were, uneasy in their minds because no punishment had fallen upon Crosbie, – no vengeance had overtaken him in consequence of his great sin. How little did they know about it! Could he have been prosecuted and put into prison, with hard labour, for twelve months, the punishment would not have been heavier. He would, in that case, at any rate, have been saved from Lady Alexandrina.

'George and his wife are coming up to town; couldn't we ask them to come to us for a week or so?' said his wife to him, as soon as they were in the fly together, going home.

'No,' shouted Crosbie; 'we will do no such thing.' There was not another word said on the subject, – nor on any other subject till they got home. When they reached their house Alexandrina had a headache, and went up to her room immediately. Crosbie threw himself into a chair before the remains of a fire in the dining-room, and resolved that he would cut the whole De Courcy family altogether. His wife, as his wife, should obey him. She should obey him – or else leave him and go her way by herself, leaving him to go his way. There was an income of twelve hundred a year. Would it not be a fine thing for him if he could keep six hundred for himself and return to his old manner of life. All his old comforts of course he would not have, – nor the old esteem and regard of men. But the luxury of a club dinner he might enjoy. Unembarrassed evenings might be his, – with liberty to him to pass them as he pleased. He knew many men who were separated from their wives, and who seemed to be as happy as their neighbours. And then he remembered how ugly Alexandrina had been this evening, wearing a great tinsel coronet full of false stones, with a cold in her head which had reddened her nose. There had, too, fallen upon her in these married days a certain fixed dreary dowdiness. She certainly was very plain! So he said to himself, and then he went to bed. I myself am inclined to think that his punishment was sufficiently severe.

The next morning his wife still complained of headache, so that he breakfasted alone. Since that positive refusal which he had given to her proposition for inviting her brother, there had not been much conversation between them. 'My head is splitting, and Sarah shall bring some tea and toast up to me, if you will not mind it.'

He did not mind it in the least, and ate his breakfast by himself, with more enjoyment than usually attended that meal.

It was clear to him that all the present satisfaction of his life must come to him from his office work. There are men who find it difficult to live without some source of daily comfort, and he was such a man. He could hardly endure his life unless there was some page in it on which he could look with gratified eyes. He had always liked his work, and he now determined that he would like it better than ever. But in order that he might do so it was necessary that he should have much of his own way. According to the theory of his office, it was incumbent on him as Secretary simply to take the orders of the Commissioners, and see that they were executed; and to such work as this his predecessor had strictly confined himself. But he had already done more than this, and had conceived the ambition of holding the Board almost under his thumb. He flattered himself that he knew his own work and theirs better than they knew either, and that by a little management he might be their master. It is not impossible that such might have been the case had there been no fracas at the Paddington station; but, as we all know, the dominant cock of the farmyard must be ever dominant. When he shall once have had his wings so smeared with mud as to give him even the appearance of adversity, no other cock will ever respect him again. Mr. Optimist and Mr. Butterwell knew very well that their secretary had been cudgelled, and they could not submit themselves to a secretary who had been so treated.

'Oh, by-the-by, Crosbie,' said Butterwell, coming into his room, soon after his arrival at his office on that day of his solitary breakfast, 'I want to say just a few words to you.' And Butterwell turned round and closed the door, the lock of which had not previously been fastened. Crosbie, without much thinking, immediately foretold himself the nature of the coming conversation.

'Do you know—' said Butterwell, beginning.

'Sit down, won't you?' said Crosbie, seating himself as he spoke. If there was to be a contest, he would make the best fight he could. He would show a better spirit here than he had done on the railway platform. Butterwell did sit down, and felt as he did so, that the very motion of sitting took away some of his power. He ought to have sent for Crosbie into his own room. A man when he wishes to reprimand another, should always have the benefit of his own atmosphere.

'I don't want to find any fault,' Butterwell began.

'I hope you have not any cause,' said Crosbie.

'No, no; I don't say that I have. But we think at the Board—'

'Stop, stop, Butterwell. If anything unpleasant is coming, it had better come from the Board. I should take it in better spirit; I should, indeed.'

'What takes place at the Board must be official.'

'I shall not mind that in the least. I should rather like it than otherwise.'

'It simply amounts to this, – that we think you are taking a little too much on yourself. No doubt, it's a fault on the right side, and arises from your wishing to have the work well done.'

'And if I don't do it, who will?' asked Crosbie.

'The Board is very well able to get through all that appertains to it. Come, Crosbie, you and I have known each other a great many years, and it would be a pity that we should have any words. I have come to you in this way because it would be disagreeable to you to have any question raised officially. Optimist isn't given to being very angry, but he was downright angry yesterday. You had better take what I say in good part, and go along a little quieter.'

But Crosbie was not in a humour to take anything quietly. He was sore all over, and prone to hit out at everybody that he met. 'I have done my duty to the best of my ability, Mr. Butterwell,' he said, 'and I believe I have done it well. I believe I know my duty here as well as any one can teach me. If I have done more than my share of work, it is because other people have done less than theirs.' As he spoke, there was a black cloud upon his brow, and the Commissioner could perceive that the Secretary was very wrathful.

'Oh! very well,' said Butterwell, rising from his chair. 'I can only, under such circumstances, speak to the Chairman, and he will tell you what he thinks at the Board. I think you're foolish; I do, indeed. As for myself, I have only meant to act kindly by you.' After that, Mr. Butterwell took himself off.

On the same afternoon, Crosbie was summoned into the Board-room in the usual way, between two and three. This was a daily occurrence, as he always sat for about an hour with two out of the three Commissioners, after they had fortified themselves with a biscuit and a glass of sherry. On the present occasion, the usual amount of business was transacted, but it was done in a manner which made Crosbie feel that they did not all stand together on their usual footing. The three Commissioners were all there. The Chairman gave his directions in a solemn, pompous voice, which was by no means usual to him when he was in a good humour. The Major said little or nothing; but there was a gleam of satisfied sarcasm in his eye. Things were going wrong at the Board, and he was pleased. Mr. Butterwell was exceedingly civil in his demeanour, and rather more than ordinarily brisk. As soon as the regular work of the day was over, Mr.

Optimist shuffled about on his chair, rising from his seat, and then sitting down again. He looked through a lot of papers close to his hand, peering at them over his spectacles. Then he selected one, took off his spectacles, leaned back in his chair, and began his little speech.

'Mr. Crosbie,' he said, 'we are all very much gratified, – very much gratified, indeed, – by your zeal and energy in the service.'

'Thank you, sir,' said Crosbie; 'I am fond of the service.'

'Exactly, exactly; we all feel that. But we think that you, – if I were to say take too much upon yourself, I should say, perhaps, more than we mean.'

'Don't say more than you mean, Mr. Optimist.' Crosbie's eyes, as he spoke, gleamed slightly with his momentary triumph; as did also those of Major Fiasco.

'No, no, no,' said Mr. Optimist; 'I would say rather less than more to so very good a public servant as yourself. But you, doubtless, understand me?'

'I don't think I do quite, sir. If I have not taken too much on me, what is it that I have done that I ought not to have done?'

'You have given directions in many cases for which you ought first to have received authority. Here is an instance,' and the selected paper was at once brought out.

It was a matter in which the Secretary had been manifestly wrong according to written law, and he could not defend it on its own merits.

'If you wish me,' said he, 'to confine myself exactly to the positive instructions of the office, I will do so; but I think you will find it inconvenient.'

'It will be far the best,' said Mr. Optimist.

'Very well,' said Mr. Crosbie, 'it shall be done.' And he at once determined to make himself as unpleasant to the three gentlemen in the room as he might find it within his power to do. He could make himself very unpleasant, but the unpleasantness would be as much to him as to them.

Nothing would now go right with him. He could look in no direction for satisfaction. He sauntered into Sebright's, as he went home, but he could not find words to speak to any one about the little matters of the day. He went home, and his wife, though she was up, complained still of her headache.

'I haven't been out of the house all day,' she said, 'and that has made it worse.'

'I don't know how you are to get out if you won't walk,' he answered.

Then there was no more said between them till they sat down to their meal.

Had the squire at Allington known all, he might, I think, have been satisfied with the punishment which Crosbie had encountered.

CHAPTER XLIX

PREPARATIONS FOR GOING

'MAMMA, read that letter.'

It was Mrs. Dale's eldest daughter who spoke to her, and they were alone together in the parlour at the Small House. Mrs. Dale took the letter and read it very carefully. She then put it back into its envelope and returned it to Bell.

'It is, at any rate, a good letter, and, as I believe, tells the truth.'

'I think it tells a little more than the truth, mamma. As you say, it is a well-written letter. He always writes well when he is in earnest. But yet—'

'Yet what, my dear?'

'There is more head than heart in it.'

'If so, he will suffer the less; that is, if you are quite resolved in the matter.'

'I am quite resolved, and I do not think he will suffer much. He would not, I suppose, have taken the trouble to write like that, if he did not wish this thing.'

'I am quite sure that he does wish it, most earnestly; and that he will be greatly disappointed.'

'As he would be if any other scheme did not turn out to his satisfaction; that is all.'

The letter, of course, was from Bell's cousin Bernard, and containing the strongest plea he was able to make in favour of his suit for her hand. Bernard Dale was better able to press such a plea by letter than by spoken words. He was a man capable of doing anything well in the doing of which a little time for consideration might be given to him; but he had not in him that power of passion which will force a man to eloquence in asking for that which he desires to obtain. His letter on this occasion was long, and well argued. If there was little in it of passionate love, there was much of pleasant flattery. He told Bell how advantageous to both their families their marriage would be; he declared to her that his own feeling in the matter had been rendered stronger by absence; he alluded without boasting to his past career of life as her best guarantee for his future conduct; he explained to her that if this marriage could be arranged there need be no further question as to his aunt removing with Lily from the Small House; and he told her that his affection for herself was the absorbing passion of his existence. Had the letter been written with the view of obtaining from a third person a favourable verdict as to his suit, it would have been a very good letter indeed; but there was not a word in it that could stir the heart of such a girl as Bell Dale.

'Answer him kindly,' Mrs. Dale said.

'As kindly as I know how,' said Bell. 'I wish you would write the letter, mamma.'

'I fear that would not do. What I should say would only tempt him to try again.'

Mrs. Dale knew very well, – had known for some months past, – that Bernard's suit was hopeless. She felt certain, although the matter had not been discussed between them, that whenever Dr. Crofts might choose to come again and ask for her daughter's hand he would not be refused. Of the two men she probably liked Dr. Crofts the best; but she liked them both, and she could not but remember that the one, in a worldly point of view, would be a very poor match, whereas the other would, in all respects, be excellent. She would not, on any account, say a word to influence her daughter, and knew, moreover, that no word which she could say would influence her; but she could not divest herself of some regret that it should be so.

'I know what you would wish, mamma,' said Bell.

'I have but one wish, dearest, and that is for your happiness. May God preserve you from any such fate as Lily's. When I tell you to write kindly to your cousin, I simply mean that I think him to have deserved a kind reply by his honesty.'

'It shall be as kind as I can make it, mamma; but you know what the lady says in the play, – how hard it is to take the sting from that word "no".' Then Bell walked out alone for a while, and on her return got her desk and wrote her letter. It was very firm and decisive. As for that wit which should pluck the sting 'from such a sharp and waspish word as "no",' I fear she had it not. 'It will be better to make him understand that I, also, am in earnest,' she said to herself; and in this frame of mind she wrote her letter. 'Pray do not allow yourself to think that what I have said is unfriendly,' she added, in a postscript. 'I know how good you are, and I know the great value of what I refuse; but in this matter it must be my duty to tell you the simple truth.'

It had been decided between the squire and Mrs. Dale that the removal from the Small House to Guestwick was not to take place till the first of May. When he had been made to understand that Dr. Crofts had thought it injudicious that Lily should be taken out of their present house in March, he had used all the eloquence of which he was master to induce Mrs. Dale to consent to abandon her project. He had told her that he had always considered that house as belonging, of right, to some other of the family than himself; that it had always been so inhabited, and that no squire of Allington had for years past taken rent for it. 'There is no favour conferred, – none at all,' he had said; but speaking nevertheless in his usual sharp, ungenial tone.

'There is a favour, a great favour, and great generosity,' Mrs. Dale had replied. 'And I have never been too proud to accept it; but when I tell you that we think we shall be happier at Guestwick, you will not refuse to let us go. Lily has had a great blow in that house, and Bell feels that she is running counter to your wishes on her behalf, – wishes that are so very kind!'

'No more need be said about that. All that may come right yet, if you will remain where you are.'

But Mrs. Dale knew that 'all that' could never come right, and persisted. Indeed, she would hardly have dared to tell her girls that she had yielded to the squire's entreaties. It was just then, at that very time, that the squire was, as it were, in treaty with the earl about Lily's fortune; and he did feel it hard that he should be opposed in such a way by his own relatives at the moment when he was behaving towards them with so much generosity. But in his arguments about the house he said nothing of Lily, or her future prospects.

They were to move on the first of May, and one week of April was already past. The squire had said nothing further on the matter after the interview with Mrs. Dale to which allusion has just been made. He was vexed and sore at the separation, thinking that he was ill-used by the feeling which was displayed by this refusal. He had done his duty by them, as he thought; indeed more than his duty, and now they told him that they were leaving him because they could no longer bear the weight of an obligation conferred by his hands. But in truth he did not understand them; nor did they understand him. He had not been hard in his manner, and had occasionally domineered, not feeling that his position, though it gave him all the privileges of a near and a dear friend, did not give him the authority of a father or a husband. In that matter of Bernard's proposed marriage he had spoken as though Bell should have considered his wishes before she refused her cousin. He had taken upon himself to scold Mrs. Dale, and had thereby given offence to the girls, which they at the time had found it utterly impossible to forgive.

But they were hardly better satisfied in the matter than was he; and now that the time had come, though they could not bring themselves to go back from their demand, almost felt that they were treating the squire with cruelty. When their decision had been made, – while it had been making, – he had been stern and hard to them. Since that he had been softened by Lily's misfortune, and softened also by the anticipated loneliness which would come upon him when they should be gone from his side. It was hard upon him that they should so treat him when he was doing his best for them all! And they also felt this, though they did not know the extent to which he was anxious to go in serving them. When they had sat round the fire planning the scheme of their removal, their hearts had been hardened against him, and they had resolved to assert their independence. But now, when the time for action had come, they felt that their grievances against him had already been in a great measure assuaged. This tinged all that they did with a certain sadness; but still they continued their work.

Who does not know how terrible are those preparations for house-moving; – how infinite in number are the articles which must be packed, how inexpressibly uncomfortable is the period of packing, and how poor and tawdry is the aspect of one's belongings while they are thus in a state of dislocation? Nowadays people who understand the world, and have money commensurate with their understanding, have learned the way of shunning all these disasters, and of leaving the work to the hands of persons paid for doing it. The crockery is left in the cupboards, the books on the shelves, the wine in the bins, the curtains on their poles, and the family that is understanding goes for a fortnight to Brighton. At the end of that time the crockery is comfortably settled in other cupboards, the books on other shelves, the wine in other bins, the curtains are hung on other poles, and all is arranged. But Mrs. Dale and her daughters understood nothing of such a method of moving as this. The assistance of the village carpenter in filling certain cases that he had made was all that they knew how to obtain beyond that of their own two servants. Every article had to pass through the hands of some one of the family; and as they felt almost overwhelmed by the extent of the work to be done, they began it much sooner than was necessary, so that it became evident as they advanced in their work, that they would have to pass a dreadfully dull, stupid, uncomfortable week at last, among their boxes and cases, in all the confusion of dismantled furniture.

At first an edict had gone forth that Lily was to do nothing. She was an invalid, and was to be petted and kept quiet. But this edict soon fell to the ground, and Lily worked harder than either her mother or her sister. In truth she was hardly an invalid any longer, and would not submit to an invalid's treatment. She felt herself that for the present constant occupation could alone save her from the misery of looking back, – and she had conceived an idea that the harder that occupation was, the better it would be for her. While pulling down the books, and folding the linen, and turning out from their old hiding-places the small long-forgotten properties of the household, she would be as gay as ever she had been in old times. She would talk over her work standing with flushed cheek and laughing eyes among the dusty ruins around her, till for a moment her mother would think that all was well within her. But then at other moments, when the reaction came, it would seem as though nothing were well. She could not sit quietly over the fire, with quiet rational work in her hands, and chat in a rational quiet way. Not as yet could she do so. Nevertheless it was well with her, – within her own bosom. She had declared to herself that she would conquer her misery, – as she had also declared to herself during her illness that her misfortune should not kill her, – and she was in the way to conquer it. She told herself that the world was not over for her because her sweet hopes had been frustrated. The wound had been deep and very sore, but the flesh of the patient had been sound and healthy, and her blood pure. A physician having knowledge in such cases would have declared, after long watching of her symptoms, that a cure was probable. Her mother was the physician who watched her with the closest eyes; and she, though she was sometimes driven to doubt, did hope, with stronger hope from day to day, that her child might live to remember the story of her love without abiding agony.

That nobody should talk to her about it, – that had been the one stipulation which she had seemed to make, not sending forth a request to that effect among her friends in so many words, but showing by certain signs that such was her stipulation. A word to that effect she had spoken to her uncle, – as may be remembered, which word had been regarded with the closest obedience. She had gone out into her little world very soon after the news of Crosbie's falsehood had reached her, – first to church and then among the people of the village, resolving to carry herself as though no crushing weight had fallen upon her. The village people had understood it all, listening to her and answering her without the proffer of any outspoken parley.

'Lord bless 'ee,' said Mrs. Crump, the postmistress, – and Mrs. Crump was supposed to have the sourest temper in Allington, – 'whenever I look at thee, Miss Lily, I thinks that surely thee is the beautifulest young 'ooman in all these parts.'

'And you are the crossest old woman,' said Lily, laughing, and giving her hand to the postmistress.

'So I be,' said Mrs. Crump. 'So I be.' Then Lily sat down in the cottage and asked after her ailments. With Mrs. Hearn it was the same. Mrs. Hearn, after that first meeting which has been already mentioned, petted and caressed her, but spoke no further word of her misfortune. When Lily called a second time upon Mrs. Boyce, which she did boldly by herself, that lady did begin one other word of commiseration. 'My dearest Lily, we have all been made so unhappy—' So far Mrs. Boyce got, sitting close to Lily and striving to look into her face; but Lily, with a slightly heightened colour, turned sharp round upon one of the Boyce girls, tearing Mrs. Boyce's commiseration into the smallest shreds. 'Minnie,' she said, speaking quite loud, almost with girlish ecstasy, 'what do you think Tartar did yesterday? I never laughed so much in my life.' Then she told a ludicrous story about a very vulgar terrier which belonged to the squire. After that even Mrs. Boyce made no further attempt. Mrs. Dale and Bell both understood that such was to be the rule – the rule even to them. Lily would speak to them occasionally on the matter, – to one of them at a time, beginning with some almost single word of melancholy resignation, and then would go on till she opened her very bosom before them; but no such conversation was ever begun by them. But now, in these busy days of the packing, that topic seemed to have been banished altogether.

'Mamma,' she said, standing on the top rung of a house-ladder, from which position she was handing down glass out of a cupboard, 'are you sure that these things are ours? I think some of them belong to the house.'

'I'm sure about that bowl at any rate, because it was my mother's before I was married.'

'Oh, dear, what should I do if I were to break it? Whenever I handle anything very precious I always feel inclined to throw it down and smash it. Oh! it was as nearly gone as possible, mamma; but that was your fault.'

'If you don't take care, you'll be nearly gone yourself. Do take hold of something.'

'Oh, Bell, here's the inkstand for which you've been moaning for three years.'

'I haven't been moaning for three years; but who could have put it up there?'

'Catch it,' said Lily; and she threw the bottle down on to a pile of carpets.

At this moment a step was heard in the hall, and the squire entered through the open door of the room. 'So you're all at work,' said he.

'Yes, we're at work,' said Mrs. Dale, almost with a tone of shame. 'If it is to be done it is as well that it should be got over.'

'It makes me wretched enough,' said the squire. 'But I didn't come to talk about that. I've brought you a note from Lady Julia De Guest, and I've had one from the earl. They want us all to go there and stay the week after Easter.'

Mrs. Dale and the girls, when this very sudden proposition was made to them, all remained fixed in their places, and, for a moment, were speechless. Go and stay a week at Guestwick Manor! The whole family! Hitherto the intercourse between the Manor and the Small House had been confined to morning calls, very far between. Mrs. Dale had never dined there, and had latterly even deputed the calling to her daughters. Once Bell had dined there with her uncle, the squire, and once Lily had gone over with her uncle Orlando.

Even this had been long ago, before they were quite brought out, and they had regarded the occasion with the solemn awe of children. Now, at this time of their flitting into some small mean dwelling at Guestwick, they had previously settled among themselves that that affair of calling at the Manor might be allowed to drop. Mrs. Eames never called, and they were descending to the level of Mrs. Eames. 'Perhaps we shall get game sent to us, and that will be better,' Lily had said. And now, at this very moment of their descent in life, they were all asked to go and stay a week at the Manor! Stay a week with Lady Julia! Had the Queen sent the Lord Chamberlain down to bid them all go to Windsor Castle it could hardly have startled them more at the first blow. Bell had been seated on the folded carpet when her uncle had entered, and now had again sat herself in the same place. Lily was still standing at the top of the ladder, and Mrs. Dale was at the foot with one hand on Lily's dress. The squire had told his story very abruptly, but he was a man who, having a story to tell, knew nothing better than to tell it out abruptly, letting out everything at the first moment.

'Wants us all!' said Mrs. Dale. 'How many does the all mean?' Then she opened Lady Julia's note and read it, not moving from her position at the foot of the ladder.

'Do let me see, mamma,' said Lily; and then the note was handed up to her. Had Mrs. Dale well considered the matter she might probably have kept the note to herself for a while, but the whole thing was so sudden that she had not considered the matter well.

'My Dear Mrs. Dale (the letter ran),

'I send this inside a note from my brother to Mr. Dale. We particularly want you and your two girls to come to us for a week from the seventeenth of this month. Considering our near connection we ought to have seen more of each other than we have done for years past, and of course it has been our fault. But it is never too late to amend one's ways; and I hope you will receive my confession in the true spirit of affection in which it is intended, and that you will show your goodness by coming to us. I will do all I can to make the house pleasant to your girls, for both of whom I have much real regard.

'I should tell you that John Eames will be here for the same week. My brother is very fond of him, and thinks him the best young man of the day. He is one of my heroes, too, I must confess.

'Very sincerely yours,
'Julia De Guest.'

Lily, standing on the ladder, read the letter very attentively. The squire meanwhile stood below speaking a word or two to his sister-in-law and niece. No one could see Lily's face, as it was turned away towards the window, and it was still averted when she spoke. 'It is out of the question that we should go, mamma; – that is, all of us.'

'Why out of the question?' said the squire.

'A whole family!' said Mrs. Dale.

'That is just what they want,' said the squire.

'I should like of all things to be left alone for a week,' said Lily, 'if mamma and Bell would go.'

'That wouldn't do at all,' said the squire. 'Lady Julia specially wants you to be one of the party.'

The thing had been badly managed altogether. The reference in Lady Julia's note to John Eames had explained to Lily the whole scheme at once, and had so opened her eyes that all the combined influence of the Dale and De Guest families could not have dragged her over to the Manor.

'Why not do?' said Lily. 'It would be out of the question a whole family going in that way, but it would be very nice for Bell.'

'No, it would not,' said Bell.

'Don't be ungenerous about it, my dear,' said the squire, turning to Bell; 'Lady Julia means to be kind. But, my darling,' and the squire turned again towards Lily, addressing her, as was his wont in these days, with an affection that was almost vexatious to her; 'but, my darling, why should you not go? A change of scene like that will do you all the good in the world, just when you are getting well. Mary, tell the girls that they ought to go.'

Mrs. Dale stood silent, again reading the note, and Lily came down from the ladder. When she reached the floor she went directly up to her uncle, and taking his hand turned him round with herself towards one of the windows, so that they stood with their backs to the room. 'Uncle,' she said, 'do not be angry with me. I can't go;' and then she put up her face to kiss him.

He stooped and kissed her and still held her hand. He looked into her face and read it all. He knew well, now, why she could not go; or, rather, why she herself thought that she could not go. 'Cannot you, my darling?' he said.

'No, uncle. It is very kind, – very kind; but I cannot go. I am not fit to go anywhere.'

'But you should get over that feeling. You should make a struggle.'

'I am struggling, and I shall succeed; but I cannot do it all at once. At any rate I could not go there. You must give my love to Lady Julia, and not let her think me cross. Perhaps Bell will go.'

What would be the good of Bell's going – or the good of his putting himself out of the way, by a visit which would of itself be so tiresome to him, if the one object of the visit could not be carried out? The earl and his sister had planned the invitation with the express intention of bringing Lily and Eames together. It seemed that Lily was firm in her determination to resist this intention; and, if so, it would be better that the whole thing should fall to the ground. He was very vexed, and yet he was not angry with her. Everybody lately had opposed him in everything. All his intended family arrangements had gone wrong. But yet he was seldom angry respecting them. He was so accustomed to be thwarted that he hardly expected success. In this matter of providing Lily with a second lover, he had not come forward of his own accord. He had been appealed to by his neighbour the earl, and had certainly answered the appeal with much generosity. He had been induced to make the attempt with eagerness, and a true desire for its accomplishment; but in this, as in all his own schemes, he was met at once by opposition and failure.

'I will leave you to talk it over among yourselves,' he said. 'But, Mary, you had better see me before you send your answer. If you will come up by and by, Ralph shall take the two notes over together in the afternoon.' So saying, he left the Small House, and went back to his own solitary home.

'Lily, dear,' said Mrs. Dale, as soon as the front door had been closed, 'this is meant for kindness to you, – for most affectionate kindness.'

'I know it, mamma; and you must go to Lady Julia, and must tell her that I know it. You must give her my love. And, indeed, I do love her now. But—'

'You won't go, Lily?' said Mrs. Dale, beseechingly.

'No, mamma; certainly I will not go.' Then she escaped out of the room by herself, and for the next hour neither of them dared to go to her.

CHAPTER L

MRS. DALE IS THANKFUL FOR A GOOD THING

On that day they dined early at the Small House, as they had been in the habit of doing since the packing had commenced. And after dinner Mrs. Dale went through the gardens, up to the other house, with a written note in her hand. In that note she had told Lady Julia, with many protestations of gratitude, that Lily was unable to go out so soon after her illness, and that she herself was obliged to stay with Lily. She explained also, that the business of moving was in hand, and that, therefore, she could not herself accept the invitation. But her other daughter, she said, would be very happy to accompany her uncle to Guestwick Manor. Then, without closing her letter, she took it up to the squire in order that it might be decided whether it would or would not suit his views. It might be well that he would not care to go to Lord De Guest's with Bell alone.

'Leave it with me,' he said; 'that is, if you do not object.'

'Oh dear, no!'

'I'll tell you the plain truth at once, Mary. I shall go over myself with it, and see the earl. Then I will decline it or not, according to what passes between me and him. I wish Lily would have gone.'

'Ah! she could not.'

'I wish she could. I wish she could. I wish she could.' As he repeated the words over and over again, there was an eagerness in his voice that filled Mrs. Dale's heart with tenderness towards him.

'The truth is,' said Mrs. Dale, 'she could not go there to meet John Eames.'

'Oh, I know,' said the squire: 'I understand it. But that is just what we want her to do. Why should she not spend a week in the same house with an honest young man whom we all like.'

'There are reasons why she would not wish it.'

'Ah, exactly; the very reasons which should make us induce her to go there if we can. Perhaps I had better tell you all. Lord De Guest has taken him by the hand, and wishes him to marry. He has promised to settle on him an income which will make him comfortable for life.'

'That is very generous; and I am delighted to hear it, – for John's sake.'

'And they have promoted him at his office.'

'Ah! then he will do well.'

'He will do very well. He is private secretary now to their head man. And, Mary, so that she, Lily, should not be empty-handed if this marriage can be arranged, I have undertaken to settle a hundred a year on her, – on her and her children, if she will accept him. Now you know it all. I did not mean to tell you; but it is as well that you should have the means of judging. That other man was a villain. This man is honest. Would it not be well that she should learn to like him? She always did like him, I thought, before that other fellow came down here among us.'

'She has always liked him – as a friend.'

'She will never get a better lover.'

Mrs. Dale sat silent, thinking over it all. Every word that the squire said was true. It would be a healing of wounds most desirable and salutary; an arrangement advantageous to them all; a destiny for Lily most devoutly to be desired, –

if only it were possible. Mrs. Dale firmly believed that if her daughter could be made to accept John Eames as her second lover in a year or two all would be well. Crosbie would then be forgotten or thought of without regret, and Lily would become the mistress of a happy home. But there are positions which cannot be reached, though there be no physical or material objection in the way. It is the view which the mind takes of a thing which creates the sorrow that arises from it. If the heart were always malleable and the feelings could be controlled, who would permit himself to be tormented by any of the reverses which affection meets? Death would create no sorrow; ingratitude would lose its sting; and the betrayal of love would do no injury beyond that which it might entail upon worldly circumstances. But the heart is not malleable; nor will the feelings admit of such control.

'It is not possible for her,' said Mrs. Dale. 'I fear it is not possible. It is too soon.'

'Six months,' pleaded the squire.

'It will take years, – not months,' said Mrs. Dale.

'And she will lose all her youth.'

'Yes; he has done all that by his treachery. But it is done, and we cannot now go back. She loves him yet as dearly as she ever loved him.'

Then the squire muttered certain words below his breath, – ejaculations against Crosbie, which were hardly voluntary; but even as involuntary ejaculations were very improper. Mrs. Dale heard them, and was not offended either by their impropriety or their warmth. 'But you can understand,' she said, 'that she cannot bring herself to go there.' The squire struck the table with his fist, and repeated his ejaculations. If he could only have known how very disagreeable Lady Alexandrina was making herself, his spirit might, perhaps, have been less vehemently disturbed. If, also, he could have perceived and understood the light in which an alliance with the De Courcy family was now regarded by Crosbie, I think that he would have received some consolation from that consideration. Those who offend us are generally punished for the offence they give; but we so frequently miss the satisfaction of knowing that we are avenged! It is arranged, apparently, that the injurer shall be punished, but that the person injured shall not gratify his desire for vengeance.

'And will you go to Guestwick yourself?' asked Mrs. Dale.

'I will take the note,' said the squire, 'and will let you know to-morrow. The earl has behaved so kindly that every possible consideration is due to him. I had better tell him the whole truth, and go or stay, as he may wish. I don't see the good of going. What am I to do at Guestwick Manor? I did think that if we had all been there it might have cured some difficulties.'

Mrs. Dale got up to leave him, but she could not go without saying some word of gratitude for all that he had attempted to do for them. She well knew what he meant by the curing of difficulties. He had intended to signify that had they lived together for a week at Guestwick the idea of flitting from Allington might possibly have been abandoned. It seemed now to Mrs. Dale as though her brother-in-law were heaping coals of fire on her head in return for that intention. She felt half-ashamed of what she was doing, almost acknowledging to herself that she should have borne with his sternness in return for the benefits he had done to her daughters. Had she not feared their reproaches she would, even now, have given way.

'I do not know what I ought to say to you for your kindness.'

'Say nothing, – either for my kindness or unkindness; but stay where you are, and let us live like Christians together, striving to think good and not evil.' These were kind, loving words, showing in themselves a spirit of love and forbearance; but they were spoken in a harsh, unsympathizing voice, and the speaker, as he uttered them, looked gloomily at the fire. In truth the squire, as he spoke, was half-ashamed of the warmth of what he said.

'At any rate I will not think evil,' Mrs. Dale answered, giving him her hand. After that she left him, and returned home. It was too late for her to abandon her project of moving and remain at the Small House; but as she went across the garden she almost confessed to herself that she repented of what she was doing.

In these days of the cold early spring, the way from the lawn into the house, through the drawing-room window, was not as yet open, and it was necessary to go round by the kitchen-garden on to the road, and thence in by the front door; or else to pass through the back door, and into the house by the kitchen. This latter mode of entrance Mrs. Dale now adopted; and as she made her way into the hall Lily came upon her, with very silent steps, out from the parlour, and arrested her progress. There was a smile upon Lily's face as she lifted up her finger as if in caution, and no one looking at her would have supposed that she was herself in trouble. 'Mamma,' she said, pointing to the drawing-room door, and speaking almost in a whisper, 'you must not go in there; come into the parlour.'

'Who's there? Where's Bell?' and Mrs. Dale went into the parlour as she was bidden. 'But who is there?' she repeated.

'He's there!'

'Who is he?'

'Oh, mamma, don't be a goose! Dr. Crofts is there, of course. He's been nearly an hour. I wonder how he is managing, for there is nothing on earth to sit upon but the old lump of a carpet. The room is strewed about with crockery, and Bell is such a figure! She has got on your old checked apron, and when he came in she was rolling up the fire-irons in brown paper. I don't suppose she was ever in such a mess before. There's one thing certain, – he can't kiss her hand.'

'It's you are the goose, Lily.'

'But he's in there certainly, unless he has gone out through the window, or up the chimney.'

'What made you leave them?'

'He met me here, in the passage, and spoke to me ever so seriously. "Come in," I said, "and see Bell packing the pokers and tongs." "I will go in," he said, "but don't come with me." He was ever so serious, and I'm sure he had been thinking of it all the way along.'

'And why should he not be serious?'

'Oh, no, of course he ought to be serious; but are you not glad, mamma? I am so glad. We shall live alone together, you and I; but she will be so close to us! My belief is that he'll stay there for ever unless somebody does something. I have been so tired of waiting and looking out for you. Perhaps he's helping her to pack the things. Don't you think we might go in; or would it be ill-natured?'

'Lily, don't be in too great a hurry to say anything. You may be mistaken, you know; and there's many a slip between the cup and the lip.'

'Yes, mamma, there is,' said Lily, putting her hand inside her mother's arm, 'that's true enough.'

'Oh, my darling, forgive me,' said the mother, suddenly remembering that the use of the old proverb at the present moment had been almost cruel.

'Do not mind it,' said Lily, 'it does not hurt me, it does me good; that is to say, when there is nobody by except yourself. But, with God's help, there shall be no slip here, and she shall be happy. It is all the difference between one thing done in a hurry, and another thing done with much thinking. But they'll remain there for ever if we don't go in. Come, mamma, you open the door.'

Then Mrs. Dale did open the door, giving some little premonitory notice with the handle, so that the couple inside might be warned of approaching footsteps. Crofts had not escaped, either through the window or up the chimney, but was seated in the middle of the room on an empty box, just opposite to Bell, who was seated upon the lump of carpeting. Bell still wore the checked apron as described by her sister. What might have been the state of her hands I will not pretend to say; but I do not believe that her lover had found anything amiss with them. 'How do you do, doctor,' said Mrs. Dale, striving to use her accustomed voice, and to look as though there were nothing of special importance in his visit. 'I have just come down from the Great House.'

'Mamma,' said Bell, jumping up, 'you must not call him doctor any more.'

'Must I not? Has any one undoctored him?'

'Oh, mamma, you understand,' said Bell.

'I understand,' said Lily, going up to the doctor, and giving him her cheek to kiss, 'he is to be my brother, and I mean to claim him as such from this moment. I expect him to do everything for us, and not to call a moment of his time his own.'

'Mrs. Dale,' said the doctor, 'Bell has consented that it shall be so, if you will consent.'

'There is but little doubt of that,' said Mrs. Dale.

'We shall not be rich—' began the doctor.

'I hate to be rich,' said Bell. 'I hate even to talk about it. I don't think it quite manly even to think about it; and I'm sure it isn't womanly.'

'Bell was always a fanatic in praise of poverty,' said Mrs. Dale.

'No; I'm no fanatic. I'm very fond of money earned. I would like to earn some myself if I knew how.'

'Let her go out and visit the lady patients,' said Lily. 'They do in America.'

Then they all went into the parlour and sat round the fire talking as though they were already one family. The proceeding, considering the nature of it, – that a young lady, acknowledged to be of great beauty and known to be of good birth, had on the occasion been asked and given in marriage, – was carried on after a somewhat humdrum fashion, and in a manner that must be called commonplace. How different had it been when Crosbie had made his offer! Lily for the time had been raised to a pinnacle, – a pinnacle which might be dangerous, but which was, at any rate, lofty. With what a pretty speech had Crosbie been greeted! How it had been felt by all concerned that the fortunes of the Small House were in the ascendant, – felt, indeed, with some trepidation, but still with much inward triumph. How great had been the occasion, forcing Lily almost to lose herself in wonderment at what had occurred! There was no great occasion now, and no wonderment. No one, unless it was Crofts, felt very triumphant. But they were all very happy, and were sure that there was safety in their happiness. It was but the other day that one of them had been thrown rudely to the ground through the treachery of a lover, but yet none of them feared treachery from this lover. Bell was as sure of her lot in life as though she were already being taken home to her modest house in Guestwick. Mrs. Dale looked upon the man as her son, and the party of four as they sat round the fire grouped themselves as though they formed one family.

But Bell was not seated next to her lover. Lily, when she had once accepted Crosbie, seemed to think that she could never be too near to him. She had been in no wise ashamed of her love, and had shown it constantly by some little caressing motion of her hand, leaning on his arm, looking into his face, as though she were continually desirous of some palpable assurance of his presence. It was not so at all with Bell. She was happy in loving and in being loved, but she required no overt testimonies of affection. I do not think it would have made her unhappy if some sudden need had required that Crofts should go to India and back before they were married. The thing was settled, and that was enough for her. But on the other hand, when he spoke of the expediency of an immediate marriage, she raised no difficulty. As her mother was about to go into a new residence, it might be as well that that residence should be fitted to the wants of two persons instead of three. So they talked about chairs and tables, carpets and kitchens, in a most unromantic, homely, useful manner! A considerable portion of the furniture in the house they were now about to leave belonged to the squire, – or to the house rather, as they were in the habit of saying. The older and more solid things, – articles of household stuff that stand the wear of half a century, – had been in the Small House when they came to it. There was, therefore, a question of buying new furniture for a house in Guestwick, – a question not devoid of importance to the possessor of so moderate an income as that owned by Mrs. Dale. In the first month or two they were to live in lodgings, and their goods were to be stored in some friendly warehouse. Under such circumstances would it not be well that Bell's marriage should be so arranged that the lodging question might not be in any degree complicated by her necessities? This was the last suggestion made by Dr. Crofts, induced no doubt by the great encouragement he had received.

'That would be hardly possible,' said Mrs. Dale. 'It only wants three weeks; – and with the house in such a condition!'

'James is joking,' said Bell.

'I was not joking at all,' said the doctor.

'Why not send for Mr. Boyce, and carry her off at once on a pillion* behind you?' said Lily. 'It's just the sort of thing for primitive people to do, like you and Bell. All the same, Bell, I do wish you could have been married from this house.'

'I don't think it will make much difference,' said Bell.

'Only if you would have waited till summer we would have had such a nice party on the lawn. It sounds so ugly, being married from lodgings; doesn't it, mamma?'

'It doesn't sound at all ugly to me,' said Bell.

'I shall always call you Dame Commonplace when you're married,' said Lily.

Then they had tea, and after tea, Dr. Crofts got on his horse and rode back to Guestwick.

'Now may I talk about him?' said Lily, as soon as the door was closed behind his back.

'No; you may not.'

'As if I hadn't know it all along! And wasn't it hard to bear that you should have scolded me with such pertinacious austerity, and that I wasn't to say a word in answer.'

'I don't remember the austerity,' said Mrs. Dale.

'Nor yet Lily's silence,' said Bell.

'But it's all settled now,' said Lily, 'and I'm downright happy. I never felt more satisfaction, – never, Bell!'

'Nor did I,' said her mother; 'I may truly say that I thank God for this good thing.'

CHAPTER LI

JOHN EAMES DOES THINGS WHICH HE OUGHT NOT TO HAVE DONE

JOHN EAMES succeeded in making his bargain with Sir Raffle Buffle. He accepted the private secretaryship on the plainly expressed condition that he was to have leave of absence for a fortnight towards the end of April. Having arranged this he took an affectionate leave of Mr. Love, who was really much affected at parting with him, discussed valedictory pots of porter in the big room, over which many wishes were expressed that he might be enabled to compass the length and breadth of old Huffle's feet, uttered a last cutting joke at Mr. Kissing as he met that gentleman hurrying through the passages with an enormous ledger in his hands, and then took his place in the comfortable arm-chair which FitzHoward had been forced to relinquish.

'Don't tell any of the fellows,' said Fitz, 'but I'm going to cut the concern altogether. My governor wouldn't let me stop here in any other place than that of private secretary.'

'Ah, your governor is a swell,' said Eames.

'I don't know about that,' said FitzHoward. 'Of course he has a good deal of family interest. My cousin is to come in for St. Bungay at the next election, and then I can do better than remain here.'

'That's a matter of course,' said Eames. 'If my cousin were Member for St. Bungay, I'd never stand anything east of Whitehall.'

'And I don't mean,' said FitzHoward. 'This room, you know, is all very nice; but it is a bore coming into the City every day. And then one doesn't like to be rung for like a servant. Not that I mean to put you out of conceit with it.'

'It will do very well for me,' said Eames. 'I never was very particular.' And so they parted, Eames assuming the beautiful arm-chair and the peril of being asked to carry Sir Raffle's shoes, while FitzHoward took the vacant desk in the big room till such time as some member of his family should come into Parliament for the borough of St. Bungay.

But Eames, though he drank the porter, and quizzed FitzHoward, and gibed at Kissing, did not seat himself in his new arm-chair without some serious thoughts. He was aware that his career in London had not hitherto been one on which he could look back with self-respect. He had lived with friends whom he did not esteem; he had been idle, and sometimes worse than idle; and he had allowed himself to be hampered by the pretended love of a woman for whom he had never felt any true affection, and by whom he had been cozened out of various foolish promises which even yet were hanging over his head. As he sat with Sir Raffle's notes before him, he thought almost with horror of the men and women in Burton Crescent. It was now about three years since he had first known Cradell, and he shuddered as he remembered how very poor a creature was he whom he had chosen for his bosom friend. He could not make for himself those excuses which we can make for him. He could not tell himself that he had been driven by circumstances to choose a friend, before he had learned to know what were the requisites for which he should look. He had lived on terms of closest intimacy with this man for three years, and now his eyes were opening themselves to the nature of his friend's character. Cradell was in age three years his senior. 'I won't drop him,' he said to himself; 'but he is a poor creature.' He thought, too, of the Lupexes, of Miss Spruce, and of Mrs. Roper, and tried to imagine what Lily Dale would do if she found herself among such people. It would be impossible that should ever so find herself. He might as well ask her to drink at the bar of a gin-shop as to sit down in Mrs. Roper's drawing-room. If destiny had in store for him such good fortune as that of calling Lily his own, it was necessary that he should altogether alter his mode of life.

In truth his hobbledehoyhood was dropping off from him, as its old skin drops from a snake. Much of the feeling and something of the knowledge of manhood was coming on him, and he was beginning to recognize to himself that the future manner of his life must be to him a matter of very serious concern. No such thought had come near him when he first established himself in London. It seems to me that in this respect the fathers and mothers of the present generation understand but little of the inward nature of the young men for whom they are so anxious. They give them credit for so much that it is impossible they should have, and then deny them credit for so much that they possess! They expect from them when boys the discretion of men, – that discretion which comes from thinking; but will not give them credit for any of that power of thought which alone can ultimately produce good conduct. Young men are generally thoughtful, – more thoughtful than their seniors; but the fruit of their thought is not as yet there. And then so little is done for the amusement of lads who are turned loose into London at nineteen or twenty. Can it be that any mother really expects her son to sit alone evening after evening in a dingy room drinking bad tea, and reading good books? And yet it seems that mothers do so expect, – the very mothers who talk about the thoughtlessness of youth! O ye mothers who from year to year see your sons launched forth upon the perils of the world, and who are so careful with your good advice, with under flannel shirting, with books of devotion and tooth-powder, does it never occur to you that provisions should be made for amusement, for dancing, for parties, for the excitement and comfort of women's society? That excitement your sons will have, and if it be not provided by you of one kind, will certainly be provided by themselves of another kind. If I were a mother sending lads out into the world, the matter most in my mind would be this, – to what houses full of nicest girls could I get them admission, so that they might do their flirting in good company.

Poor John Eames had been so placed that he had been driven to do his flirting in very bad company, and he was now fully aware that it had been so. It wanted but two days to his departure for Guestwick Manor, and as he sat breathing a while after the manufacture of a large batch of Sir Raffle's notes, he made up his mind that he would give Mrs. Roper notice before he started, that on his return to London he would be seen no more in Burton Crescent. He would break his bonds altogether asunder, and if there should be any penalty for such breaking he would pay it in what best manner he might be able. He acknowledged to himself that he had been behaving badly to Amelia, confessing, indeed, more sin in that respect than he had in truth committed; but this, at any rate, was clear to him, that he must put himself on a proper footing in that quarter before he could venture to speak to Lily Dale.

As he came to a definite conclusion on this subject the little handbell which always stood on Sir Raffle's table was sounded, and Eames was called into the presence of the great man. 'Ah,' said Sir Raffle, leaning back in his arm-chair, and stretching himself after the great exertions which he had been making – 'Ah, let me see! You are going out of town the day after to-morrow.'

'Yes, Sir Raffle, the day after to-morrow.'

'Ah! it's a great annoyance, – a very great annoyance. But on such occasions I never think of myself. I never have done so, and don't suppose I ever shall. So you're going down to my old friend De Guest?'

Eames was always angered when his new patron Sir Raffle talked of his old friendship with the earl, and never gave the Commissioner any encouragement. 'I am going down to Guestwick,' said he.

'Ah! yes; to Guestwick Manor? I don't remember that I was ever there. I daresay I may have been, but one forgets those things.'

'I never heard Lord De Guest speak of it.'

'Oh, dear, no. Why should his memory be better than mine? Tell him, will you, how very glad I shall be to renew our old intimacy. I should think nothing of running down to him for a day or two in the dull time of the year, – say in September or October. It's rather a coincidence our both being interested about you, – isn't it?'

'I'll be sure to tell him.'

'Mind you do. He's one of our most thoroughly independent noblemen, and I respect him very highly. Let me see; didn't I ring my bell? What was it I wanted? I think I rang my bell.'

'You did ring your bell.'

'Ah, yes; I know. I am going away, and I wanted my— would you tell Rafferty to bring me – my boots?' Whereupon Johnny rang the bell – not the little handbell, but the other bell. 'And I shan't be here to-morrow,' continued Sir Raffle. 'I'll thank you to send my letters up to the square; and if they should send down from the Treasury; – but the Chancellor would write, and in that case you'll send up his letter at once by a special messenger, of course.'

'Here's Rafferty,' said Eames, determined that he would not even sully his lips with speaking of Sir Raffle's boots.

'Oh, ah, yes; Rafferty, bring me my boots.'

'Anything else to say?' asked Eames.

'No, nothing else. Of course you'll be careful to leave everything straight behind you.'

'Oh, yes; I'll leave it all straight.' Then Eames withdrew, so that he might not be present at the interview between Sir Raffle and his boots. 'He'll not do,' said Sir Raffle to himself. 'He'll never do. He's not quick enough, – has no go in him. He's not man enough for the place. I wonder why the earl has taken him by the hand in that way.'

Soon after the little episode of the boots Eames left his office, and walked home alone to Burton Crescent. He felt that he had gained a victory in Sir Raffle's room, but the victory there had been easy. Now he had another battle on his hands, in which, as he believed, the achievement of victory would be much more difficult. Amelia Roper was a person much more to be feared than the Chief Commissioner. He had one strong arrow in his quiver on which he would depend, if there should come to him the necessity of giving his enemy a death-wound. During the last week she had been making powerful love to Cradell, so as to justify the punishment of desertion from a former lover. He would not throw Cradell in her teeth if he could help it; but it was incumbent on him to gain a victory, and if the worst should come to the worst, he must use such weapons as destiny and the chance of war had given him.

He found Mrs. Roper in the dining-room as he entered, and immediately began his work. 'Mrs. Roper,' he said, 'I'm going out of town the day after to-morrow.'

'Oh, yes, Mr. Eames, we know that. You're going as a visitor to the noble mansion of the Earl De Guest.'

'I don't know about the mansion being very noble, but I'm going down into the country for a fortnight. When I come back—'

'When you come back, Mr. Eames, I hope you'll find your room a deal more comfortable. I know it isn't quite what it should be for a gentleman like you, and I've been thinking for some time past—'

'But, Mrs. Roper, I don't mean to come back here any more. It's just that that I want to say to you.'

'Not come back to the crescent!'

'No, Mrs. Roper. A fellow must move sometimes, you know; and I'm sure I've been very constant to you for a long time.'

'But where are you going, Mr. Eames?'

'Well; I haven't just made up my mind as yet. That is, it will depend on what I may do, – on what friends of mine may say down in the country. You'll not think I'm quarrelling with you, Mrs. Roper.'

'It's them Lupexes as have done it,' said Mrs. Roper, in her deep distress.

'No, indeed, Mrs. Roper, nobody has done it.'

'Yes, it is; and I'm not going to blame you, Mr. Eames. They've made the house unfit for any decent young gentleman like you. I've been feeling that all along; but it's hard upon a lone woman like me, isn't it, Mr. Eames?'

'But, Mrs. Roper, the Lupexes have had nothing to do with my going.'

'Oh, yes, they have; I understand it all. But what could I do, Mr. Eames? I've been giving them warning every week for the last six months; but the more I give them warning, the more they won't go. Unless I were to send for a policeman, and have a row in the house—'

'But I haven't complained of the Lupexes, Mrs. Roper.'

'You wouldn't be quitting without any reason, Mr. Eames. You are not going to be married in earnest, are you, Mr. Eames?'

'Not that I know of.'

'You may tell me; you may, indeed. I won't say a word, – not to anybody. It hasn't been my fault about Amelia. It hasn't really.'

'Who says there's been any fault?'

'I can see, Mr. Eames. Of course it didn't do for me to interfere. And if you had liked her, I will say I believe she'd have made as good a wife as any young man ever took; and she can make a few pounds go farther than most girls. You can understand a mother's feelings; and if there was to be anything, I couldn't spoil it; could I, now?'

'But there isn't to be anything.'

'So I've told her for months past. I'm not going to say anything to blame you; but young men ought to be very particular; indeed they ought.' Johnny did not choose to hint to the disconsolate mother that it also behoved young women to be very particular, but he thought it. 'I've wished many a time, Mr. Eames, that she had never come here; indeed I have. But what's a mother to do? I couldn't put her outside the door.' Then Mrs. Roper raised her apron up to her eyes, and began to sob.

'I'm very sorry if I've made any mischief,' said Johnny.

'It hasn't been your fault,' continued the poor woman, from whom, as her tears became uncontrollable, her true feelings forced themselves and the real outpouring of her feminine nature. 'Nor it hasn't been my fault. But I knew what it would come to when I saw how she was going on; and I told her so. I knew you wouldn't put up with the likes of her.'

'Indeed, Mrs. Roper, I've always had a great regard for her, and for you too.'

'But you weren't going to marry her. I've told her so all along, and I've begged her not to do it, – almost on my knees I have; but she wouldn't be said by me. She never would. She's always been that wilful that I'd sooner have her away from me than with me. Though she's a good young woman in the house, – she is, indeed, Mr. Eames; – and there isn't a pair of hands in it that works so hard; but it was no use my talking.'

'I don't think any harm has been done.'

'Yes, there has; great harm. It has made the place not respectable. It's the Lupexes is the worst. There's Miss Spruce, who has been with me for nine years, – ever since I've had the house, – she's been telling me this morning that she means to go into the country. It's all the same thing. I understand it. I can see it. The house isn't respectable, as it should be; and your mamma, if she were to know all, would have a right to be angry with me. I did mean to be respectable, Mr. Eames; I did indeed.'

'Miss Spruce will think better of it.'

'You don't know what I've had to go through. There's

none of them pays, not regular, – only she and you. She's been like the Bank of England, has Miss Spruce.'

'I'm afraid I've not been very regular, Mrs. Roper.'

'Oh, yes, you have. I don't think of a pound or two more or less at the end of a quarter, if I'm sure to have it some day. The butcher, – he understands one's lodgers just as well as I do, – if the money's really coming, he'll wait; but he won't wait for such as them Lupexes, whose money's nowhere. And there's Cradell; would you believe it, that fellow owes me eight and twenty pounds!'

'Eight and twenty pounds!'

'Yes, Mr. Eames, eight and twenty pounds! He's a fool. It's them Lupexes as have had his money. I know it. He don't talk of paying, and going away. I shall be just left with him and the Lupexes on my hands; and then the bailiffs may come and sell every stick about the place. I won't say nay to them.' Then she threw herself into the old horsehair armchair, and gave way to her womanly sorrow.

'I think I'll go upstairs, and get ready for dinner,' said Eames.

'And you must go away when you come back?' said Mrs. Roper.

'Well, yes, I'm afraid I must. I meant you to have a month's warning from to-day. Of course I shall pay for the month.'

'I don't want to take any advantage; indeed, I don't. But I do hope you'll leave your things. You can have them whenever you like. If Chumpend knows that you and Miss Spruce are both going, of course he'll be down upon me for his money.' Chumpend was the butcher. But Eames made no answer to this piteous plea. Whether or no he could allow his old boots to remain in Burton Crescent for the next week or two, must depend on the manner in which he might be received by Amelia Roper this evening.

When he came down to the drawing-room, there was no one there but Miss Spruce. 'A fine day, Miss Spruce,' said he.

'Yes, Mr. Eames, it is a fine day for London; but don't you think the country air is very nice?'

'Give me the town,' said Johnny, wishing to say a good word for poor Mrs. Roper, if it were possible.

'You're a young man, Mr. Eames; but I'm only an old woman. That makes a difference,' said Miss Spruce.

'Not much,' said Johnny, meaning to be civil. 'You don't like to be done any more than I do.'

'I like to be respectable, Mr. Eames. I always have been respectable, Mr. Eames.' This the old woman said almost in a whisper, looking anxiously to see that the door had not been opened to other listening ears.

'I'm sure Mrs. Roper is very respectable.'

'Yes; Mrs. Roper is respectable, Mr. Eames; but there are some here that— Hush-sh-sh!' And the old lady put her finger up to her lips. The door opened and Mrs. Lupex swam into the room.

'How d'ye do, Miss Spruce? I declare you're always first. It's to get a chance of having one of the young gentlemen to yourself, I believe. What's the news in the city to-day, Mr Eames? In your position now of course you hear all the news.'

'Sir Raffle Buffle has got a new pair of shoes. I don't know that for certain, but I guess it from the time it took him to put them on.'

'Ah! now you're quizzing. That's always the way with you gentlemen when you get a little up in the world. You don't think women are worth talking to then, unless just for a joke or so.'

'I'd a great deal sooner talk to you, Mrs. Lupex, than I would to Sir Raffle Buffle.'

'It's all very well for you to say that. But we women know what such compliments as those mean; – don't we, Miss Spruce? A woman that's been married five years as I have – or I may say six, – doesn't expect much attention from young men. And though I was young when I married – young in years, that is, – I'd seen too much and gone through too much to be young in heart.' This she said almost in a whisper; but Miss Spruce heard it, and was confirmed in her belief that Burton Crescent was no longer respectable.

'I don't know what you were then, Mrs. Lupex,' said Eames; 'but you're young enough now for anything.'

'Mr. Eames, I'd sell all that remains of my youth at a

cheap rate, – at a very cheap rate, if I could only be sure of—'

'Sure of what, Mrs. Lupex?'

'The undivided affection of the one person that I loved. That is all that is necessary to a woman's happiness.'

'And isn't Lupex—'

'Lupex! But, hush, never mind. I should not have allowed myself to be betrayed into an expression of feeling. Here's your friend, Mr. Cradell. Do you know I sometimes wonder what you find in that man to be so fond of him.' Miss Spruce saw it all, and heard it all, and positively resolved upon moving herself to those two small rooms at Dulwich.

Hardly a word was exchanged between Amelia and Eames before dinner. Amelia still devoted herself to Cradell, and Johnny saw that that arrow, if it should be needed, would be a strong weapon. Mrs. Roper they found seated at her place at the dining-table, and Eames could perceive the traces of her tears. Poor woman! Few positions in life could be harder to bear than hers! To be ever tugging at others for money that they could not pay; to be ever tugged at for money which she could not pay; to desire respectability for its own sake, but to be driven to confess that it was a luxury beyond her means; to put up with disreputable belongings for the sake of lucre, and then not to get the lucre, but be driven to feel that she was ruined by the attempt! How many Mrs. Ropers there are who from year to year sink down and fall away, and no one knows whither they betake themselves! One fancies that one sees them from time to time at the corners of the streets in battered bonnets and thin gowns, with the tattered remnants of old shawls upon their shoulders, still looking as though they had within them a faint remembrance of long-distant respectability. With anxious eyes they peer about, as though searching in the streets for other lodgers. Where do they get their daily morsels of bread, and their poor cups of thin tea, – their cups of thin tea, with perhaps a pennyworth of gin added to it, if Providence be good! Of this state of things Mrs. Roper had a lively appreciation, and now, poor woman, she feared that she was reaching it, by the aid of the Lupexes. On the present occasion she carved her joint of meat in silence, and sent out her slices to the good guests that would leave her, and to the bad guests that would remain, with apathetic impartiality. What was the use now of doing favour to one lodger or disfavour to another? Let them take their mutton, – they who would pay for it and they who would not. She would not have the carving of many more joints in that house if Chumpend acted up to all the threats which he had uttered to her that morning.

The reader may, perhaps, remember the little back room behind the dining parlour. A description was given in some former pages of an interview which was held between Amelia and her lover. It was in that room that all the interviews of Mrs. Roper's establishment had their existence. A special room for interviews is necessary in all households of a mixed nature. If a man lives alone with his wife, he can have his interviews where he pleases. Sons and daughters, even when they are grown up, hardly create the necessity of an interview-chamber, though some such need may be felt if the daughters are marriageable and independent in their natures. But when the family becomes more complicated than this, if an extra young man be introduced, or an aunt comes into residence, or grown-up children by a former wife interfere with the domestic simplicity, then such accommodation becomes quite indispensable. No woman would think of taking in lodgers without such a room; and this room there was at Mrs. Roper's, very small and dingy, but still sufficient, – just behind the dining parlour and opposite to the kitchen stairs. Hither, after dinner, Amelia was summoned. She had just seated herself between Mrs. Lupex and Miss Spruce, ready to do battle with the former because she would stay, and with the latter because she would go, when she was called out by the servant girl.

'Miss Mealyer, Miss Mealyer, – sh – sh – sh!' And Amelia, looking round, saw a large red hand beckoning to her. 'He's down there,' said Jemima, as soon as her young mistress had joined her, 'and wants to see you most partic'lar.'

'Which of 'em?' asked Amelia, in a whisper.

'Why, Mr. Heames, to be sure. Don't you go and have anythink to say to the other one, Miss Mealyer, pray don't; he ain't no good; he ain't indeed.'

Amelia stood still for a moment on the landing, calculating whether it would be well for her to have the interview, or well to decline it. Her objects were two; – or, rather, her object was in its nature twofold. She was, naturally, anxious to drive John Eames to desperation; and anxious also, by some slight added artifice, to make sure of Cradell if Eames's desperation did not have a very speedy effect. She agreed with Jemima's criticism in the main, but she did not go quite so far as to think that Cradell was no good at all. Let it be Eames, if Eames were possible; but let the other string be kept for use if Eames were not possible. Poor girl! in coming to this resolve she had not done so without agony. She had a heart, and with such power as it gave her, she loved John Eames. But the world had been hard to her; knocking her about hither and thither unmercifully; threatening, as it now threatened, to take from her what few good things she enjoyed. When a girl is so circumstanced she cannot afford to attend to her heart. She almost resolved not to see Eames on the present occasion, thinking that he might be made the more desperate by such refusal, and remembering also that Cradell was in the house and would know of it.

'He's there a-waiting, Miss Mealyer. Why don't yer come down?' and Jemima plucked her young mistress by the arm.

'I am coming,' said Amelia. And with dignified steps she descended to the interview.

'Here she is, Mr. Heames,' said the girl. And then Johnny found himself alone with his lady-love.

'You have sent for me, Mr. Eames,' she said, giving her head a little toss, and turning her face away from him. 'I was engaged upstairs, but I thought it uncivil not to come down to you as you sent for me so special.'

'Yes, Miss Roper, I did want to see you very particularly.'

'Oh, dear!' she exclaimed, and he understood fully that the exclamation referred to his having omitted the customary use of her Christian name.

'I saw your mother before dinner, and I told her that I am going away the day after to-morrow.'

'We all know about that; – to the earl's, of course!' And then there was another chuck of her head.

'And I told her also that I had made up my mind not to come back to Burton Crescent.'

'What! leave the house altogether!'

'Well; yes. A fellow must make a change sometimes, you know.'

'And where are you going, John?'

'That I don't know as yet.'

'Tell me the truth, John; are you going to be married? Are you – going – to marry – that young woman, – Mr. Crosbie's leavings? I demand to have an answer at once. Are you going to marry her?'

He had determined very resolutely that nothing she might say should make him angry, but when she thus questioned him about 'Crosbie's leavings' he found it very difficult to keep his temper.

'I have not come,' said he, 'to speak to you about any one but ourselves.'

'That put-off won't do with me, sir. You are not to treat any girl you may please in that sort of way; – oh, John!' Then she looked at him as though she did not know whether to fly at him and cover him with kisses, or to fly at him and tear his hair.

'I know I haven't behaved quite as I should have done,' he began.

'Oh, John!' and she shook her head. 'You mean, then, to tell me that you are going to marry her?'

'I mean to say nothing of the kind. I only mean to say that I am going away from Burton Crescent.'

'John Eames, I wonder what you think will come to you! Will you answer me this; have I had a promise from you, – a distinct promise, over and over again, or have I not?'

'I don't know about a distinct promise—'

'Well, well! I did think that you was a gentleman that would not go back from your word. I did think that. I did think that you would never put a young lady to the necessity of bringing forward her own letters to prove that she is not expecting more than she has a right! You don't know! And that, after all that has been between us! John Eames!' And again it seemed to him as though she were about to fly.

'I tell you that I know I haven't behaved well. What more can I say?'

'What more can you say? Oh, John! to ask me such a question! If you were a man you would know very well what more to say. But all you private secretaries are given to deceit, as the sparks fly upward. However, I despise you, – I do, indeed. I despise you.'

'If you despise me, we might as well shake hands and part at once. I daresay that will be best. One doesn't like to be despised, of course; but sometimes one can't help it.' And then he put out his hand to her.

'And is this to be the end of all?' she said, taking it.

'Well, yes; I suppose so. You say I'm despised.'

'You shouldn't take up a poor girl in that way for a sharp word, – not when she is suffering as I am made to suffer. If you only think of it, – think what I have been expecting!' And now Amelia began to cry, and to look as though she were going to fall into his arms.

'It is better to tell the truth,' he said; 'isn't it?'

'But it shouldn't be the truth.'

'But it is the truth. I couldn't do it. I should ruin myself and you too, and we should never be happy.'

'I should be happy – very happy indeed.' At this moment the poor girl's tears were unaffected, and her words were not artful. For a minute or two her heart, – her actual heart, – was allowed to prevail.

'It cannot be, Amelia. Will you not say good-bye?'

'Good-bye,' she said, leaning against him as she spoke.

'I do so hope you will be happy,' he said. And then, putting his arm round her waist, he kissed her; which he certainly ought not to have done.

When the interview was over, he escaped out into the crescent, and as he walked down through the squares, – Woburn Square, and Russell Square, and Bedford Square, – towards the heart of London, he felt himself elated almost to a state of triumph. He had got himself well out of his difficulties, and now he would be ready for his love-tale to Lily.

CHAPTER LII

THE FIRST VISIT TO THE GUESTWICK BRIDGE

WHEN John Eames arrived at Guestwick Manor, he was first welcomed by Lady Julia. 'My dear Mr. Eames,' she said, 'I cannot tell you how glad we are to see you.' After that she always called him John, and treated him throughout his visit with wonderful kindness. No doubt that affair of the bull had in some measure produced this feeling; no doubt, also, she was well disposed to the man who she hoped might be accepted as a lover by Lily Dale. But I am inclined to think that the fact of his having beaten Crosbie had been the most potential cause of this affection for our hero on the part of Lady Julia. Ladies, – especially discreet old ladies, such as Lady Julia De Guest, – are bound to entertain pacific theories, and to condemn all manner of violence. Lady Julia would have blamed any one who might have advised Eames to commit an assault upon Crosbie. But, nevertheless, deeds of prowess are still dear to the female heart, and a woman, be she ever so old and discreet, understands and appreciates the summary justice which may be done by means of a thrashing. Lady Julia, had she been called upon to talk of it, would undoubtedly have told Eames that he had committed a fault in striking Mr. Crosbie; but the deed had been done, and Lady Julie became very fond of John Eames.

'Vickers shall show you your room, if you like to go upstairs; but you'll find my brother close about the house if you choose to go out; I saw him not half an hour since.' But John seemed to be well satisfied to sit in the arm-chair over the fire, and talk to his hostess; so neither of them moved.

'And now that you're a private secretary, how do you like it?'

'I like the work well enough; only I don't like the man,

Lady Julia. But I shouldn't say so, because he is such an intimate friend of your brother's.'

'An intimate friend of Theodore's! – Sir Raffle Buffle!' Lady Julia stiffened her back and put on a serious face, not being exactly pleased at being told that the Earl De Guest had any such intimate friend.

'At any rate he tells me so about four times a day, Lady Julia. And he particularly wants to come down here next September.' –

'Did he tell you that, too?'

'Indeed he did. You can't believe what a goose he is! Then his voice sounds like a cracked bell; it's the most disagreeable voice you ever heard in your life. And one has always to be on one's guard lest he should make one do something that is – is – that isn't quite the thing for a gentleman. You understand; – what the messenger ought to do.

'You shouldn't be too much afraid of your own dignity.'

'No, I'm not. If Lord De Guest were to ask me to fetch him his shoes, I'd run to Guestwick and back for them and think nothing of it, just because I know he's my friend. He'd have a right to send me. But I'm not going to do such things as that for Sir Raffle Buffle.'

'Fetch him his shoes!'

'That's what FitzHoward had to do, and he didn't like it.'

'Isn't Mr. FitzHoward nephew to the Duchess of St. Bungay?'

'Nephew, or cousin, or something.'

'Dear me!' said Lady Julia, 'what a horrible man!' And in this way John Eames and her ladyship became very intimate.

There was no one at dinner at the Manor that day but the earl and his sister and their single guest. The earl when he came in was very warm in his welcome, slapping his young friend on the back, and poking jokes at him with a good-humoured if not brilliant pleasantry.

'Thrashed anybody lately, John?'

'Nobody to speak of,' said Johnny.

'Brought your nightcap down for your out-o'-doors nap?'

'No; but I've got a grand stick for the bull,' said Johnny.

'Ah! that's no joke now, I can tell you,' said the earl. 'We had to sell him, and it half broke my heart. We don't know what had come to him, but he became quite unruly after that; – knocked Darvell down in the straw-yard! It was a very bad business, – a very bad business, indeed! Come, go and dress. Do you remember how you came down to dinner that day? I shall never forget how Crofts stared at you. Come, you've only got twenty minutes, and you London fellows always want an hour.'

'He's entitled to some consideration now he's a private secretary,' said Lady Julia.

'Bless us all! yes; I forgot that. Come, Mr. Private Secretary, don't stand on the grandeur of your neck-tie to-day, as there's nobody here but ourselves. You shall have an opportunity to-morrow.'

Then Johnny was handed over to the groom of the chambers, and exactly in twenty minutes he reappeared in the drawing-room.

As soon as Lady Julia had left them after dinner, the earl began to explain his plan for the coming campaign. 'I'll tell you now what I have arranged,' said he. 'The squire is to be here to-morrow with his eldest niece, – your Miss Lily's sister, you know.'

'What, Bell?'

'Yes, with Bell, if her name is Bell. She's a very pretty girl, too. I don't know whether she's not the prettiest of the two, after all.'

'That's a matter of opinion.'

'Just so, Johnny; and do you stick to your own. They're coming here for three or four days. Lady Julia did ask Mrs. Dalea and Lily. I wonder whether you'll let me call her Lily?'

'Oh, dear! I wish I might have the power of letting you.'

'That's just the battle that you've got to fight. But the mother and younger sister wouldn't come. Lady Julia says it's all right; – that, as a matter of course, she wouldn't come when she heard you were to be here. I don't quite understand it. In my days the young girls were ready enough to go where they knew they'd meet their lovers, and I never thought any the worse of them for it.'

'It wasn't because of that,' said Eames.

'That's what Lady Julia says, and I always find her to be right in things of that sort. And she says you'll have a better chance in going over there than you would here, if she were in the same house with you. If I was going to make love to a girl, of course I'd sooner have her close to me, – staying in the same house. I should think it the best fun in the world. And we might have had a dance, and all that kind of thing. But I couldn't make her come, you know.'

'Oh, no; of course not.'

'And Lady Julia thinks that it's best as it is. You must go over, you know, and get the mother on your side, if you can. I take it, the truth is this; – you mustn't be angry with me, you know, for saying it.'

'You may be sure of that.'

'I suppose she was fond of that fellow, Crosbie. She can't be very fond of him now, I should think, after the way he has treated her; but she'll find a difficulty in making her confession that she really likes you better than she ever liked him. Of course that's what you'll want her to say.'

'I want her to say that she'll be my wife, – some day.'

'And when she has agreed to the some day, then you'll begin to press her to agree to your day; – eh, sir? My belief is you'll bring her round. Poor girl! why should she break her heart when a decent fellow like you will only be too glad to make her a happy woman?' And in this way the earl talked to Eames till the latter almost believed that the difficulties were vanishing from out of his path. 'Could it be possible,' he asked himself, as he went to bed, 'that in a fortnight's time Lily Dale should have accepted him as her future husband?' Then he remembered that day on which Crosbie, with the two girls, had called at his mother's house, when in the bitterness of his heart, he had sworn to himself that he would always regard Crosbie as his enemy. Since then the world had gone well with him; and he had no longer any very bitter feeling against Crosbie. That matter had been arranged on the platform of the Paddington Station. He felt that if Lily would now accept him he could almost shake hands with Crosbie. The episode in his life and in Lily's would have been painful; but he would learn to look back upon that without regret, if Lily could be taught to believe that a kind fate had at last given her to the better of her two lovers. 'I'm afraid she won't bring herself to forget him,' he had said to the earl. 'She'll only be too happy to forget him,' the earl had answered, 'if you can induce her to begin the attempt. Of course it is very bitter at first; – all the world knew about it; but, poor girl, she is not to be wretched for ever, because of that. Do you go about your work with some little confidence, and I doubt not but what you'll have your way. You have everybody in your favour, – the squire, her mother, and all.' While such words as these were in his ears how could he fail to hope and to be confident? While he was sitting cozily over his bedroom fire he resolved that it should be as the earl had said. But when he got up on the following morning, and stood shivering as he came out of his bath, he could not feel the same confidence. 'Of course I shall go to her,' he said to himself, 'and make a plain story of it. But I know what her answer will be. She will tell me that she cannot forget him.' Then his feelings towards Crosbie were not so friendly as they had been on the previous evening.

He did not visit the Small House on that, his first day. It had been thought better that he should first meet the squire and Bell at Guestwick Manor, so he postponed his visit to Mrs. Dale till the next morning.

'Go when you like,' said the earl. 'There's the brown cob for you to do what you like with him while you are here.'

'I'll go and see my mother,' said John; 'but I won't take the cob to-day. If you'll let me have him to-morrow, I'll ride to Allington.' So he walked off to Guestwick by himself.

He knew well every yard of the ground over which he went, remembering every gate and stile and greensward from the time of his early boyhood. And now as he went along through his old haunts, he could not but look back and think of the thoughts which had filled his mind in his earlier wanderings. As I have said before, in some of these pages, no walks taken by the man are so crowded with thought as those taken by the boy. He had been early taught to understand that the world to him would be very hard; that he had nothing to look to but his own exertions, and that those exertions would not, unfortunately, be backed by any great cleverness of his own. I do not know that anybody had told him that he was a fool; but he had come to understand, partly through his own modesty, and partly, no doubt, through the somewhat obtrusive diffidence of his mother, that he was less sharp than other lads. It is probably true that he had come to his sharpness later in life than is the case with many young men. He had not grown on the sunny side of the wall. Before that situation in the Income-tax Office had fallen in his way, very humble modes of life had offered themselves, – or, rather, had not offered themselves for his acceptance. He had endeavoured to become an usher at a commercial seminary, not supposed to be in a very thriving condition; but he had been, luckily, found deficient in his arithmetic. There had been some chance of his going into the leather-warehouse of Messrs. Basil and Pigskin, but those gentlemen had required a premium, and any payment of that kind had been quite out of his mother's power. A country attorney, who had known the family for years, had been humbly solicited, the widow almost kneeling before him with tears, to take Johnny by the hand and make a clerk of him; but the attorney had discovered that Master Johnny Eames was not supposed to be sharp, and would have none of him. During those days, those gawky, gainless, unadmired days, in which he had wandered about the lanes of Guestwick as his only amusement, and had composed hundreds of rhymes in honour of Lily Dale which no human eye but his own had ever seen, he had come to regard himself as almost a burden upon the earth. Nobody seemed to want him. His own mother was very anxious; but her anxiety seemed to him to indicate a continual desire to get rid of him. For hours upon hours he would fill his mind with castles in the air, dreaming of wonderful successes in the midst of which Lily Dale always reigned as a queen. He would carry on the same story in his imagination from month to month, almost contenting himself with such ideal happiness. Had it not been for the possession of that power, what comfort could there have been to him in his life? There are lads of seventeen who can find happiness in study, who can busy themselves in books and be at their ease among the creations of other minds. These are they who afterwards become well-informed men. It was not so with John Eames. He had never been studious. The perusal of a novel was to him in those days a slow affair; and of poetry he read but little, storing up accurately in his memory all that he did read. But he created for himself his own romance, though to the eye a most unromantic youth; and he wandered through the Guestwick woods with many thoughts of which they who knew him best knew nothing. All this he thought of now as, with devious steps, he made his way towards his old home; – with very devious steps, for he went backwards through the woods by a narrow path which led right away from the town down to a little water-course, over which stood a wooden foot-bridge with a rail. He stood on the centre of the plank, at a spot which he knew well, and rubbing his hand upon the rail, cleansed it for the space of a few inches of the vegetable growth produced by the spray of the water. There, rudely carved in the wood, was still the word LILY. When he cut those letters she had been almost a child. 'I wonder whether she will come here with me and let me show it to her,' he said to himself. Then he took out his knife and cleared the cuttings of the letters, and having done so, leaned upon the rail, and looked down upon the running water. How well things in the world had gone for him! How well! And yet what would it all be if Lily would not come to him? How well the world had gone for him! In those days when he stood there carving the girl's name everybody had seemed to regard him as a heavy burden, and he had so regarded himself. Now he was envied by many, respected by many, taken by the hand as a friend by those high in the world's esteem. When he had come near the Guestwick Mansion in his old walks, – always, however, keeping at a great distance lest the grumpy old lord should be down upon him and scold him, – he had little dreamed that he and the grumpy old lord would ever be together on such familiar terms, that he would tell to that lord more of his private thoughts than to any other living being; yet it had come to that. The grumpy old lord had now told him that that gift of money was to be his whether Lily Dale accepted him or no. 'Indeed, the thing's done,' said the grumpy lord, pulling out from his pocket certain papers, 'and you've got to receive the dividends as they become due.' Then, when Johnny had expostulated, – as, indeed, the circumstances had left him no alternative but to expostulate, – the earl had roughly bade him hold his tongue, telling him that he would have to fetch Sir Raffle's boots directly he got back to London. So the conversation had quickly turned itself away to Sir Raffle, whom they had both ridiculed with much satisfaction. 'If he finds his way down here in September, Master Johnny, or in any other month either, you may fit my head with a foolscap. Not remember, indeed! Is it not wonderful that any man should make himself so mean a fool?' All this was thought over again, as Eames leaned upon the bridge. He remembered every word, and remembered many other words, – earlier words, spoken years ago, filling him with desolation as to the prospects of his life. It had seemed that his friends had united in prophesying that the outlook into the world for him was hopeless, and that the earning of bread must be for ever beyond his power. And now his lines had fallen to him in very pleasant places, and he was among those whom the world had determined to caress. And yet, what would it all be if Lily would not share his happiness? When he had carved that name on the rail, his love for Lily had been an idea. It had now become a reality which might probably be full of pain. If it were so, – if such should be the result of his wooing, – would not those dreamy old days have been better than these – the days of his success?

It was one o'clock by the time that he reached his mother's house, and he found her and his sister in a troubled and embarrassed state. 'Of course you know, John,' said his mother, as soon as their first embraces were over, 'that we are going to dine at the Manor this evening?' But he did not know it, neither the earl nor Lady Julia having said anything on the subject. 'Of course we are going,' said Mrs. Eames, 'and it was so very kind. But I've never been out to such a house for so many years, John, and I do feel in such a twitter. I dined there once, soon after we were married; but I never have been there since that.'

'It's not the earl I mind, but Lady Julia,' said Mary Eames.

'She's the most good-natured woman in the world,' said Johnny.

'Oh, dear; people say she is so cross!'

'That's because people don't know her. If I was asked who is the kindest-hearted woman I know in the world, I think I should say Lady Julia De Guest. I think I should.'

'Ah! but then they're so fond of you,' said the admiring mother. 'You saved his lordship's life, – under Providence.'

'That's all bosh, mother. You ask Dr. Crofts. He knows them as well as I do.'

'Dr. Crofts is going to marry Bell Dale,' said Mary; and then the conversation was turned from the subject of Lady Julia's perfections, and the awe inspired by the earl.

'Crofts going to marry Bell!' exclaimed Eames, thinking almost with dismay of the doctor's luck in thus getting himself accepted all at once, while he had been suing with the constancy almost of a Jacob.*

'Yes,' said Mary; 'and they say that she has refused her cousin Bernard, and that, therefore, the squire is taking away the house from them. You know they're all coming into Guestwick.'

'Yes, I know they are. But I don't believe that the squire is taking away the house.'

'Why should they come, then? Why should they give up such a charming place as that?'

'Rent-free!' said Mrs. Eames.

'I don't know why they should come away, but I can't believe the squire is turning them out; at any rate not for that reason.' The squire was prepared to advocate John's suit, and therefore John was bound to do battle on the squire's behalf.

'He is a very stern man,' said Mrs. Eames, 'and they say that since that affair of poor Lily's he has been more cross than ever with them. As far as I know, it was not Lily's fault.'

'Poor Lily!' said Mary. 'I do pity her. If I was her I should hardly know how to show my face; I shouldn't, indeed.'

'And why shouldn't she show her face?' said John, in an angry tone. 'What has she done to be ashamed of? Show her face indeed! I cannot understand the spite which one woman will sometimes have to another.'

'There is no spite, John; and it's very wrong of you to say so,' said Mary, defending herself. 'But it's a very unpleasant thing for a girl to be jilted. All the world knows that she was engaged to him.'

'And all the world knows—' But he would not proceed to declare that all the world knew also that Crosbie had been well thrashed for his baseness. It would not become him to mention that even before his mother and sister. All the world did know it; all the world that cared to know anything of the matter; – except Lily Dale herself. Nobody had ever yet told Lily Dale of that occurrence at the Paddington Railway Station, and it was well for John that her friends and his had been so discreet.

'Oh, of course you are her champion,' said Mary. 'And I didn't mean to say anything unkind. Indeed I didn't. Of course it was a misfortune.'

'I think it was the best piece of good fortune that could have happened to her, not to marry a d— scoundrel like—'

'Oh, John!' exclaimed Mrs. Eames.

'I beg your pardon, mother. But it isn't swearing to call such a man as that a d— scoundrel.' And he particularly emphasized the naughty word, thinking that thereby he would add to its import, and take away from its naughtiness. 'But we won't talk any more about him. I hate the man's very name. I hated him the first moment that I saw him, and knew that he was a blackguard from his look. And I don't believe a word about the squire having been cross to them. Indeed I know he has been the reverse of cross. So Bell is going to marry Dr. Crofts!'

'There is no doubt on earth about that,' said Mary. 'And they say that Bernard Dale is going abroad with his regiment.'

Then John discussed with his mother his duties as private secretary, and his intention of leaving Mrs. Roper's house. 'I suppose it isn't nice enough for you now, John,' said his mother.

'It never was very nice, mother, to tell you the truth. There were people there—. But you mustn't think I am turning up my nose because I'm getting grand. I don't want to live any better than we all lived at Mrs. Roper's; but she took in persons that were not agreeable. There is a Mr. and Mrs. Lupex there.' Then he described something of their life in Burton Crescent, but did not say much about Amelia Roper. Amelia Roper had not made her appearance at Guestwick, as he had once feared that she would do; and therefore it did not need that he should at present make known to his mother that episode in his life.

When he got back to the Manor House he found that Mr. Dale and his niece had arrived. They were both sitting with Lady Julia when he went into the morning room, and Lord De Guest was standing over the fire talking to them. Eames as he came among them felt terribly conscious of his position, as though all there were aware that he had been brought down from London on purpose to make a declaration of love; – as, indeed, all of them were aware of that fact. Bell, though no one had told her so in direct words, was as sure of it as the others.

'Here comes the prince of matadores,' said the earl.

'No, my lord; you're the prince. I'm only your first follower.' Though he could contrive that his words should be gay, his looks were sheepish, and when he gave his hand to the squire it was only by a struggle that he could bring himself to look straight into the old man's face.

'I'm very glad to see you, John,' said the squire, 'very glad indeed.'

'And so am I,' said Bell. 'I have been so happy to hear that you have been promoted at your office, and so is mamma.'

'I hope Mrs. Dale is quite well,' said he; – 'and Lily.' The word had been pronounced, but it had been done with so manifest an effort that all in the room were conscious of it, and paused as Bell prepared her little answer.

'My sister has been very ill, you know, – with scarlatina. But she has recovered with wonderful quickness, and is nearly well again now. She will be so glad to see you if you will go over.'

'Yes; I shall certainly go over,' said John.

'And now shall I show you your room, Miss Dale?' said Lady Julia. And so the party was broken up, and the ice had been broken.

CHAPTER LIII

LOQUITUR HOPKINS*

THE squire had been told that his niece Bell had accepted Dr. Crofts, and he had signified a sort of acquiescence in the arrangement, saying that if it were to be so, he had nothing to say against Dr. Crofts. He spoke this in a melancholy tone of voice, wearing on his face that look of subdued sorrow which was now almost habitual to him. It was to Mrs. Dale that he spoke on the subject. 'I could have wished that it might have been otherwise,' he said, 'as you are well aware. I had family reasons for wishing that it might be otherwise. But I have nothing to say against it. Dr. Crofts, as her husband, shall be welcome to my house.' Mrs. Dale, who had expected much worse than this, began to thank him for his kindness, and to say that she also would have preferred to see her daughter married to her cousin. 'But in such a matter the decision should be left entirely to the girl. Don't you think so?'

'I have not a word to say against her,' he repeated. Then Mrs. Dale left him, and told her daughter that her uncle's manner of receiving the news had been, for him, very gracious. 'You were his favourite, but Lily will be so now,' said Mrs. Dale.

'I don't care a bit about that; – or, rather, I do care, and think it will be in every way better. But as I, who am the naughty one, will go away, and as Lily, who is the good one, will remain with you, doesn't it almost seem a pity that you should be leaving the house?'

Mrs. Dale thought it was almost a pity, but she could not say so now. 'You think Lily will remain,' she said.

'Yes, mamma; I feel sure she will.'

'She was always very fond of John Eames; – and he is doing so well.'

'It will be of no use, mamma. She is fond of him, – very fond. In a sort of a way she loves him – so well, that I feel sure she never mentions his name without some inward reference to her old childish thoughts and fancies. If he had come before Mr. Crosbie it would have all been well with her. But she cannot do it now. Her pride would prevent her, even if her heart permitted it. Oh! dear; it's very wrong of me to say so, after all that I have said before; but I almost wish you were not going. Uncle Christopher seems to be less hard than he used to be; and as I was the sinner, and as I am disposed of—'

'It is too late now, my dear.'

'And we should neither of us have the courage to mention it to Lily,' said Bell.

On the following morning the squire sent for his sister-in-law as it was his wont to do when necessity came for any discussion on matters of business. This was perfectly understood between them, and such sending was not taken as indicating any lack of courtesy on the part of Mr. Dale. 'Mary,' he said, as soon as Mrs. Dale was seated, 'I shall do for Bell exactly what I have proposed to do for Lily. I had intended more than that once, of course. But then it would all have gone into Bernard's pocket; as it is, I shall make no difference between them. They shall each have a hundred a year, – that is, when they marry. You had better tell Crofts to speak to me.'

'Mr. Dale, he doesn't expect it. He does not expect a penny.'

'So much the better for him; and, indeed, so much the better for her. He won't make her the less welcome to his home because she brings some assistance to it.'

'We have never thought of it, – any of us. The offer has come so suddenly that I don't know what I ought to say.'

'Say – nothing. If you choose to make me a return for it—; but I am only doing what I conceive to be my duty, and have no right to ask for a kindness in return.'

'But what kindness can we show you, Mr. Dale?'

'Remain in that house.' In saying these last words he spoke as though he were again angry, – as though he were again laying down the law to them, – as though he were telling her of a duty which was due to him and incumbent on her. His voice was as stern and his face as acid as ever. He said that he was asking for a kindness; but surely no man ever asked for kindness in a voice so peremptory. 'Remain in that house.' Then he turned himself in towards his table as though he had no more to say.

But Mrs. Dale was beginning, now at last, to understand something of his mind and real character. He could be affectionate and forbearing in his giving; but when asking, he could not be otherwise than stern. Indeed, he could not ask; he could only demand.

'We have done so much now,' Mrs. Dale began to plead.

'Well, well, well. I did not mean to speak about that. Things are unpacked easier than they are packed. But, however— Never mind. Bell is to go with me this afternoon to Guestwick Manor. Let her be up here at two. Grimes can bring her box round, I suppose.'

'Oh, yes; of course.'

'And don't be talking to her about money before she starts. I had rather you didn't; – you understand. But when you see Crofts, tell him to come to me. Indeed, he'd better come at once, if this thing is to go on quickly.'

It may easily be understood that Mrs. Dale would disobey the injunctions contained in the squire's last words. It was quite out of the question that she should return to her daughters and not tell them the result of her morning's interview with their uncle. A hundred a year in the doctor's modest household would make all the difference between plenty and want, between modest plenty and endurable want. Of course she told them, giving Bell to understand that she must dissemble so far as to pretend ignorance of the affair.

'I shall thank him at once,' said Bell; 'and tell him that I did not at all expect it, but am not too proud to accept it.'

'Pray don't, my dear; not just now. I am breaking a sort of promise in telling you at all, – only I could not keep it to myself. And he has so many things to worry him! Though he says nothing about it now, he has half broken his heart about you and Bernard.' Then, too, Mrs. Dale told the girls what request the squire had just made, and the manner in which he had made it. 'The tone of his voice as he spoke brought tears into my eyes. I almost wish we had not done anything.'

'But, mamma,' said Lily, 'what difference can it make to him? You know that our presence near him was always a trouble to him. He never really wanted us. He liked to have Bell there when he thought that Bell would marry his pet.'

'Don't be unkind, Lily.'

'I don't mean to be unkind. Why shouldn't Bernard be his pet? I love Bernard dearly, and always thought it the best point in uncle Christopher that he was fond of him. I knew, you know, that it was no use. Of course I knew it, as I understood all about – somebody else. But Bernard is his pet.'

'He's fond of you all, in his own way,' said Mrs. Dale.

'But is he fond of you? – that's the question,' said Lily. 'We could have forgiven him anything done to us, and have put up with any words he might have spoken to us, because he regards us as children. His giving a hundred a year to Bell won't make you comfortable in this house if he still domineers over you. If a neighbour be neighbourly, near neighbourhood is very nice. But uncle Christopher has not been neighbourly. He has wanted to be more than an uncle to us, on condition that he might be less than a brother to you. Bell and I have always felt that his regard on such terms was not worth having.'

'I almost feel that we have been wrong,' said Mrs. Dale; 'but in truth I never thought that the matter would be to him one of so much moment.'

When Bell had gone, Mrs. Dale and Lily were not disposed to continue with much energy the occupation on which they had all been employed for some days past. There had been life and excitement in the work when they had first commenced their packing, but now it was grown wearisome, dull, and distasteful. Indeed so much of it was done that but little was left to employ them, except those final strappings and fastenings, and that last collection of odds and ends which could not be accomplished till they were absolutely on the point of starting. The squire had said that unpacking would be easier than packing, and Mrs. Dale, as she wandered about among the hampers and cases, began to consider whether the task of restoring all the things to their old places would be very disagreeable. She said nothing of this to Lily, and Lily herself, whatever might be her thoughts, made no such suggestion to her mother.

'I think Hopkins will miss us more than any one else,' she said. 'Hopkins will have no one to scold.'

Just at that moment Hopkins appeared at the parlour window, and signified his desire for a conference.

'You must come round,' said Lily. 'It's too cold for the window to be opened. I always like to get him into the house, because he feels himself a little abashed by the chairs and tables; or, perhaps, it is the carpet that is too much for him. Out on the gravel-walks he is such a terrible tyrant, and in the greenhouse he almost tramples upon one!'

Hopkins, when he did appear at the parlour door, seemed by his manner to justify Lily's discretion. He was not at all masterful in his tone or bearing, and seemed to pay to the chairs and tables all the deference which they could have expected.

'So you be going in earnest, ma'am,' he said, looking down at Mrs. Dale's feet.

As Mrs. Dale did not answer him at once, Lily spoke: – 'Yes, Hopkins, we are going in a very few days, now. We shall see you sometimes, I hope, over at Guestwick.'

'Humph!' said Hopkins. 'So you be really going! I didn't think it'd ever come to that, miss; I didn't indeed, – and no more it oughtn't; but of course it isn't for me to speak.'

'People must change their residence sometimes, you

know,' said Mrs. Dale, using the same argument by which Eames had endeavoured to excuse his departure to Mrs. Roper.

'Well, ma'am; it ain't for me to say anything. But this I will say, I've lived here about t' squire's place, man and boy, jist all my life, seeing I was born here, as you knows, Mrs. Dale; and of all the bad things I ever see come about the place, this is a sight the worst.'

'Oh, Hopkins!'

'The worst of all, ma'am; the worst of all! It'll just kill t' squire! There's ne'ery doubt in the world about that. It'll be the very death of t' old man.'

'That's nonsense, Hopkins,' said Lily.

'Very well, miss. I don't say but what it is nonsense; only you'll see. There's Mr. Bernard, – he's gone away; – and by all accounts he never did care very much for the place. They all say he's a-going to the Hingies. And Miss Bell is going to be married, – which is all proper, in course; why shouldn't she? And why shouldn't you, too, Miss Lily?'

'Perhaps I shall, some day, Hopkins.'

'There's no day like the present, Miss Lily. And I do say this, that the man as pitched into him would be the man for my money.' This, which Hopkins spoke in the excitement of the moment, was perfectly unintelligible to Lily, and Mrs. Dale, who shuddered as she heard him, said not a word to call for any explanation. 'But,' continued Hopkins, 'that's all as it may be, Miss Lily, and you'll be in the hands of Providence, – as is others.'

'Exactly so, Hopkins.'

'But why should your mamma be all for going away? She ain't going to marry no one. Here's the house, and there's she, and there's t' squire; and why should she be for going away? So much going away all at once can't be for any good. It's just a breaking up of everything, as though nothing wasn't good enough for nobody. I never went away, and I can't abide it.'

'Well, Hopkins; it's settled now,' said Mrs. Dale, 'and I'm afraid it can't be unsettled.'

'Settled; – well. Tell me this: do you expect, Mrs. Dale,

that he's to live there all alone by hisself without any one to say a cross word to, – unless it be me or Dingles; for Jolliffe's worse than nobody, he's so mortial cross hisself. Of course he can't stand it. If you goes away, Mrs. Dale, Mister Bernard, he'll be squire in less than twelve months. He'll come back from the Hinges, then, I suppose?'

'I don't think my brother-in-law will take it in that way, Hopkins.'

'Ah, ma'am, you don't know him, – not as I knows him; – all the ins and outs and crinks and crannies of him. I knows him as I does the old apple-trees that I've been a-handling for forty year. There's a deal of bad wood about them old cankered trees, and some folk say they ain't worth the ground they stand on; but I know where the sap runs, and when the fruit-blossom shows itself I know where the fruit will be the sweetest. It don't take much to kill one of them old trees, – but there's life in 'm yet if they be well handled.'

'I'm sure I hope my brother's life may be long spared to him,' said Mrs. Dale.

'Then don't be taking yourself away, ma'am, into them gashly lodgings at Guestwick. I says they are gashly for the likes of a Dale. It is not for me to speak, ma'am, of course. And I only came up now just to know what things you'd like with you out of the greenhouse.'

'Oh, nothing, Hopkins, thank you,' said Mrs. Dale.

'He told me to put up for you the best I could pick, and I means to do it;' and Hopkins, as he spoke, indicated by a motion of his head that he was making reference to the squire.

'We shan't have any place for them,' said Lily.

'I must send a few, miss, just to cheer you up a bit. I fear you'll be very dolesome there. And the doctor, – he ain't got what you can call a regular garden, but there is a bit of a place behind.'

'But we wouldn't rob the dear old place,' said Lily.

'For the matter of that what does it signify? T' squire'll be that wretched he'll turn sheep in here to destroy the place, or he'll have the garden ploughed. You see if he don't. As for the place, the place is clean done for if you leave it. You don't suppose he'll go and let the Small House to strangers. T' squire ain't one of that sort any ways.'

'Ah me!' exclaimed Mrs. Dale, as soon as Hopkins had taken himself off.

'What is it, mamma? He's a dear old man, but surely what he says cannot make you really unhappy.'

'It is so hard to know what one ought to do. I did not mean to be selfish, but it seems to me as though I were doing the most selfish thing in the world.'

'Nay, mamma; it has been anything but selfish. Besides, it is we that have done it; not you.'

'Do you know, Lily, that I also have that feeling as to breaking up one's old mode of life of which Hopkins spoke. I thought that I should be glad to escape from this place, but now that the time has come I dread it.'

'Do you mean that you repent?'

Mrs. Dale did not answer her daughter at once, fearing to commit herself by words which could not be retracted. But at last she said, 'Yes, Lily; I think I do repent. I think that it has not been well done.'

'Then let it be undone,' said Lily.

The dinner-party at Guestwick Manor on that day was not very bright, and yet the earl had done all in his power to make his guests happy. But gaiety did not come naturally to his house, which, as will have been seen, was an abode very unlike in its nature to that of the other earl at Courcy Castle. Lady De Courcy at any rate understood how to receive and entertain a house full of people, though the practice of doing so might give rise to difficult questions in the privacy of her domestic relations. Lady Julia did not understand it; but then Lady Julia was never called upon to answer for the expense of extra servants, nor was she asked about twice a week who the — was to pay the wine-merchant's bill? As regards Lord De Guest and the Lady Julia themselves, I think they had the best of it; but I am bound to admit, with reference to chance guests, that the house was dull. The people who were now gathered at the earl's table could hardly have been expected to be very sprightly when in company with each other. The squire was not a man much given to general society, and was unused to amuse a table full of people. On the present occasion he sat next to Lady Julia, and from time to time muttered a few words to her about the state of the country. Mrs. Eames was terribly afraid of everybody there, and especially of the earl, next to whom she sat, and whom she continually called 'my lord,' showing by her voice as she did so that she was almost alarmed by the sound of her own voice. Mr. and Mrs. Boyce were there, the parson sitting on the other side of Lady Julia, and the parson's wife on the other side of the earl. Mrs. Boyce was very studious to show that she was quite at home, and talked perhaps more than any one else; but in doing so she bored the earl most exquisitely, so that he told John Eames the next morning that she was worse than the bull. The parson ate his dinner, but said little or nothing between the two graces. He was a heavy, sensible, slow man, who knew himself and his own powers. 'Uncommon good stewed beef,' he said, as he went home; 'why can't we have our beef stewed like that?' 'Because we don't pay our cook sixty pounds a year,' said Mrs. Boyce. 'A woman with sixteen pounds can stew beef as well as a woman with sixty,' said he; 'she only wants looking after.' The earl himself was possessed of a sort of gaiety. There was about him a lightness of spirit which often made him an agreeable companion to one single person. John Eames conceived him to be the most sprightly old man of his day, – an old man with the fun and frolic almost of a boy. But this spirit, though it would show itself before John Eames, was not up to the entertainment of John Eames's mother and sister, together with the squire, the parson, and the parson's wife of Allington. So that the earl was overweighted and did not shine on this occasion at his own dinner-table. Dr. Crofts, who had also been invited, and who had secured the place which was now peculiarly his own, next to Bell Dale, was no doubt happy enough; as, let us hope, was the young lady also; but they added very little to the general hilarity of the company. John Eames was seated between his own sister and the parson, and did not at all enjoy his position. He had a full view of the doctor's felicity, as the happy pair sat opposite to him, and conceived himself to be hardly treated by Lily's absence.

The party was certainly very dull, as were all such dinners at Guestwick Manor. There are houses which, in their every-day course, are not conducted by any means in a sad or unsatisfactory manner, – in which life, as a rule, runs along merrily enough; but which cannot give a dinner-party; or, I might rather say, should never allow themselves to be allured into the attempt. The owners of such houses are generally themselves quite aware of the fact, and dread the dinner which they resolve to give quite as much as it is dreaded by their friends. They know that they prepare for their guests an evening of misery, and for themselves certain long hours of purgatory which are hardly to be endured. But they will do it. Why that long table, and all those super-numerary glasses and knives and forks, if they are never to be used? That argument produces all this misery; that and others cognate to it. On the present occasion, no doubt, there were excuses to be made. The squire and his niece had been invited on special cause, and their presence would have been well enough. The doctor added in would have done no harm. It was good-natured, too, that invitation given to Mrs. Eames and her daughter. The error lay in the parson and his wife. There was no necessity for their being there, nor had they any ground on which to stand, except the party-giving ground. Mr. and Mrs. Boyce made the dinner-party, and destroyed the social circle. Lady Julia knew that she had been wrong as soon as she had sent out the note.

Nothing was said on that evening which has any bearing on our story. Nothing, indeed, was said which had any bear-ing on anything. The earl's professed object had been to bring the squire and young Eames together; but people are never brought together on such melancholy occasions. Though they sip their port in close contiguity, they are poles asunder in their minds and feelings. When the Guestwick fly came for Mrs. Eames, and the parson's pony phaeton came for him and Mrs. Boyce, a great relief was felt. but the misery of those who were left had gone too far to allow of any reaction on that evening. The squire yawned, and the earl yawned, and then there was an end of it for that night.

CHAPTER LIV

THE SECOND VISIT TO THE GUESTWICK BRIDGE

Bell had declared that her sister would be very happy to see John Eames if he would go over to Allington, and he had replied that of course he would go there. So much having been, as it were, settled, he was able to speak of his visit as a matter of course at the breakfast table, on the morning after the earl's dinner-party. 'I must get you to come round with me, Dale, and see what I am doing to the land,' the earl said. And then he proposed to order saddle-horses. But the squire preferred walking, and in this way they were disposed of soon after breakfast.

John had it in his mind to get Bell to himself for half an hour and hold a conference with her; but it either happened that Lady Julia was too keen in her duties as a hostess, or else, as was more possible, Bell avoided the meeting. No opportunity for such an interview offered itself, though he hung about the drawing-room all the morning. 'You had better wait for luncheon, now,' Lady Julia said to him about twelve. But this he declined; and taking himself away hid himself about the place for the next hour and a half. During this time he considered much whether it would be better for him to ride or walk. If she should give him any hope, he could ride back triumphant as a field-marshal. Then the horse would be delightful to him. But if she should give him no hope, – if it should be his destiny to be rejected utterly on that morning, – then the horse would be terribly in the way of his sorrow. Under such circumstances what could he do but roam wide about across the fields, resting when he might choose to rest, and running when it might suit him to run. 'And she is not like other girls,' he thought to himself. 'She won't care for my boots being dirty.' So at last he elected to walk.

'Stand up to her boldly, man,' the earl had said to him. 'By George, what is there to be afraid of? It's my belief they'll give most to those who ask for most. There's nothing sets 'em against a man like being sheepish.' How the earl knew so much, seeing that he had not himself given signs of any success in that walk of life, I am not prepared to say. But Eames took his advice as being in itself good, and resolved to act upon it. 'Not that any resolution will be of any use,' he said to himself, as he walked along. 'When the moment comes I know that I shall tremble before her, and I know that she'll see it; but I don't think it will make any difference in her.'

He had last seen her on the lawn behind the Small House, just at that time when her passion for Crosbie was at the strongest. Eames had gone thither impelled by a foolish desire to declare to her his hopeless love, and she had answered him by telling him that she loved Mr. Crosbie better than all the world besides. Of course she had done so, at that time; but, nevertheless, her manner of telling him had seemed to him to be cruel. And he also had been cruel. He had told her that he hated Crosbie, – calling him 'that man,' and assuring her that no earthly consideration should induce him to go into 'that man's house.' Then he had walked away moodily wishing him all manner of evil. Was it not singular that all the evil things which he, in his mind, had meditated for the man, had fallen upon him. Crosbie had lost his love! He had so proved himself to be a villain that his name might not be so much as mentioned! He had been ignominiously thrashed! But what good would all this be if his image were still dear to Lily's heart? 'I told her that I loved her then,' he said to himself, 'though I had no right to do so. At any rate I have a right to tell her now.'

When he reached Allington he did not go in through the village and up to the front of the Small House by the cross street, but turned by the church gate and passed over the squire's terrace, and by the end of the Great House through the garden. Here he encountered Hopkins. 'Why, if that b'aint Mr. Eames!' said the gardener. 'Mr. John, may I make so bold!' and Hopkins held out a very dirty hand, which Eames of course took, unconscious of the cause of this new affection.

'I'm just going to call at the Small House, and I thought I'd come this way.'

'To be sure; this way, or that way, or any way, who's so welcome, Mr. John! I envies you; I envies you more than I envies any man. If I could a got him by the scruff of the neck, I'd a treated him just like any wermin; – I would, indeed! He was wermin! I ollays said it. I hated him ollays; I did indeed, Mr. John, from the first moment when he used to be nigging away at them foutry balls, knocking them in among the rhododendrons, as though there weren't no flower blossoms for next year. He never looked at one as though one were a Christian; did he, Mr. John?'

'I wasn't very fond of him myself, Hopkins.'

'Of course you weren't very fond of him. Who was? – only she, poor young lady. She'll be better now, Mr. John, a deal better. He wasn't a wholesome lover, – not like you are. Tell me, Mr. John, did you give it him well when you got him? I heard you did; – two black eyes, and all his face one mash of gore!' And Hopkins, who was by no means a young man, stiffly put himself into a fighting attitude.

Eames passed on over the little bridge, which seemed to be in a state of fast decay, unattended to by any friendly carpenter, now that the days of its use were so nearly at an end; and on into the garden, lingering on the spot where he had last said farewell to Lily. He looked about as though he expected still to find her there; but there was no one to be seen in the garden, and no sound to be heard. As every step brought him nearer to her whom he was seeking, he became more and more conscious of the hopelessness of his errand. Him she had never loved, and why should he venture to hope that she would love him now? He would have turned back had he not been aware that his promise to others required that he should persevere. He had said that he would do this thing, and he would be as good as his word. But he hardly ventured to hope that he might be successful In this frame of mind he slowly made his way up across the lawn.

'My dear, there is John Eames,' said Mrs. Dale, who had first seen him from the parlour window.

'Don't go, mamma.'

'I don't know; perhaps it will be better that I should.'

'No, mamma, no; what good can it do? It can do no good. I like him as well as I can like any one. I love him dearly. But it can do no good. Let him come in here, and be very kind to him; but do not go away and leave us. Of course I knew he would come and I shall be very glad to see him.'

Then Mrs. Dale went round to the other room, and admitted her visitor through the window of the drawing-room. 'We are in terrible confusion, John, are we not?'

'And so you are really going to live in Guestwick?'

'Well, it looks like it, does it not? But, to tell you a secret, – only it must be a secret; you must not mention it at Guestwick Manor; even Bell does not know; – we have half made up our minds to unpack all our things and stay where we are.'

Eames was so intent on his own purpose, and so fully occupied with the difficulty of the task before him, that he could hardly receive Mrs. Dale's tidings with all the interest which they deserved. 'Unpack them all again,' he said. 'That will be very troublesome. Is Lily with you, Mrs. Dale?'

'Yes, she is in the parlour. Come and see her.' So he followed Mrs. Dale through the hall, and found himself in the presence of his love.

'How do you do, John?' 'How do you do, Lily?' We all know the way in which such meetings are commenced. Each longed to be tender and affectionate to the other, – each in a different way; but neither knew how to throw any tenderness into this first greeting. 'So you're staying at the Manor House,' said Lily.

'Yes; I'm staying there. Your uncle and Bell came yesterday afternoon.'

'Have you heard about Bell?' said Mrs. Dale.

'Oh, yes; Mary told me. I'm so glad of it. I always liked Dr. Crofts very much. I have not congratulated her, because I didn't know whether it was a secret. But Crofts was there last night, and if it is a secret he didn't seem to be very careful about keeping it.'

'It is no secret,' said Mrs. Dale. 'I don't know that I am

fond of such secrets.' But as she said this, she thought of Crosbie's engagement, which had been told to every one, and of its consequences.

'Is it to be soon?' he asked.

'Well, yes; we think so. Of course nothing is settled.'

'It was such fun,' said Lily. 'James, who took a year or two to make his proposal, wanted to be married the next day afterwards.'

'No, Lily; not quite that.'

'Well, mamma, it was very nearly that. He thought it could all be done this week. It has made us so happy, John! I don't know anybody I should so much like for a brother. I'm very glad you like him; – very glad. I hope you'll be friends always.' There was some little tenderness in this, – as John acknowledged to himself.

'I'm sure we shall, – if he likes it. That is, if I ever happen to see him. I'll do anything for him I can if he ever comes up to London. Wouldn't it be a good thing, Mrs. Dale, if he settled himself in London?'

'No, John; it would be a very bad thing. Why should he wish to rob me of my daughter?'

Mrs. Dale was speaking of her eldest daughter; but the very allusion to any such robbery covered John Eames's face with a blush, made him hot up to the roots of his hair, and for the moment silenced him.

'You think he would have a better career in London?' said Lily, speaking under the influence of her superior presence of mind.

She had certainly shown defective judgment in desiring her mother not to leave them alone; and of this Mrs. Dale soon felt herself aware. The thing had to be done, and no little precautionary measure, such as this of Mrs. Dale's enforced presence, would prevent it. Of this Mrs. Dale was well aware; and she felt, moreover, that John was entitled to an opportunity of pleading his own cause. It might be that such opportunity would avail him nothing, but not the less should he have it of right, seeing that he desired it. But yet Mrs. Dale did not dare to get up and leave the room. Lily had asked her not to do so, and at the present period of their lives all Lily's requests were sacred. They continued for some time to talk of Crofts and his marriage; and when that subject was finished, they discussed their own probable, – or, as it seemed now, improbable, – removal to Guestwick. 'It's going too far, mamma,' said Lily, 'to say that you think we shall not go. It was only last night that you suggested it. The truth is, John, that Hopkins came in and discoursed with the most wonderful eloquence. Nobody dared to oppose Hopkins. He made us almost cry; he was so pathetic.'

'He has just been talking to me, too,' said John, 'as I came through the squire's garden.'

'And what has he been saying to you?' said Mrs. Dale.

'Oh, I don't know; not much.' John, however, remembered well, at this moment, all that the gardener had said to him. Did she know of that encounter between him and Crosbie? and if she did know of it, in what light did she regard it?

They had sat thus for an hour together, and Eames was not as yet an inch nearer to his object. He had sworn to himself that he would not leave the Small House without asking Lily to be his wife. It seemed to him as though he would be guilty of falsehood towards the earl if he did so. Lord De Guest had opened his house to him, and had asked all the Dales there, and had offered himself up as a sacrifice at the cruel shrine of a serious dinner-party, to say nothing of that easier and lighter sacrifice which he had made in a pecuniary point of view, in order that this thing might be done. Under such circumstances Eames was too honest a man not to do it, let the difficulties in his way be what they might.

He had sat there for an hour, and Mrs. Dale still remained with her daughter. Should he get up boldly and ask Lily to put on her bonnet and come out into the garden? As the thought struck him, he rose and grasped at his hat. 'I am going to walk back to Guestwick,' said he.

'It was very good of you to come so far to see us.'

'I was always fond of walking,' he said. 'The earl wanted me to ride, but I prefer being on foot when I know the country, as I do here.'

'Have a glass of wine before you go.'

'Oh, dear, no. I think I'll go back through the squire's fields, and out on the road at the white gate. The path is quite dry now.'

'I dare say it is,' said Mrs. Dale.

'Lily, I wonder whether you would come as far as that with me.' As the request was made Mrs. Dale looked at her daughter almost beseechingly. 'Do, pray do,' said he; 'it is a beautiful day for walking.'

The path proposed lay right across the field into which Lily had taken Crosbie when she made her offer to let him off from his engagement. Could it be possible that she should ever walk there again with another lover? 'No, John,' she said; 'not to-day, I think. I am almost tired, and I had rather not go out.'

'It would do you good,' said Mrs. Dale.

'I don't want to be done good to, mamma. Besides, I should have to come back by myself.'

'I'll come back with you,' said Johnny.

'Oh, yes; and then I should have to go again with you. But, John, really I don't wish to walk to-day.' Whereupon John Eames again put down his hat.

'Lily,' said he; and then he stopped. Mrs. Dale walked away to the window, turning her back upon her daughter and visitor. 'Lily, I have come over here on purpose to speak to you. Indeed, I have come down from London only that I might see you.'

'Have you, John?'

'Yes, I have. You know well all that I have got to tell you. I loved you before he ever saw you; and now that he has gone, I love you better than I ever did. Dear Lily!' and he put out his hand to her.

'No, John; no,' she answered.

'Must it be always no?'

'Always no to that. How can it be otherwise? You would not have me marry you while I love another!'

'But he is gone. He has taken another wife.'

'I cannot change myself because he is changed. If you are kind to me you will let that be enough.'

'But you are so unkind to me!'

'No, no; oh, I would wish to be so kind to you! John, here; take my hand. It is the hand of a friend who loves you, and will always love you. Dear John, I will do anything, – everything for you but that.'

'There is only one thing,' said he, still holding her by the hand, but with his face turned from her.

'Nay; do not say so. Are you worse off than I am? I could not have that one thing, and I was nearer to my heart's longings than you have ever been. I cannot have that one thing; but I know that there are other things, and I will not allow myself to be broken-hearted.'

'You are stronger than I am,' he said.

'Not stronger, but more certain. Make yourself as sure as I am, and you, too, will be strong. Is it not so, mamma?'

'I wish it could be otherwise; – I wish it could be otherwise! If you can give him any hope—'

'Mamma!'

'Tell me that I may come again, – in a year,' he pleaded.

'I cannot tell you so. You may not come again, – not in this way. Do you remember what I told you before, in the garden; that I loved him better than all the world besides? It is still the same. I still love him better than all the world. How, then, can I give you any hope?'

'But it will not be so for ever, Lily.'

'For ever! Why should he not be mine as well as hers when that for ever comes? John, if you understand what it is to love, you will say nothing more of it. I have spoken to you more openly about this than I have ever done to anybody, even to mamma, because I have wished to make you understand my feelings. I should be disgraced in my own eyes if I admitted the love of another man, after – after—. It is to me almost as though I had married him. I am not blaming him, remember. These things are different with a man.'

She had not dropped his hand, and as she made her last speech was sitting in her old chair with her eyes fixed upon the ground. She spoke in a low voice, slowly, almost with difficulty; but still the words came very clearly, with a clear, distinct voice which caused them to be remembered with accuracy, both by Eames and Mrs. Dale. To him it seemed to be impossible that he should continue his suit after such a declaration. To Mrs. Dale they were terrible words, speaking of a perpetual widowhood, and telling of an amount of suffering greater even than that which she had anticipated. It was true that Lily had never said so much to her as she had now said to John Eames, or had attempted to make so clear an exposition of her own feelings. 'I should be disgraced in my own eyes if I admitted the love of another man!' They were terrible words, but very easy to be understood. Mrs. Dale had felt, from the first, that Eames was coming too soon, that the earl and the squire together were making an effort to cure the wound too quickly after its infliction; that time should have been given to her girl to recover. But now the attempt had been made, and words had been forced from Lily's lips, the speaking of which would never be forgotten by herself.

'I knew that it would be so,' said John.

'Ah, yes; you know it, because your heart understands my heart. And you will not be angry with me, and say naughty, cruel words, as you did once before. We will think of each other, John, and pray for each other; and will always love one another. When we do meet let us be glad to see each other. No other friend shall ever be dearer to me than you are. You are so true and honest! When you marry I will tell your wife what an infinite blessing God has given her.'

'You shall never do that.'

'Yes, I will. I understand what you mean; but yet I will.'

'Good-bye, Mrs. Dale,' he said.

'Good-bye, John. If it could have been otherwise with her, you should have had all my best wishes in the matter. I would have loved you dearly as my son; and I will love you now.' And she put up her lips and kissed his face.

'And so will I love you,' said Lily, giving him her hand again. He looked longingly into her face as though he had thought it possible that she also might kiss him: then he pressed her hand to his lips, and without speaking any further farewell, took up his hat and left the room.

'Poor fellow!' said Mrs. Dale.

'They should not have let him come,' said Lily. 'But they don't understand. They think that I have lost a toy, and they mean to be good-natured, and to give me another.' Very shortly after that Lily went away by herself, and sat alone for hours; and when she joined her mother again at tea-time, nothing further was said of John Eames's visit.

He made his way out by the front door, and through the churchyard, and in this way on to the field through which he had asked Lily to walk with him. He hardly began to think of what had passed till he had left the squire's house behind him. As he made his way through the tombstones he paused and read one, as though it interested him. He stood a moment under the tower looking up at the clock, and then pulled out his own watch as though to verify the one by the other. He made, unconsciously, a struggle to drive away from his thoughts the facts of the last scene, and for some five or ten minutes he succeeded. He said to himself a word or two about Sir Raffle and his letters, and laughed inwardly as he remembered the figure of Rafferty bringing in the knight's shoes. He had gone some half mile upon his way before he ventured to stand still and tell himself that he had failed in the great object of his life.

Yes; he had failed: and he acknowledged to himself, with bitter reproaches, that he had failed, now and for ever. He told himself that he had obtruded upon her in her sorrow with an unmannerly love, and rebuked himself as having been not only foolish but ungenerous. His friend the earl had been wont, in his waggish way, to call him the conquering hero, and had so talked him out of his common sense as to have made him almost think that he would be successful in his suit. Now, as he told himself that any such success must have been impossible, he almost hated the earl for having brought him to this condition. A conquering hero, indeed! How should he manage to sneak back among them all at the Manor House, crestfallen and abject in his misery? Everybody knew the errand on which he had gone, and everybody must know of his failure. How could he have been such a fool as to undertake such a task under the eyes of so many lookers-on? Was it not the case that he had so fondly expected success, as to think only of his triumph in returning, and not of his more probable disgrace? He had allowed others to make a fool of him, and had so made a fool of himself that now all hope and happiness were over for him. How could he escape at once out of the country, – back to London? How could he get away without saying a word further to any one? That was the thought that at first occupied his mind.

He crossed the road at the end of the squire's property, where the parish of Allington divides itself from that of Abbott's Guest in which the earl's house stands, and made his way back along the copse which skirted the field where they had encountered the bull, into the high woods at the back of the park. Ah, yes; it had been well for him that he had not come out on horseback. That ride home along the high road and up to the Manor House stables would, under his present circumstances, have been almost impossible to him. As it was, he did not think it possible that he should return to his place in the earl's house. How could he pretend to maintain his ordinary demeanour under the eyes of those two old men? It would be better for him to get home to his mother, – to send a message from thence to the Manor, and then to escape back to London. So thinking, but with no resolution made, he went on through the woods, and down from the hill back towards the town till he again came to the little bridge over the brook. There he stopped and stood awhile with his broad hand spread over the letters he had cut in those early days, so as to hide them from his sight. 'What an ass I have been, – always and ever!' he said to himself.

It was not only of his late disappointment that he was thinking, but of his whole past life. He was conscious of his hobbledehoyhood, – of that backwardness on his part in assuming manhood which had rendered him incapable of making himself acceptable to Lily before she had fallen into the clutches of Crosbie. As he thought of this he declared to himself that if he could meet Crosbie again he would again thrash him, – that he would so belabour him as to send him out of the world, if such sending might possibly be done by fair beating, regardless whether he himself might be called upon to follow him. Was it not hard that for the two of them, – for Lily and for him also, – there should be such punishment because of the insincerity of that man? When he had thus stood upon the bridge for some quarter of an hour, he took out his knife, and, with deep, rough gashes in the wood, cut out Lily's name from the rail.

He had hardly finished, and was still looking at the chips as they were being carried away by the stream, when a gentle step came close up to him, and turning round, he saw that Lady Julia was on the bridge. She was close to him, and had already seen his handiwork. 'Has she offended you, John?' she said.

'Oh, Lady Julia!'

'Has she offended you?'

'She has refused me, and it is all over.'

'It may be that she has refused you, and that yet it need not be all over. I am sorry that you have cut out the name, John. Do you mean to cut it out from your heart?'

'Never. I would if I could, but I never shall.'

'Keep to it as to a great treasure. It will be a joy to you in after years, and not a sorrow. To have loved truly, even though you shall have loved in vain, will be a consolation when you are as old as I am. It is something to have had a heart.'

'I don't know. I wish that I had none.'

'And, John; – I can understand her feeling now; and indeed, I thought all through that you were asking her too soon; but the time may yet come when she will think better of your wishes.'

'No, no; never. I begin to know her now.'

'If you can be constant in your love you may win her yet. Remember how young she is; and how young you both are. Come again in two years' time, and then, when you have won her, you shall tell me that I have been a good old woman to you both.'

'I shall never win her, Lady Julia.' As he spoke these last words the tears were running down his cheeks, and he was weeping openly in presence of his companion. It was well for him that she had come upon him in his sorrow. When he once knew that she had seen his tears, he could pour out to her the whole story of his grief; and as he did so she led him back quietly to the house.

CHAPTER LV

NOT VERY FIE FIE AFTER ALL

It will perhaps be remembered that terrible things had been foretold as about to happen between the Hartletop and Omnium families. Lady Dumbello had smiled whenever Mr. Plantagenet Palliser had spoken to her. Mr. Palliser had confessed to himself that politics were not enough for him, and that Love was necessary to make up the full complement of his happiness. Lord Dumbello had frowned latterly when his eyes fell on the tall figure of the duke's heir; and the duke himself, – that potentate, generally so mighty in his silence, – the duke himself had spoken. Lady De Courcy and Lady Clandidlem were, both of them, absolutely certain that the thing had been fully arranged. I am, therefore, perfectly justified in stating that the world was talking about the loves, – the illicit loves, – of Mr. Palliser and Lady Dumbello.

And the talking of the world found its way down to that respectable country parsonage in which Lady Dumbello had been born, and from which she had been taken away to those noble halls she now graced by her presence. The talking of the world was heard at Plumstead Episcopi, where still lived Archdeacon Grantly, the lady's father; and was heard also at the deanery of Barchester, where lived the lady's aunt and grandfather. By whose ill-mannered tongue the rumour was spread in these ecclesiastical regions it boots not now to tell. But it may be remembered that Courcy Castle was not far from Barchester, and that Lady De Courcy was not given to hide her lights under a bushel.

It was a terrible rumour. To what mother must not such a rumour respecting her daughter be very terrible? In no mother's ears could it have sounded more frightfully than it did in those of Mrs. Grantly. Lady Dumbello, the daughter, might be altogether worldly; but Mrs. Grantly had never been more than half worldly. In one moiety of her character, her habits, and her desires, she had been wedded to things good in themselves, – to religion, to charity, and to honest-hearted uprightness. It is true that the circumstances of her life had induced her to serve both God and Mammon, and that, therefore, she had gloried greatly in the marriage of her daughter with the heir of a marquis. She had revelled in the aristocratic elevation of her child, though she continued to dispense books and catechisms with her own hands to the children of the labourers of Plumstead Episcopi. When Griselda first became Lady Dumbello the mother feared somewhat lest her child should find herself unequal to the exigencies of her new position. But the child had proved herself more than equal to them, and had mounted up to a dizzy height of success, which brought to the mother great glory and great fear also. She delighted to think that her Griselda was great even among the daughters of marquises; but she trembled as she reflected how deadly would be the fall from such a height – should there ever be a fall!

But she had never dreamed of such a fall as this! She would have said, – indeed, she often had said, – to the archdeacon that Griselda's religious principles were too firmly fixed to be moved by outward worldly matters; signifying, it may be, her conviction that that teaching of Plumstead Episcopi had so fastened her daughter into a groove, that all the future teaching of Hartlebury would not suffice to undo the fastenings. When she had thus boasted no such idea as that of her daughter running from her husband's house had ever come upon her; but she had alluded to vices of a nature kindred to that vice, – to vices into which other aristocratic ladies sometimes fell, who had been less firmly grooved; and her boastings had amounted to this, – that she herself had so successfully served God and Mammon together, that her child might go forth and enjoy all worldly things without risk of damage to things heavenly. Then came upon her this rumour. The archdeacon told her in a hoarse whisper that he had been recommended to look to it, that it was current through the world that Griselda was about to leave her husband.

'Nothing on earth shall make me believe it,' said Mrs. Grantly. But she sat alone in her drawing-room afterwards and trembled. Then came her sister, Mrs. Arabin, the dean's wife, over to the parsonage, and in half-hidden words told the same story. She had heard it from Mrs. Proudie, the bishop's wife. 'That woman is as false as the father of falsehoods,' said Mrs. Grantly. But she trembled the more; and as she prepared her parish work, could think of nothing but her child. What would be all her life to come, what would have been all that was past of her life, if this thing should happen to her? She would not believe it; but yet she trembled the more as she thought of her daughter's exaltation, and remembered that such things had been done in that world to which Griselda now belonged. Ah! would it not have been better for them if they had not raised their heads so high! And she walked out alone among the tombs of the neighbouring churchyard, and stood over the grave in which had been laid the body of her other daughter. Could it be that the fate of that one had been the happier?

Very few words were spoken on the subject between her and the archdeacon, and yet it seemed agreed among them that something should be done. He went up to London, and saw his daughter, – not daring, however, to mention such a subject. Lord Dumbello was cross with him, and very uncommunicative. Indeed both the archdeacon and Mrs. Grantly had found that their daughter's house was not comfortable to them, and as they were sufficiently proud among their own class they had not cared to press themselves on the hospitality of their son-in-law. But he had been able to perceive that all was not right in the house in Carlton Gardens. Lord Dumbello was not gracious with his wife, and there was something in the silence, rather than in the speech, of men, which seemed to justify the report which had reached him.

'He is there oftener than he should be,' said the archdeacon. 'And I am sure of this, at least, that Dumbello does not like it.'

'I will write to her,' said Mrs. Grantly at last. 'I am still her mother; – I will write to her. It may be that she does not know what people say of her.'

And Mrs. Grantly did write.

PLUMSTEAD, April 186—.

'Dearest Griselda,

'It seems sometimes that you have been moved so far away from me that I have hardly a right to concern myself more in the affairs of your daily life, and I know that it is impossible that you should refer to me for advice or sympathy, as you would have done had you married some gentleman of our own standing. But I am quite sure that my child does not forget her mother, or fail to look back upon her mother's love; and that she will allow me to speak to her if she be in trouble, as I would to any other child whom I had loved and cherished. I pray God that I may be wrong in supposing that such trouble is near you. If I am so you will forgive me my solicitude.

'Rumours have reached us from more than one quarter that— Oh! Griselda, I hardly know in what words to conceal and yet to declare that which I have to write. They say that you are intimate with Mr. Palliser, the nephew of the duke, and that your husband is much offended. Perhaps I had better tell you all, openly, cautioning you not to suppose that I have believed it. They say that it is thought that you are going to put yourself under Mr. Palliser's protection. My dearest child, I think you can imagine with what an agony I write these words, – with what terrible grief I must have been oppressed before I could have allowed myself to entertain the thoughts which have produced them. Such things are said openly in Barchester, and your father, who has been in town and has seen you, feels himself unable to tell me that my mind may be at rest.

'I will not say to you a word as to the injury in a worldly point of view which would come to you from any rupture with your husband. I believe that you can see what would be the effect of so terrible a step quite as plainly as I can show it you. You would break the heart of your father, and send your mother to her grave; – but it is not even on that

that I may most insist. It is this, – that you would offend your God by the worst sin that a woman can commit, and cast yourself into a depth of infamy in which repentance before God is almost impossible, and from which escape before man is not permitted.

'I do not believe it, my dearest, dearest child, – my only living daughter; I do not believe what they have said to me. But as a mother I have not dared to leave the slander unnoticed. If you will write to me and say that it is not so, you will make me happy again, even though you should rebuke me for my suspicion.

'Believe that at all times, and under all circumstances, I am still your loving mother, as I was in other days.

'Susan Grantly.'

We will now go back to Mr. Palliser as he sat in his chambers at the Albany, thinking of his love. The duke had cautioned him, and the duke's agent had cautioned him; and he, in spite of his high feeling of independence, had almost been made to tremble. All his thousands a year were in the balance, and perhaps everything on which depended his position before the world. But, nevertheless, though he did tremble, he resolved to persevere. Statistics were becoming dry to him, and love was very sweet. Statistics, he thought, might be made as enchanting as ever, if only they could be mingled with love. The mere idea of loving Lady Dumbello had seemed to give a salt to his life of which he did not now know how to rob himself. It is true that he had not as yet enjoyed many of the absolute blessings of love, seeing that his conversations with Lady Dumbello had never been warmer than those which have been repeated in these pages; but his imagination had been at work; and now that Lady Dumbello was fully established at her house in Carlton Gardens, he was determined to declare his passion on the first convenient opportunity. It was sufficiently manifest to him that the world expected him to do so, and that the world was already a little disposed to find fault with the slowness of his proceedings.

He had been once at Carlton Gardens since the season

had commenced, and the lady had favoured him with her sweetest smile. But he had only been half a minute alone with her, and during that half-minute had only time to remark that he supposed she would now remain in London for the season.

'Oh, yes,' she had answered, 'we shall not leave till July.' Nor could he leave till July, because of the exigencies of his statistics. He therefore had before him two, if not three, clear months in which to manœuvre, to declare his purposes, and prepare for the future events of his life. As he resolved on a certain morning that he would say his first tender word to Lady Dumbello that very night, in the drawing-room of Lady De Courcy, where he knew that he should meet her, a letter came to him by the post. He well knew the hand and the intimation which it would contain. It was from the duke's agent, Mr. Fothergill, and informed him that a certain sum of money had been placed to his credit at his bankers. But the letter went further, and informed him also that the duke had given his agent to understand that special instructions would be necessary before the next quarterly payment could be made. Mr. Fothergill said nothing further, but Mr. Palliser understood it all. He felt his blood run cold round his heart; but, nevertheless, he determined that he would not break his word to Lady De Courcy that night.

And Lady Dumbello received her letter also on the same morning. She was being dressed as she read it, and the maidens who attended her found no cause to suspect that anything in the letter had excited her ladyship. Her ladyship was not often excited, though she was vigilant in exacting from them their utmost cares. She read her letter, however, very carefully, and as she sat beneath the toilet implements of her maidens thought deeply of the tidings which had been brought to her. She was angry with no one; – she was thankful to no one. She felt no special love for any person concerned in the matter. Her heart did not say, 'Oh, my lord and husband!' or, 'Oh, my lover!' or, 'Oh, my mother, the friend of my childhood!' But she became aware that matter for thought had been brought before her, and she did think. 'Send my love to Lord Dumbello,' she said, when the operations were nearly completed, 'and tell him that I shall be glad to see him if he will come to me while I am at breakfast.'

'Yes, my lady.' And then the message came back: 'His lordship would be with her ladyship certainly.'

'Gustavus,' she said, as soon as she had seated herself discreetly in her chair, 'I have had a letter from my mother, which you had better read;' and she handed to him the document. 'I do not know what I have done to deserve such suspicions from her; but she lives in the country, and has probably been deceived by ill-natured people. At any rate you must read it, and tell me what I should do.'

We may predicate from this that Mr. Palliser's chance of being able to shipwreck himself upon that rock was but small, and that he would, in spite of himself, be saved from his uncle's anger. Lord Dumbello took the letter and read it very slowly, standing, as he did so, with his back to the fire. He read it very slowly, and his wife, though she never turned her face directly upon his, could perceive that he became very red, that he was fluttered and put beyond himself, and that his answer was not ready. She was well aware that his conduct to her during the last three months had been much altered from his former usages; that he had been rougher with her in his speech when alone, and less courteous in his attention when in society; but she had made no complaint or spoken a word to show him that she had marked the change. She had known, moreover, the cause of his altered manner, and having considered much, had resolved that she would live it down. She had declared to herself that she had done no deed and spoken no word that justified suspicion, and therefore she would make no change in her ways, or show herself to be conscious that she was suspected. But now, – having her mother's letter in her hand, – she could bring him to an explanation without making him aware that she had ever thought that he had been jealous of her. To her, her mother's letter was a great assistance. It justified a scene like this, and enabled her to fight her battle after her own fashion. As for eloping with any Mr. Palliser, and giving up the position which she had won; – no, indeed! She had been fastened in her grooves too well for that! Her mother, in entertaining any fear on such a subject, had shown herself to be ignorant of the solidity of her daughter's character.

'Well, Gustavus,' she said at last. 'You must say what answer I shall make, or whether I shall make any answer.' But he was not even yet ready to instruct her. So he unfolded the letter and read it again, and she poured out for herself a cup of tea.

'It's a very serious matter,' said he.

'Yes, it is serious; I could not but think such a letter from my mother to be serious. Had it come from any one else I doubt whether I should have troubled you; unless, indeed, it had been from any as near to you as she is to me. As it is, you cannot but feel that I am right.'

'Right! Oh, yes, you are right, – quite right to tell me; you should tell me everything. D— them!' But whom he meant to condemn he did not explain.

'I am above all things averse to cause you trouble,' she said. 'I have seen some little things of late—'

'Has he ever said anything to you?'

'Who, – Mr. Palliser? Never a word.'

'He has hinted at nothing of this kind?'

'Never a word. Had he done so, I must have made you understand that he could not have been allowed again into my drawing-room.' Then again he read the letter, or pretended to do so.

'Your mother means well,' he said.

'Oh, yes, she means well. She has been foolish to believe the tittle-tattle that has reached her, – very foolish to oblige me to give you this annoyance.'

'Oh, as for that, I'm not annoyed. By Jove, no. Come, Griselda, let us have it all out; other people have said this, and I have been unhappy. Now, you know it all.'

'Have I made you unhappy?'

'Well, no; not you. Don't be hard upon me when I tell you the whole truth. Fools and brutes have whispered things that have vexed me. They may whisper till the devil fetches them, but they shan't annoy me again. Give me a kiss, my girl.' And he absolutely put out his arms and embraced her. 'Write a good-natured letter to your mother, and ask her to come up for a week in May. That'll be the best thing; and then she'll understand. By Jove, it's twelve o'clock. Good-bye.'

Lady Dumbello was well aware that she had triumphed, and that her mother's letter had been invaluable to her. But it had been used, and therefore she did not read it again. She ate her breakfast in quiet comfort, looking over a milliner's French circular as she did so; and then, when the time for such an operation had fully come, she got to her writing-table and answered her mother's letter.

'Dear Mamma (she said),

'I thought it best to show your letter at once to Lord Dumbello. He said that people would be ill-natured, and seemed to think that the telling of such stories could not be helped. As regards you, he was not a bit angry, but said that you and papa had better come to us for a week about the end of next month. Do come. We are to have rather a large dinner-party on the 23rd. His Royal Highness is coming, and I think papa would like to meet him. Have you observed that those very high bonnets have all gone out: I never liked them; and as I had got a hint from Paris, I have been doing my best to put them down. I do hope nothing will prevent your coming.

'Your affectionate daughter,
'G. Dumbello.

'Carlton Gardens, Wednesday.'

Mrs. Grantley was aware, from the moment in which she received the letter, that she had wronged her daughter by her suspicions. It did not occur to her to disbelieve a word that was said in the letter, or an inference that was implied. She had been wrong, and rejoiced that it was so. But nevertheless there was that in the letter which annoyed and irritated her, though she could not explain to herself the cause of her annoyance. She had thrown all her heart into that which she had written, but in the words which her child had written not a vestige of heart was to be found. In that reconciling of God and Mammon which Mrs. Grantly had carried on so successfully in the education of her daughter, the organ had not been required, and had become withered, if not defunct, through want of use.

'We will not go there, I think,' said Mrs. Grantly, speaking to her husband.

'Oh dear, no; certainly not. If you want to go to town at all, I will take rooms for you. And as for his Royal Highness—! I have a great respect for his Royal Highness, but I do not in the least desire to meet him at Dumbello's table.'

And so that matter was settled, as regarded the inhabitants of Plumstead Episcopi.

And whither did Lord Dumbello betake himself when he left his wife's room in so great a hurry at twelve o'clock? Not to the Park, nor to Tattersall's, nor to a committee-room of the House of Commons, nor yet to the bow-window of his club. But he went straight to a great jeweller's in Ludgate-hill, and there purchased a wonderful green necklace, very rare and curious, heavy with green sparkling drops, with three rows of shining green stones embedded in chaste gold, – a necklace amounting almost to a jewelled cuirass* in weight and extent. It had been in all the exhibitions, and was very costly and magnificent. While Lady Dumbello was still dressing in the evening this was brought to her with her lord's love, as his token of renewed confidence, and Lady Dumbello, as she counted the sparkles, triumphed inwardly, telling herself that she had played her cards well.

But while she counted the sparkles produced by her full reconciliation with her lord, poor Plantagenet Palliser was still trembling in his ignorance. If only he could have been allowed to see Mrs. Grantly's letter, and the lady's answer, and the lord's present! But no such seeing was vouchsafed to him, and he was carried off in his brougham to Lady De Courcy's house, twittering with expectant love, and trembling with expectant ruin. To this conclusion he had come at any rate, that if anything was to be done, it should be done now. He would speak a word of love, and prepare his future in accordance with the acceptance it might receive.

Lady De Courcy's rooms were very crowded when he

arrived there. It was the first great crush party of the season, and all the world had been collected into Portman Square. Lady De Courcy was smiling as though her lord had no teeth, as though her eldest son's condition was quite happy, and all things were going well with the De Courcy interests. Lady Margaretta was there behind her, bland without and bitter within; and Lady Rosina also, at some further distance, reconciled to this world's vanity and finery because there was to be no dancing. And the married daughters of the house were there also, striving to maintain their positions on the strength of their undoubted birth, but subjected to some snubbing by the lowness of their absolute circumstances. Gazebee was there, happy in the absolute fact of his connexion with an earl, and blessed with the consideration that was extended to him as an earl's son-in-law. And Crosbie, also, was in the rooms, – was present there, though he had sworn to himself that he would no longer dance attendance on the countess, and that he would sever himself away from the wretchedness of the family. But if he gave up them and their ways, what else would then be left to him? He had come, therefore, and now stood alone, sullen, in a corner, telling himself that all was vanity. Yes; to the vain all will be vanity; and to the poor of heart all will be poor.

Lady Dumbello was there in a small inner room, seated on a couch to which she had been brought on her first arrival at the house, and on which she would remain till she departed. From time to time some very noble or very elevated personage would come before her and say a word, and she would answer that elevated personage with another word; but nobody had attempted with her the task of conversation. It was understood that Lady Dumbello did not converse, – unless it were occasionally with Mr. Palliser.

She knew well that Mr. Palliser was to meet her there. He had told her expressly that he should do so, having inquired, with much solicitude, whether she intended to obey the invitation of the countess. 'I shall probably be there,' she had said, and now had determined that her mother's letter and her husband's conduct to her should not cause her to break her word. Should Mr. Palliser 'forget' himself, she would know how to say a word to him as she had known how to say a word to her husband. Forget himself! She was very sure that Mr. Palliser had been making up his mind to forget himself for some months past.

He did come to her, and stood over her, looking unutterable things. His unutterable things, however, were so looked, that they did not absolutely demand notice from the lady. He did not sigh like a furnace, nor open his eyes upon her as though there were two suns in the firmament above her head, nor did he beat his breast or tear his hair. Mr. Palliser had been brought up in a school which delights in tranquillity, and never allows its pupils to commit themselves either to the sublime or to the ridiculous. He did look an unutterable thing or two; but he did it with so decorous an eye, that the lady, who was measuring it all with great accuracy, could not, as yet, declare that Mr. Palliser had 'forgotten himself'.

There was room by her on the couch, and once or twice, at Hartlebury, he had ventured so to seat himself. On the present occasion, however, he could not do so without placing himself manifestly on her dress. She would have known how to fill a larger couch even than that, – as she would have known, also, how to make room, – had it been her mind to do so. So he stood still over her, and she smiled at him. Such a smile! It was cold as death, flattering no one, saying nothing hideous in its unmeaning, unreal grace. Ah! how I hate the smile of a woman who smiles by rote! It made Mr. Palliser feel very uncomfortable; – but he did not analyse it, and persevered.

'Lady Dumbello,' he said, and his voice was very low, 'I have been looking forward to meeting you here.'

'Have you, Mr. Palliser? Yes; I remember that you asked me whether I was coming.'

'I did. Hm – Lady Dumbello!' and he almost trenched upon the outside verge of that schooling which had taught him to avoid both the sublime and the ridiculous. But he had not forgotten himself as yet, and so she smiled again.

'Lady Dumbello, in this world in which we live, it is so hard to get a moment in which we can speak.' He had thought that she would move her dress, but she did not.

'Oh, I don't know,' she said; 'one doesn't often want to say very much, I think.'

'Ah, no; not often, perhaps. But when one does want! How I do hate these crowded rooms!' Yet, when he had been at Hartlebury he had resolved that the only ground for him would be the crowded drawing-room of some large London house. 'I wonder whether you ever desire anything beyond them?'

'Oh, yes,' said she; 'but I confess that I am fond of parties.'

Mr. Palliser looked round and thought that he saw that he was unobserved. He had made up his mind as to what he would do, and he was determined to do it. He had in him none of that readiness which enables some men to make love and carry off their Dulcineas at a moment's notice, but he had that pluck which would have made himself disgraceful in his own eyes if he omitted to do that as to the doing of which he had made a solemn resolution. He would have preferred to do it sitting, but, faute de mieux,* seeing that a seat was denied to him, he would do it standing.

'Griselda,' he said, – and it must be admitted that his tone was not bad. The word sank softly into her ear, like small rain upon moss, and it sank into no other ear. 'Griselda!'

'Mr. Palliser!' said she; – and though she made no scene, though she merely glanced upon him once, he could see that he was wrong.

'May I not call you so?'

'Certainly not. Shall I ask you to see if my people are there?' He stood a moment before her hesitating. 'My carriage, I mean.' As she gave the command she glanced at him again, and then he obeyed her orders.

When he returned she had left her seat; but he heard her name announced on the stairs, and caught a glance of the back of her head as she made her way gracefully down through the crowd. He never attempted to make love to her again, utterly disappointing the hopes of Lady De Courcy, Mrs. Proudie, and Lady Clandidlem.

As I would wish those who are interested in Mr. Palliser's fortunes to know the ultimate result of this adventure, and as we shall not have space to return to his affairs in this little history, I may, perhaps, be allowed to press somewhat forward, and tell what Fortune did for him before the close of that London season. Everybody knows that in that spring, Lady Glencora MacCluskie was brought out before the world, and it is equally well known that she, as the only child of the late Lord of the Isles, was the great heiress of the day. It is true that the hereditary possession of Skye, Staffa, Mull, Arran, and Bute went, with the title, to the Marquis of Auldreekie, together with the counties of Caithness and Ross-shire. But the property in Fife, Aberdeen, Perth, and Kincardineshire, comprising the greater part of those counties, and the coal-mines in Lanark, as well as the enormous estate within the city of Glasgow, were unentailed, and went to the Lady Glencora. She was a fair girl, with bright blue eyes and short wavy flaxen hair, very soft to the eye. The Lady Glencora was small in stature, and her happy round face lacked, perhaps, the highest grace of female beauty. But there was ever a smile upon it, at which it was very pleasant to look; and the intense interest with which she would dance, and talk, and follow up every amusement that was offered her, was very charming. The horse she rode was the dearest love; oh! she loved him so dearly! And she had a little dog that was almost as dear as the horse. The friend of her youth, Sabrina Scott, was – oh, such a girl! And her cousin, the little Lord of the Isles, the heir of the marquis, was so gracious and beautiful that she was always covering him with kisses. Unfortunately he was only six, so that there was hardly a possibility that the properties should be brought together.

But Lady Glencora, though she was so charming, had even in this, her first outset upon the world, given great uneasiness to her friends, and caused the Marquis of Auldreekie to be almost wild with dismay. There was a terribly handsome man about town, who had spent every shilling that anybody would give him, who was very fond of brandy, who was known, but not trusted, at Newmarket, who was said to be deep in every vice, whose father would not speak to him; – and with him the Lady Glencora was never tired of dancing. One morning she had told her cousin the marquis, with a flashing eye, – for the round blue eye could flash, – that Burgo Fitzgerald was more sinned against than sinning. Ah me! what was a guardian marquis, anxious for the fate of the family property, to do under such circumstances as that?

But before the end of the season the marquis and the duke were both happy men, and we will hope that the Lady Glencora also was satisfied. Mr. Plantagenet Palliser had danced with her twice, and had spoken his mind. He had an interview with the marquis, which was pre-eminently satisfactory, and everything was settled. Glencora no doubt told him how she had accepted that plain gold ring from Burgo Fitzgerald, and how she had restored it; but I doubt whether she ever told him of that wavy lock of golden hair which Burgo still keeps in his receptacle for such treasures.

'Plantagenet,' said the duke, with quite unaccustomed warmth, 'in this, as in all things, you have shown yourself to be everything that I could desire. I have told the marquis that Matching Priory, with the whole estate, should be given over to you at once. It is the most comfortable country-house I know. Glencora shall have The Horns as her wedding present.'

But the genial, frank delight of Mr. Fothergill pleased Mr. Palliser the most. The heir of the Pallisers had done his duty, and Mr. Fothergill was unfeignedly a happy man.

CHAPTER LVI

SHOWING HOW MR. CROSBIE BECAME AGAIN A HAPPY MAN

It has been told in the last chapter how Lady De Courcy gave a great party in London in the latter days of April, and it may therefore be thought that things were going well with the De Courcys; but I fear the inference would be untrue. At any rate, things were not going well with Lady Alexandrina, for she, on her mother's first arrival in town, had rushed to Portman-square with a long tale of her sufferings.

'Oh, mamma! you would not believe it; but he hardly ever speaks to me.'

'My dear, there are worse faults in a man than that.'

'I am alone there all the day. I never get out. He never offers to get me a carriage. He asked me to walk with him once last week, when it was raining. I saw that he waited till the rain began. Only think, I have not been out three evenings this month, – except to Amelia's; and now he says he won't go there any more, because a fly is so expensive. You can't believe how uncomfortable the house is.'

'I thought you chose it, my dear.'

'I looked at it, but, of course, I didn't know what a house ought to be. Amelia said it wasn't nice, but he would have it. He hates Amelia. I'm sure of that, for he says everything he can to snub her and Mr. Gazebee. Mr. Gazebee is as good as he, at any rate. What do you think? He has given Richard warning to go. You never saw him, but he was a very good servant. He has given him warning, and he is not talking of getting another man. I won't live with him without somebody to wait upon me.'

'My dearest girl, do not think of such a thing as leaving him.'

'But I will think of it, mamma. You do not know what my life is in that house. He never speaks to me, – never. He comes home before dinner at half-past six, and when he has just shown himself he goes to his dressing-room. He is always silent at dinner-time, and after dinner he goes to sleep. He breakfasts always at nine, and goes away at half-past nine, though I know he does not get to his office till eleven. If I want anything, he says that it cannot be afforded. I never thought before that he was stingy, but I am sure now that he must be a miser at heart.'

'It is better so than a spendthrift, Alexandrina.'

'I don't know that it is better. He could not make me more unhappy than I am. Unhappy is no word for it. What can I do, shut up in such a house as that by myself from nine o'clock in the morning till six in the evening? Everybody knows what he is, so that nobody will come to see me. I tell you fairly, mamma, I will not stand it. If you cannot help me, I will look for help elsewhere.'

It may be said that things were not going well with that branch of the De Courcy family. Nor, indeed, was it going well with some other branches. Lord Porlock had married, not having selected his partner for life from the choicest cream of the aristocratic circles, and his mother, while endeavouring to say a word in his favour, had been so abused by the earl that she had been driven to declare that she could no longer endure such usage. She had come up to London in direct opposition to his commands, while he was fastened to his room by gout; and had given her party in defiance of him, so that people should not say, when her back was turned, that she had slunk away in despair.

'I have borne it,' she said to Margaretta, 'longer than any other woman in England would have done. While I thought that any of you would marry—'

'Oh, don't talk of that, mamma,' said Margaretta, putting a little scorn into her voice. She had not been quite pleased that even her mother should intimate that all her chance was over, and yet she herself had often told her mother that she had given up all thought of marrying.

'Rosina will go to Amelia's,' the countess continued; 'Mr. Gazebee is quite satisfied that it should be so, and he will take care that she shall have enough to cover her own expenses. I propose that you and I, dear, shall go to Baden-Baden.'

'And about money, mamma?'

'Mr. Gazebee must manage it. In spite of all that your father says, I know that there must be money. The expense will be much less so than in our present way.'

'And what will papa do himself?'

'I cannot help it, my dear. No one knows what I have had to bear. Another year of it would kill me. His language has become worse and worse, and I fear every day that he is going to strike me with his crutch.'

Under all these circumstances it cannot be said that the De Courcy interests were prospering.

But Lady De Courcy, when she had made up her mind to go to Baden-Baden, had by no means intended to take her youngest daughter with her. She had endured for years, and now Alexandrina was unable to endure for six months. Her chief grievance, moreover, was this, – that her husband was silent. The mother felt that no woman had a right to complain much of any such sorrow as that. If her earl had sinned only in that way, she would have been content to have remained by him till the last!

And yet I do not know whether Alexandrina's life was not quite as hard as that of her mother. She barely exceeded the truth when she said that he never spoke to her. The hours with her in her new comfortless house were very long, – very long and very tedious. Marriage with her had by no means been the thing that she had expected. At home, with her mother, there had always been people around her, but they had not always been such as she herself would have chosen for her companions. She had thought that, when married, she could choose and have those about her who were congenial to her; but she found that none came to her. Her sister, who was a wiser woman than she, had begun her married life with a definite idea, and had carried it out; but this poor creature found herself, as it were, stranded. When once she had conceived it in her heart to feel anger against her husband, – and she had done so before they had been a week together, – there was no love to bring her back to him again. She did not know that it behoved her to look pleased when he entered the room, and to make him think that his presence gave her happiness. She became gloomy before she reached her new house, and never laid her gloom aside. He would have made a struggle for some domestic comfort, had any seemed to be within his reach. As it was, he struggled for domestic propriety, believing that he might so best bolster up his present lot in life. But the task became harder and harder to him, and the gloom became denser and more dense. He did not think of her unhappiness, but of his own; as she did not think of his tedium, but of hers. 'If this be domestic felicity!' he would say to himself, as he sat in his arm-chair, striving to fix his attention upon a book.

'If this be the happiness of married life!' she thought, as she remained listless, without even the pretence of a book, behind her teacups. In truth she would not walk with him, not caring for such exercise round the pavement of a London square; and he had resolutely determined that she should not run into debt for carriage hire. He was not a curmudgeon with his money; he was no miser. But he had found that in marrying an earl's daughter he had made himself a poor man, and he was resolved that he would not also be an embarrassed man.

When the bride heard that her mother and sister were about to escape to Baden-Baden, there rushed upon her a sudden hope that she might be able to accompany the flight. She would not be parted from her husband, or at least not so parted that the world should suppose that they had quarrelled. She would simply go away and make a long visit, – a very long visit. Two years ago a sojourn with her mother and Margaretta at Baden-Baden would not have offered to her much that was attractive; but now, in her eyes, such a life seemed to be a life in Paradise. In truth, the tedium of those hours in Princess Royal Crescent had been very heavy.

But how could she contrive that it should be so? That conversation with her mother had taken place on the day preceding the party, and Lady De Courcy had repeated it with dismay to Margaretta.

'Of course he would allow her an income,' Margaretta had coolly said.

'But, my dear, they have been married only ten weeks.'

'I don't see why people are to be made absolutely wretched because they are married,' Margaretta answered. 'I don't want to persuade her to leave him, but if what she says is true, it must be very uncomfortable.'

Crosbie had consented to go to the party in Portman-square, but had not greatly enjoyed himself on that festive occasion. He had stood about moodily, speaking hardly a word to any one. His whole aspect of life seemed to have been altered during the last few months. It was here, in such spots as this that he had been used to find his glory. On such occasions he had shone with peculiar light, making envious the hearts of many who watched the brilliance of his career as they stood around in dull quiescence. But now no one in those rooms had been more dull, more silent, or less courted than he; and yet he was established there as the son-in-law of that noble house. 'Rather slow work, isn't it?' Gazebee had said to him, having, after many efforts, succeeded in reaching his brother-in-law in a corner. In answer to this Crosbie had only grunted. 'As for myself,' continued Gazebee, 'I would a deal sooner be at home with my paper and slippers. It seems to me these sort of gatherings don't suit married men.' Crosbie had again grunted, and had then escaped into another corner.

Crosbie and his wife went home together in a cab, – speechless both of them. Alexandrina hated cabs, – but she had been plainly told that in such vehicles, and in such vehicles only, could she be allowed to travel. On the following morning he was at the breakfast-table punctually by nine, but she did not make her appearance till after he had gone to his office. Soon after that, however, she was away to her mother and her sister; but she was seated grimly in her drawing-room when he came in to see her, on his return to his house. Having said some word which might be taken for a greeting, he was about to retire; but she stopped him with a request that he would speak to her.

'Certainly,' said he. 'I was only going to dress. It is nearly the half-hour.'

'I won't keep you very long, and if dinner is a few minutes late it won't signify. Mamma and Margaretta are going to Baden-Baden.'

'To Baden-Baden, are they?'

'Yes; and they intend to remain there – for a considerable time.' There was a little pause, and Alexandrina found it necessary to clear her voice and to prepare herself for further speech by a little cough. She was determined to make her proposition, but was rather afraid of the manner in which it might be first received.

'Has anything happened at Courcy Castle?' Crosbie asked.

'No; that is, yes; there may have been some words between papa and mamma; but I don't quite know. That, however, does not matter now. Mamma is going, and purposes to remain there for the rest of the year.'

'And the house in town will be given up.'

'I suppose so, but that will be as papa chooses. Have you any objection to my going with mamma?'

What a question to be asked by a bride of ten weeks' standing! She had hardly been above a month with her husband in her new house, and she was now asking permission to leave it, and to leave him also, for an indefinite number of months – perhaps for ever. But she showed no excitement as she made her request. There was neither sorrow, nor regret, nor hope in her face. She had not put on half the animation which she had once assumed in asking for the use, twice a week, of a carriage done up to look as though it were her own private possession. Crosbie had then answered her with great sternness, and she had wept when his refusal was made certain to her. But there was to be no weeping now. She meant to go, – with his permission if he would accord it, and without it if he should refuse it. The question of money was no doubt important, but Gazebee should manage that, – as he managed all those things.

'Going with them to Baden-Baden?' said Crosbie. 'For how long?'

'Well; it would be no use unless it were for some time.'

'For how long a time do you mean, Alexandrina? Speak out what you really have to say. For a month?'

'Oh, more than that.'

'For two months, or six, or as long as they may stay there?'

'We could settle that afterwards, when I am there.' During all this time she did not once look into his face, though he was looking hard at her throughout.

'You mean,' said he, 'that you wish to go away from me.'

'In one sense it would be going away, certainly.'

'But in the ordinary sense? is it not so? When you talk of going to Baden-Baden for an unlimited number of months, have you any idea of coming back again?'

'Back to London, you mean?'

'Back to me, – to my house, – to your duties as a wife? Why

cannot you say at once what it is you want? You wish to be separated from me?'

'I am not happy here, – in this house.'

'And who chose the house? Did I want to come here? But it is not that. If you are not happy here, what could you have in any other house to make you happy?'

'If you were left alone in this room for seven or eight hours at a time, without a soul to come to you, you would know what I mean. And even after that, it is not much better. You never speak to me when you are here.'

'Is it my fault that nobody comes to you? The fact is, Alexandrina, that you will not reconcile yourself to the manner of life which is suitable to my income. You are wretched because you cannot have yourself driven round the Park. I cannot find you a carriage, and will not attempt to do so. You may go to Baden-Baden, if you please; – that is, if your mother is willing to take you.'

'Of course I must pay my own expenses,' said Alexandrina. But to this he made no answer on the moment. As soon as he had given his permission he had risen from his seat and was going, and her last words only caught him in the doorway. After all, would not this be the cheapest arrangement that he could make? As he went through his calculations he stood up with his elbow on the mantel-piece in his dressing-room. He had scolded his wife because she had been unhappy with him: but had he not been quite as unhappy with her? Would it not be better that they should part in this quiet, half-unnoticed way; – that they should part and never again come together? He was lucky in this, that hitherto had come upon them no prospect of any little Crosbie to mar the advantages of such an arrangement. If he gave her four hundred a year, and allowed Gazebee two more towards the paying off of incumbrances, he would still have six on which to enjoy himself in London. Of course he could not live as he had lived in those happy days before his marriage, nor, independently of the cost, would such a mode of life be within his reach. But he might go to his club for his dinners; he might smoke his cigar in luxury; he would not be bound to that wooden home which, in spite of all his resolutions, had become almost unendurable to him. So he made his calculations, and found that it would be well that his bride should go. He would give over his house and furniture to Gazebee, allowing Gazebee to do as he would about that. To be once more a bachelor, in lodgings, with six hundred a year to spend on himself, seemed to him now such a prospect of happiness that he almost became light-hearted as he dressed himself. He would let her go to Baden-Baden.

There was nothing said about it at dinner, nor did he mention the subject again till the servant had left the tea-things on the drawing-room table. 'You can go with your mother if you like it,' he then said.

'I think it will be best,' she answered.

'Perhaps it will. At any rate you shall suit yourself.'

'And about money?'

'You had better leave me to speak to Gazebee about that.'

'Very well. Will you have some tea?' And then the whole thing was finished.

On the next day she went after lunch to her mother's house, and never came back again to Princess Royal Crescent. During that morning she packed up those things which she cared to pack herself, and sent her sisters there, with an old family servant, to bring away whatever else might be supposed to belong to her. 'Dear, dear,' said Amelia, 'what trouble I had in getting these things together for them, and only the other day. I can't but think she's wrong to go away.'

'I don't know,' said Margaretta. 'She has not been so lucky as you have in the man she has married. I always felt that she would find it difficult to manage him.'

'But, my dear, she has not tried. She has given up at once. It isn't management that was wanting. The fact is that when Alexandrina began she didn't make up her mind to the kind of thing she was coming to. I did. I knew it wasn't to be all party-going and that sort of thing. But I must own that Crosbie isn't the same sort of man as Mortimer. I don't think I could have gone on with him. You might as well have those small books put up; he won't care about them.' And in this way Crosbie's house was dismantled.

She saw him no more, for he made no farewell visit to the house in Portman-square. A note had been brought to him at his office: 'I am here with mamma, and may as well say good-bye now. We start on Tuesday. If you wish to write, you can send your letters to the housekeeper here. I hope you will make yourself comfortable, and that you will be well. Yours affectionately, A. C.' He made no answer to it, but went that day and dined at his club.

'I haven't seen you this age,' said Montgomerie Dobbs.

'No. My wife is going abroad with her mother, and while she is away I shall come back here again.'

There was nothing more said to him, and no one ever made any inquiry about his domestic affairs. It seemed to him now as though he had no friend sufficiently intimate with him to ask him after his wife or family. She was gone, and in a month's time he found himself again in Mount Street, – beginning the world with five hundred a year, not six. For Mr. Gazebee, when the reckoning came, showed him that a larger income at the present moment was not possible for him. The countess had for a long time refused to let Lady Alexandrina go with her on so small a pittance as four hundred and fifty; – and then were there not the insurances to be maintained?

But I think he would have consented to accept his liberty with three hundred a year, – so great to him was the relief.

CHAPTER LVII

LILIAN DALE VANQUISHES HER MOTHER

Mrs. Dale had been present during the interview in which John Eames had made his prayer to her daughter, but she had said little or nothing on that occasion. All her wishes had been in favour of the suitor, but she had not dared to express them, neither had she dared to leave the room. It had been hard upon him to be thus forced to declare his love in the presence of a third person, but he had done it, and had gone away with his answer. Then, when the thing was over, Lily, without any communion with her mother, took herself off, and was no more seen till the evening hours had come on, in which it was natural that they should be together again. Mrs. Dale, when thus alone, had been able to think of nothing but this new suit for her daughter's hand. If only it might be accomplished! If any words from her to Lily might be efficacious to such an end! And yet, hitherto, she had been afraid almost to utter a word.

She knew that it was very difficult. She declared to herself over and over that he had come too soon, – that the attempt had been made too quickly after that other shipwreck. How was it possible that the ship should put to sea again at once, with all her timbers so rudely strained? Now that the attempt had been made, now that Eames had uttered his request and been sent away with an answer, she felt that she must at once speak to Lily on the subject, if ever she were to speak upon it. She thought that she understood her child and all her feelings. She recognized the violence of the shock which must be encountered before Lily could be brought to acknowledge such a change in her heart. But if the thing could be done, Lily would be a happy woman. When once done it would be in all respects a blessing. And if it were not done, might not Lily's life be blank, lonely, and loveless to the end? Yet when Lily came down in the evening, with some light, half-joking word on her lips, as was usual to her, Mrs. Dale was still afraid to venture upon her task.

'I suppose, mamma, we may consider it as a settled thing that everything must be again unpacked, and that the lodging scheme will be given up.'

'I don't know that, my dear.'

'Oh, but I do – after what you said just now. What geese everybody will think us!'

'I shouldn't care a bit for that, if we didn't think ourselves geese, or if your uncle did not think us so.'

'I believe he would think we were swans. If I had ever thought he would be so much in earnest about it, or that he would ever have cared about our being here, I would never have voted for going. But he is so strange. He is affectionate when he ought to be angry, and ill-natured when he ought to be gentle and kind.'

'He has, at any rate, given us reason to feel sure of his affection.'

'For us girls, I never doubted it. But, mamma, I don't think I could face Mrs. Boyce. Mrs. Hearn and Mrs. Cramp would be very bad, and Hopkins would come down upon us terribly when he found that we had given way. But Mrs. Boyce would be worse than any of them. Can't you fancy the tone of her congratulations?'

'I think I should survive Mrs. Boyce.'

'Ah, yes; because we should have to go and tell her. I know your cowardice of old, mamma; don't I? And Bell wouldn't care a bit, because of her lover. Mrs. Boyce will be nothing to her. It is I that must bear it all. Well, I don't mind; I'll vote for staying if you will promise to be happy here. Oh, mamma, I'll vote for anything if you will be happy.'

'And will you be happy?'

'Yes, as happy as the day is long. Only I know we shall never see Bell. People never do see each other when they live just at that distance. It's too near for long visits, and too far for short visits. I'll tell you what; we might make arrangements each to walk half-way, and meet at the corner of Lord De Guest's wood. I wonder whether they'd let us put up a seat there. I think we might have a little house and carry sandwiches and a bottle of beer. Couldn't we see something of each other in that way?'

Thus it came to be the fixed idea of both of them that they would abandon their plan of migrating to Guestwick, and on this subject they continued to talk over their tea-table; on that evening Mrs. Dale ventured to say nothing about John Eames.

But they did not even yet dare to commence the work of reconstructing their old home. Bell must come back before they would do that, and the express assent of the squire must be formally obtained. Mrs. Dale must, in a degree, acknowledge herself to have been wrong, and ask to be forgiven for her contumacy.

'I suppose the three of us had better go up in sackcloth, and throw ashes on our foreheads as we meet Hopkins in the garden,' said Lily, 'and then I know he'll heap coals of fire on our heads by sending us an early dish of peas. And Dingles would bring us in a pheasant, only that pheasants don't grow in May.'

'If the sackcloth doesn't take an unpleasanter shape than that, I shan't mind it.'

'That's because you've got no delicate feelings. And then uncle Christopher's gratitude!'

'Ah! I shall feel that.'

'But, mamma, we'll wait till Bell comes home. She shall decide. She is going away, and therefore she'll be free from prejudice. If uncle offers to paint the house, – and I know he will, – then I shall be humbled to the dust.'

But yet Mrs. Dale had said nothing on the subject which was nearest to her heart. When Lily in pleasantry had accused her of cowardice, her mind had instantly gone off to that other matter and she had told herself that she was a coward. Why should she be afraid of offering her counsel to her own child? It seemed to her as though she had neglected some duty in allowing Crosbie's conduct to have passed away without hardly a word of comment on it between herself and Lily. Should she not have forced upon her daughter's conviction the fact that Crosbie had been a villain, and as such should be discarded from her heart? As it was, Lily had spoken the simple truth when she told John Eames that she was dealing more openly with him on that affair of her engagement than she had ever dealt, even with her mother. Thinking of this as she sat in her own room that night, before she allowed herself to rest, Mrs. Dale resolved that on the next morning she would endeavour to make Lily see as she saw and think as she thought.

She let breakfast pass by before she began her task, and even then she did not rush at it at once. Lily sat herself down to her work when the teacups were taken away, and Mrs. Dale went down to her kitchen as was her wont. It was nearly eleven before she seated herself in the parlour, and even then she got her work-box before her and took out her needle.

'I wonder how Bell gets on with Lady Julia,' said Lily.

'Very well, I'm sure.'

'Lady Julia won't bite her, I know, and I suppose her dismay at the tall footmen has passed off by this time.'

'I don't know that they have any tall footmen.'

'Short footmen then, – you know what I mean; all the noble belongings. They must startle one at first, I'm sure, let one determine ever so much not to be startled. It's a very mean thing, no doubt, to be afraid of a lord merely because he is a lord; yet I'm sure I should be afraid at first, even of Lord De Guest, if I were staying in the house.'

'It's well you didn't go then.'

'Yes, I think it is. Bell is of a firmer mind, and I dare say she'll get over it after the first day. But what on earth does she do there? I wonder whether they mend their stockings in such a house as that.'

'Not in public, I should think.'

'In very grand houses they throw them away at once, I suppose. I've often thought about it. Do you believe the Prime Minister ever has his shoes sent to a cobbler?'

'Perhaps a regular shoemaker will condescend to mend a Prime Minister's shoes.'

'You do think they are mended then? But who orders it? Does he see himself when there's a little hole coming, as I do? Does an archbishop allow himself so many pair of gloves in a year?'

'Not very strictly, I should think.'

'Then I suppose it comes to this, that he has a new pair whenever he wants them. But what constitutes the want? Does he ever say to himself that they'll do for another Sunday? I remember the bishop coming here once, and he had a hole at the end of his thumb. I was going to be confirmed, and I remember thinking that he ought to have been smarter.'

'Why didn't you offer to mend it?'

'I shouldn't have dared for all the world.'

The conversation had commenced itself in a manner that

did not promise much assistance to Mrs. Dale's project. When Lily got upon any subject, she was not easily induced to leave it, and when her mind had twisted itself in one direction, it was difficult to untwist it. She was now bent on a consideration of the smaller social habits of the high and mighty among us, and was asking her mother whether she supposed that the royal children ever carried halfpence in their pockets, or descended so low as fourpenny-bits.

'I suppose they have pockets like other children,' said Lily.

But her mother stopped her suddenly, –

'Lily, dear, I want to say something to you about John Eames.'

'Mamma, I'd sooner talk about the Royal Family just at present.'

'But, dear, you must forgive me if I persist. I have thought much about it, and I'm sure you will not oppose me when I am doing what I think to be my duty.'

'No, mamma; I won't oppose you, certainly.'

'Since Mr. Crosbie's conduct was made known to you, I have mentioned his name in your hearing very seldom.'

'No, mamma, you have not. And I have loved you so dearly for your goodness to me. Do not think that I have not understood and known how generous you have been. No other mother ever was so good as you have been. I have known it all, and thought of it every day of my life, and thanked you in my heart for your trusting silence. Of course, I understand your feelings. You think him bad and you hate him for what he has done.'

'I would not willingly hate any one, Lily.'

'Ah, but you do hate him. If I were you, I should hate him; but I am not you, and I love him. I pray for his happiness every night and morning, and for hers. I have forgiven him altogether, and I think that he was right. When I am old enough to do so without being wrong, I will go to him and tell him so. I should like to hear of all his doings and all his success, if it were only possible. How, then, can you and I talk about him? It is impossible. You have been silent and I have been silent, – let us remain silent.'

'It is not about Mr. Crosbie that I wish to speak. But I

think you ought to understand that conduct such as his will be rebuked by all the world. You may forgive him, but you should acknowledge—.'

'Mamma, I don't want to acknowledge anything; – not about him. There are things as to which a person cannot argue.' Mrs. Dale felt that this present matter was one as to which she could not argue. 'Of course, mamma,' continued Lily, 'I don't want to oppose you in anything, but I think we had better be silent about this.'

'Of course I am thinking only of your future happiness.'

'I know you are; but pray believe me that you need not be alarmed. I do not mean to be unhappy. Indeed, I think I may say I am not unhappy; of course I have been unhappy, – very unhappy. I did think that my heart would break. But that has passed away, and I believe I can be as happy as my neighbours. We're all of us sure to have some troubles, as you used to tell us when we were children.'

Mrs. Dale felt that she had begun wrong, and that she would have been able to make better progress had she omitted all mention of Crosbie's name. She knew exactly what it was that she wished to say, — what were the arguments which she desired to expound before her daughter; but she did not know what language to use, or how she might best put her thoughts into words. She paused for a while, and Lily went on with her work as though the conversation was over. But the conversation was not over.

'It was about John Eames, and not about Mr. Crosbie, that I wished to speak to you.'

'Oh, mamma!'

'My dear, you must not hinder me in doing what I think to be a duty. I heard what he said to you and what you replied, and of course I cannot but have my mind full of the subject. Why should you set yourself against him in so fixed a manner?'

'Because I love another man.' These words she spoke out loud, in a steady, almost dogged tone, with a certain show of audacity, – as though aware that the declaration was unseemly, but resolved that, though unseemly, it must be made.

'But, Lily, that love, from its very nature, must cease; or, rather, such love is not the same as that you felt when you thought that you were to be his wife.'

'Yes, it is. If she died, and he came to me in five years' time, I would still take him. I should think myself constrained to take him.'

'But she is not dead, nor likely to die.'

'That makes no difference. You don't understand me, mamma.'

'I think I do, and I want you to understand me also. I know how difficult is your position; I know what your feelings are; but I know this also, that if you could reason with yourself, and bring yourself in time to receive John Eames as a dear friend—'

'I did receive him as a dear friend. Why not? He is a dear friend. I love him heartily, – as you do.'

'You know what I mean?'

'Yes, I do; and I tell you it is impossible.'

'If you would make the attempt, all this misery would soon be forgotten. If once you could bring yourself to regard him as a friend, who might become your husband, all this would be changed, – and I should see you happy!'

'You are strangely anxious to be rid of me, mamma!'

'Yes, Lily; – to be rid of you in that way. If I could see you put your hand in his as his promised wife, I think that I should be the happiest woman in the world.'

'Mamma, I cannot make you happy in that way. If you really understood my feelings, my doing as you propose would make you very unhappy. I should commit a great sin, – the sin against which women should be more guarded than against any other. In my heart I am married to that other man. I gave myself to him, and loved him, and rejoiced in his love. When he kissed me I kissed him again, and I longed for his kisses. I seemed to live only that he might caress me. All that time I never felt myself to be wrong, – because he was all in all to me. I was his own. That has been changed, – to my great misfortune; but it cannot be undone or forgotten. I cannot be the girl I was before he came here. There are things that will not have themselves buried and put out of sight, as though they had never been. I am as you are, mamma, – widowed. But you have your daughter, and I have my mother. If you will be contented, so will I.' Then she got up and threw herself on her mother's neck.

Mrs. Dale's argument was over now. To such an appeal as that last made by Lily no rejoinder on her part was possible. After that she was driven to acknowledge to herself that she must be silent. Years as they rolled on might make a change, but no reasoning could be of avail. She embraced her daughter, weeping over her, – whereas Lily's eyes were dry. 'It shall be as you will,' Mrs. Dale murmured.

'Yes, as I will. I shall have my own way; shall I not? That is all I want; to be a tyrant over you, and make you do my bidding in everything, as a well-behaved mother should do. But I won't be stern in my orderings. If you will only be obedient, I will be so gracious to you! There's Hopkins again. I wonder whether he has come to knock us down and trample upon us with another speech.'

Hopkins knew very well to which window he must come, as only one of the rooms was at the present time habitable. He came up to the dining-room, and almost flattened his nose against the glass.

'Well, Hopkins,' said Lily, 'here we are.' Mrs. Dale had turned her face away, for she knew that the tears were still on her cheek.

'Yes, miss, I see you. I want to speak to your mamma, miss.'

'Come round,' said Lily, anxious to spare her mother the necessity of showing herself at once. 'It's too cold to open the window; come round, and I'll open the door.'

'Too cold!' muttered Hopkins, as he went. 'They'll find it a deal colder in lodgings at Guestwick.' However, he went round through the kitchen, and Lily met him in the hall.

'Well, Hopkins, what is it? Mamma has got a headache.'

'Got a headache, has she? I won't make her headache no worse. It's my opinion that there's nothing for a headache so good as fresh air. Only some people can't abear to be blowed upon, not for a minute. If you don't let down the lights in a greenhouse more or less every day, you'll never get any plants, – never; – and it's just the same with the grapes. Is I to go back and say as how I couldn't see her?'

'You can come in if you like; only be quiet, you know.'

'Ain't I ollays quiet, Miss? Did anybody ever hear me rampage? If you please, ma'am, the squire's come home.'

'What, home from Guestwick? Has he brought Miss Bell?'

'He ain't brought none but hisself, 'cause he come on horseback; and it's my belief he's going back almost immediate. But he wants you to come to him, Mrs. Dale.'

'Oh, yes, I'll come at once.'

'He bade me say with his kind love. I don't know whether that makes any difference.'

'At any rate, I'll come, Hopkins.'

'And I ain't to say nothing about the headache?'

'About what?' said Mrs. Dale.

'No, no, no,' said Lily. 'Mamma will be there at once. Go and tell my uncle, there's a good man,' and she put up her hand and backed him out of the room.

'I don't believe she's got no headache at all,' said Hopkins, grumbling, as he returned through the back premises. 'What lies gentlefolks do tell! If I said I'd a headache when I ought to be out among the things, what would they say to me? But a poor man mustn't never lie, nor yet drink, nor yet do nothing.' And so he went back with his message.

'What can have brought your uncle home?' said Mrs. Dale.

'Just to look after the cattle, and to see that the pigs are not all dead. My wonder is that he should ever have gone away.'

'I must go up to him at once.'

'Oh, yes, of course.'

'And what shall I say about the house?'

'It's not about that, – at least I think not. I don't think he'll speak about that again till you speak to him.'

'But if he does?'

'You must put your trust in Providence. Declare you've got a bad headache, as I told Hopkins just now; only you would throw me over by not understanding. I'll walk with you down to the bridge.' So they went off together across the lawn.

But Lily was soon left alone, and continued her walk, waiting for her mother's return. As she went round and round the gravel paths, she thought of the words that she had said to her mother. She had declared that she also was widowed. 'And so it should be,' she said, debating the matter with herself. 'What can a heart be worth if it can be transferred hither and thither as circumstances and convenience and comfort may require? When he held me here in his arms' – and, as the thoughts ran through her brain, she remembered the very spot on which they had stood – 'oh, my love!' she had said to him then as she returned his kisses – 'oh, my love, my love, my love!' 'When he held me here in his arms, I told myself that it was right, because he was my husband. He has changed, but I have not. It might be that I should have ceased to love him, and then I should have told him so. I should have done as he did.' But, as she came to this, she shuddered, thinking of the Lady Alexandrina. 'It was very quick,' she said, still speaking to herself; 'very, very. But then men are not the same as women.' And she walked on eagerly, hardly remembering where she was, thinking over it all, as she did daily; remembering every little thought and word of those few eventful months in which she had learnt to regard Crosbie as her husband and master. She had declared that she had conquered her unhappiness; but there were moments in which she was almost wild with misery. 'Tell me to forget him!' she said. 'It is the one thing which will never be forgotten.'

At last she heard her mother's step coming down across the squire's garden, and she took up her post at the bridge.

'Stand and deliver,' she said, as her mother put her foot upon the plank. 'That is, if you've got anything worth delivering. Is anything settled?'

'Come up to the house,' said Mrs. Dale, 'and I'll tell you all.'

CHAPTER LVIII

THE FATE OF THE SMALL HOUSE

THERE was something in the tone of Mrs. Dale's voice, as she desired her daughter to come up to the house, and declared that her budget of news should be opened there, which at once silenced Lily's assumed pleasantry. Her mother had been away fully two hours, during which Lily had still continued her walk round the garden, till at last she had become impatient for her mother's footstep. Something serious must have been said between her uncle and her mother during those long two hours. The interviews to which Mrs. Dale was occasionally summoned at the Great House did not usually exceed twenty minutes, and the upshot would be communicated to the girls in a turn or two round the garden; but in the present instance Mrs. Dale positively declined to speak till she was seated within the house.

'Did he come over on purpose to see you, mamma?'

'Yes, my dear, I believe so. He wished to see you, too; but I asked his permission to postpone that till after I had talked to you.'

'To see me, mamma? About what?'

'To kiss you, and bid you love him; solely for that. He has not a word to say to you that will vex you.'

'Then I will kiss him, and love him, too.'

'Yes, you will when I have told you all. I have promised him solemnly to give up all idea of going to Guestwick. So that is over.'

'Oh, oh! And we may begin to unpack at once? What an episode in one's life!'

'We may certainly unpack, for I have pledged myself to him; and he is to go into Guestwick himself and arrange about the lodgings.'

'Does Hopkins know it?'

'I should think not yet.'

'Nor Mrs. Boyce! Mamma, I don't believe I shall be able to survive this next week. We shall look such fools! I'll tell you what we'll do; – it will be the only comfort I can have; – we'll go to work and get everything back into its place before Bell comes home, so as to surprise her.'

'What! in two days?'

'Why not? I'll make Hopkins come and help, and then he'll not be so bad. I'll begin at once and go to the blankets and beds because I can undo them myself.'

'But I haven't told you all; and, indeed, I don't know how to make you understand what passed between us. He is very unhappy about Bernard; Bernard has determined to go abroad, and may be away for years.'

'One can hardly blame a man for following up his profession.'

'There was no blaming. He only said that it was very sad for him that, in his old age, he should be left alone. This was before there was any talk about our remaining. Indeed he seemed determined not to ask that again as a favour. I could see that in his eye, and I understood it from his tone. He went on to speak of you and Bell, saying how well he loved you both; but that, unfortunately, his hopes regarding you had not been fulfilled.'

'Ah, but he shouldn't have had hopes of that sort.'

'Listen, my dear, and I think that you will not feel angry with him. He said that he felt his house had never been pleasant to you. Then there followed words which I could not repeat, even if I could remember them. He said much about myself, regretting that the feeling between us had not been more kindly. "But my heart," he said, "has ever been kinder than my words." Then I got up from where I was seated, and going over to him, I told him that we would remain here.'

'And what did he say?'

'I don't know what he said. I know that I was crying, and that he kissed me. It was the first time in his life. I know that he was pleased, – beyond measure pleased. After a while he became animated, and talked of doing ever so many things. He promised that very painting of which you spoke.'

'Ah, yes, I knew it; and Hopkins will be here with the peas before dinner-time to-morrow, and Dingles with his shoulders smothered with rabbits. And then Mrs. Boyce! Mamma, he didn't think of Mrs. Boyce; or, in very charity of heart, he would still have maintained his sadness.'

'Then he did not think of her; for when I left him he was not at all sad. But I haven't told you half yet.'

'Dear me, mamma; was there more than that?'

'And I've told it all wrong; for what I've got to tell now was said before a word was spoken about the house. He brought it in just after what he said about Bernard. He said that Bernard would, of course, be his heir.'

'Of course he will.'

'And that he should think it wrong to encumber the property with any charges for you girls.'

'Mamma, did any one ever—'

'Stop, Lily, stop; and make your heart kinder towards him if you can.'

'It is kind; only I hate to be told that I'm not to have a lot of money, as though I had ever shown a desire for it. I have never envied Bernard his man-servant, or his maid-servant, or his ox, or his ass, or anything that is his. To tell the truth I didn't even wish it to be Bell's because I knew well that there was somebody she would like a great deal better than ever she could like Bernard.'

'I shall never get to the end of my story.'

'Yes, you will, mamma, if you persevere.'

'The long and the short of it is this, that he has given Bell three thousand pounds, and has given you three thousand also.'

'But why me, mamma?' said Lily, and the colour of her cheeks became red as she spoke. There should if possible be nothing more said about John Eames; but whatever might or might not be the necessity of speaking, at any rate, let there be no mistake. 'But why me, mamma?'

'Because, as he explained to me, he thinks it right to do the same by each of you. The money is yours at this moment, – to buy hair-pins with, if you please. I had no idea that he could command so large a sum.'

'Three thousand pounds! The last money he gave me was half-a-crown, and I thought he was so stingy! I particularly wanted ten shillings. I should have liked it so much better now if he had given me a nice new five-pound note.'

'You'd better tell him so.'

'No; because then he'd give me that too. But with five pounds I should have the feeling that I might do what I liked with it; – buy a dressing-case, and a thing for a squirrel to run round in. But nobody ever gives girls money like that, so that they can enjoy it.'

'Oh, Lily; you ungrateful child!'

'No, I deny it. I'm not ungrateful. I'm very grateful, because his heart was softened – and because he cried and kissed you. I'll be ever so good to him! But how I'm to thank him for giving me three thousand pounds, I cannot think. It's a sort of thing altogether beyond my line of life. It sounds like something that's to come to me in another world, but which I don't want quite yet. I am grateful, but with a misty, mazy sort of gratitude. Can you tell me how soon I shall have a new pair of Balmoral boots* because of this money? If that were brought home to me I think it would enliven my gratitude.'

The squire, as he rode back to Guestwick, fell again from that animation, which Mrs. Dale had described, into his natural sombre mood. He thought much of his past life, declaring to himself the truth of those words in which he had told his sister-in-law that his heart had ever been kinder than his words. But the world, and all those nearest to him in the world, had judged him always by his words rather than by his heart. They had taken the appearance, which he could not command or alter, rather than the facts, of which he had been the master. Had he not been good to all his relations? – and yet was there one among them that cared for him? 'I'm almost sorry that they are going to stay,' he said to himself; – 'I know that I shall disappoint them.' Yet when he met Bell at the Manor House he accosted her cheerily, telling her with much appearance of satisfaction that that flitting into Guestwick was not to be accomplished.

'I am so glad,' said she. 'It is long since I wished it.'

'And I do not think your mother wishes it now.'

'I am sure she does not. It was all a misunderstanding

from the first. When some of us could not do all that you wished, we thought it better—' Then Bell paused, finding that she would get herself into a mess if she persevered.

'We will not say any more about it,' said the squire. 'The thing is over, and I am very glad that it should be so pleasantly settled. I was talking to Dr. Crofts yesterday.'

'Were you, uncle?'

'Yes; and he is to come and stay with me the day before he is married. We have arranged it all. And we'll have the breakfast up at the Great House. Only you must fix the day. I should say some time in May. And, my dear, you'll want to make yourself fine; here's a little money for you. You are to spend that before your marriage, you know.' Then he shambled away, and as soon as he was alone, again became sad and despondent. He was a man for whom we may predicate some gentle sadness and continued despondency to the end of his life's chapter.

We left John Eames in the custody of Lady Julia, who had overtaken him in the act of erasing Lily's name from the railing which ran across the brook. He had been premeditating an escape home to his mother's house in Guestwick, and thence back to London, without making any further appearance at the Manor House. But as soon as he heard Lady Julia's step, and saw her figure close upon him, he knew that his retreat was cut off from him. So he allowed himself to be led away quietly up to the house. With Lady Julia herself he openly discussed the whole matter, – telling her that his hopes were over, his happiness gone, and his heart half-broken. Though he would perhaps have cared but little for her congratulations in success, he could make himself more amenable to consolation and sympathy from her than from any other inmate in the earl's house. 'I don't know what I shall say to your brother,' he whispered to her, as they approached the side door at which she intended to enter.

'Will you let me break it to him? After that he will say a few words to you of course, but you need not be afraid of him.'

'And Mr. Dale?' said Johnny. 'Everybody has heard about it. Everybody will know what a fool I have made myself.'

She suggested that the earl should speak to the squire, assured him that nobody would think him at all foolish, and then left him to make his way up to his own bedroom. When there he found a letter from Cradell, which had been delivered in his absence; but the contents of that letter may best be deferred to the next chapter. They were not of a nature to give him comfort or to add to his sorrow.

About an hour before dinner there was a knock at his door, and the earl himself, when summoned, made his appearance in the room. He was dressed in his usual farming attire, having been caught by Lady Julia on his first approach to the house, and had come away direct to his young friend, after having been duly trained in what he ought to say by his kind-hearted sister. I am not, however, prepared to declare that he strictly followed his sister's teaching in all that he said upon the occasion.

'Well, my boy,' he began, 'so the young lady has been perverse.'

'Yes, my lord. That is. I don't know about being perverse. It is all over.'

'That's as may be, Johnny. As far as I know, not half of them accept their lovers the first time of asking.'

'I shall not ask her again.'

'Oh, yes, you will. You don't mean to say you are angry with her for refusing you.'

'Not in the least. I have no right to be angry. I am only angry with myself for being such a fool, Lord De Guest. I wish I had been dead before I came down here on this errand. Now I think of it, I know there are so many things which ought to have made me sure how it would be.'

'I don't see that at all. You come down again, — let me see, – it's May now. Say you come when the shooting begins in September. If we can't get you leave of absence in any other way, we'll make old Buffle come too. Only, by George, I believe he'd shoot us all. But never mind; we'll manage that. You keep up your spirits till September, and then we'll fight the battle in another way. The squire shall get up a little party for the bride, and my lady Lily must go then. You shall meet her so; and then we'll shoot over the squire's land. We'll bring you together so; you see if we don't. Lord bless me! Refused once! My belief is, that in these days a girl thinks nothing of a man till she has refused him half a dozen times.'

'I don't think Lily is at all like that.'

'Look here, Johnny. I have not a word to say against Miss Lily. I like her very much, and think her one of the nicest girls I know. When she's your wife, I'll love her dearly, if she'll let me. But she's made of the same stuff as other girls, and will act in the same way. Things have gone a little astray among you, and they won't right themselves all in a minute. She knows now what your feelings are, and she'll go on thinking of it, till at last you'll be in her thoughts more than that other fellow. Don't tell me about her becoming an old maid, because at her time of life she has been so unfortunate as to come across a false-hearted man like that. It may take a little time; but if you'll carry on and not be down-hearted, you'll find it will all come right in the end. Everybody doesn't get all that they want in a minute. How I shall quiz you about all this when you have been two or three years married!'

'I don't think I shall ever be able to ask her again; and I feel sure, if I do, that her answer will be the same. She told me in so many words—; but never mind, I cannot repeat her words.'

'I don't want you to repeat them; nor yet to heed them beyond their worth. Lily Dale is a very pretty girl; clever, too, I believe, and good, I'm sure; but her words are not more sacred than those of other men or women. What she has said to you now, she means, no doubt; but the minds of men and women are prone to change, especially when such changes are conducive to their own happiness.'

'At any rate I'll never forget your kindness, Lord De Guest.'

'And there is one other thing I want to say to you, Johnny. A man should never allow himself to be cast down by anything, – not outwardly, to the eyes of other men.'

'But how is he to help it?'

'His pluck should prevent him. You were not afraid of a

roaring bull, nor yet of that man when you thrashed him at the railway station. You've pluck enough of that kind. You must now show that you've that other kind of pluck. You know the story of the boy who would not cry though the wolf was gnawing him underneath his frock. Most of us have some wolf to gnaw us somewhere; but we are generally gnawed beneath our clothes, so that the world doesn't see; and it behoves us so to bear it that the world shall not suspect. The man who goes about declaring himself to be miserable will be not only miserable, but contemptible as well.'

'But the wolf hasn't gnawed me beneath my clothes; everybody knows it.'

'Then let those who do know it learn that you are able to bear such wounds without outward complaint. I tell you fairly that I cannot sympathize with a lackadaisical lover.'

'I know that I have made myself ridiculous to everybody. I wish I had never come here. I wish you had never seen me.'

'Don't say that, my dear boy; but take my advice for what it is worth. And remember what it is that I say; with your grief I do sympathize, but not with any outward expression of it; – not with melancholy looks, and a sad voice, and an unhappy gait. A man should always be able to drink his wine and seem to enjoy it. If he can't, he is so much less of a man than he would be otherwise, – not so much more, as some people seem to think. Now get yourself dressed, my dear fellow, and come down to dinner as though nothing had happened to you.'

As soon as the earl was gone John looked at his watch and saw that it still wanted some forty minutes to dinner. Fifteen minutes would suffice for him to dress, and therefore there was time sufficient for him to seat himself in his arm-chair and think over it all. He had for a moment been very angry when his friend had told him that he could not sympathize with a lackadaisical lover. It was an ill-natured word. He felt it to be so when he heard it, and so he continued to think during the whole of the half-hour that he sat in that chair. But it probably did him more good than any word that the earl had ever spoken to him, – or any other word that he could have used. 'Lackadaisical! I'm not lackadaisical,' he said to himself, jumping up from his chair, and instantly sitting down again. 'I didn't say anything to him. I didn't tell him. Why did he come to me?' And yet, though he endeavoured to abuse Lord De Guest in his thoughts, he knew that Lord De Guest was right, and that he was wrong. He knew that he had been lackadaisical, and was ashamed of himself; and at once resolved that he would henceforth demean himself as though no calamity had happened to him. 'I've a good mind to take him at his word, and drink wine till I'm drunk.' Then he strove to get up his courage by a song.

If she be not fair for me,
*What care I how—**

'But I do care. What stuff it is a man writing poetry and putting into it such lies as that! Everybody knows that he did care, – that is, if he wasn't a heartless beast.'

But nevertheless, when the time came for him to go down into the drawing-room he did make the effort which his friend had counselled, and walked into the room with less of that hang-dog look than the earl and Lady Julia had expected. They were both there, as was also the squire, and Bell followed him in less than a minute.

'You haven't seen Crofts to-day, John, have you?' said the earl.

'No; I haven't been anywhere his way!'

'His way! His ways are every way, I take it. I wanted him to come and dine, but he seemed to think it improper to eat two dinners in the same house two days running. Isn't that his theory, Miss Dale?'

'I'm sure I don't know, Lord De Guest. At any rate, it isn't mine.'

So they went to their feast, and before his last chance was over John Eames found himself able to go through the pretence of enjoying his roast mutton.

There can, I think, be no doubt that in all such calamities as that which he was now suffering, the agony of the misfortune is much increased by the conviction that the facts of the case are known to those round about the sufferer. A most warm-hearted and intensely-feeling young gentleman might, no doubt, eat an excellent dinner after being refused by the girl of his devotions, provided that he had reason to believe that none of those in whose company he ate it knew anything of his rejection. But the same warm-hearted and intensely-feeling young gentleman would find it very difficult to go through the ceremony with any appearance of true appetite or gastronomic enjoyment, if he were aware that all his convives knew all the facts of his little misfortune. Generally, we may suppose a man in such condition goes to his club for his dinner, or seeks consolation in the shades of some adjacent Richmond or Hampton Court. There he meditates on his condition in silence, and does ultimately enjoy his little plate of whitebait, his cutlet and his moderate pint of sherry. He probably goes alone to the theatre, and, in his stall, speculates with a somewhat bitter sarcasm on the vanity of the world. Then he returns home, sad indeed, but with a moderated sadness, and as he puffs out the smoke of his cigar at the open window, – with perhaps the comfort of a little brandy-and-water at his elbow, – swears to himself that, 'By Jove, he'll have another try for it.' Alone, a man may console himself, or among a crowd of unconscious mortals; but it must be admitted that the position of John Eames was severe. He had been invited down there to woo Lily Dale, and the squire and Bell had been asked to be present at the wooing. Had it all gone well, nothing could have been nicer. He would have been the hero of the hour, and everybody would have sung for him his song of triumph. But everything had not gone well, and he found it very difficult to carry himself otherwise than lackadaisically. On the whole, however, his effort was such that the earl gave him credit for his demeanour, and told him when parting with him for the night that he was a fine fellow, and that everything should go right with him yet.

'And you mustn't be angry with me for speaking harshly to you,' he said.

'I wasn't a bit angry.'

'Yes, you were; and I rather meant that you should be. But you mustn't go away in dudgeon.'

He stayed at the Manor House one day longer, and then he returned to his room at the Income-tax Office, to the disagreeable sound of Sir Raffle's little bell, and the much more disagreeable sound of Sir Raffle's big voice.

CHAPTER LIX

JOHN EAMES BECOMES A MAN

EAMES, when he was half way up to London in the railway carriage, took out from his pocket a letter and read it. During the former portion of his journey he had been thinking of other things; but gradually he had resolved that it would be better for him not to think more of those other things for the present, and therefore he had recourse to his letter by way of dissipating his thoughts. It was from Cradell, and ran as follows: –

'INCOME-TAX OFFICE, May —, 186—.

'My Dear John, –

'I hope the tidings which I have to give you will not make you angry, and that you will not think I am untrue to the great friendship which I have for you because of that which I am now going to tell you. There is no *man* – [and the word was underscored] – there is no *man* whose regard I value so highly as I do yours; and though I feel that you can have no just ground to be displeased with me after all that I have heard you say on many occasions, nevertheless, in matters of the heart it is very hard for one person to understand the sentiments of another, and when the affections of a lady are concerned, I know that quarrels will sometimes arise.'

Eames, when he had got so far as this, on the first perusal of the letter, knew well what was to follow. 'Poor Caudle!' he said to himself, 'he's hooked, and he'll never get himself off the hook again.'

'But let that be as it may, the matter has now gone too far for any alteration to be made by me; nor would any mere earthly inducement suffice to change me. The claims of friendship are very strong, *but those of love are paramount.* Of course I know all that has passed between you and Amelia Roper. Much of this I had heard from you before, but the rest she has now told me with that pure-minded honesty which is the most remarkable feature in her character. She has confessed that at one time she felt attached to you, and that she was induced by your perseverance to allow you to regard her as your fiancy. [Fancy-girl he probably conceived to be the vulgar English for the elegant term which he used.] But all that must be over between you now. *Amelia has promised to be mine* – [this also was underscored] – and mine I intend that she shall be. That you may find in the kind smiles of L. D. consolation for any disappointment at which this may occasion you, is the ardent wish of your true friend,

'Joseph Cradell.'

'P.S. – Perhaps I had better tell you the whole. Mrs. Roper has been in some trouble about her house. She is a little in arrears with her rent, and some bills have not been paid. As she explained that she has been brought into this by those dreadful Lupexes I have consented to take the house into my own hands, and have given bills to one or two tradesmen for small amounts. Of course she will take them up, but it was the credit that was wanting. She will carry on the house, but I shall, in fact, be the proprietor. I suppose it will not suit you now to remain here, but don't you think I might make it comfortable enough for some of our fellows; say, half a dozen, or so? That is Mrs. Roper's idea, and I certainly think it is not a bad one. Our first efforts must be to get rid of the Lupexes. Miss Spruce goes next week. In the meantime we are all taking our meals up in our own rooms, so that there is nothing for the Lupexes to eat. But they don't seem to mind that, and still keep the sitting-room and best bedroom. We mean to lock them out after Tuesday, and send all their boxes to the public-house.'

Poor Cradell! Eames, as he threw himself back upon his seat and contemplated the depth of misfortune into which his friend had fallen, began to be almost in love with his own position. He himself was, no doubt, a very miserable fellow. There was only one thing in life worth living for, and that he could not get. He had been thinking for the last three days of throwing himself before a locomotive steam-engine, and was not quite sure that he would not do it yet; but, nevertheless, his place was a place among the gods as compared to that which poor Cradell had selected for himself. To be not only the husband of Amelia Roper, but to have been driven to take upon himself as his bride's fortune the whole of his future mother-in-law's debts! To find himself the owner of a very indifferent lodging-house; – the owner as regarded all responsibility, though not the owner as regarded any possible profit! And then, above and almost worse than all the rest, to find himself saddled with the Lupexes in the beginning of his career! Poor Cradell indeed!

Eames had not taken his things away from the lodging-house before he left London, and therefore determined to drive to Burton Crescent immediately on his arrival, not with the intention of remaining there, even for a night, but that he might bid them farewell, speak his congratulations to Amelia, and arrange for his final settlement with Mrs. Roper. It should have been explained in the last chapter that the earl had told him before parting with him that his want of success with Lily would make no difference as regarded money. John had, of course, expostulated, saying that he did not want anything, and would not, under his existing circumstances, accept anything; but the earl was a man who knew how to have his own way, and in this matter did have it. Our friend, therefore, was a man of wealth when he returned to London, and could tell Mrs. Roper that he would send her a cheque for her little balance as soon as he reached his office.

He arrived in the middle of the day, – not timing his return at all after the usual manner of Government clerks, who generally manage to reach the metropolis not more than half an hour before the moment at which they are bound to show themselves in their seats. But he had come back two days before he was due, and had run away from the country as though London in May to him were much pleasanter than the woods and fields. But neither had London nor the woods and fields any influence on his return. He had gone down that he might throw himself at the feet of Lily Dale, – gone down, as he now confessed to himself, with hopes almost triumphant, and he had returned because Lily Dale would not have him at her feet. 'I loved him, – him, Crosbie, – better than all the world besides. It is still the same. I still love him better than all the world.' Those were the words which had driven him back to London; and having been sent away with such words as those, it was little matter to him whether he reached his office a day or two sooner or later. The little room in the city, even with the accompaniment of Sir Raffle's bell and Sir Raffle's voice, would be now more congenial to him than Lady Julia's drawing-room. He would therefore present himself to Sir Raffle on that very afternoon, and expel some interloper from his seat. But he would first call in Burton Crescent and say farewell to the Ropers.

The door was opened for him by the faithful Jemima. 'Mr. Heames, Mr. Heames! ho dear, ho dear!' and the poor girl, who had always taken his side in the adventures of the lodging-house, raised her hands on high and lamented the fate which had separated her favourite from its fortunes. 'I suppose you knows it all, Mister Johnny?' Mister Johnny said that he believed he did know it all, and asked for the mistress of the house. 'Yes, sure enough, she's at home. She don't dare stir out much, 'cause of them Lupexes. Ain't this a pretty game? No dinner and no nothink! Them boxes is Miss Spruce's. She's a-going now, this minute. You'll find 'em all upstairs in the drawen-room.' So upstairs into the drawing-room he went, and there he found the mother and daughter, and with them Miss Spruce, tightly packed up in her bonnet and shawl. 'Don't, mother,' Amelia was saying; 'what's the good of going on in that way? If she chooses to go, let her go.'

'But she's been with me now so many years,' said Mrs. Roper, sobbing; 'and I've always done everything for her! Haven't I, now, Sally Spruce?' It struck Eames immediately, that, though he had been an inmate in the house for two years, he had never before heard that maiden lady's Christian name. Miss Spruce was the first to see Eames as he entered the room. It is probable that Mrs. Roper's pathos might have produced some answering pathos on her part had she remained unobserved, but the sight of a young man brought her back to her usual state of quiescence. 'I'm only an old woman,' said she; 'and here's Mr. Eames come back again.'

'How d'ye do, Mrs. Roper? how d'ye do,—Amelia? how d'ye do, Miss Spruce?' and he shook hands with them all.

'Oh, laws,' said Mrs. Roper, 'you have given me such a start!'

'Dear me, Mr. Eames; only think of your coming back in that way,' said Amelia.

'Well, what way should I come back? You didn't hear me knock at the door, that's all. So Miss Spruce is really going to leave you?'

'Isn't it dreadful, Mr. Eames? Nineteen years we've been together; – taking both houses together, Miss Spruce, we have, indeed.' Miss Spruce, at this point, struggled very hard to convince John Eames that the period in question had in truth extended over only eighteen years, but Mrs. Roper was authoritative, and would not permit it. 'It's nineteen years if it's a day. No one ought to know dates if I don't, and there isn't one in the world understands her ways unless it's me. Haven't I been up to your bedroom every night, and with my own hand given you—' But she stopped herself, and was too good a woman to declare before a young man what had been the nature of her nightly ministrations to her guest.

'I don't think you'll be so comfortable anywhere else, Miss Spruce,' said Eames.

'Comfortable! of course she won't,' said Amelia. 'But if I was mother I wouldn't have any more words about it.'

'It isn't the money I'm thinking of, but the feeling of it,' said Mrs. Roper. 'The house will be so lonely like. I shan't know myself; that I shan't. And now that things are all settled so pleasantly, and that the Lupexes must go on Tuesday—I'll tell you what, Sally; I'll pay for the cab myself, and I'll start off to Dulwich by the omnibus to-morrow, and settle it all out of my own pocket. I will indeed. Come; there's the cab. Let me go down, and send him away.'

'I'll do that,' said Eames. 'It's only sixpence, off the stand,' Mrs. Roper called to him as he left the room. But the cabman got a shilling, and John, as he returned, found Jemima in the act of carrying Miss Spruce's boxes back to her room. 'So much the better for poor Caudle,' said he to himself. 'As he has gone into the trade it's well that he should have somebody that will pay him.'

Mrs. Roper followed Miss Spruce up the stairs and Johnny was left with Amelia. 'He's written to you, I know,' said she, with her face turned a little away from him. She was certainly very handsome, but there was a hard, cross, almost sullen look about her, which robbed her countenance of all its pleasantness. And yet she had no intention of being sullen with him.

'Yes,' said John. 'He has told me how it's all going to be.'

'Well?' she said.

'Well?' said he.

'Is that all you've got to say?'

'I'll congratulate you, if you'll let me.'

'Psha; – congratulations! I hate such humbug. If you've no feelings about it, I'm sure that I've none. Indeed I don't know what's the good of feelings. They never did me any good. Are you engaged to marry L. D.?'

'No, I am not.'

'And you've nothing else to say to me?'

'Nothing, – except my hopes for your happiness. What else can I say? You are engaged to marry my friend Cradell, and I think it will be a happy match.'

She turned away her face further from him, and the look of it became even more sullen. Could it be possible that at such a moment she still had a hope that he might come back to her?

'Good-bye, Amelia,' he said, putting out his hand to her.

'And this is to be the last of you in this house!'

'Well, I don't know about that. I'll come and call upon you, if you'll let me, when you're married.'

'Yes,' she said, 'that there may be rows in the house, and

noise, and jealousy, – as there have been with that wicked woman upstairs. Not if I know it, you won't! John Eames, I wish I'd never seen you. I wish we might have both fallen dead when we first met. I didn't think ever to have cared for a man as I have cared for you. It's all trash and nonsense and foolery; I know that. It's all very well for young ladies as can sit in drawing-rooms all their lives, but when a woman has her way to make in the world it's all foolery. And such a hard way too to make as mine is!'

'But it won't be hard now.'

'Won't it? But I think it will. I wish you would try it. Not that I'm going to complain. I never minded work, and as for company, I can put up with anybody. The world's not to be all dancing and fiddling for the likes of me. I know that well enough. But—' and then she paused.

'What's the "but" about, Amelia?'

'It's like you to ask me; isn't it?' To tell the truth he should not have asked her. 'Never mind. I'm not going to have any words with you. If you've been a knave I've been a fool, and that's worse.'

'But I don't think I have been a knave.'

'I've been both,' said the girl; 'and both for nothing. After that you may go. I've told you what I am, and I'll leave you to name yourself. I didn't think it was in me to have been such a fool. It's that that frets me. Never mind, sir; it's all over now, and I wish you good-bye.'

I do not think that there was the slightest reason why John should have again kissed her at parting, but he did so. She bore it, not struggling with him; but she took his caress with sullen endurance. 'It'll be the last,' she said. 'Good-bye, John Eames.'

'Good-bye, Amelia. Try to make him a good wife and then you'll be happy.' She turned up her nose at this, assuming a look of unutterable scorn. But she said nothing further, and then he left the room. At the parlour door he met Mrs. Roper, and had his parting words with her.

'I am so glad you came,' said she. 'It was just that word you said that made Miss Spruce stay. Her money is so ready, you know! And so you've had it all out with her about

Cradell. She'll make him a good wife, she will indeed; much better than you've been giving her credit for.'

'I don't doubt she'll be a very good wife.'

'You see, Mr. Eames, it's all over now, and we understand each other; don't we? It made me very unhappy when she was setting her cap at you; it did indeed. She is my own daughter, and I couldn't go against her; – could I? But I knew it wasn't in any way suiting. Laws, I know the difference. She's good enough for him any day of the week, Mr. Eames.'

'That she is, – Saturdays or Sundays,' said Johnny, not knowing exactly what he ought to say.

'So she is; and if he does his duty by her she won't go astray in hers by him. And as for you, Mr. Eames, I am sure I've always felt it an honour and a pleasure to have you in the house; and if ever you could use a good word in sending to me any of your young men, I'd do by them as a mother should; I would indeed. I know I've been to blame about those Lupexes, but haven't I suffered for it, Mr. Eames? And it was difficult to know at first; wasn't it? And as to you and Amelia, if you would send any of your young men to try, there couldn't be anything more of that kind, could there? I know it hasn't all been just as it should have been; – that is as regards you; but I should like to hear you say that you've found me honest before you went. I have tried to be honest, I have indeed.'

Eames assured her that he was convinced of her honesty, and that he had never thought of impugning her character either in regard to those unfortunate people, the Lupexes, or in reference to other matters. 'He did not think,' he said, 'that any young men would consult him as to their lodgings; but if he could be of any service to her, he would.' Then he bade her good-bye, and having bestowed half-a-sovereign on the faithful Jemima, he took a long farewell of Burton Crescent. Amelia had told him not to come and see her when she should be married, and he had resolved that he would take her at her word. So he walked off from the Crescent, not exactly shaking the dust from his feet, but resolving that he would know no more either of its dust or of its dirt.

Dirt enough he had encountered there certainly, and he was now old enough to feel that the inmates of Mrs. Roper's house had not been those among whom a resting-place for his early years should judiciously have been sought. But he had come out of the fire comparatively unharmed, and I regret to say that he felt but little for the terrible scorchings to which his friend had been subjected and was about to subject himself. He was quite content to look at the matter exactly as it was looked at by Mrs. Roper. Amelia was good enough for Joseph Cradell – any day of the week. Poor Cradell, of whom in these pages after this notice no more will be heard! I cannot but think that a hard measure of justice was meted out to him, in proportion to the extent of his sins. More weak and foolish than our friend and hero he had been, but not to my knowledge more wicked. But it is to the vain and foolish that the punishments fall; – and to them they fall so thickly and constantly that the thinker is driven to think that vanity and folly are of all sins those which may be the least forgiven. As for Cradell I may declare that he did marry Amelia, that he did, with some pride, take the place of master of the house at the bottom of Mrs. Roper's table, and that he did make himself responsible for all Mrs. Roper's debts. Of his future fortunes there is not space to speak in these pages.

Going away from the Crescent Eames had himself driven to his office, which he reached just as the men were leaving it, at four o'clock. Cradell was gone, so that he did not see him on that afternoon; but he had an opportunity of shaking hands with Mr. Love, who treated him with all the smiling courtesy due to an official big-wig, – for a private secretary, if not absolutely a big-wig, is semi-big, and entitled to a certain amount of reverence; – and he passed Mr. Kissing in the passage, hurrying along as usual with a huge book under his arm. Mr. Kissing, hurried as he was, stopped his shuffling feet; but Eames only looked at him, hardly honouring him with the acknowledgement of a nod of his head. Mr. Kissing, however, was not offended; he knew that the private secretary of the First Commissioner had been the guest of an earl; and what more than a nod could be expected from him? After that John made his way into the august presence of Sir Raffle, and found that great man putting on his shoes in the presence of FitzHoward. FitzHoward blushed; but the shoes had not been touched by him, as he took occasion afterwards to inform John Eames.

Sir Raffle was all smiles and civility. 'Delighted to see you back, Eames: am, upon my word; though I and FitzHoward have got on capitally in your absence; haven't we, FitzHoward?'

'Oh, yes,' drawled FitzHoward. 'I haven't minded it for a time, just while Eames has been away.'

'You're much too idle to keep at it, I know; but your bread will be buttered for you elsewhere, so it doesn't signify. My compliments to the Duchess when you see her.' Then FitzHoward went. 'And how's my dear old friend?' asked Sir Raffle, as though of all men living Lord De Guest were the one for whom he had the strongest and the oldest love. And yet he must have known that John Eames knew as much about it as he did himself. But there are men who have the most lively gratification in calling lords and marquises their friends, though they know that nobody believes a word of what they say, – even though they know how great is the odium they incur, and how lasting is the ridicule which their vanity produces. It is a gentle insanity which prevails in the outer courts of every aristocracy; and as it brings with itself considerable annoyance and but a lukewarm pleasure, it should not be treated with too keen a severity.

'And how's my dear old friend?' Eames assured him that his dear old friend was all right, that Lady Julia was all right, that the dear old place was all right. Sir Raffle now spoke as though the 'dear old place' were quite well known to him. 'Was the game doing pretty well? Was there a promise of birds?' Sir Raffle's anxiety was quite intense, and expressed with almost familiar affection. 'And, by the by, Eames, where are you living at present?'

'Well, I'm not settled. I'm at the Great Western Railway Hotel at this moment.'

'Capital house, very; only it's expensive if you stay there

the whole season.' Johnny had no idea of remaining there beyond one night, but he said nothing as to this. 'By the by, you might as well come and dine with us to-morrow. Lady Buffle is most anxious to know you. There'll be one or two with us. I did ask my friend Dumbello, but there's some nonsense going on in the House, and he thinks that he can't get away.' Johnny was more gracious than Lord Dumbello, and accepted the invitation. 'I wonder what Lady Buffle will be like?' he said to himself, as he walked away from the office.

He had turned into the Great Western Hotel, not as yet knowing where to look for a home; and there we will leave him, eating his solitary mutton-chop at one of those tables which are so comfortable to the eye, but which are so comfortless in reality. I speak not now with reference to the excellent establishment which has been named, but to the nature of such tables in general. A solitary mutton-chop in an hotel coffee-room is not a banquet to be envied by any god; and if the mutton-chop be converted into soup, fish, little dishes, big dishes, and the rest, the matter becomes worse and not better. What comfort are you to have, seated alone on that horsehair chair, staring into the room and watching the waiters as they whisk about their towels? No one but an Englishman has ever yet thought of subjecting himself to such a position as that! But here we will leave John Eames, and in doing so I must be allowed to declare that only now, at this moment, has he entered on his manhood. Hitherto he has been a hobbledehoy, – a calf, as it were, who had carried his calfishness later into life than is common with calves; but who did not, perhaps, on that account, give promise of making a worse ox than the rest of them. His life hitherto, as recorded in these pages, had afforded him no brilliant success, had hardly qualified him for the rôle of hero which he has been made to play. I feel that I have been ir fault in giving such prominence to a hobbledehoy, and that I should have told my story better had I brought Mr. Crosbie more conspicuously forward on my canvas. He at any rate has gotten to himself a wife – as a hero always should do; whereas I must leave my poor friend Johnny without any matrimonial prospects.

It was thus that he thought of himself as he sat moping over his solitary table in the hotel coffee-room. He acknowledged to himself that he had not hitherto been a man; but at the same time he made some resolution which, I trust, may assist him in commencing his manhood from this date.

CHAPTER LX

CONCLUSION

It was early in June that Lily went up to her uncle at the Great House, pleading for Hopkins, – pleading that to Hopkins might be restored all the privileges of head gardener at the Great House. There was some absurdity in this, seeing that he had never really relinquished his privileges; but the manner of the quarrel had been in this wise.

There was in those days, and had been for years, a vexed question between Hopkins and Jolliffe the bailiff on the matter of – stable manure. Hopkins had pretended to the right of taking what he required from the farmyard, without asking leave of any one. Jolliffe in return had hinted, that if this were so, Hopkins would take it all. 'But I can't eat it,' Hopkins had said. Jolliffe merely grunted, signifying by the grunt, as Hopkins thought, that though a gardener couldn't eat a mountain of manure fifty feet long and fifteen high – couldn't eat in the body, – he might convert it into things edible for his own personal use. And so there had been a great feud. The unfortunate squire had of course been called on to arbitrate, and having postponed his decision by every contrivance possible to him, had at last been driven by Jolliffe to declare that Hopkins should take nothing that was not assigned to him. Hopkins, when the decision was made known to him by his master, bit his old lips, and turned round upon his old heel, speechless. 'You'll find it's so at all other places,' said the squire, apologetically. 'Other places!' sneered Hopkins. Where would he find other gardeners like himself? It is hardly necessary to declare that from that moment he resolved that he would abide by no such order. Jolliffe on the next morning informed the squire that the order had been broken, and the squire fretted and fumed, wishing that Jolliffe were well buried under the mountain in question. 'If they all is to do as they like,' said Jolliffe, 'then nobody won't care for nobody.' The squire understood that an order if given must be obeyed, and therefore, with many inner groanings of the spirit, resolved that war must be waged against Hopkins.

On the following morning he found the old man himself wheeling a huge barrow of manure round from the yard into the kitchen-garden. Now, on ordinary occasions, Hopkins was not required to do with his own hands work of that description. He had a man under him who hewed wood, and carried water, and wheeled barrows, – one man always, and often two. The squire knew when he saw him that he was sinning, and bade him stop upon his road.

'Hopkins,' he said, 'why didn't you ask for what you wanted, before you took it?' The old man put down the barrow on the ground, looked up in his master's face, spat into his hands, and then again resumed his barrow. 'Hopkins, that won't do,' said the squire. 'Stop where you are.'

'What won't do?' said Hopkins, still holding the barrow from the ground, but not as yet progressing.

'Put it down, Hopkins,' and Hopkins did put it down. 'Don't you know that you are flatly disobeying my orders?'

'Squire, I've been here about this place going on nigh seventy years.'

'If you've been going on a hundred and seventy it wouldn't do that there should be more than one master. I'm the master here, and I intend to be so to the end. Take that manure back into the yard.'

'Back into the yard?' said Hopkins, very slowly.

'Yes; back into the yard.'

'What, – afore all their faces?'

'Yes; you've disobeyed me before all their faces?'

Hopkins paused a moment, looking away from the squire, and shaking his head as though he had need of deep thought, but by the aid of deep thought had come at last to a right conclusion. Then he resumed the barrow, and putting himself almost into a trot, carried away his prize into the kitchen-garden. At the pace which he went it would have been beyond the squire's power to stop him, nor would Mr. Dale have wished to come to a personal encounter with his servant. But he called after the man in dire wrath that if he were not obeyed the disobedient servant should rue the consequences for ever. Hopkins, equal to the occasion, shook his head as he trotted on, deposited his load at the foot of the cucumber-frames, and then at once returning to his master, tendered to him the key of the greenhouse.

'Master,' said Hopkins, speaking as best he could with his scanty breath, 'there it is; – there's the key; of course I don't want no warning, and doesn't care about my week's wages. I'll be out of the cottage afore night, and as for the work'us, I suppose they'll let me in at once, if your honour'll give 'em a line.'

Now as Hopkins was well known by the squire to be the owner of three or four hundred pounds, the hint about the workhouse must be allowed to have been melodramatic.

'Don't be a fool,' said the squire, almost gnashing his teeth.

'I know I've been a fool,' said Hopkins, 'about that 'ere doong; my feelings has been too much for me. When a man's feelings has been too much for him, he'd better just take hisself off, and lie in the work'us till he dies.' And then he again tendered the key. But the squire did not take the key, and so Hopkins went on. 'I s'pose I'd better just see to the lights and the like of that, till you've suited yourself, Mr. Dale. It 'ud be a pity all them grapes should go off, and they, as you may say, all one as fit for the table. It's a long way the best crop I ever seen on 'em. I've been that careful with 'em that I haven't had a natural night's rest, not since February. There ain't nobody about this place as understands grapes, nor yet anywhere nigh that could be got at. My lord's head man is wery ignorant; but even if he knew ever so, of course he couldn't come here. I suppose I'd better keep the key till you're suited, Mr. Dale.'

Then for a fortnight there was an interregnum in the gardens, terrible in the annals of Allington. Hopkins lived in his cottage indeed, and looked most sedulously after the grapes. In looking after the grapes, too, he took the greenhouses under his care; but he would have nothing to do with the outer gardens, took no wages, returning the amount sent to him back to the squire, and insisted with everybody that he had been dismissed. He went about with some terrible horticultural implement always in his hand, with which it was said that he intended to attack Jolliffe; but Jolliffe prudently kept out of his way.

As soon as it had been resolved by Mrs. Dale and Lily that the flitting from the Small House at Allington was not to be accomplished, Lily communicated the fact to Hopkins.

'Miss,' said he, 'when I said them few words to you and your mamma, I knew that you would listen to reason.'

This was no more than Lily had expected; that Hopkins should claim the honour of having prevailed by his arguments was a matter of course.

'Yes,' said Lily; 'we've made up our minds to stay. Uncle wishes it.'

'Wishes it! Laws, miss; it ain't only wishes. And we all wishes it. Why, now, look at the reason of the thing. Here's this here house—'

'But, Hopkins, it's decided. We're going to stay. What I want to know is this; can you come at once and help me to unpack?'

'What! this very evening, as is—'

'Yes, now; we want to have the things about again before they come back from Guestwick.'

Hopkins scratched his head and hesitated, not wishing to yield to any proposition that could be considered as childish; but he gave way at last, feeling that the work itself was a good work. Mrs. Dale also assented, laughing at Lily for her folly as she did so, and in this way the things were unpacked very quickly, and the alliance between Lily and Hopkins became, for the time, very close. This work of unpacking and resettling was not yet over, when the battle of the manure broke out, and therefore it was that Hopkins, when his feelings had become altogether too much for him 'about the doong,' came at last to Lily, and laying down at her feet all the weight and all the glory of his sixty odd years of life, implored her to make matters straight for him. 'It's been a killing me, miss, so it has; to see the way they've been a cutting that 'sparagus. It ain't cutting at all. It's just hocking it up; – what is fit, and what isn't, all together. And they've been a-putting the plants in where I didn't mean 'em, though they know'd I didn't mean 'em. I've stood by, miss, and said never a word. I'd a died sooner. But, Miss Lily, what my sufferings have been, 'cause of my feelings getting the better of me about that – you know, miss – nobody will ever tell; – nobody – nobody – nobody.' Then Hopkins turned away and wept.

'Uncle,' said Lily, creeping close up against his chair, 'I want to ask you a great favour.'

'A great favour. Well, I don't think I shall refuse you anything at present. It isn't to ask another earl to the house, – is it?'

'Another earl!' said Lily.

'Yes; haven't you heard? Miss Bell has been here this morning, insisting that I should have over Lord De Guest and his sister for the marriage. It seems that there was some scheming between Bell and Lady Julia.'

'Of course you'll ask them.'

'Of course I must. I've no way out of it. It'll be all very well for Bell, who'll be off to Wales with her lover; but what am I to do with the earl and Lady Julia, when they're gone? Will you come and help me?'

In answer to this, Lily of course promised that she would come and help. 'Indeed,' said she, 'I thought we were all asked up for the day. And now for my favour. Uncle, you must forgive poor Hopkins.'

'Forgive a fiddlestick!' said the squire.

'No, but you must. You can't think how unhappy he is.'

'How can I forgive a man who won't forgive me. He goes prowling about the place doing nothing; and he sends me back his wages, and he looks as though he were going to murder some one; and all because he wouldn't do as he was told. How am I to forgive such a man as that?'

'But, uncle, why not?'

'It would be his forgiving me. He knows very well that he may come back whenever he pleases; and, indeed, for the matter of that he has never gone away.'

'But he is so very unhappy.'

'What can I do to make him happier?'

'Just go down to his cottage and tell him that you forgive him.'

'Then he'll argue with me.'

'No; I don't think he will. He is too much down in the world for arguing now.'

'Ah! you don't know him as I do. All the misfortunes in the world wouldn't stop that man's conceit. Of course I'll go if you ask me, but it seems to me that I'm made to knock under to everybody. I hear a great deal about other people's feelings, but I don't know that mine are very much thought of.' He was not altogether in a happy mood, and Lily almost regretted that she had persevered; but she did succeed in carrying him off across the garden to the cottage, and as they went together she promised him that she would think of him always, – always. The scene with Hopkins cannot be described now, as it would take too many of our few remaining pages. It resulted, I am afraid I must confess, in nothing more triumphant to the squire than a treaty of mutual forgiveness. Hopkins acknowledged, with much self-reproach, that his feelings had been too many for him; but then, look at his provocation! He could not keep his tongue from that matter, and certainly said as much in his own defence as he did in confession of his sins. The substantial triumph was altogether his, for nobody again ever dared to interfere with his operations in the farmyard. He showed his submission to his master mainly by consenting to receive his wages for the two weeks which he had passed in idleness.

Owing to this little accident, Lily was not so much oppressed by Hopkins as she had expected to be in that matter of their altered plans; but this salvation did not extend to

Mrs. Hearn, to Mrs. Crump, or, above all, to Mrs. Boyce. They, all of them, took an interest more or less strong in the Hopkins controversy; but their interest in the occupation of the Small House was much stronger, and it was found useless to put Mrs. Hearn off with the gardener's persistent refusal of his wages, when she was big with inquiry whether the house was to be painted inside, as well as out. 'Ah,' said she, 'I think I'll go and look at lodgings at Guestwick myself, and pack up some of my beds.' Lily made no answer to this, feeling that it was a part of that punishment which she had expected. 'Dear, dear,' said Mrs. Crump to the two girls; 'well, to be sure, we should 'a been lone without 'ee, and mayhap we might a got worse in your place; but why did 'ee go and fasten up all your things in them big boxes, just to unfasten 'em all again?'

'We changed our minds, Mrs. Crump,' said Bell, with some severity.

'Yees, I know ye changed your mindses. Well, it's all right for loiks o' ye, no doubt; but if we changes our mindses, we hears of it.'

'So, it seems, do we!' said Lily. 'But never mind, Mrs. Crump. Do you send us our letters up early, and then we won't quarrel.'

'Oh, letters! Drat them for letters. I wish there weren't no sich things. There was a man here yesterday with his imperence. I don't know where he come from, – down from Lun'on, I b'leeve: and this was wrong, and that was wrong, and everything wrong; and then he said he'd have me discharged the sarvice.'

'Dear me, Mrs. Crump; that wouldn't do at all.'

'Discharged the sarvice! Tuppence farden a day. So I told 'un to discharge hisself, and take all the old bundles and things away upon his shoulders. Letters indeed! What business have they with post-missusses, if they cannot pay 'em better nor tuppence farden a day?' And in this way, under the shelter of Mrs. Crump's storm of wrath against the inspector who had visited her, Lily and Bell escaped much that would have fallen upon their own heads; but Mrs. Boyce still remained. I may here add, in order that Mrs. Crump's history may be carried on to the farthest possible point, that she was not 'discharged the sarvice,' and that she still receives her twopence farthing a day from the Crown. 'That's a bitter old lady,' said the inspector to the man who was driving him. 'Yes, sir; they all say the same about she. There ain't none of 'em get much change out of Mrs. Crump.'

Bell and Lily went together also to Mrs. Boyce's. 'If she makes herself very disagreeable, I shall insist upon talking of your marriage,' said Lily.

'I've not the slightest objection,' said Bell; 'only I don't know what there can be to say about it. Marrying the doctor is such a very commonplace sort of thing.'

'Not a bit more commonplace than marrying the parson,' said Lily.

'Oh, yes, it is. Parsons' marriages are often very grand affairs. They come in among county people. That's their luck in life. Doctors never do; nor lawyers. I don't think lawyers ever get married in the country. They're supposed to do it up in London. But a country doctor's wedding is not a thing to be talked about much.'

Mrs. Boyce probably agreed on this view of the matter, seeing that she did not choose the coming marriage as her first subject of conversation. As soon as the two girls were seated she flew away immediately to the house, and began to express her very great surprise, – her surprise and her joy also, – at the sudden change which had been made in their plans. 'It is so much nicer, you know,' said she, 'that things should be pleasant among relatives.'

'Things always have been tolerably pleasant with us,' said Bell.

'Oh, yes; I'm sure of that. I've always said it was quite a pleasure to see you and your uncle together. And when we heard about your all having to leave—'

'But we didn't have to leave, Mrs. Boyce. We were going to leave because we thought mamma would be more comfortable in Guestwick; and now we're not going to leave, because we've all "changed our mindses," as Mrs. Crump calls it.'

'And is it true the house is going to be painted?' asked Mrs. Boyce.

'I believe it is true,' said Lily.

'Inside and out?'

'It must be done some day,' said Bell.

'Yes, to be sure; but I must say it is generous of the squire. There's such a deal of wood-work about your house. I know I wish the Ecclesiastical Commissioners would paint ours; but nobody ever does anything for the clergy. I'm sure I'm delighted you're going to stay. As I said to Mr. Boyce, what should we ever have done without you? I believe the squire had made up his mind that he would not let the place.'

'I don't think he ever has let it.'

'And if there was nobody in it, it would all go to rack and ruin; wouldn't it? Had your mamma to pay anything for the lodgings she engaged at Guestwick?'

'Upon my word, I don't know. Bell can tell better about that than I, as Dr. Crofts settled it. I suppose Dr. Crofts tells her everything.' And so the conversation was changed, and Mrs. Boyce was made to understand that whatever further mystery there might be, it would not be unravelled on that occasion.

It was settled that Dr. Crofts and Bell should be married about the middle of June, and the squire determined to give what grace he could to the ceremony by opening his own house on the occasion. Lord De Guest and Lady Julia were invited by special arrangement betwen her ladyship and Bell, as has been before explained. The colonel also with Lady Fanny came up from Torquay on the occasion, this being the first visit made by the colonel to his paternal roof for many years. Bernard did not accompany his father. He had not yet gone abroad, but there were circumstances which made him feel that he would not find himself comfortable at the wedding. The service was performed by Mr. Boyce, assisted, as the *County Chronicle* very fully remarked, by the Reverend John Joseph Jones, M.A., late of Jesus College, Cambridge, and curate of St. Peter's, Northgate, Guestwick; the fault of which little advertisement was this, – that as none of the readers of the paper had patience to get beyond the Reverend John Joseph Jones, the fact of Bell's marriage with Dr. Crofts was not disseminated as widely as might have been wished.

The marriage went off very nicely. The squire was upon his very best behaviour, and welcomed his guests as though he really enjoyed their presence there in his halls. Hopkins, who was quite aware that he had been triumphant, decorated the old rooms with mingled flowers and greenery with an assiduous care which pleased the two girls mightily. And during this work of wreathing and decking there was one little morsel of feeling displayed which may as well be told in these last lines. Lily had been encouraging the old man while Bell for a moment had been absent.

'I wish it had been for thee, my darling!' he said; 'I wish it had been for thee!'

'It is much better as it is, Hopkins,' she answered solemnly.

'Not with him, though,' he went on, 'not with him. I wouldn't a hung a bough for him. But with t'other one.'

Lily said no word further. She knew that the man was expressing the wishes of all around her. She said no word further, and then Bell returned to them.

But no one at the wedding was so gay as Lily, – so gay, so bright, and so wedding-like. She flirted with the old earl till he declared that he would marry her himself. No one seeing her that evening, and knowing nothing of her immediate history, would have imagined that she herself had been cruelly jilted some six or eight months ago. And those who did know her could not imagine that what she then suffered had hit her so hard, that no recovery seemed possible for her. But though no recovery, as she herself believed, was possible for her – though she was as a man whose right arm had been taken from him in battle, still all the world had not gone with that right arm. The bullet which had maimed her sorely had not touched her life, and she scorned to go about the world complaining either by word or look of the injury she had received. 'Wives when they have lost their husbands still eat and laugh,' she said to herself, 'and he is not dead like that.' So she resolved that she would be happy, and I here declare that she not only seemed to carry out her resolution, but that she did carry it out in very truth. 'You're a dear good man, and I know you'll be good to her,' she said to Crofts just as he was about to start with his bride.

'I'll try, at any rate,' he answered.

'And I shall expect you to be good to me too. Remember you have married the whole family; and, sir, you mustn't believe a word of what that bad man says in his novels about mothers-in-law. He has done a great deal of harm, and shut half the ladies in England out of their daughters' homes.'

'He shan't shut Mrs. Dale out of mine.'

'Remember he doesn't. Now, good-bye.' So the bride and bridegroom went off, and Lily was left to flirt with Lord De Guest.

Of whom else it is necessary that a word or two should be said before I allow the weary pen to fall from my hand? The squire, after much inward struggling on the subject, had acknowledged to himself that his sister-in-law had not received from him that kindness which she had deserved. He had acknowledged this, purporting to do his best to amend his past errors; and I think I may say that his efforts in that line would not be received ungraciously by Mrs. Dale. I am inclined therefore to think that life at Allington, both at the Great House and at the Small, would soon become pleasanter than it used to be in former days. Lily soon got the Balmoral boots, or, at least, soon learned that the power of getting them as she pleased had devolved upon her from her uncle's gift; so that she talked even of buying the squirrel's cage; but I am not aware that her extravagance led her as far as that.

Lord De Courcy was left suffering dreadfully from gout and ill temper at Courcy Castle. Yes, indeed! To him in his latter days life did not seem to offer much that was comfortable. His wife had now gone from him, and declared positively to her son-in-law that no earthly consideration should ever induce her to go back again; – 'not if I were to starve!' she said. By which she intended to signify that she would be firm in her resolve, even though she should thereby lose her carriage and horses. Poor Mr. Gazebee went down to Courcy, and had a dreadful interview with the earl; but matters were at last arranged, and her ladyship remained at Baden-Baden in a state of semi-starvation. That is to say, she had but one horse to her carriage.

As regards Crosbie, I am inclined to believe that he did again recover his power at his office. He was Mr. Butterwell's master, and the master also of Mr. Optimist, and the major. He knew his business, and could do it, which was more, perhaps, than might fairly be said of any of the other three. Under such circumstances he was sure to get in his hand, and lead again. But elsewhere his star did not recover its ascendancy. He dined at his club almost daily, and there were those with whom he habitually formed some little circle. But he was not the Crosbie of former days, – the Crosbie known in Belgravia and in St. James's Street. He had taken his little vessel bravely out into the deep waters, and had sailed her well while fortune stuck close to him. But he had forgotten his nautical rules, and success had made him idle. His plummet and lead had not been used, and he had kept no look-out ahead. Therefore the first rock he met shivered his bark to pieces. His wife, the Lady Alexandrina, is to be seen in the one-horse carriage with her mother at Baden-Baden.

NOTES

1. *entail*: a means of restricting the inheritance of property, ordinarily to the male line.

2. *luck of Edenhall*: a reference to a legend, celebrated in at least two popular ballads, which speaks of a company of fairies warning that the good fortune of the estate, located near Penrith in Cumberland, is dependent upon the safety of an enamelled goblet, kept in a case dating from the time of Henry IV or Henry V.

13. *Damon to any Pythias*: Damon and Pythias are inseparable friends famous in Greek literature.

22. *meo periculo*: colloquial classical Latin for 'at my own risk'.

24. *Rhadamanthine moralists*: extremely severe moralists, after Rhadamanthus, son of Zeus and Europa and one of the judges of souls in the lower world.

26. *tire-woman*: a wardrobe woman in a theatre or a lady's maid.

28. *succedaneum*: substitute.

42. *deshabille*: careless dress, negligée.

46. *L. S. D.*: abbreviations for pounds, shillings, and pence.

63. *Lotharios*: Lothario is a famous seducer in the play *The Fair Penitent* (1703) by Nicholas Rowe.

64. *merino*: a soft wool dress.

68. *gaiters*: leather leg-coverings.

82. *ha-ha*: a sunken fence.

100. *some Greek kalends*: a time that will never arrive, from the fact that the Greeks did not reckon time by calendars.

113. *cambric*: thin, finely-woven linen.

130. *the time of King John*: reigned 1199–1216.

155. *ménage*: household.

160. *my thoughts ... shall be like those of Ruth*: a reference to Ruth 1: 16–17, which begins, 'And Ruth said, Intreat me not to leave thee, *or* to return from following after thee: for whither thou goest, I will go'.

169. *warden*: The story of Mr. Harding as warden of Hiram's Hospital is the dominant subject of *The Warden* and is central in *Barchester Towers*.

169. *Nolo decanari*: 'I do not wish to be dean.'

180. *tanner*: sixpence.

191. *blue-books*: government publications, reports, and manuals.

192. *Damon and Pythias*: see note for p. 13.

192. *syllabub*: a desert made with curdled cream and wine.

197. *ennuyé*: bored.

205. *take me up*: interrupt, bring up short.

229. *sward*: a grassy spot.

230. *adamant*: imaginary stones of impenetrable hardness.

235. *Falernian*: a wine from the district of Campania celebrated by Horace.

237. *Pawkins's*: Perhaps another reference to Dickens's *Martin Chuzzlewit*, this time to Mr. Pawkins's boarding-house in America.

251. *old Nestor*: legendary Greek hero, a wise elder counsellor.

261. *deshabille*: see note for p. 42.

265. *dog-cart*: a light, two-wheeled carriage.

267. *Lothario, Don Juan,* and ... *Lovelace*: notorious rakes and seducers: for Lothario, see note for p. 63; Lovelace is the villain-hero of Samuel Richardson's *Clarissa* (1747–8).

281. *corn-chandlers*: retailers of grain.

281. *Newmarket or Homburg*: Newmarket is a famous English race-track. Homburg was such a notorious gambling centre that it was occupied in 1849 by Austrian troops in order to close its gaming-tables. Play soon began again, however, and continued unchecked until 1872, when the Prussian government finally took action.

289. *gig*: a light two-wheeled, one horse carriage.

303. *Lord Eldon*: John Scott, first Earl of Eldon (1751–1838) and lord chancellor, whose maxim, according to the *DNB*, was 'that a lawyer should live like a hermit and work like a horse'.

305. *the Daily Jupiter*: In other Barsetshire novels, most prominently in *Barsetshire Towers*, an obvious reference to the *Times*. In this novel, however, the *Times* is also mentioned.

312. *Egyptia conjux*: a form of *coniunx Aegyptia* from Ovid's *Metamorphoses* xv. 826, a pejorative reference to Cleopatra as mistress of Marc Antony. Lupex is a play on *lupa* (female wolf and also whore) and *lupanar* (whorehouse).

313. *Mr. Todgers*: A slightly miscalculated reference to Mrs. Todgers' boarding-house in Dickens's *Martin Chuzzlewit*. There is no longer a Mr. Todgers.

339. *dies non*: short for *dies non juridicus*, a holiday, or literally a day on which courts may not convene nor any legal business be conducted.

342. *The show of fat beasts in London*: annual cattle shows.

344. *negus*: a beverage made of wine, hot water, sugar, and spices.

346. *arrowroot*: a medicinal food made from a starch prepared from Maranta tubers.

347. *senna and salts*: a cathartic compound.

349. *as far apart as Dives and Lazarus*: a reference to Luke 16 and the parable of the rich man (*dives* is Latin for 'rich') and Lazarus the beggar, between whom, after death, there is 'a great gulf fixed'.

361. *retricked our beams*: restored our lives; a reference either to Milton's 'Lycidas', ll. 166–71 or, more probably, to Elizabeth

Barrett Browning's *Prometheus Bound*: 'The sun [shall] dispense with retrickt beams the morning-frosts' (l. 28).

368. *Mr. Smith's book-stall*: W. H. Smith, famous bookseller.

372. *Væ Victis*: 'woe to the conquered', the proverbial remark made by the chieftan of the Gauls after the defeat of the Romans at the battle of Allia, probably in 387 B.C. The source is in Livy, v. 48.9.

388. *show the white feather*: a mark of cowardice; the phrase refers to a belief that a white feather in the plumage of a gamecock is a sign of a poor fighter.

411. *Lady-day*: March 25, the Feast of the Annunciation of the Virgin.

429. *hecatombs*: ironic reference to Greek sacrifices to the Gods.

432. *appanages*: natural accompaniments, wedding clothes.

440. *Palissy ware*: after the famous French potter Bernard Palissy (1510–89).

445. *maternal devotion ... of the pelican*: in legend and iconographic representation, the pelican is supposed to draw its own blood with its beak to feed its young.

472. *ad valorem*: in proportion to the value.

473. *ignis fatuus*: phosphorescent light that flits over swamp ground, a deception.

474. *Amaryllis in the shade*: Amaryllis is a shepherdess in Virgil; the reference is to Milton's 'Lycidas' (1637): 'Were it not better done as others use/ To sport with Amaryllis in the shade' (ll. 67–8).

482. *whatever is, is right*: a reference to Pope's 'An Essay on Man' (1733): 'And, spite of pride, in erring reason's spite,/ One truth is clear: Whatever IS, is RIGHT' (ll. 293–4).

484. *Paul and Virginia*: Jacques Henri Bernardin de Saint-Pierre's *Paul et Virginie* (1789).

491. *rissole*: a meat- or fish-filled croquette.

552. *pillion*: a cushion for an extra rider behind the saddle on a horse.

575. *constancy ... of a Jacob*: a reference to *Genesis* 29: 1–30, where Jacob, tricked into marrying Leah, has to wait seven years extra to marry his beloved Rachel.

578. *Loquitur*: a stage direction meaning 'begins to speak'.

609. *cuirass*: a breastplate or a close-fitting bodice.

612. *faute de mieux*: undertaken for lack of something better.

637. *Balmoral boots*: heavy, front-laced walking shoes.

642. *If she be not fair ...*: A reference to a sonnet from George Wither's *Fair Virtue* (1622): 'If she be not so to me,/ What care I how fair she be?'

WHO'S WHO

A.S. *The American Senator*
B.T. *Barchester Towers*
C.Y.F.H. *Can You Forgive Her?*
D.C. *The Duke's Children*
D.T. *Doctor Thorne*
E.D. *The Eustace Diamonds*
F.P. *Framley Parsonage*
L.C.B. *The Last Chronicle of Barset*
P.F. *Phineas Finn*
P.M. *The Prime Minister*
P.R. *Phineas Redux*
W. *The Warden*

Arabin, The Rev. Francis, Dean of Barchester, 169; m. Eleanor Bold (Harding), y.d. of Septimus Harding, q.v., 169.
See also *W.*, *B.T.*, *D.T.*, *F.P.*, *L.C.B.*

Boggs, clerk at the General Committee Office, 375.

Bold, John, the first husband of Eleanor Arabin, q.v., 169.
See also *W.*

Boyce, The Rev. Mr., vicar of Allington, 10; m. —— 91; many children, among them Minnie, 539; Dick, 82; Jane, Charles, Florence, and Bessy, 91.
See also *L.C.B.*

Brock, Lord, Prime Minister, 308.
See also *C.Y.F.H.*

Buffle, Sir Raffle, Chief Commissioner of the Income-Tax Office, formerly of the General Committee, 302.
See also *L.C.B.*

Bushers, the constable arresting Johnny Eames at Paddington Station, 369.

Butterwell, Mr., member of the General Committee; formerly secretary to the General Committee; succeeded as secretary by Adolphus Crosbie, q.v., 301.

Chumpend, butcher who serves Mrs. Roper, q.v., 560.

Clandidlem, Lady, guest at the house party of the Countess De Courcy, q.v., 183.

Colepepper, friend of Lord Porlock, 181.

Connor, Mrs., housekeeper at Guestwick Manor, 213.

CRADELL, JOSEPH ('Caudle'), civil-service clerk; friend of Johnny Eames, q.v.; lodger at the boardinghouse operated by Mrs. Roper, q.v.; m. Amelia Roper, q.v.; his mother, the widow of a barrister, 39.
See also *L.C.B.*

CROFTS, DR. JAMES, physician at Guestwick; m. Isabella Dale, q.v., 29.
See also *L.C.B.*

CROSBIE, ADOLPHUS, clerk in the General Committee Office and later secretary to the General Committee; m. Lady Alexandrina De Courcy, q.v., 10.
See also *L.C.B.*